T0213203

ENGINE OILS
AND
AUTOMOTIVE
LUBRICATION

MECHANICAL ENGINEERING

A Series of Textbooks and Reference Books

Editor

L. L. Faulkner

*Columbus Division, Battelle Memorial Institute
and Department of Mechanical Engineering
The Ohio State University
Columbus, Ohio*

Additional Volumes in Preparation

Mechanical Engineering Software

Spring Design with an IBM PC, Al Dietrich

Mechanical Design Failure Analysis: With Failure Analysis System Software for the IBM PC, David G. Ullman

ENGINE OILS AND AUTOMOTIVE LUBRICATION

EDITED BY

WILFRIED J. BARTZ

Technische Akademie Esslingen
Ostfildern, Germany

CRC Press
Taylor & Francis Group
Boca Raton London New York

CRC Press is an imprint of the
Taylor & Francis Group, an **informa** business

CRC Press
Taylor & Francis Group
6000 Broken Sound Parkway NW, Suite 300
Boca Raton, FL 33487-2742

First issued in paperback 2019

© 1993 by Taylor & Francis Group, LLC
CRC Press is an imprint of Taylor & Francis Group, an Informa business

No claim to original U.S. Government works

ISBN-13: 978-0-8247-8807-0 (hbk)
ISBN-13: 978-0-367-40270-9 (pbk)

**Visit the Taylor & Francis Web site at
http://www.taylorandfrancis.com**

**and the CRC Press Web site at
http://www.crcpress.com**

Preface

Lubricants and lubrication techniques are indispensable in the automobile industry. Owing to the special operating conditions, characterized by high temperatures, loads and speeds, lubricants have to cover extreme requirements. Therefore, the necessary properties of these lubricants require appropriate classification, production and formulation, testing and application as well as expert disposal.

This book deals with the state of the art in the field of automotive lubrication, particularly engine lubrication. The different topics are covered by experts from the mineral oil, additive and automobile industries as well as from research institutes, thus providing high standard expert knowledge in any specific area of automotive and engine lubrication. Experts from several countries contributed to this book.

In different chapters the following topics are covered:
— Film thickness in engine bearings
— Base oils for automotive lubricants
— Additives and mechanism of effectiveness
— Engine oils and their evaluation
— Sludge deposits in gasoline engines
— Special aspects of engine lubrication
— Two-Stroke-Engine Oils
— Tractor lubrication
— Gear lubrication
— Lubricant influence on ceramic and seal materials

This book is characterized by the fact that experts from all over the world gathered and summarized their knowledge, resulting in a general but nevertheless comprehensive presentation of all major aspects of automotive and engine lubrication.

This book might be useful to all who are active in the field of automobile tribology and lubrication. Experts from the mineral oil and additive industries can also find new points of view to supplement their knowledge, as will junior scientists and engineers who are introducing themselves to this field of tribology and lubrication engineering. In addition, with the aid of this book a great number of students of tribology may gain a great deal of useful information.

Prof. W.J. Bartz

Table of Contents

3.13 Evaluation of the Antiwear Performance of Aged Oils through Tribological and Physicochemical Tests 359

G. Monteil, A.M. Merillon and J. Lonchampt
C. Roques-Carmes

3.14 Mathematic Model for the Thickening Power of Viscosity Index Improvers. Application in Engine Oil Formulations 383

H. Bourgognon and C. Rodes, C. Neveu and F. Huby

Sludge Deposits in Gasoline Cars

Special Aspects of Engine Lubrication

Two-Stroke Engine Oils

5. Tractor Lubrication

1. Oil Film Thickness in Engine Bearings

1.1 Measurement of Oil Film Thickness in Big-End Bearings and Its Relevance to Engine Oil Viscosity Classifications

T.W. Bates, Shell Research Ltd., Chester, Great Britain
M.A. Vickars, Esso Research Centre, Abingdon, Great Britain

Summary

Minimum oil film thickness (MOFT) measurements have been carried out in big-end bearings of V-6 and in-line four cylinder gasoline engines during engine operation. MOFT decreases with increasing crankshaft speed above 2000 r/min. The most severe, practical, steady-state operation is high-speed cruising. Maximum shear rates are in the region of 10^7 s^{-1} at 4000 r/min. The dynamic viscosities at a shear rate of 10^6 s^{-1} correlate significantly better with monograde MOFT data than with multigrade data; the correlation parameters for mono- and multigrade data are also significantly different. Although the dynamic viscosity measurement correlates with multigrade data better than the low-shear-rate kinematic viscosity, the differences are not always significant at the 95 % confidence level. Some other rheological parameter or combination of parameters may be better than either kinematic or dynamic viscosities.

1.1.1 Introduction

Until recently, the most widely used method of measuring journal bearing performance in operating gasoline or diesel engines was to measure bearing weight loss at the end of a test which inevitably involved operating under severe conditions of load and speed in order to obtain metal-to-metal contact (1.2). Such tests are the ultimate arbiter for assessing long-term durability performance. They are less than ideal, however, for studying the effects of oil rheology on journal bearing performance because of: (a) poor test repeatability; and (b) complications arising from operating under boundary/mixed lubrication conditions where chemical effects of the dispersant/inhibitor package complicate interpretation of the results in terms of oil rheology (1,2). Poor repeatability also makes bearing weight loss experiments unsatisfactory for studying the effects of bearing design on performance.

A preferred assessment of journal bearing performance is measurement of the oil film thickness during engine operation.. The latter is critical since it ensures the bearing is subject to relevant dynamic loadings (as opposed to the use of tactically-loaded journal bearing rigs). Moreover, operation of the bearing under full-film hydrodynamic loading ensures only oil rheology effects are involved. Considerable progress has been made in the last ten years in instrumenting bearings of engines to allow measurement of oil film thickness during operation. Most of the activity has centred around main bearings (3-9) . Big-end bearings, however, are of more interest because the combination of smaller bearing area and severe dynamic loadings make them more prone to field failure than main bearings. The problem of making electrical connections to the reciprocating big-end has been overcome by the use of light-weight, mechanical scissor linkages to support the wires and prevent their premature breakage (10-14).

As part of an extended CEC programme (15-18) into the effects of oil viscometry on engine performance, the CEC Project Group PL-33 has initiated a study of the relationship between oil rheology and oil film thickness in the big-end bearings of two different engines. This study was initiated partly in response to a request from CCMC for further information on the role of high-temperature, high-shear viscosity (HTHSV) and other rheological properties on bearing performance, and partly to provide information which could be used as input to the debate on HTHSV classifications and limits in the SAE Viscosity Classification J300. There is considerable interest at the present time in a better characterization of the high-temperature performance of lubricants than that provided by the present low-shear, kinematic-viscosity limits in SAE J300. This interest stems from the trend towards lower viscosity oils for easier low-temperature starting and improved fuel economy. There is a concern that journal bearing durability problems may arise due to possible limitations of the kinematic viscosity as a guide to oil-rheological effects and its lack of relevance in terms of temperature and shear rate to critical areas of the engine. In journal bearings for instance shear rates in the region 10^5 s^{-1} to 10^7 s^{-1} occur.

1.1.2 Experimental

1.1.2.1 Engines

The connecting-rod, big-end bearings of two gasoline engines were instrumented. One engine was an in-line, four-cylinder, 2.3 litre, fuel-injected unit (hereinafter referred to as the 2.3l L-4 engine). The ungrooved big-end bearings were of lead/bronze on a steel backing. There was a squirt hole in the upper shell for lubrication of the bore. The bearing corresponding to cylinder number 1 (pulley-end) was instrumented. The other engine was a 60°, V-6, fuel-injected unit of 2.8 litre capacity (referred to as the 2.8l V-6 engine). The ungrooved bearing material was copper/lead on a steel backing with an overlay of lead/tin. The cylinder of the instrumented bearing was in the middle of the bank. Special big-end bearing shells without the normal squirt hole in the upper shell were obtained from the engine manufacturer. Values of 40 µm and 25 µm were used for the radial

3

clearances of the instrumented big-end bearings of the 2.3 litre and 2.8 litre engines, respectively. These are estimated values based on pre-build information. Since the bearings have not been dismantled their present radial clearances are not known precisely. Table 1.1.1 gives further information on the engines and the bearing dimensions.

The engines were installed in a test cell, speed and torque being controlled by a dynamometer. Oil temperature was controlled at the gallery by passing the oil through an external oil cooler and electric heater.

Table 1.1.1: Test-engine data

		2.3l L-4	2.8l V-6
Big-end bearing dimensions:			
length	, mm	21.5	16.5
diameter	, mm	48.5	54.0
area	, cm^2	32.8	28.0
length :diameter ratio		0.44	0.30
shell thickness	, mm	1.8	1.4
Maximum brake horsepower	, kW	100 at 5100 r/min	110 at 5800 r/min
Maximum torque	, N · m	205 at 3500 r/min	216 at 4000 r/min
Bore diameter	, mm	95.5	93
Stroke	, mm	80.3	68.5
Compression ratio		9.0:1	9.2:1
Radial clearance★	, μm	40	25

★ Estimated for the instrumented big-end bearing

1.1.2.2 Oil Film Thickness Measurement

The oil film thickness in the big-end bearing during engine operation was calculated from measurement of the total capacitance, C, of the oil film in the bearing. For a cylindrical bearing and shaft (i. e. no distortion) and in the absence of cavitation (i. e. uniform dielectric constant of the oil in the bearing), it can be shown (3) that:

$$\text{MOFT} = R\{1 - [1-(kAe/RC)^2]^{0.5}\} . \tag{1}$$

Here MOFT is the minimum oil film thickness at a given crankshaft position (see Fig. 1.1.1), R is the radial clearance, k the permittivity of free space, A the area of the bearing and e the dielectric constant of the oil at the temperature and pressure in the bearing.

4

In order to measure capacitance of the oil film in the bearing, the bearing shell was electrically insulated from the connecting rod by replacing ca. 200 μm of metal in the bearing housing by an equal thickness of alumina ceramic applied by plasma spraying (see. Fig. 1.1.1). The journal was earthed by means of copper braid held in tension over a pulley on the end of the crankshaft. The bearing shells and the journal now act as a cylindrical capacitor with the oil film as dielectric.

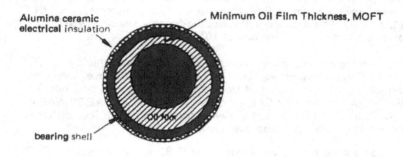

Figure 1.1.1:   Electrical insulation of the big-end bearing shell for the measurement of the minimum oil film thickness, MOFT, as a function of crankangle

The oil film capacitance was measured continuously as a function of crankshaft angle by a capacitive divider circuit (3) in the case of the 2.3l L-4 engine and by a transformer ratio arm bridge circuit for the 2.8l V-6 engine (12). A check was made (19) that both techniques gave the same capacitance by replacing the ratio-arm circuit by a capacitive divider circuit for the 2.8l V-6 engine. The ac voltages applied to the bearing capacitor had frequencies of 100 kHz and 20 kHz for the 2.3 litre and 2.8 litre engines, respectively. The output voltage was used to calculate a capacitance by use of a voltage/capacitance calibration relationship established by replacing the bearing by a series of fixed capacitors. Further details of the electrical circuits and the data acquisition systems are given elsewhere (3,4,12-14).

Electrical connections were made to the reciprocating big-end bearing by screened cables supported by a light-weight, aluminium-alloy, scissor linkage. The linkages were custom built for each engine by T&N Technology, Cawston, England and allowed operation at speeds up to 4000 r/min for periods in excess of 200 hours before wire breakage occured. Connections of the wires to the bearing were arranged so that replacement of broken wires could be carried out without removal of the big-end bearing cap, thereby ensuring continuity of results. In the case of the 2.8l V-6 engine, theoretical calculations (20) established that the linkage did not contribute significantly to loading in the bearing.

5

1.1.2.3 Oils

A series of five mono- and sixteen multigrade oils were specially blended for this programme. The oils are listed in Table 1.1.2 along with their viscosities and dielectric constants. Viscosities were measured at $100°C$, $130°C$ and $150°C$ at both low- and high-shear rates (see Table 1.1.2). The low-shear-rate viscosities were measured in an Ubbelohde viscometer (ASTM Procedure D 445) which yields kinematic viscosities, $V_h(T)$, at a temperature T. The high-shear-rate viscosities were measured in a Ravenfield tapered plug viscometer by CEC procedure L-36-A-87. This procedure provides the dynamic viscosity, $V_d(T, 10^6)$, at a temperature T and a fixed shear rate of 10^6 s^{-1}.

The oils all contained the same (commercial) dispersant/inhibitor SF/CC performance package. Conventional base oils from a single source were used, all fully-formulated oils being blended from the same batches of base oils. The monograde oils covered the four SAE viscosity grades SAE 20, SAE 30, SAE 40 and SAE 50. The sixteen multigrade oils were formulated from four different, commercial, viscosity index (VI) improver types:

namely styrene-isoprene (S-I), styrene-butadiene (S-B), olefin copolymer (OCP) and polymethacrylate (PMA). The base oil and VI improver concentrations were determined by the respective supplier of the VI improver. Each VI improver type was formulated into the following SAE grades: 10W30, 10W40, 15W40 and 20W50.

The dielectric constants of the fresh oils were measured by Southwest Research Institute, San Antonio, Texas using a brass, cylindrical capacitor and an excitation signal of 100 kHz. Results at $100°C$ are shown in Table 1.1.2. The mean dielectric constant is 2.13, there being little difference between the oils. A value of 2.1 was used throughout these studies, the effects of temperature between $100°C$ and $150°C$ (13,14) and pressure between atmospheric and 100 MPa (21) being small (i. e. ca. 1 %). Changes of 1 % in dielectric constant alter MOFT as calculated by Equation (1) by about 1 %.

1.1.3 Results

Voltage/crankangle curves were averaged over sixteen, $720°$, engine cycles in order to minimise the effects of cycle-to-cycle variations. The repeatability is not very sensitive to the number of cycles averaged, provided it is not less than about eight; sixteen is a somewhat arbitrary, convenient number.

Measurements were taken after about one hour of engine operation to allow temperature equilibrium around the engine to be established. MOFT was determined as a function of crankangle for each of the mono- and multigrade oils at the conditions shown in Table 1.1.3. One of the SAE 40 oils, RL 153, was used for a study of the effects of engine speed and torque on MOFT; the engine conditions used are shown in Table 1.1.4.

Table 1.1.2: Viscometric and dielectric constant data for mono- and multigrade oils

Oil Code	SAE Grade	VII Type	CCS[a] Pa·s	$V_k(T)$, mm²/s			$V_d(T,10^6)$, mPa·s			e[c]
				100°C	130°C	150°C	100°C	130°C	150°C	100°C
RL151	SAE 20	None	nd[b]	7.98	4.41	3.25	6.48	3.57	2.62	2.10
RL152	SAE 30	None	nd	9.74	5.28	3.83	8.00	4.30	3.11	2.12
RL153	SAE 40	None	nd	12.98	6.74	4.76	11.00	5.47	3.81	2.12
RL154	SAE 40	None	nd	14.33	7.30	5.26	11.74	5.97	4.20	2.13
RL155	SAE 50	None	nd	18.65	9.12	6.33	15.65	7.52	5.14	2.15
RL156	10W30	S-I	32.0	11.43	6.46	4.78	7.14	4.16	3.12	2.11
RL157	10W40	S-I	32.2	15.21	8.49	6.24	8.54	4.99	3.74	2.10
RL158	15W40	S-I	31.4	14.83	8.11	5.85	9.05	5.16	3.82	2.11
RL159	20W50	S-I	42.8	18.27	9.46	6.78	12.20	6.44	4.60	2.12
RL160	10W30	S-B	28.6	11.10	6.31	4.62	7.72	4.41	3.28	2.12
RL161	10W40	S-B	30.1	14.78	8.13	5.88	9.66	5.49	4.06	2.12
RL162	15W40	S-B	30.0	14.39	7.65	5.33	9.82	5.54	4.08	2.12
RL163	20W50	S-B	43.9	18.08	9.21	6.47	12.28	6.62	4.77	2.13
RL164	10W30	OCP	31.7	11.11	6.27	4.66	7.06	4.17	3.15	2.10
RL165	10W40	OCP	35.7	14.61	8.10	6.06	8.55	5.02	3.77	2.11
RL166	15W40	OCP	34.1	14.83	8.05	5.93	9.32	5.32	3.94	2.12
RL167	20W50	OCP	45.2	17.96	9.26	6.73	12.00	6.43	4.62	2.13
RL168	10W30	PMA	35.0	11.16	6.44	4.83	7.36	4.34	3.27	2.16
RL169	10W40	PMA	35.0	14.70	8.61	6.37	8.72	5.15	3.88	2.19
RL170	15W40	PMA	33.5	14.56	8.28	6.22	9.31	5.24	3.86	2.17
RL171	20W50	PMA	40.8	18.26	9.74	7.25	12.10	6.60	4.78	2.17

a) Cold cranking simulator viscosities. 10 WX, 15 WX and 20 WX measured at −20°C, −15°C, −10°C respectively.
b) nd = not determined.
c) dielectric constant.

Table 1.1.3: Engine conditions used to measure MOFT of mono- and multigrade oils

Engine	Engine Condition	Crankshaft Speed r/min	Torque		Gallery Temperature °C
			N · m	% max.	
2.3 litre L-4	I	2000	45	25	100
	II	2000	108	60	100
	III	3000	120	60	100
	IV	3000	120	60	130
2.8 litre V-6	V	2500	100	40	100
	VI	2500	100	40	130
	VII	3000	100	40	100

1.1.3.1 Precision

The value of MOFT obtained by the total capacitance technique can be inde-
terminate within a factor of two (13-14) due to uncertainty in assigning a
precise value to the radial clearance in Equation (1) (see section dealing with the
enignes). The assumptions of no-cavitation and no-distortion made in deriving
Equation (1) also introduce further uncertainties in the absolute magnitude of
MOFT (8,20). These factors must be borne in mind when comparing oil film
thickness given in this paper with theoretical predictions.

The method, however, provides excellent precision in terms of the magnitude
of MOFT. Thus MOFT values for repeat tests carried out sequentially are usually
within 1 %: excellent repeatability is also obtained between measurements
made on different days for the same oil at the same engine condition, e. g.
such repeat tests have a standard deviation of less than 3 % for the 2.8l V-6
engine. MOFT values for in-house reference oils showed no drift over the period
in which measurements were made.

1.1.3.2 MOFT/crankangle curves — general features

Table 1.1.4 shows the results of the various speed/torque studies on the SAE 40
oil, RL 153. In this table, $(MOFT)_o$ is the minimum value of MOFT, as determined from the MOFT/crankangle curve (see Fig. 1.1.2), G_{max} is the maximum
shear rate (i. e. the shear rate of $(MOFT)_o$ — see later) and $_m$ is the crankangle
locating $(MOFT)_o$. (Note that 0° is top-dead-centre of the firing stroke of the
cylinder corresponding to that of the instrumented bearing.) A $(MOFT)_o$ "result" as reported in this paper is the mean of two sixteen cycle averages in
the case of the 2.3l L-4 engine and of five such averages for the 2.8l V-6 engine.
The individual averages were taken over a five to ten minute period.

Table 1.1.4: Effect of torque and crankshaft speed on $(MOFT)_o$, G_{max}, and h_m. SAE 40 oil RL 153 at a gallery temperature of $100°C$

Torque N · m	Speed r/min	2.8l V-6			2.3l L-4		
		$(MOFT)_o$ μm	$10^6 G_{max}$ /s	m degrees	$(MOFT)_o$ μm	$10^6 G_{max}$ /s	m degrees
45	1500	2.16	1.7	239	2.52	1.7	67
45	2000	1.77	2.6	225	2.22	2.0	243
45	2500	1.33	6.2	414	1.88	3.2	261
45	3000	1.05	9.6	401	1.62	4.7	272
45	3500	0.83	14.3	407	nd	nd	nd
45	4000	0.56	24.3	403	nd	nd	nd
120	1500	nd	nd	nd	1.40	3.1	59
120	2000	1.28	3.6	222	1.81	2.1	156
120	2500	1.14	5.0	222	nd	nd	nd
120	3000	0.96	7.3	227	1.46	5.2	270
120	3500	0.77	15.5	406	nd	nd	nd
120	4000	0.39	34.5	408	nd	nd	nd
10	2500	1.29	6.4	413	2.04	3.0	264
80	2500	1.26	5.0	244	1.75	3.4	258
120	2500	1.14	5.3	233	1.66	3.6	258
160	2500	1.00	5.9	230	nd	nd	nd
175	2500	nd	nd	nd	1.54	3.7	247

nd = not determined

Fig. 1.1.2 shows a comparison of MOFT/crankangle curves for the two engines for an SAE 40 oil at 2500 r/min and 80 N · m torque. These curves are representative of those obtained at different speed/torque conditions and for different viscosities. The curves for the two big-end bearings are remarkably similar. Thus there is a maximum in each of the firing, exhaust and induction strokes and a minimum in each of the exhaust, induction and compression strokes. The crankangles locating the maxima and minima are very close for the two engines, as are their relative heights and depths. The magnitude of MOFT is close for the two bearings over some of the $720°$ cycle; $(MOFT)_o$ is, however, higher by some 40 % for the 2.3l L-4 big-end bearing than for that of the 2.8l V-6 (i. e. 1.75 μm vs. 1.26 μm). It is not possible to state whether this difference is real or is due to uncertainties about the correct values of the radial clearance to be used in Equation (1) - see previous section.

Thus if a radial clearance of 20 μm, instead of 25 μm, was assumed in Equation (1) for the 2.8l V-6 engine, the calculated value of $(MOFT)_o$ would be increased to 1.59 μm. This value would be obtained for the 2.3l L-4 engine by using a radial clearance of 44 μm instead of 40 μm. Uncertainties of $\pm$5 μm are quite possible for the radial clearances of the big-end bearings since they have not been dismantled.

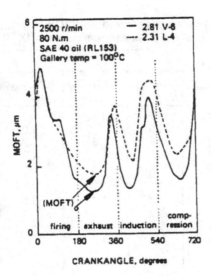

Figure 1.1.2: Minimum oil film thickness vs. crankangle for the big-end bearings of the 2.8 litre V-6 and 2.3 litre L-4 engines showing the definition of (MOFT)$_o$

1.1.3.3 Effect of Speed and Torque

Figs. 1.1.3a and 1.1.3b show the effects of crankshaft speed on the MOFT/crankangle curves for the big-end bearings of the 2.8l V-6 and 2.3l L-4 engines at a constant torque of 45 N · m; Fig. 1.1.3c shows results for the 2.3l L-4 at the higher torque of 120 N · m. The effects of speed are qualitatively similar for the two bearings. Thus MOFT *decreases* with *increasing* speed in the exhaust, induction and compression strokes. The effects of speed in the firing stroke are complicated by crossovers in this region; thus MOFT *decreases* with *increasing* speed at a crankangle of about 60° from top-dead-centre of the firing stroke; at 30°, however, MOFT passes through a maximum with increasing speed and at about 0° tends to *decrease* with *increasing* speed. For the 2.8l V-6 engine, a shallow minimum develops in the firing stroke as speed decreases; however, this minimum never becomes the lowest minimum (see Fig. 1.1.3a). No such minimum develops with decreasing speed in the firing stroke for the 2.3l L-4 engine; instead at 1500 r/min the minimum in the exhaust stroke develops into a trough extending from about 60° to about 300°; at this speed, (MOFT)$_o$ in fact is located in the firing stroke.

10

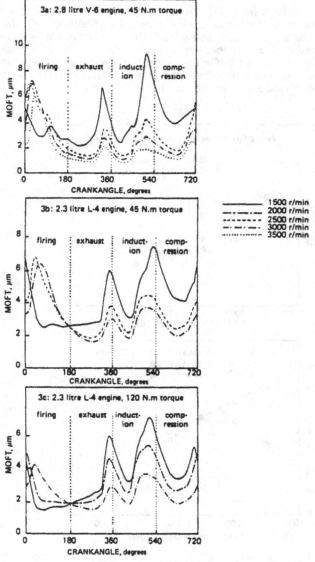

Figure 1.1.3: Effect of crankshaft speed on MOFT/crankangle curves for SAE 40 oil RL 153 at a gallery temperature of 100°C

As seen in Fig. 1.1.4a, $(MOFT)_0$ decreases monotonically with increasing speed for the big-end bearing of the 2.8l V-6 engine for speeds above 1500 r/min at 45 N · m torque and at speeds above 2000 r/min at 120 N · m torque (these are the lowest speeds at which the engine was run at each torque). This decrease arises from the location of $(MOFT)_0$ in either the exhaust or induction strokes for this engine. As shown in Table 1.1.4 and Fig. 1.1.3a, $(MOFT)_0$ shifts from the exhaust to the induction stroke with increasing speed.

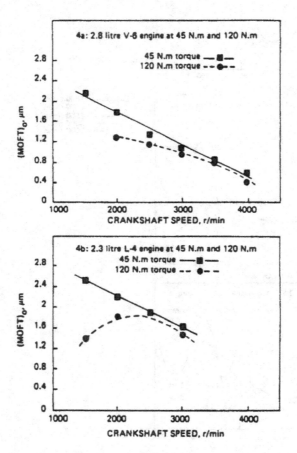

Figure 1.1.4: Dependence of $(MOFT)_0$ on crankshaft speed for SAE 40 oil RL 153 at constant torque. Gallery temperature = 100°C

At a torque of 45 N · m, $(MOFT)_o$ for the big-end bearing of the 2.3l L-4 engine also decreases with increasing speed, as shown in Fig. 1.1.4b. At the higher torque of 120 N · m, on the other hand, $(MOFT)_o$ passes through a maximum just above 2000 r/min. This effect arises from the previously mentioned shift of $(MOFT)_o$ at 1500 r/min to the firing stroke where the high gas pressures at 120 N · m are sufficient to depress the minimum oil film thickness in this region to below the minima in the exhaust stroke at the higher speeds (compare Figs. 1.1.3a and 1.1.3c).

The maximum in the data for the 2.3l L-4 engine at 120 N · m suggests that this curve passes through the origin. Indeed this must be so since at zero speed there is no basis for hydrodynamic film formation and $(MOFT)_o$ must be zero. All the curves in Fig. 1.1.4 must therefore have a maximum; experimentally, it is below 1500 r/min at low torques. It is apparent from Figs. 1.1.4a and 1.1.4b that $(MOFT)_o$ decreases monotonically with increasing speed for speeds above about 2000 r/min for both bearings, there being no evidence of a levelling off at 4000 r/min (the maximum speed presently studied) for the 2.8l V-6 engine. The presence of a maximum and the monotonic decrease in $(MOFT)_o$ with increasing speed are in accordance with expectations; thus initially as speed increases, $(MOFT)_o$ will increase due to the increase in oil film pressure; inertial forces are, however, also increasing for a big-end bearing with increasing speed and eventually these will overwhelm the oil film pressure effects and result in a decrease in $(MOFT)_o$ with increasing speed. The most severe steady-state operating conditions for these big-end bearings are therefore low-speed (<1500 r/min) lugging and high-speed cruising (>4000 r/min) at any torque. Since the latter is likely to be more frequently encountered in practice, further data at speeds above 4000 r/min. are desirable to determine if the effects of oil rheology on $(MOFT)_o$ are the same at say 6000 r/min as they are at the lower speeds studied here. Concern about the durability of the mechanical scissor linkage at these speeds has so far prevented such measurements being attempted.

Figs. 1.1.5a and 1.1.5b show typical results of the effect of torque on the MOFT/crankangle curves. As expected, MOFT decreases with increasing torque in the firing stroke, the effect being quite dramatic. After about the midpoint of the exhaust stroke, however, the effects of gas pressure on MOFT have decreased considerably; as a consequence, $(MOFT)_o$ which occurs in the induction and compression strokes, decreases only slightly with increasing torque.

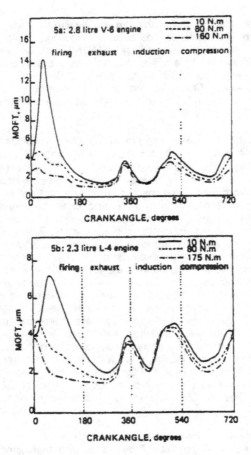

Figure 1.1.5: Effect of torque on MOFT/crankangle curves for SAE 40 oil RL 153 at a constant crankshaft speed of 2500 r/min. Gallery temperature = 100°C

1.1.3.4 Shear rates in the bearing

For a big-end bearing, the maximum shear rate, G_{max}, depends not only on speed and $(MOFT)_o$ but also on the crankangle, $_m$, locating $(MOFT)_o$. The latter dependence arises because the throw of the crank has an effect on the relative velocities of the shaft and bearing. The value of G_{max} for any given $(MOFT)_o/m$ can be calculated from the following expression (14):

$$G_{max} = [r_j/(MOFT)_o](2 \pi N/60)(1 + d\beta_b/d\beta_a), \quad (2) \tag{2}$$

14

where r_j is the journal radius, N the rotational speed of the crankshaft, β_b and β_a are the angles between the centre-line of the cylinder and lines joining the centres of the big- and little-ends and big-end and main bearings, respectively. The differential in Equation (2) can be evaluated by a numerical procedure using the following relationship:

$$\beta_b = \sin^{-1}\left[(r/1)\sin\beta_a\right],$$

where 1 is the distance between the centres of the big- and little-ends and r is the radius of the crank.

Values of G_{max} for the SAE 40 oil RL 153 are shown in Table 1.1.4 for the two big-end bearings at different speed/torque conditions. G_{max} increases with increasing crankshaft speed due to the direct dependence on speed and the inverse dependence on $(MOFT)_O$ which decreases with increasing speed above 2000 r/min. It is seen that G_{max} exceeds 4×10^6 s^{-1} for speeds above 2500 r/min for both engines. At the highest speeds evaluated in this study (4000 r/min for the 2.8l V-6 engine), G_{max} is in the region of 3×10^7 s^{-1}. Even assuming that the absolute values of $(MOFT)_O$ are low by a factor of two (see earlier discussion on precision), maximum shear rates at speeds in the region of 4000 r/min will be close to 10^7 s^{-1}. A crankshaft speed of 4000 r/min is by no means excessive for many modern engines for which peak torque may occur above 4000 r/min and peak power above 6000 r/min. For big-end bearings therefore, maximum shear rates in the region of 10^7 s^{-1} would seem to be normal for high-speed cruising. Such shear rates are a factor of ten higher than the value currently considered typical in journal bearings of operating engines. Correlations of $(MOFT)_O$ for multigrade oils with viscosities measured at 10^6 s^{-1} must therefore be treated with caution, recognising that the rate of change of viscosity with increasing shear rate depends on the chemical type, molecular weight and concentration in the oil of the VI improver. As a consequence, rankings obtained at 10^6 s^{-1} will not necessarily be a guide to the situation actually pertained in the bearing at, say, 10^7 s^{-1}

1.1.3.5 Effect of viscosity

$(MOFT)_O$ values were related to viscosity, separately, for each of the five classes of oil - namely monogrades, S-I, S-B, OCP and PMA. Correlations were found for each of the engine conditions for which measurements were made. Thus there was a total of fifteen correlations for the 2.8l V-6 engine and twenty for the 2.3l L-4 engine. Values of $(MOFT)_O$ were related to both the low-shear, kinematic viscosity at $100°C$ and the high-shear dynamic viscosity at $150°C$ and 10^6 s^{-1}; these viscosities are denoted by $V_k(100°C)$ and $V_d(150°C, 10^6)$, respectively. They were chosen because the former is the only high-temperature viscosity specified in the SAE Viscosity Classification J300 whilst the latter is under active discussion as a possible alternative, more relevant, measure of a lubricant's performance in journal bearings.

The indices of determination, R^2, are shown in Table 1.1.5 for the two engines. The main conclusions to be drawn from the results in this table are:

a) For the monogrades, $V_k(100°C)$ and $V_d(150°C, 10^6)$ are equally valid predictors of $(MOFT)_o$. The R^2 values are consistently high, eleven of the fourteen regressions having values of 0.99 with lowest being 0.95. That there is little to choose between $V_k(100°C)$ and $V_d(150°C, 10^6)$ for monograde (Newtonian) oils is to be expected since definitively viscosity is independent of shear rate for such fluids; hence the two viscosities are simply related via density. The well-defined dependence on viscosity for monograde oils provides a solid baseline against which to evaluate the more-rheologically-complex multigrade oils.

b) For multigrade oils, $V_d(150°C, 10^6)$ gives better correlations than $V_k(100°C)$, R^2 *always* being greater for the former.

c) For both engines, $V_d(150°C, 10^6)$ correlates monograde data somewhat better than it does data for *individual* multigrades (i. e. those based on a given VI improver type).

d) $V_d(150°C, 10^6)$ does not correlate the *combined* mono- and multigrade oils as well as it does the monograde or individual multigrade oils. Thus R^2 is in the range 0.56 to 0.90 for the combined data compared with 0.74 to 0.99 for the individual data.

An alternative method of analysis is to normalise the $(MOFT)_o$ values by dividing by $(MOFT)_o$ for some reference oil. Previous results (14, 20) have shown that such normalisation removes the effect of engine condition on $(MOFT)_o$, thereby greatly increasing the number of data points available for each correlation.

The value of $(MOFT)_o$ for the SAE 40 oil RL 153 was (arbitrarily) chosen as the reference for normalisation. The $(MOFT)_o$ values for each of the remaining oils at a given speed/torque condition were divided by that for RL 153 at 100°C at the same speed/torque condition. The normalised $(MOFT)_o$ values, which will be denoted by H_o, were then correlated against viscosity at the applicable gallery temperature (i. e. 100°C for 130°C). In this way, the data can be condensed into two correlations (mono- and multigrade) for each engine for any given viscosity parameter.

The dependence of normalised minimum oil film thickness on viscosity was represented by the relation

$$H_o = A + BV^{0.5}, \tag{3}$$

where V is either the low-shear kinematic viscosity, $V_k(T)$, at the gallery temperature T, or the high-shear dynamic viscosity, $V_d(T,10^6)$. The square root of viscosity was used because it gave somewhat better correlations than those based on the first power of viscosity. There is also theoretical justification (20) for the square-root model. The conclusions would not be altered, however, if V rather than $V^{0.5}$ was used.

16

Table 1.1.5:   Correlation coefficients, R^2, for the model $(MOFT)_0 = a + bV$

		2.3l L-4				2.8l V-6		
	n[a)	I[b)	II	III	IV	V	VI	VII
$V_d(150°C, 10^6)$								
Monogrades	5	0.99	0.99	0.99	0.99	0.99	0.99	0.95
S-I	4	0.99	0.98	0.99	0.99	0.92	0.95	0.94
S-B	4	0.99	0.95	0.94	0.98	0.87	0.97	0.89
OCP	4	0.98	0.97	0.97	0.96	0.74	0.99	0.93
PMA	4	0.97	0.95	0.94	0.98	0.79	0.93	0.94
Mono- + multigrades	21	0.89	0.90	0.87	0.89	0.68	0.56	0.65
$V_k(100°C)$								
Monogrades	5	0.99	0.99	0.99	0.99	0.98	0.99	0.94
S-I	4	0.95	0.91	0.94	0.93	0.81	0.86	0.85
S-B	4	0.98	0.94	0.95	0.97	0.85	0.96	0.90
OCP	4	0.95	0.92	0.92	0.92	0.66	0.96	0.88
PMA	4	0.93	0.89	0.88	0.93	0.73	0.89	0.90
Mono- + multigrades	21	0.90	0.86	0.74	0.81	0.52	0.43	0.48

a) n is the number of oils in the correlation
b) I,II,III, etc. denote engine conditions as described in Table 1.1.3

Figs. 1.1.6a and 1.1.6b show plots of H_0 against the square root of $V_d(T,10^6)$ for the monograde oils for the 2.3 litre and 2.8 litre engines, respectively; results for the multigrade oils are shown in Figs. 1.1.7a and 1.1.7b. The data were analysed as follows:

The goodness-of-fit for each of the models represented in Figs. 1.1.6 and 1.1.7 was measured by its residual mean square, RMS; the smaller the RMS, the better the agreement between the data points and the model. For each model, the RMS was obtained by dividing the residual sum-of-squares by the degrees of freedom. The residual sum-of-squares was determined by summing, across all data points, the square of the difference between the observed value of H_0 and the value predicted by the model. The degrees of freedom (DF) were cal-culated as the number of data points minus two (the number of coefficients,

A and B, in the model). When comparing mono- and multigrade data sets, allowance must be made for the larger number of data points for the (normalised) multigrade oils (64 vs. 20 for the 2.3l engine and 48 vs. 15 for the 2.8l engine). In such cases, comparison of the RMS values for the mono- and multigrade models, using F-tests, was used to assess whether the models differed significantly in their goodness-of-fit. All significant differences reported here are significant at the 95 % confidence level (2-sided).

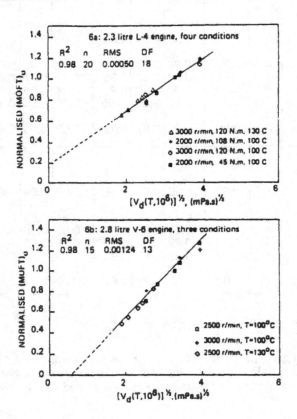

Figure 1.1.6: Dependence of normalised $(MOFT)_0$ on square root dynamic viscosity at $10^6 s^{-1}$ for monograde oils at different engine conditions

18

Prior to comparing mono- and multigrade data sets, the data were analysed to determine if there was an effect of engine condition. For both mono- and multigrade oils in the 2.8l V-6 engine, there were no significant difference between the regression lines for the three engine conditions used. There were significant differences between some of the regression equation for the four engine conditions used for the 2.3l L-4 engine. However, because the equations frequently fit the data almost perfectly, some of these differences are so small as to be of no practical interest (see Fig. 1.1.6a, for example). There were no significant interactions between engine conditions and oil type so that the effects of engine conditions were consistent across oil types and vice versa.

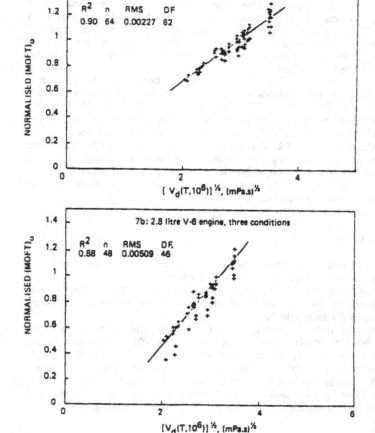

Figure 1.1.7: Dependence of normalised $(MOFT)_o$ on square root dynamic viscosity at $10^6 s^{-1}$ for multigrade oils at different engine conditions

19

Table 1.1.6 summarises the RMS and degrees of freedom, as well as the R^2 values and number of data points, n, for the mono- and multigrade oils in the two engines, using both V_k and V_d. Also shown are the coefficients A and B in Equation (3). The following conclusions are drawn:

a) For $V_d(10^6)$, the fit to the monograde data (Fig. 1.1.6) is significantly better than that to the multigrade data when the different VI improver chemistries are combined, as seen in Fig. 1.1.7. In other words, shilst a HTHSV at a fixed shear rate of 10^6 s^{-1} correlates monograde $(MOFT)_o$ data well, it is less satisfactory for the more-rheologically-complex multigrade oils.

b) For $V_d(10^6)$, the regression equation for monograde oils is significantly different from that for the multigrade oils for both engines. HTHSV at a fixed shear rate of 10^6 s^{-1} is, therefore, deficient as a parameter for representing both mono- and multigrade data. This deficiency must partly arise from the fact that 10^6 s^{-1} is a factor of five to ten lower than the actual maximum shear rate in the bearing. The viscosity at G_{max} will *always* be lower than that at 10^6 s^{-1} for multigrade oils. (For monograde oils, of course, viscosity is independent of shear rate.) In the absence of viscosities at the relevant shear rate a meaningful comparison of the relative performance of mono- and multigrade oils is not possible.

b) For the 2.3l L-4 bearing, $V_d(10^6)$ provides a better fit (at the 95 % confidence level) to the multigrade data than does the low-shear kinematic viscosity (RMS = 0.00227 vs. 0.00390 with 62 DF). Although $V_d(10^6)$ provides a better correlation than V_k for the multigrade data in the 2.8l V-6 bearing, the difference is only significant at the 85 % confidence level (RMS = 0.00509 vs. 0.00784 with 46 DF). Thus whilst $V_d(10^6)$ correlates multigrade data better than V_k, the former is by no means greatly superior. It is probably best to say that either parameter has a first order effect on $(MOFT)_o$ but that some other parameter (e. g. V_d at G_{max}) or combination of parameters (6) may be significantly better.

1.1.4 Conclusions

From the measurement of minimum oil film thickness in the big-end bearings of two different operating engines, it is concluded that:
1. The general shape of the MOFT/crankangle curve (i. e. the number, crankangle positions and relative heights of the maxima and minima) is independent of the big-end bearing design.
2. The minimum oil film thickness, $(MOFT)_o$, in big-end bearings decreases with increasing crankshaft speed for speeds above 2000 r/min and is not strongly dependent on torque. The most severe steady-state operating condition for big-end bearings of gasoline engines is high-speed cruising, independent of load. Low-speed lugging is also severe but since this requires wide-open-throttle operating at speeds of 1500 r/min or below, it is assumed that it occurs much less frequently in service than high-speed cruising.

Table 1.1.6: Statistical parameters for models based on Equation (3)

	n	[$V_d(T \cdot 10^6)$]$^{0.5}$					[$V_k(T)$]$^{0.5}$				
		RMS	DF	R²	A	B	RMS	DF	R²	A	B
a) 2.8l V-6											
Monogrades	15	0.00124	13	0.98	−0.19	0.37	0.00101	13	0.98	−0.23	0.34
Multigrades	48	0.00509	46	0.88	−0.43	0.43	0.00784	46	0.82	−0.33	0.34
b) 2.3l L-4											
Monogrades	20	0.00050	18	0.98	0.22	0.22	0.00053	18	0.98	0.18	0.23
Multigrades	64	0.00227	62	0.90	0.22	0.00	0.00390	62	0.83	0.04	0.26

3. The maximum shear rates in big-end bearings are in the region of 10^7 s^{-1} rather than 10^6 s^{-1} for high-speed operation.
4. Dynamic viscosity at 10^6 s^{-1} is deficient as a parameter for correlating MOFT data of multigrade oils formulated with different VI improvers or for correlating mono- and multigrade data simultaneously. The adoption of $V_d(150°C,10^6)$ in existing or proposed viscosity classifications, therefore, should be viewed as a compromise which attempts to represent the range of shear rate conditions in the lubricated contacts in an operating engine.
5. Although dynamic viscosity correlates $(MOFT)_0$ data for multigrade oils better than the low-shear kinematic viscosity, the difference is not always significant at the 95% confidence level. Some other parameter (e. g. the dynamic viscosity at the maximum shear rate) or combination of parameters could correlate mono- and multigrade data better.
6. Further studies aimed at elucidating the rheological parameters which determine bearing oil film thickness are required. Such studies require measurement of rheological properties at shear rates well above 10^6 s^{-1}. Oil film thickness measurements in big-end bearings under both steady and non-steady state operation at crankshaft speeds above 4000 r/min will also be necessary in order to relate MOFT measurements to field practice.

Acknowledgements

The authors thank the following members of the CEC PL-33 Group for supplying the viscosity index improvers and formulations for the oils used in this study: Shell Research Ltd, Röhm and Haas, Lubrizol and Paramins. The contributions of Roger Park and Phil Normington of Esso Research Centre, Abingdon in carrying out the oil film thickness measurements on the 2.3l L-4 engine are greatly appreciated.

Measurements on the 2.8l V-6 engine were made possible by the efforts of Peter Evans, Clive Sims, Brian Spencer and Stephen Benwell of Shell Research Ltd, Thornton. Statistical analyses were carried out by Geoff Morris of Shell Research Ltd.

1.1.5 References

(1) Rhodes, R.B.; Henderson, B.M.: "The Effects of Engine Oil Viscosity and Composition on Bearing Wear", SAE Paper No. 811224 (1981).

(2) Lonstrup, T.F.; Smith Jr., M.F.: "Engine Oil and Bearing Wear", SAE Paper No. 810330 (1981).

(3) Craig, R.C.; King, W.H.; Appeldoorn, J.K.: "Oil Film Thickness in a Bearing of a Fired Engine-Part II", SAE Paper No. 821250 (1982).

(4) Girshick, F.; Craig, R.C.: "Oil Film Thickness in a Bearing of a Fired Engine-Part III", SAE Paper 831691 (1983).

(5) Spearot, J.A.; Murphy, C.K.; Rosenberg, R.C.: "Measuring the Effect of Oil Viscosity on Oil Film Thickness in Journal Bearings", SAE Paper No. 831689 (1983).

(6) Bates, T.W.; Williamson, B.P.; Spearot, J.A.; Murphy, C.K.: "A Correlation Between Engine Oil Rheology and Oil Film Thickness in Engine Journal Bearings", SAE Paper No. 860376 (1986).

(7) Schilowitz, A.M.; Waters, J.L.: "Oil Film Thickness in a Bearing of a Fired Engine-Part IV", SAE Paper No. 861561 (1986).

(8) Spearot, J.A.; Murphy, C.K.: "A Comparison of the Total Capacitance and Total Resistance Techniques for Measuring the Thickness of Journal Bearing Oil Films in an Operating Engine", SAE Paper No. 880680 (1988).

(9) Deysarkar, A.: "Bearing Oil Film Thickness of Single and Multigrade oils-Part 1", SAE Paper No. 880681 (1988).

(10) Campbell, J.; Love, P.P.; Rafique, S.O.: "Bearings for Reciprocating Machinery: A Review of the Present State of Theoretical, Experimental and Service Knowledge", Proc. I. Mech. E., 182, Part 3A, pp51–74 (1967).

(11) Dobson, G.R.; Pike, W.C.: "Predicting Viscosity Related Performance of Engine Oils", I. Mech. E. Conf. Publ., p65 (1982).

(12) Bates, T.W.; Evans, P.G.: "Effect of Oil Rheology on Journal Bearing Performance: Part 1", Proc. JSLE Int. Tribology Conf., Tokyo, July 8–10, p445 (1985).

(13) Bates, T.W.; Evans, P.G.; Benwell, S.: "Effect of Oil Rheology on Journal Bearing Performance: Part 2", 4th SAE Int. Pacific Conf. on Automotive Engineering", Melbourne, Nov 8–14 (1988).

(14) Bates, T.W.; Benwell, S.: "Effect of Oil Rheology on Journal Bearing Performance", SAE Paper No. 880679 (1988).

(15) Cassiani Ingoni, A.A.; DiLelio, G.F.; Eberan-Eberhorst, C.G.A.: "European Activity Concerning Engine Oil Viscosity Classification", SAE Paper No. 770374 (1977).

(16) Eberan-Eberhorst, C.G.A.; Dilelio, G.F.; Cassiani Ingoni, A.A.: "European Activity Concerning Engine Oil Viscosity Classification-Part II", SAE Paper No. 780377 (1978).

(17) Cassiani Ingoni, A.A.; DiLelio, G.F.; van Os, N.; Vickars, M.A.: "European Activity Concerning Engine Oil Viscosity Classification-Part III", SAE Paper No. 800546 (1980).

(18) Cassiani Ingoni, A.A.; DiLelio, G.F.; van Os, N.; Vickars, M.A.: "Measurements of Viscosity at High Temperature, High Shear — CEC CL 23 Achievements", Paper EL/4/1 presented at CEC Symposium, Rome, Italy, June 3–5, 1981.

(19) Evans, P.G.: Shell Research Ltd., private communication.

(20) Bates, T.W.; Fantino, B.; Launay, L.; Frene, J.: "Oil Film Thickness in an Elastic Connecting-Rod Bearing: Comparison Between Theory and Experiment", to be presented at ASME-STLE Tribology Conference, Atlanta, U.S.A. (1989).

(21) Scaife, W.G.S.; Lyons, C.G.R.: "Dielectric Permittivity and pvT Data of Some n-Alkanes", Proc. R. Soc. Lond., a370, 193 (1980).

23

1.2 Does the Automotive Industry Need a Standard Engine Test to Measure Journal Bearing Oil Film Thickness?

J.A. Spearot
General Motors Research Laboratories, Warren, USA

1.2.1 Summary

Automotive manufacturers are dissatisfied with the high-temperature portion of the SAE Viscosity Classification J300. This dissatisfaction occurs because the low-shear, 100°C conditions at which viscosities are specified are not representative of the conditions to which oils are subjected at operating temperatures in an engine. Recents efforts to modify the high-temperature viscosity specifications have dealt primarily with developing correlations between oil film thickness measured in the journal bearings of operating engines and viscosities values measured at different high-temperature, high-shear rate conditions in laboratory viscometers. However, it has also been suggested that the industry develop a standard engine test for measuring journal bearing oil film thickness, and that oils be "classified" on the basis of film thickness rather than viscosity. This paper reviews the methods which have been developed to measure bearing film thickness, the correlations which have been calculated between film thickness and viscosity, and the conclusions drawn from such efforts. Reasons for relying on high-temperature, high-shear viscosity specifications rather than a standard industry bearing film thickness test are offered. In addition, the merits and drawbacks of developing an industry bearing distress test are also discussed.

1.2.2 Introduction

Concern over the long-term durability of automotive journal bearings has increased during the past decade due to two competing factors. First, engine oil temperatures have generally increased because of efforts to generate more power from small displacement, 4- and 6-cylinder engines and because of cooling limitations created by sloping, aerodynamic hood designs. As oil temperatures have increased, some have considered specifying higher oil viscosity grades in order to generate the same level of hydrodynamic protection provided by the engine oil in earlier years. At the same time, however, government-mandated standards in the U.S. have forced automotive engineers to consider the use of lower viscosity oils to maximize corporate average fuel economy. Because long-term durability is still a primary design goal, engine and bearing designs have been modified to account for the use of such lower viscosity oils.

At the same time that engine oil viscosity has reemerged as an important issue with regard to engine design and performance, the deficiencies in the present method for specifying the high-temperature viscometric properties of engine oils have become more apparent. Currently, the SAE Engine Oil Viscosity Classification System, J300, uses kinematic viscosity measured at 100°C to define the high-temperature properties of engine oils. This specification has a history which dates from the earliest attempts within SAE (circa 1911) to categorize oils according to their viscosities. Since only Newtonian, single-grade, mineral oil-based lubricants were available at that time, the complexities of describing the viscometric properties of non-Newtonian, polymer-containing, multigrade oils were of no concern. In today's engine bearings, with the lubricant subjected to shear rates approaching 10^7 s^{-1} , and temperatures exceeding 150°C, the non-Newtonian character of the oil is of great importance. Kinematic viscosity measurements at 100°C provide no information regarding the viscosity of the oil under these severe engine operating conditions.

Engine manufacturers both in the U.S. and in Europe have felt sufficiently concerned about the lack of adequate viscosity definition to set their own service and factory-fill requirements on engine oil viscosity measured at 10^6 s^{-1} and 150°C. In addition, in Europe the CCMC has issued an industry-wide requirement of 3.5 cP or greater for the viscosity of its new G4 and G5 engine oil grades measured at these same conditions. In the U.S., the issuance of industry-wide standards on high-temperature, high-shear (HTHS) viscosity have been delayed by debate in the relationship between such measurements and engine operation or performance.

In 1985, the American Society for Testing and Materials, ASTM, published a review of oil rheological effects on engine performance (1) as they were understood at that time. Generally, it was agreed that bearing durability is a function of both the rheological properties of an oil as well as the chemical characteristics of the additive package. It was stated that separating the effects of rheological and chemical factors on bearing wear would be very difficult. In contrast, it was suggested that bearing oil film thickness (BOFT) appears to be affected only by the rheological properties of the oil, and it was proposed that to determine the correct conditions at which to specify oil viscosity, development of measurement techniques for determining BOFT should be pursued.

This paper reviews the development of bearing oil film thickness measurement techniques which have occurred during the decade of the 1980's. In addition, it describes some of the latest findings with regard to the effects of oil rheological properties on BOFT in engines. Some have suggested that the relationship between oil rheological properties and bearing film thickness is so complex that no single viscosity value measured at a particular temperature and shear rate will adequately define bearing performance. Instead, they argue for a BOFT "engine sequence test" to define the effect of oil on bearing operation. The concept of a BOFT engine test is compared with simply including HTHS viscosity limits in SAE J300, and the merits of each prosposal are discussed.

1.2.3 Development of Methods for Measuring BOFT

As has been described (1), early attempts to measure bearing oil film thickness in operating engines and laboratory rigs dealt primarily with the use of discrete electrical probes. These probes were positioned at different locations around the bearing and measured the thickness of the oil film by measuring the change in some electrical property (eddy currents or capacitance, for example) due to the variation of the film. Discrete probes suffer from several experimental drawbacks. First, they are difficult to install and "zero" with respect to the surface of the bearing. They are fragile and subject to physical deterioration during bearing operation. And finally, the electrical properties of the probes can change in a complicated fashion with temperature at different operating conditions, making analysis of the data more complex.

In 1978, Bassoli, Cornetti, and Belei (2) suggested a technique for assessing the general operating conditions of an engine crankshaft. Instead of using discrete probes, they proposed a method by which the entire oil film in a bearing could be used as a "sensor". Specifically, their technique included electrically insulating each of the journal bearings from the engine block, and applying an electric potential between the bearings and the crankshaft. They assessed the operating condition of the crankshaft by determining the overall electrical resistance produced by the oil films between the crankshaft and bearings. In addition, they formulated equations which predicted that, if the electrical conductivity of the oil in the bearing were known, measurements of the oil film electrical resistance could be used to calculate a specific value for the bearing oil film thickness.

Spearot, Murphy, and Rosenberg (3) used this concept to determine the oil film thickness in the front main bearing of a 3.8l V-6 engine by measuring the electrical resistance and oil conductivity provided by a series of single- and multigrade oils. At approximately the same time, Craig, King and Appeldoorn (4) using a modification of the technique, measured BOFT values in a main bearing for a series of single-grade oils by determining the dielectric constant of the oil and the electrical capacitance of the oil film. Later Girshick and Craig (5) continued this work by measuring film thickness values for multigrade oils using the capacitance technique. The flexibility of this general experimental method was demonstrated by Schilowitz and Waters (6) who measured oil film thickness in the front main bearing of an engine in an accelerating vehicle. In addition, Bates and Benwell (7), and later Bates and Vickars (8), used a mechanical linkage to provide electrical connections to the connecting rod bearings of an engine. Using the capacitance technique, the BOFT of such "big-end" bearings have been measured for both single- and multigrade oil formulations.

1.2.4 Experimental Methods for Making BOFT Measurements

All of the references cited have used similar methods for preparing the engines for making bearing film thickness measurements. Whether film thickness values are to be determined from resistance or capacitance measurements, whether they are to be determined in main or connecting rod bearings, the first requirement is to electrically insulate the test bearing from the rest of the engine as shown in Fig. 1.2.1. Two methods have been used to accomplish this task. In one instance, a sufficient amount of metal is removed from the inside diameter of the bearing housing and replaced with a thin (nominally 0.1 mm thick) plastic insulating sheet (2, 3, 4, 5, 6). In the second instance, the metal removed from the diameter of the bearing housing is replaced with a comparable thickness of an "alumina ceramic" coating (7, 8). In either case the bearing can be sufficiently insulated.

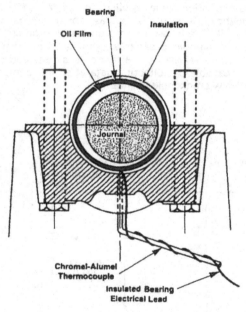

Figure 1.2.1: Engine Test Bearing for Making Resistance or Capacitance Film Thickness Measurement

In addition to insulating the bearing, a second requirement of these methods is to electrically ground the crankshaft. Most references have accomplished this task by simply draping a braided metal wire, one end of which is connected to ground, over the portion of the shaft extending outside of the engine. By connecting electrical leads to the side of the test bearing and to "ground", an electrical circuit can be completed.

Depending on whether measurements of electrical resistance or capacitance are to be used in calculating film thickness values, two different methods for applying a voltage across the bearing are employed. Spearot and Murphy (9) reviewed both methods and described the techniques for applying voltages in detail. In the case of resistance measurements, a low voltage, DC signal, is used as shown in Fig. 1.2.2, and the resistance of the oil film, R_0, in the bearing is calculated directly from circuit analysis. The value of the self-generated voltage, ζ, produced during bearing rotation can be determined from measurements of bearing voltage without the excitation signal applied. In the case of capacitance measurements, a high frequency, AC signal is used as shown in Fig. 1.2.3, and the capacitance, C_0, of the oil film is calculated from calibration curves generated by using the circuit with known capacitors in place of the test bearing. The output voltage produced by the bearing in either case is measured by means of a data acquisition system capable of collecting data at speeds comparable to the time required to rotate a few crankangle degrees.

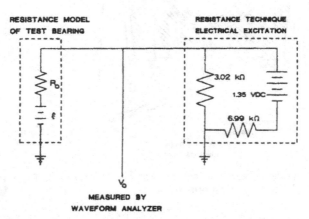

Figure 1.2.2: Total Resistance Measurement Method, Reprinted with permission, Copyright 1988, Society of Automotive Engineers, Inc.

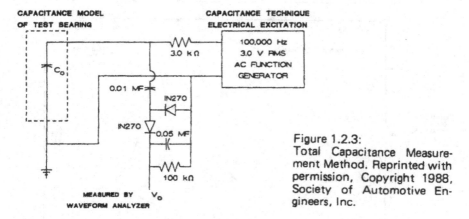

Figure 1.2.3:
Total Capacitance Measurement Method. Reprinted with permission, Copyright 1988, Society of Automotive Engineers, Inc.

Having measured the resistance or capacitance of the total oil film at any instant, the minimum film thickness, h_m, in the bearing can be calculated directly from either Equation 1 or 2, respectively.

$$h_m = \delta \left[1 - \sqrt{1 - \left[\frac{2\pi W R_B R_o \sigma}{\delta} \right]^2} \right] \qquad (1)$$

$$h_m = \delta \left[1 - \sqrt{1 - \left[\frac{2\pi W R_B D_o k}{\delta C_o} \right]^2} \right] \qquad (2)$$

In these equations, W is the width, R_B is the radius, and δ is the radial clearance of the test bearing. D_o is the dielectric constant, σ is the electrical conductivity of the oil, and k is the permittivity of free space.

By coupling the film thickness values provided by Equations 1 or 2 with a measure of angular rotation of the crankshaft, plots of minimum film thickness in the bearing versus crankangle during a single combustion cycle can be constructed using either method as shown in Fig. 1.2.4 for a 2.5l L-4 engine. It has been demonstrated (9), that the resistance and the capacitance technique can provide comparable film thickness values when used in the same engine at a fixed set of conditions with the same oil. However, statistical analyses have also demonstrated that it is easier to measure film thickness with greater precision using the capacitance technique than it is using the resistance technique. The explanation for this finding is that the dielectric constant of the oil, D_o, is much less a function of temperature and oil chemistry, than is oil conductivity, σ. The precision with which dielectric constants can be measured is reflected in greater precision in calculated values of h_m.

29

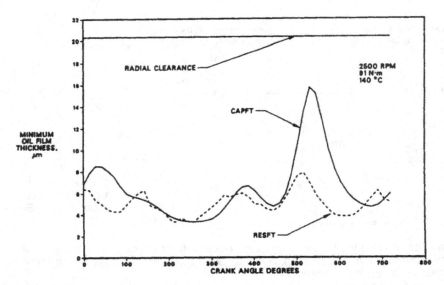

Figure 1.2.4: Minimum Bearing Oil Film Thickness for an Oil During a Single Combustion Cycle as Determined by Both the Total Capacitance and Total Resistance Methods. Reprinted with permission, Copyright 1988, Society of Automotive Engineers, Inc.

Although the precision of minimum oil film thickness values calculated has been shown to be very good when determined using the capacitance method (9), there are still sources of error in the development of Equations 1 and 2 which could affect the accuracy of h_m values. It is assumed in the derivation of these euqations that the geometry of both the bearing and the journal are circular. It is known that this is not true for the bearing; the two bearing shells form a slight oval. However, the deviation from circular geometry is not expected to greatly influence the calculation of h_m as long as the minimum film thickness does not occur near the bearing "split line".

A more critical concern involves the possible presence of cavitation in the bearing. It has been shown (9) that the presence of cavitation can influence the values of h_m calculated either by the capacitance or the resistance methods. Because the dielectric constant of air is closer to that of D_o than the electrical conductivity of air is to that of σ, cavitation affects the values determined with the capacitance technique less than it does the resistance method. However, the presence of cavitation in the bearing will increase the calculated values of h_m over the actual film thickness in the bearing no matter which method is used. The greater the extent of cavitation in the bearing, the greater will be the effect on calculated values of h_m. At the same time, the smaller the value of h_m (greater eccentricity), the less effect any fixed amount of cavitation will have on the calculated film thickness.

30

1.2.5 Effects of Oil Rheology on Bearing Performance

Single-Grade Oils

An example of how oil rheological properties affect film thickness in an engine journal bearing can be readily provided by a series of single-grade oils as shown in Fig. 1.2.5. For the front main bearing of a 3.8l V-6 engine at the conditions indicated, the effect of viscosity grade in going from an SAE 20 to an SAE 40 oil is to merely shift the minimum film thickness curve to higher values. In interpreting the curves in Fig. 1.2.5, it is important to remember that every point along each curve is a minimum film thickness at a particular time during the combustion process (in this instance the data are referenced to the combustion process in cylinder #1, the closest cylinder to the front main bearing). At some point during the cycle, this locus of minima, forms an absolute minimum. It is this absolute minimum which is of interest because it defines the point at which the bearing and the journal come closest to touching during the combustion process (in this instance the data are referenced to the combustion process in cylinder #1, the closest cylinder to the front main bearing). At some point during the cycle, this locus of minima forms an absolute minimum. It is this absolute minimum which is of interest because it defines the point at which the bearing and the journal come closest to touching during and load, and the type of bearing being analyzed. Presumably because of greater inertial loads, con-rod or "big-end" bearings have an absolute minimum film thickness which falls within the induction or exhaust strokes of the combustion cycle more often than during the power stroke (8).

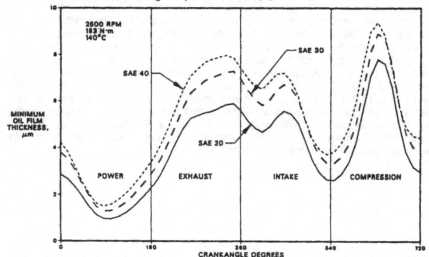

Figure 1.2.5: Effect of Single-Grade Oil Viscosity on Bearing Film Thickness Reprinted with Permission, Copyright 1989, Society of Automotive Engineers, Inc.

31

Many published studies have demonstrated that BOFT values correlate well with a variety of different measures of single-grade oil viscosity. In some studies (3, 5), linear correlations were developed between film thickness and kinematic viscosity. In some (3, 5), linear correlations have been developed between film thickness and high-temperature, high-shear viscosity (defined in this work as viscosity measured at $150°C$ and 10^6 s^{-1}). In other studies (7, 8), conducted over wider temperature ranges, excellent linear correlations have been developed between film thickness and the square root of viscosity measured either in kinematic viscometers or at HTHS conditions. For a range of engine speeds and loads, Deysarkar (10) has demonstrated a good correlation for single-grade oils between film thickness in a main bearing and the square root of a Sommerfeld Number, S, defined as:

$$S = \frac{N \cdot \eta}{P} \tag{3}$$

where N is the engine speed in RPM, η is the HTHS viscosity in cP, and P is the output torque of the engine in N · m. As shown in Table 1.2.1, for all of these correlations the Indices of Determination (R^2 values) range from a low of 0.71 to a high of 0.99 demonstrating a clear relationship between the viscometric properties of these single-grade oils and journal bearing film thickness in operating engines.

These correlations all appear valid because of three reasons. First, single-grade oils are essentially Newtonian, and it is of little consequence to the correlations what shear rate exists in the bearing or in the laboratory viscometer used to measure their viscosity. Second, since the heat transfer characteristics of these mineral oil-based fluids are all approximately the same, and since viscous heating is a uniform function of oil viscosity, any temperature rise in the bearing is hidden in the film thickness/viscosity correlations. Finally, over the range of pressures generated in a hydrodynamic journal bearing, the pressure-viscosity coefficients of the oils are constant and roughly equivalent. Thus any pressure effects are constant across the sampling of test oils investigated. If single-grade, Newtonian oils were all that were commercially available, these data demonstrate that current SAE J300 kinematic viscosity specifications would adequately predict bearing performance.

Table 1.2.1: Correlations between Bearing Oil Film Thickness and Single-Grade Oil Viscosity

Authors	Reference	Bearing	Correlating Variable	Index of Determination R^2	Comments
Spearot, Murphy, and Rosenberg	3	Main	η_{Kin} (100°C)	0.71	
Spearot, Murphy and Rosenberg	3	Main	η_{HTHS}	0.73	
Girshick and Craig	5	Main	η_{Kin} (100°C)	----	(a)
Girshick and Craig	5	Main	η_{HTHS}	----	(a)
Bates and Benwell	7	Con-Rod	$(\eta_{Kin})^{0.5}$	0.98	(b)
Bates and Vickars	8	Con-Rod	$(\eta_{Kin})^{0.5}$	0.98	(c)
Bates and Vickars	8	Con-Rod	$(\eta_{HTHS})^{0.5}$	0.98	(d)
Deysarkar	10	Main	$S^{0.5}$	0.90	(e)

Comments
(a) R^2 value for BOFT correlation with single-grade oil viscosity not specified, but qualitatively the correlation appears excellent.
(b) Viscosity calculated from kinematic viscosity times density measured at bearing temperatures.
(c) Viscosity calculated at bearing temperatures.
(d) Viscosity calculated at 10^6 s^{-1} and bearing temperatures.
(e) Sommerfeld Numbers calculated using η_{HTHS}.

Multigrade Oils

As apparently simple and straightforward as the interpretation of BOFT data for single-grade oil is, the analysis of similar data for multigrade oils is complex and ambiguous. As shown in Table 1.2.2, for some of the same studies which demonstrated consistent correlations between bearing film thickness and viscosity in the case of single-grade oils, a wide disparity of results are documented for multigrade formulations. Indices of Determination range from 0.0 to 0.90 for differenct correlations. In addition to providing lower values of R^2 in every case, the regression equations between BOFT and multigrade oil viscosity are statistically different from regressions between BOFT and single-grade oil viscosity. Since the engine is not aware of the formulation differences between single- and multigrade oils, the lack of agreement between correlations involving these two classes of lubricants must be due to either 1) measuring their viscosities at temperatures and shear rates which are not representative of bearing operation or 2) the effect of some other engine or oil variable (possibly oil elasticity) which has yet to be taken into consideration.

Recently, Spearot, Murphy, and Deysarkar (11) developed techniques for obtaining better estimates of the temperatures and shear rates to which oils are subjected in the front main bearing of a 3.8l engine. By assuming that viscous heating in the bearing is minimized when extremely low-viscous, Newtonian oils are evaluated at low sump oil temperature, a "true" BOFT versus Sommerfeld Number curve in the presumed absence of viscous heating is generated for the test bearing. Using this curve and the measured BOFT values for higher viscosity oils at other engine conditions, the "true" Sommerfeld Number for the bearing can be estimated. Since the engine speed and load are known, the viscosity in the bearing can be calculated and using the well known Walther Equation, ASTM D446, the temperature corresponding to this viscosity can be determined.

Using the difference in temperature between the oil in the bearing and the oil in the sump as a dependent variable, a multi-variable, linear regression technique was used to develop an equation to describe this temperature difference in terms of sump temperature, sump oil viscosity, engine speed, engine load, and minimum oil film thickness in the bearing. As shown in Fig. 1.2.6, the temperature difference predicted from the regression equation agrees reasonably well with that calculated from the previously described Sommerfeld approximations. More importantly, both analyses predict that under certain engine operating conditions, the temperature of the oil in the bearing can be more than twenty degrees greater than that in the sump. Thus, assuming the oil in the bearing to have the same temperature as that in the sump can be a gross miscalculation.

34

Table 1.2.2: Correlations between Bearing Oil Film Thickness and Multigrade Oil Viscosity

Authors	Reference	Bearing	Correlating Variable	Index of Determination R^2	Comments
Spearot, Murphy, and Rosenberg	3	Main	η_{Kin} (100°C)	0.00	(a)
Spearot, Murphy, and Rosenberg	3	Main	η_{HTHS}	0.01	(a)
Girshick and Craig	5	Main	η_{Kin} (100°C)	0.22	(b)
Girshick and Craig	5	Main	η_{HTHS}	0.76	(c)
Bates and Vickars	8	Con-Rod	$(\eta_{Kin})^{0.5}$	0.82 to 0.83	(d)
Bates and Vickars	8	Con-Rod	$(\eta_{HTHS})^{0.5}$	0.88 to 0.90	(d)
Deysarkar	10	Main	$S^{0.5}$	0.64	(e)

Comments
(a) R^2 value for BOFT correlation with multigrade oil viscosity calculated including data for four oils which were subsequently identified as statistical "outliers".
(b) R^2 value calculated for data set including both single- and multigrade oils.
(c) R^2 value calculated for data set including both single- and multigrade oils. Viscosity evaluated at 10^6 s^{-1} and 100°C.
(d) Viscosity evaluated at bearing temperatures. Different R^2 values correspond to different engines.
(e) Sommerfeld Numbers calculated using η_{HTHS}.

Some of the data in Fig. 1.2.6 suggest that there can be a negative temperature rise between the oil sump and the bearing. This finding is believed to be an artifact of the way these engine tests were conducted. The sump temperature was controlled by means of a heater and cooling coils independent of what speed and load conditions were applied to the engine. At some sets of conditions the sump temperature was controlled at a higher value than the engine would produce in the absence of any sump controls. When this occurred, the result was a decrease in oil temperature as the oil passed from the "hot" sump to the relatively cool engine block. This negative temperature rise could be detected by means of the assumptions described in the preceding paragraphs.

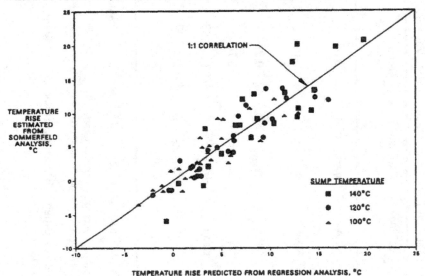

TEMPERATURE RISE PREDICTED FROM REGRESSION ANALYSIS, °C

Figure 1.2.6: A Comparison of the Temperature Rise in the Bearing Predicted by Multilinear Regression Analysis with that Calculated from Sommerfeld Approximations. Reprinted with permission, Copyright 1989, Society of Automotive Engineers, Inc.

Using the regression equation to calculate bearing oil temperatures and determining the corresponding viscosities provides the BOFT versus Sommerfeld Number curve shown in Fig. 1.2.7. The assumptions made regarding the Sommerfeld Number and the amount of viscous heating produce a high degree of correlation for this series of single-grade oils. Although the temperature regression equation developed is truly valid only for the engine bearing and operating conditions used in this work, it can be used to estimate the temperature rise in the bearing for a series of multigrade oils over the same range of operating conditions.

36

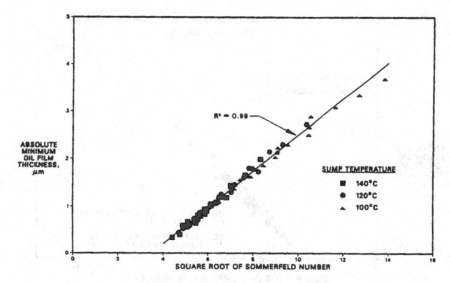

Figure 1.2.7: Absolute Minimum Oil Film Thickness as a Function of Sommer-
feld Number for Single-Grade Oils at the Estimated Bearing Oil
Temperature. Reprinted with permission, Copyright 1989,
Society of Automotive Engineers, Inc.

Spearot, Murphy, and Deysarkar (11) also calculated values for different shear
rates associated with the test bearing. The two shear rate definitions which were
thought might be related to film thickness values in a journal bearing were 1)
the maximum shear rate, and 2) the average shear rate in the loaded portion of
the bearing. The maximum shear rate occurs at the minimum film thickness
point in the bearing and thus the viscosity calculated at such conditions could
influence BOFT. The average shear rate in the loaded portion of the bearing is a
variable which reflects the shearing conditions in the entire portion of the oil
film which carries the applied load. Thus, it could also influence BOFT.

By using the temperature regression equation to calculate the temperature of the
oil in the bearing and using these different shear rate definitions, the viscosity
of a series of both single- and multigrade oils were calculated at different bearing
operating conditions. Using these viscosities to calculate values of S, regressions
between film thickness and Sommerfeld Number were constructed as shown in
Figs. 1.2.8 and 1.2.9. In each of these figures, (1) provides the raw data, and (b)
provides the linear regressions through the data.

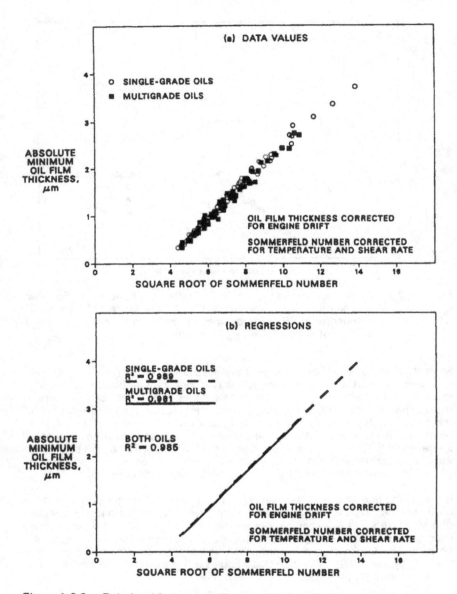

Figure 1.2.8: Relationship between Bearing Oil Film Thickness and Sommerfeld Number at an Average Shear Rate. Reprinted with permission, Copyright 1989, Society of Automotive Engineers, Inc.

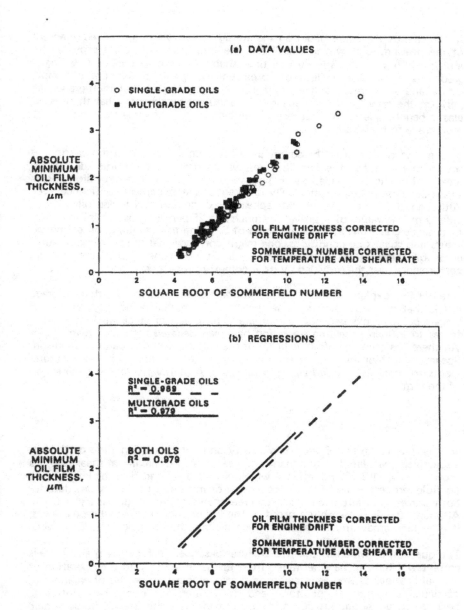

Figure 1.2.9: Relationship between Bearing Oil Film Thickness and Sommerfeld Number at the Maximum Shear Rate. Reprinted with Permission, Copyright 1989, Society of Automotive Engineers, Inc.

In the case of viscosities based on the average shear rate in the loaded quadrant of the bearing, as shown in Fig. 1.2.8, the film thickness values provided by multigrade oils are statistically indistinguishable from those provided by single-grade oils at a confidence level of 95 percent. Using this combination of temperature and shear rate conditions, it would be concluded that multigrade oils perform the same as single-grade oils in a journal bearing and that there is no elastic benefit due to the presence of the high molecular weight polymer in the multigrade formulation.

In the case of viscosities based on the maximum shear rate in the bearing, as shown in Fig. 1.2.9, the film thickness values provided by multigrade oils are greater than those provided by single-grade formulations. The regression curves based on these data are statistically different at the 95 percent confidence level. Although the differences do not appear large, certain multigrade oils, particularly at low values of S, provide as much as a 25 percent greater film thickness than single-grade oils at the same value of S. Using this combination of temperature and shear rate conditions, one might conclude that the polymer in multigrade oil formulations does provide an additional benefit to journal bearing performance over the provided by its viscometry properties.

One possible explanation for some additional benefit associated with multigrade oils is that of fluid elasticity. The question of whether or not high molecular weight polymer blended into a multigrade oil could produce sufficient elastic forces to influence bearing operation has been debated for many years. As of yet there has been no definitive proof of such elastic benefits. Bates, Williamson, Spearot, and Murphy (12) attempted to relate bearing film thickness measurements to both viscous and elastic parameters. A multivariable linear regression of the form:

$$h_m = C_0 + C_1 \eta + C_2 \theta \tag{4}$$

was used where η and θ are the viscosity and the relaxation time of the oil, respectively, caculated at the bearing temperature and maximum shear rate. As shown in Fig. 1.2.10, Equation 4 which gives a predicted film thickness based on fluid properties was fit to a collection of measured film thickness data for both single- and multigrade oils with a reasonable degree of success ($R^2 = 0.73$). Although this is not absolute proof of the importance of oil elastic properties, it does lend credibility to theories which include the influence of such effects.

The question of which of the two analyses described in Figs. 1.2.8 and 1.2.9 is correct will have to be determined from further research into the operation of journal bearings as well as research on the rheological properties of engine oils, particularly at high temperatures and shear rates. The numerical solution to bearing design equations for both non-Newtonian and elastic fluids should provide an understanding of what characteristic shear rate is required for defining bearing performance. Coincidentally, the measurement of oil viscosities and elastic properties at temperatures and shear rates representative of bearing operation should allow the determination of the relative magnitudes of these two rheological effects. Research into both of these areas is a continuing effort in several laboratories around the world.

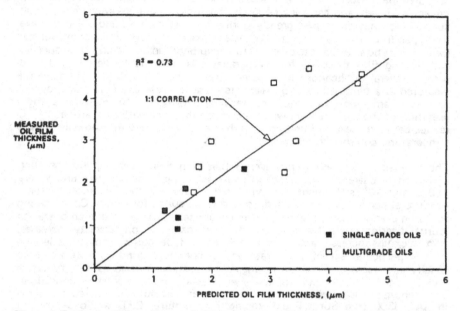

Figure 1.2.10: A Comparison of Measured and Predicted Bearing Oil Film Thicknesses from a Regression Using Viscosity and Relaxation Time Measured at 100°C and at the Maximum Shear Rate in the Bearing as Independent Variables. Reprinted with permission, Copyright 1986, Society of Automotive Engineers, Inc.

1.2.6 Does the Automobile Industry Need a Bearing Film Thickness Test?

The answer to the central question of this paper depends entirely on whether or not the film thickness in a journal bearing of an operating engine can be adequately described by one or more laboratory oil rheological properties measurements. Clearly, if relatively straightforward laboratory measurements are available, there is no justification for a more complex, harder to control, more time-consuming and expensive engine test. SAE J300 was originally developed as a table of oil viscosities from which engine designers could choose an oil suitable for use in their engines without having to evaluate a large number of oils in engine tests. The concept of such a table of oil viscometric properties is still valid and desirable.

But are the proper rheological measurements which affect bearing oil film thickness known, and can they be easily measured in the laboratory? Although our research has progressed greatly during the last decade, and we now have the capabilities to measure film thickness in bearings more easily than ever before, it must be said that the exact relationship between oil rheological properties and bearing film thickness for multigrade oils is yet to be determined. The data in several publications have demonstrated that the viscosity of engine oil measured at a temperature and shear rate close to those of an operating journal bearing is an important factor in bearing performance. No reasonable interpretation of the existing data would conclude that kinematic viscosity at 100°C relates better to bearing film thickness than does a viscosity measured at bearing temperatures and shear rates.

The problem which occurs, however, is that each engine bearing and operating condition is characterized by a different shear rate and temperature. Where 150°C and 10^6 s^{-1} might be well suited to describe the operation of some bearings at certain operating conditions, it is less suited for others. Complicating the issue even further is the fact that the characteristic shear rate of an operating journal bearing is itself a function of oil viscosity. As oil viscosity increases, film thickness increases and shear rate, in general, decreases. Thus, the selection of any particular set of shear rate and temperature conditions at which to specify the viscosity grade of an engine oil must be viewed as a compromise based on the characteristics of many engines and many operating conditions. This concept is not without precedent, however. The pumpability specifications in SAE J300 (the Borderline Pumping Temperature, BPT) are based on an average of the pumpability characteristics of a set of widely differing engines.

The advancements which have been made over the past decade in measuring the viscosity of oils at high temperatures and shear rates have also been significant. Although the maximum shear rates at which oil viscosities have been reported measured in the laboratory range from 2 to 5×10^6 s^{-1} (13, 14), less than the shear rates which can occur in con-rod bearings (6), these measurements are significant in that a decade ago viscosity determinations at 10^5 s^{-1} were considered the limits of laboratory capabilities. Regardless of the value of shear rate which is identified as being representative of bearing operation, if the shear rate is produced in an operating engine, it can be reproduced in a laboratory viscometer. Viscosity values may be determined on a relative rather than an absolute basis, but they will be determined none-the-less.

The other rheological property which may influence journal bearing operation is oil elasticity. Although the relative influence of this multigrade oil property is still to be verified, the measurements needed to define its effect in engines

42

are available, and the laboratory measurements needed to quantify its magnitude are beginning to be developed. The limited data which have been collected in recent engine bearing studies suggest that the effect of elasticity is additive to that of any viscosity effect. If this is true, it means that a viscosity classification system based only on measurements of high-temperature, high-shear viscosity would define a level of minimum performance for multigrade oils. This would provide protection for the automotive industry from shear-thinning, non-elastic multigrade oils. If it can be unequivocally demonstrated that elasticity provides an extra measure of protection for certain multigrade formulations, then recognition of this fact could also be incorporated into a system such as SAE J300 at a later date.

With regard to the problems which might be experienced in developing an engine BOFT test, it is worth noting that the development of any "standard" industry engine test is a time-consuming, arduous task. In this instance, potential problems would arise even in the selection of a test engine. Because the operating characteristics of passenger car and heavy-duty engines are radically different, it is doubtful that either segment of the industry would accept a BOFT engine test which uses the other's engine. Journal bearings in heavy-duty engines are designed differently and for different objectives as are those in passenger car applications. Even if an engine could be selected to which both segments of the industry agree, the selection of limits on film thickness to define different viscosity grades of oil in such a test would present a significant hurdle. BOFT engine tests are excellent tools for developing an understanding of how journal bearings operate in an engine. As a technique for defining different grades of engine oil, however, they should be considered as a last resort after it is demonstrated that laboratory rheological properties measurements can not do the job adequately.

Given the current data regarding the effects of oil rheological properties on bearing oil film thickness in engines, it is the opinion of General Motors Corporation, that including oil viscosities measured at a representative temperature and shear rate in the Viscosity Classification System, SAE J300 is sufficient for providing a measure of oil high-temperature, high-shear viscometric performance. The engine test procedures which have been developed for research projects designed to study oil rheological effects on bearing performance are very good, and the data collected from them will help in the selection of a temperature and shear rate which is representative of industry bearing characteristics. However, at this time, there is no compelling reason for complicating the oil qualification process with an engine test designed to measure only journal bearing film thickness.

1.2.7 Is an Industry Engine Bearing Test of any Sort Needed?

It has been demonstrated in previous engine studies that bearing wear is a function of both the rheological properties of an oil as well as its chemical characteristics (1). The rheological properties insure a sufficient oil film thickness when the engine is up to speed and operating at relatively steady conditions. The chemical characteristics protect the bearing during intermittent contacts between it and the journal which can occur during accelerations, decelerations and each time the engine is started. At General Motors, we believe the rheological properties which are required to insure suitable oil films during steady engine operation can be provided by the inclusion of HTHS viscosity specifications in SAE J300. However, we also realize that there is very little in the remainder of the oil qualification system which protects bearings based on the chemical characteristics of the oil. If the industry is to take it upon itself to develop an engine bearing performance test, then we believe that test should evaluate bearing distress and wear rather than oil film thickness. In particular, correlation between bearing wear in a laboratory engine and long-term bearing durability in the field should be the ultimate objective of such a test. The combination of viscometric specifications in SAE J300 and the ability to rank chemistries based on long-term bearing wear would provide the industry with an excellent tool for meeting the twin objectives of producing high-powered, durable engines while at the same time optimizing corporate fuel economies.

1.2.8 References

(1)   The Relationship Between High-Temperature Oil Rheology and Engine Operation: A Status Report, ASTM Data Series DS-62, ASTM, Philadelphia, PA 19103 (1985).

(2)   Bassoli, C.; Cornetti, G.; Belei, M.: "A System for Assessing the General Conditions of Lubricated Crankshafts", RIVISTA ATAT, January (1978).

(3)   Spearot, J.A.; Murphy, C.K.; Rosenberg, R.C.: "Measuring the Effect of Oil Viscosity on Oil Film Thickness in Engine Journal Bearings", SAE Paper No. 831689 (1983).

(4)   Craig, R.C.; King Jr., W.H.; Appeldoorn, J.K.: "Oil Film Thickness in Engine Bearings — The Bearing as a Capacitor", SAE Paper No. 821250 (1982).

(5)   Girshick, F.; Craig, R.C.: "Oil Film Thickness in a Bearing of a Fired Engine, Part III: The Effects of Lubricant Rheology", SAE Paper No. 831691 (1983).

(6)   Schilowitz, A.M.; Waters, J.L.: "Oil Film Thickness in a Bearing of a Fired Engine — Part IV", SAE Paper No. 861561 (1986).

(7)   Bates, T.W.; Benwell, S.: "Effect of Oil Rheology on Journal Bearing Performance", SEA Paper No. 880679 (1988).

(8)   Bates, T.W.; Vickars, M.A.: "Measurement of II Film Thickness in Big-End Bearings and Its Relevance to Engine Oil Viscosity Classifications", 3rd CED International Symposium on Performance Evaluation for Automotive Fuels and Lubricants, Paris, April 19—21 (1989).

(9)   Spearot, J.A.; Murphy, C.K.: "A Comparison of the Total Capacitance and Total Resistance Technique for Measuring the Thickness of Journal Bearing Oil Films in an Operating Engine", SAE Paper No. 880680 (1988).

(10)   Deysarkar, A.K.: "The Bearing Oil Film Thickness of Single and Multi-Grade Oils — Part I: Experimental Results in a 3.8 L Engine", SAE Paper No. 880681 (1988).

(11)   Spearot, J.A.; Murphy, C.K.; Deysarkar, A.K.: "Interpreting Experimental Bearing Oil Film Thickness Data", SAE Paper No. 892151 (1989).

(12)   Bates, T.W.; Williamson, B.; Spearot, J.A.; Murphy, C.K.: "A Correlation Between Engine Oil Rheology and Oil Film Thickness in Engine Journal Bearings", SAE Paper No. 860376 (1986).

(13)   Tolton, T.: "The Temporary Viscosity Losses of New and Sheared Multiweight Oil Formulations Characterized at Multiple Shear Rates", SAE Paper No. 890272 (1989).

(14)   Lodge, A.S.: "Multigrade Oil Elasticity and Viscosity Measurement at High Shear Rates", SAE Paper No. 872043 (1987).

2. Base Oils

2.1 Structure of Oils According to Type and Group Analysis of Oils by the Combination of Chromatographic and Spectral Methods

P. Daucik, T. Jakubik, N. Pronayova and B. Zuzi,
Slovak Technical University, Bratislava, Czechoslovakia

Summary

In this work was used the rapid preparation-analysis separation of oils on the chromatographic column with recirculation of the eluant. The basic quantitative data concerning the group composition of oils have been supplement by analysis of chromatographic fractions by means of NMR spectrometry. The analysis results of group composition of oils obtained by means of chromatographic fractionation have been completed with the results of NMR spectrometry and mass spectrometry.

2.1.1 Introduction

Knowledge of the composition of oil products is a prerequisite of optimization of technological processes and better utilization of raw materials. Therefore, the requirements for a prompt analysis of vacuum distillates and residues are still in the foreground of interest. With increasing boiling point of hydrocarbons the possibility of their resolution in a mixture decreases. Therefore, by chromatographic separation of oils, data on their group composition are obtained. In spite of the intensive development of high efficient liquid chromatography problems with detection, calibration, recovery, etc. remain unsolved during the separation of oil compounds (1-3), and therefore, also preparative separations are employed (4, 5).

Introduction of spectral methods to the analysis of oil fractions resulted in a remarkable progress. Their great advantages are reliability of results, time saving and considerable information. Structural type analysis an information on an average molecule can be obtained from infrared (6) or NMR (7, 8) spectra. Direct determinations of group composition of oils by NMR spectrometry (9) are derived from simplifying prerequisites which are not always fulfilled. The quantitative analysis by mass spectrometry (10, 11) provides more information on the group composition of oils. However, they are more demanding as for the instrument calibration, its price and considerable wearing.

The combination of separation and spectral methods always brings the most valuable information on the sample composition. The preparative-analytical methods of the chromatographic determination of the group composition of oils together with the mass and NMR spectrometry are the most efficient modes of oil analysis. The use of a simple and cheap separation method together with the analysis of fractions obtained ensures reliable quantitative and qualitative data on the sample analyzed.

2.1.2 Experimental

Chromatography:

A sample weighing 1 g was separated by a preparative-analytical method of chromatographic determination of group composition. The sample was dissolved in 5 ml of hexane. A glass column was used (Fig. 2.1.1) which was connected to a boiling flask and cooler. The eluant was evaporated through a side tube of the column and condensed in the cooler. The side tube was used for recycling the eluant on the head of the column. The column was filled with 15 g of alumina and 25 g of silica gel (bottom layer). Alumina and silica gel were thermally activated for 4 hours: alumina at 450°C and silica gel at 160°C. The eluants were n-hexane, benzene and a mixture of 34 vol. % of ethanol in benzene.

The sample dissolved in n-hexane was quantitatively poured into the chromatographic column. For the sample elution 85 ml of n-hexane were used which added "per partes" always after penetration of the previous volume into the column packing. The first three portions of the added n-hexane (3 x 5 ml) were used to wash the sample from the column walls.

Fractions with compounds of similar structures were gradually eluted from the column. Paraffinic and cyclanic (naphthenic) hydrocarbons (P + N) were determined from the first 40 ml of eluate. The remaining volume of n-hexane was collected into the boiling flask. The n-hexane was recycled three times by redistillation through the side tube of the column. From this eluate the amount of light aromates (A_H) was determined after the evaporation of n-hexane. Middle aromates (A_B) were eluted after addition of 50 ml of benzene into the column. Benzene was also recycled three times. Heavy aromates (A_{BE}) were eluted after application of 50 ml of the binary solvent (ethanol + benzene). The binary solvent was poured into the cooled column. After the flowing of the binary solvent through the column the boiling flask under it was again heated. The solvent was recycled three times.

Groups of structurally similar compounds (P + N, A_H, A_B, A_{BE}) were determined by weighing the boiling flasks after the solvents evaporation and drying up.

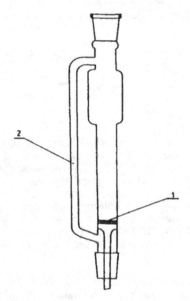

Figure 2.1.1

Mass spectrometry:

For both high (HRMS) and low (LRMS) resolution mass spectrometry we used MS 902 SAE I Manchester Co. mass spectrometer. Our results were compared with measurements on the equipments listed in the Table 2.1.8. Evaluation and elaboration were performed according to the data from the literature (10, 11).

NMR spectrometry:

The sample were analysed using the instruments VXR-300 of the Varian Company and FX-100 of the JEOL Company. The ^1H NMR spectra were measured at a frequency of 300 MHz, the pulse delay being 1 s. The ^{13}C NMR spectra were measured at the band width of 6000 Hz, frequency of 25 MHz and pulse delay of 6 s. The relaxation agent was $(AcAc)_3 Cr$. The solvent used was deuterated chloroform and the standard for the determination of the chemical shift was tetramethylsilane.

2.1.3 Results and Discussion

The preparative-analytical method for the determination of the group composition of oils was tested with samples from various stages of technological processing. Samples of a vacuum distillate, selective and hydrogenation refining were chosen. The basic characteristics of the chosen samples are shown in Table 2.1.1. The values of viscosities, densities and refractive indices show the differences in the chemical composition of the oils analyzed.

Table 2.1.1: Basic characteristics of analyzed oils

	Vacuum distillate	Raffinate	Extract	Hydrogenate
Mw	356	368	359	385
d^{20}_4 (kg m^{-3})	907.1	889.1	937.7	876.9
$\| ^{50}_d$	1.4934	1.4818	1.5572	1.4825 (20°C)
ν_{40} (mm^2 s^{-1})	26.59	22.57	88.46	57.60
ν_{100} (mm^2 s^{-1})	4.52	4.25	6.76	6.95
% C	884.86	85.12	84.71	86.35
% H	12.43	12.93	10.45	13.48
% N	0.03	–	0.19	–

The group composition determined by the above method confirms the differences in the content of saturated and aromatic fractions (Table 2.1.2). Reproducibility of the method for the determination of the basic groups eluted by hexane, benzene and binary eluant (ethanol-benzene) was verified with the samples of a refined product and extract. It has been found that the saturated fractions and the group of compounds eluted by the binary solvent can be determined with a standard deviation of 0.2 %. The groups eluted with hot hexane or benzene can be determined with a deviation of 0.4 %.

Table 2.1.2: Group composition of analyzed oils

	Vacuum distillate	Raffinate	Extract	Hydrogenate
P + N	47.9	58.5	10.8	87.2
A_H	27.4	21.9	38.0	7.1
A_B	20.0	15.7	41.8	5.1
A_{BE}	4.4	3.6	9.2	0.4
Total	99.7	99.7	99.8	99.8

At the same time, the method enables to characterize the fractions obtained by other analytical methods. An example is the determination of molecular weights and elemental analysis of the fractions of the extract in Table 2.1.3. Even this simple analysis demonstrates the differences between the individual fractions. Particularly the information on the content of heteroatoms is valuable. The content of carbon and hydrogen points to their purely paraffinic-cyclic character. The fractions eluted with hot hexane or benzene differ particularly in the sulphur and hydrogen content. With regard to the significantly different elution power of these solvents one can expect that the separation of these fractions is not influenced only by the unequal number of aromatic rings in the hydrocarbon molecules. It is quite probable that the change in the adsorption properties is also due to the different structural sulphur groups in these fractions. The fractions eluted by the binary eluant differ from the previous ones in a considerable higher content of the total nitrogen. It seems that this heteroatom causes markedly higher adsorption retention of the fractions in the column. This group elutes from the column at high eluting power of the binary solvent. In the case of the desorption of this fraction the thermal effect can be observed. There is danger of breaking the adsorbent column due to the formation of the solvent vapours in the column. Therefore, the binary solvent is poured into the cooled column.

Table 2.1.3: Elemental analysis of chromatographic fractions of the extract

Fraction	Mw	% C	% H	% S	% N
P + N	360	85.77	14.20	0.00	0.00
A_H	318	84.95	11.61	1.79	0.00
A_B	313	85.11	9.68	3.53	0.20
A_{BE}	405	84.70	10.03	2.08	1.51

Valuable data are provided by NMR spectrometry. When evaluating ^1H and ^{13}C NMR spectra the characteristic spectral lines in different regions belong to the following structural groups:

$H_{\alpha 1}$ = hydrogen in CH_2 groups in the alpha positions to the aromatic ring (chemical) shift 2.3 - 4.0)

$H_{\alpha 2}$ = hydrogen in CH_3 groups in the alpha positions to the aromatic ring (chemical shift 1.9 - 2.3)

H_β = hydrogen in CH_2 and CH_3 groups in the beta positions to the aromatic ring; hydrogen in the naphthenes in beta positions to the aromatic ring and hydrogen in the alkane groups (chemical shift 1.0 - 1.9)

H_γ = hydrogen in CH_3 groups in the gamma position to the aromatic ring; hydrogen in the groups of alkanes (chemical shift 0.5 - 1.0)

H_{mono} = hydrogen in monoaromatic hydrocarbons (chemical shift 6.6 - 7.2)

H_{di} = hydrogen in diaromatic hydrocarbons (chemical shift 7.2 - 7.8)

H_{tri} = hydrogen in triaromatic hydrocarbons (chemical shift 7.8 - 8.3)

C_{al} = carbon in aliphatic hydrocarbons (chemical shift 50 − 100)

C_{ar} = carbon in aromatic and olefinic hydrocarbons (chemical shift 100 -150)

C_1 = carbon in CH_3 groups of linear alkanes (chemical shift 13.8 − 14.2)

C_2 = carbon in CH_2 groups of linear alkanes (chemical shift 22.7 − 23.0)

C_3 = carbon in CH_2 groups of linear alkanes (chemical shift 31.8 − 32.0)

$C_{4,5}$ = carbon in CH_2 groups of linear alkanes (chemical shift 29.0 − 30.0)

The above data from ^1H and ^{13}C NMR spectra of the fractions of the vacuum distillate are shown in Tables 2.1.4 and 2.1.5. It is evident from these data that there are not any spectral lines in the range of aromatic hydrogens and carbons in the spectra of the paraffine-naphthenic fraction (P + N). The separation efficiency of the column is satisfactory for the separation of saturated and aromatic hydrocarbons. In order to evaluate the separation efficiency of the column packed with alumina and silica gel with respect to the aromatic compounds qualitative characterization of the hexane eluted fractions (A_H), benzene eluted fractions (A_B) and binary solvent eluted fractions (A_{BE}) is required. William's evaluation of ^1H NMR spectra provided data on the average molecule in the above presented fractions (Table 2.1.6). The definition of average molecular structure parameter is following:

53

R_A = number of aromatic rings per molecule
R_N = number of naphthenic rings per molecule
n = number of carbon atoms per substituent
R = number of naphthenic rings per substituent

Table 2.1.4: Structural type data of the chromatographic fractions of the vacuum distillate obtained from ^1H NMR spectra

Fraction				%			
	$H_{\alpha 1}$	$H_{\alpha 2}$	H_β	H_γ	H_{mono}	H_{di}	H_{tri}
P + N	1.36	2.02	70.08	26.54	0.00	0.00	0.00
A_H	8.78	6.91	52.13	26.33	5.32	0.53	0.00
A_B	10.16	10.57	42.28	22.76	8.94	4.47	0.81
A_{BE}	10.73	8.69	45.50	26.10	6.38	2.61	0.92

Table 2.1.5: Structural type data of the chromatographic fractions of the fractions of the vacuum distillate obtained from ^{13}C NMR spectra

Fraction	%		%			
	C_{al}	C_{ar}	C_1	C_2	C_3	$C_{4,5}$
P + N	100.00	0.00	4.42	5.09	3.32	32.74
A_H	70.97	29.03	4.52	4.52	2.82	14.12
A_B	43.48	56.52	3.57	3.57	2.38	9.82
A_{BE}	63.14	36.86	5.91	4.93	1.58	12.56

In the aromatic fractions (A_H, A_B) the content of the aromatic carbon and hydrogen increases and calculated number of aromatic rings in an average molecule R_A increases as well. In the A_{BE} fraction (eluted with a binary solvent) the content of aromatic carbon and hydrogen decreased in comparison with the previous fraction. However, for the number of aromatic rings in an average molecule the following is valid according to Table 2.1.6:
R_A of A_{BE} fraction > R_A of A_B fraction. It can be explained by molecular weights included in the calculation of the average molecule. For the molecular weights of these fractions the following is valid:
M_W of A_{BE} fraction > M_W of A_B fraction.

Table 2.1.6: Data on the structure of an average molecule of the chromato-graphic fractions of the vacuum distillate and the extract calculated from the NMR measurements by Williams (7)

Oil	Fraction	R_A	R_N	n	R
Vacuum distillate	A_H	1.31	1.11	6.00	0.39
	A_B	2.44	0.76	2.94	0.26
	A_{BE}	4.10	1.01	3.22	0.31
	A_H	0.98	0.96	7.00	0.37
Extract	A_B	2.60	0.74	4.90	0.29
	A_{BE}	3.23	0.82	4.44	0.30

The basic character of the compounds eluted after saturated fractions can be determined from the data in Table 2.1.6. The aromatic fraction eluted by hot hexane is mostly of a monoaromatic character. The fraction eluted with benzene contains mostly bi- and tri-nuclear aromates per molecule. The elemental analysis also demonstrates the presence of structural groups of sulphur in these fractions. Higher aromates are washed from the column using a binary eluant. Based on the elemental analysis they can have a character of polar compounds since they contain heteroatoms of sulphur and nitrogen in their molecules. All groups contain also a cyclanic ring and a different number of alkyls in an average molecule.

The high resolution mass spectrometric analysis of oil fractions can provide seven data in the groups of paraffinic and twelve data on the aromatic and/or sulphur aromatic groups. When analyzing oils by low resolution mass spectrometry the same groups of compounds for alkanocyclanes are determined. However, this method involves all aromatic compounds as monoaromates. The determination of a group composition of the hydrogenate by mass spectrometry is shown in Table 2.1.7. From the viewpoint of qualitative and quantitative data the mass spectrometry is the top method for the solution of research and technological problems. The results of the analysis of hydro-genate from various laboratories carried out with different instruments are presented in Table 2.1.8. It seems that the problems of calibration, geometry of the instruments, etc. cause deviations in the results. However, the method provides valuable information on the group composition and a good reproduci-bility of the given equipment. A comparison of the analysis has shown that results found on MS 902 S cannot be taken as absolutely certain. It must also be taken into account that high resolution mass spectrometers are neither designed nor built for quantitative work. With regard to some inaccuracies of the analyses and a high tear and wear of the instrument working at a good repeatability there are some possibilities for an application of noncalibrated mass spectrometers to follow the changes in composition of the oils during their technological processing.

Table 2.1.7: Group analysis of the hydrogenate by high resolution (HRMS) and low resolution (LRMS) mass spectrometry

Group	HRMS % Vol	LRMS % Wt
Paraffins	10.7	14.8
Monocycloparaffins	10.8	11.4
Dicycloparaffins	17.6	15.3
Tricycloparaffins	14.1	14.5
Tetracycloparaffins	11.3	10.2
Pentacycloparaffins	9.8	11.7
Hexacycloparaffins	6.8	5.7
Alkylbenzenes	5.8	—
Benzocycloparaffins	6.0	—
Benzodicycloparaffins	2.6	—
Naphthalenes	0.6	—
Acenaphthenes	0.4	—
Fluorenes	0	—
Phenanthrenes	1.0	—
Pyrenes	1.3	—
Chrysenes	0.6	—
Benzothiophenes	0.2	—
Dibenzothiophenes	0.2	—
Naphthobenzothiophenes	0.2	—
Total aromates	18.9	16.4

Therefore, a comparison of different methods of the determination of the group composition of oils is rather difficult (Table 2.1.9). It seems from the evaluation of various methods — when calculating the results to homogeneous values — that NMR spectrometry exaggerates the saturated fractions. It can be explained by the overlapping of the spectral lines of the CH_2 and CH_3 groups in alkyls of alkylaromatics with identical groups of aliphatic hydrocarbons. At the same time, the prerequisites for an identical number of alkyls in the particular groups of aromates are not satisfied, and either is the requirement for the low content of polyaromates. Therefore, the differences in the analysis of the vacuum distillate are more distinct. The values of the group analyses determined by chromatography are comparable with those of the low resolution mass spectrometry.

Table 2.1.8: Group composition of the hydrogenate determined by mass spectrometry using different instruments

Equipment	Method	%	Saturate	Aromatics			
				mono-	di-	tri-	poly-
MAT 731 VARIAN	HRMS	Vol	81.1	14.4	1.0	1.0	2.5
MS 902 S AEI	HRMS	Vol	88.5	8.2	0.7	1.3	0.7
MS 902 S AEI	LRMS	Wt	83.6	16.4	–	–	–
CEC 21-103 C	LRMS	Wt	87.1	12.9	–	–	–
CH 5 DF VARIAN	LRMS	Wt	92.3	7.7	–	–	–

Table 2.1.9: Group composition of the hydrogenate and the vacuum distillate determined by various methods

Oil	Method	%	Saturate	Aromatics			
				mono-	di-	tri-	poly-
Vacuum distillate	Chromatography	Wt	47.9	27.4		20.0	4.4
	MMR	Mol	85.4	12.7	1.9	–	–
	Chromatography	Wt	87.2	7.1		5.1	0.4
Hydro-genate	NMR	Mol	92.2	6.7	1.1	–	–
	HRMS	Vol	81.1	8.2	0.7	1.3	0.7
	LRMS	Wt	83.6	16.4	–	–	–

2.1.4 Conclusion

The chromatographic method of the preparative-analytical character provides quantitative data on the group composition of oils. The particular fractions can be characterized by spectral methods. Using NMR, it is possible to determine the data on an average molecule in the fractions obtained by chromatographic separation. The most convenient combination seems to be the analysis of chromatographic fractions by mass spectrometry.

The elaborated chromatographic method is advantageous for its simplicity, low price and good reproducibility. After characterization of the particular fractions it is possible to use it for routine analyses of a larger scale.

2.1.5 References

(1)   Suatoni, J.C.; Swab, R.E.: Rapid Hydrocarbon Group-Type Analysis by High Performance Liquid Chromatography. Journal of Chromatographic Sci 13 (1975) 8, 361—366.

(2)   Holstein, W.: HPLC — Gruppentrennung nach Ringklassen von Aromatischen Mineralölfractionen. Erdöl, Kohle, Erdgas, Petrochem. 40 (1987) 4, 175—177.

(3)   Matsunaga, A.; Yagi, M.: Separation of Aromatic Compounds in Lubricant Base Oils by HPLC. Anal. Chem. 50 (1978) 6, 753—756.

(4)   Radke, H.; Willsch, H.; Welte, D.H.: Preparative Hydrocarbon Group Type Determination by Automated Medium Pressure Liquid Chromatography. Anal. Chem 52 (1980) 3, 406—411.

(5)   Jewell, D.M.; Weber, J.H.; Plancher, H.; Lathan, B.R.: Ion Exchange, Coordination and Adsorption Chromatographic Separation of Heavy Ends Petroleum Distillates. Anal. Chem. 44 (1972) 8, 1391—1395.

(6)   Berthold, P.H.; Staude, B.; Bernhard, U.: Beitrag zur IR-spektroskopischen Strukturgruppenanalyse aromatenhaltiger Mineralölprodukte. Chem. Tech. 27 (1975) 4, 234—239.

(7)   Williams, R.B.: Symposium on Composition of Petroleum Oils Determination and Evaluation. ASTM Spec. Tech. Publ. 244 (1958) 163—194.

(8)   Clutter, P.R.; Petrakis, R.; Stegner, R.L.; Jensen, R.K.: Nuclear Magnetic Resonance of Petroleum Fractions. Anal. Chem. 44 (1972) 8, 1395—1405.

(9)   Netzel, D.A.; Miknis, F.P.: NMR Study of US Eastern and Western Shale Oils Produced by Pyrolysis and Hydropyrolysis. Fuel 61 (1982) 11, 1101—1104.

(10)   ASTM D 2786-71, Amer. Nat. Standard Z 11.310 — 1972, Approved May 18, 1972.

(11)   Gallegos, E.J.; Green, J.W.; Lindeman, L.P.; de Tourneau, R.L.; Teeter, R.M.: Petroleum Group-Type Analysis by High Resolution Mass Spectrometry. Anal. Chem. 39 (1967) 14, 1833—1838.

(12)   Jakubik, T.: Problems of Separation and Mass Spectrometric Analyses of Hydrogenated Oils. Doctoral Thesis. ChTF SVST, Bratislava (1976).

2.2 Dependency of Viscometric Properties on Base-Stocks Chemical Structures in Multigrade Crankcase Oils

H.H. Abou el Naga and S.A. Bendary, MISR Petroleum Co., Cairo, Egypt

2.2.1 Abstract

Dependency of some viscometric properties for multigrade crankcase oils on chemical structure variables of incorporated base-stocks have been evaluated. The neat base blends were formulated to contain increasing percentage from a bright-stock. Viscosity index improvers with different chemical structures were used in these formulations at the treating rates recommended by their suppliers. Statistical models via single and multiple regression analysis were derived to express the magnitude of this dependency.

Following viscometric properties: viscosity, viscosity increase, viscosity index, specific viscosity and thickening tendency are found highly dependent on percent content of the monoaromatics and nonhydrocarbon components.

Each of the combined chemical variables (TAR+S) and (C_A+S_m) $(C_P/C_N)^{(*)}$, has been found in strong correlation with most of the viscometric properties. Derived statistical models via these two combined variables are used in the prediction of the oils viscometric properties at high degree of confidence.

A multiple model based on the combination of the two variables (SAT) and (DAR+PAR)/(S+N+O) is found in good correlation to the change in oils viscosity.

2.2.2 Introduction

Performance properties of lubricating oils either as additive or unadditive oils, are greatly affected by the chemical structure of incorporated base-stocks. Therefore, over the last two decades different research schools have been studying the relationships between base-stocks chemical structure variables and performance properties of the formulated blends (1—14). According to these previous publications, it is possible to remark the following:

1. Tested base-stocks are included in the wide range from: neutral base oils, bright-stocks, reclaimed used oils, etc.

(*)   Refer to Appendix 1 for abbreviations key-words.

2. Formulated blends are covered by one of the following oils: unadditive oils (3,5 & 7), fully formulated crankcase oils (1, 2, 6 & 8), different types of industrial oils (9—14), etc.

3. Chemical structure variables are mainly considered by one or more of the following: saturates, total aromatics, different aromatic types (mono-, di- & polyaromatics) and non-hydrocarbon compounds (sulphur, nitrogen and oxygen compounds). These chemical structure variables have been studied either from their qualitative structures or quantitative values.

4. Performance properties are included in: oxidation stability (1—7), engine cleanliness (1, 2, 6, 8 & 14), friction properties (9 & 10) and surface properties (e.g. air release, foam tendency, demulsibility, etc. (11—14).

5. Special attention has been given to moderate refined base-stocks (either as neutral base oils or bright-stocks). Such base-stocks are containing relative high amounts from undesirable compounds (e. g. diaromatics, polyaromatics and non-hydrocarbons), therefore blends formulated from them can show relative low performance properties (1, 3 — 7, 9 & 11 — 14).

With regard to multigrade oils, base-stocks chemical structure variables are also found to play an important role in defining their viscometric properties (15, 18, 22 & 25). In general, viscometric properties of multigrade oils can be affected by one or more of the following:

1. Base-stocks general characteristics, e.g. density, viscosity, viscosity index, etc. (15, 22 & 25).

2. Base-stocks chemical structure variables, e.g. total aromaticity, content of each type of aromatic compounds, percentage and type of non-hydrocarbon compounds, etc. (18, 22 & 25).

3. Viscosity index improvers general characteristics, e.g. molecular weight, molecular weight distribution, solubility, dose percentage, etc. (17, 20 — 24 & 26 — 30).

4. Viscosity index improvers chemical structure variables, e.g. hydrocarbon component(s), oxygen containing group(s), molecular structure, molecular stereoregularity, etc. (20, 24 & 26 — 30).

5. Type of interaction between base-stocks and VI improvers (16, 18, 22 & 25).

6. Chemical structures and performance activities for incorporated multipurpose additive(s).

Some of the previous publications have considered the dependency of viscometric properties for multigrade oils on the chemical structure variables of incorporated base-stocks (16 – 29). Their most important conclusions can be summarized according to the following:

1. A base-stock of good solvent power to incorporated VI improver is usually composed of molecules similar in structure to the monomer links which form the VI improver. It is obvious that similar molecules are usually associated free (16).

2. A blend of base-stock and VI improver shows a viscosity decrease as the structural of base-stock becomes more paraffinic (19, 22 & 25).

3. Response of VI improvers to base-stocks increases with the improvement of the degree or technique of base-stocks refining (18, 22 & 25).

4. Due to the high solvation power of the aromatic compounds their content in the final lubricant blends can decide to a great extent to efficiency of the incorporated VI improvers (16, 22 & 25).

5. Styrene-based VI improvers have good efficiency in moderate to highly refined base-stocks (22, 25 & 26).

6. At low temperature, there is a distinct thickening effect for all VI improvers but with an exception to the polyalkyl-methacrylates (21, 23 – 25, 28 & 29). On the other hand, the thickening effect of polyalkyl-methacrylates at high temperatures is more pronounced than with other VI improvers (23, 25 & 28).

In few of these publications multigrade crankcase oils based on moderate refined base-stocks (22) and/or with high SAE viscosity grades (e.g. 15W-50, 20W-50, etc.), (25) have been considered. Generally, the formulate oils with high SAE viscosity grades bright-stocks are usually incorporated at relative high contents. As bright-stocks contain considerable amounts of aromatics and non-hydrocarbons, the performance properties for oils with high SAE viscosity grades are relatively lower compared to those with thinner ones. The only other approach to formulate such high SAE viscosity oils is based on using synthetic oils.

Moreover, previous publications have only studied effects of base-stocks chemical structures on their viscometric properties in a descriptive way (4, 22 & 25) and without expressing that sort of dependency on statistical bases.

The present work is aimed at finding the statistical models which can express the dependency of viscometric properties of crankcase multigrade oils with high SAE viscosity grades on the chemical structure variables of incorporated base-stocks. Therefore, different VI improvers are incorporated in crankcase formulations with step increasing in bright-stock content. As viscometric properties, in this part of the work, are only included: viscosity, viscosity increase

(i. e. difference between viscosity of VI improver containing blend and viscosity of neat blend), viscosity index and specific viscosity for fully formulated crankcase blends. Thickening tendency for each of the incorporated VI improvers is also considered.

2.2.3 Experimental

Two neutral base oils (150 N & 300 N) and one bright-stock (135 Brt) were used to formulate twelve base blends (Table 2.2.1). While bright-stock content in these blends was increased in steps from zero up to 23 % wt, the content of neutral base oil 150 N was decreased from 40 down to 17 % wt. Neutral base oil 300 N was incorporated in all formulations at a constant percentage (60 % wt). To meet the requirements of the API performance level SF/CC a multipurpose additive was incorporated in all blends at 7.9 % wt. Table 2.2.1 also includes viscosity at 40 and 100°C and the viscosity index of these base blends.

The chemical structure variables for incorporated base-stocks were measured via column chromatography and spectroscopy techniques (7, 14 & 31 − 35). The measured varibles include the percentage weight of the following: saturates (SAT), total aromatics (TAR), monoaromatics (MAR), diaromatics (DAR), polyaromatics (PAR) and total non-hydrocarbons (sulphur (S), nitrogen (N) and oxygen (O)). Percent mole of the paraffinic carbon (C_P), naphthenic carbon (C_N) and aromatic carbon (C_A) were measured via infrared spectroscopy (34 & 35). Total sulphur content was also calculated as percent mole (S_m). The key-words for abbreviations of the chemical structure variables are listed in Appendix 1. Measured chemical structure variables, as chemical groups, non-hydrocarbon elements and mole percent for the formulated base blends are listed in Table 2.2.2.

To formulate the multigrade crankcase oil samples, the following eight commercial VI improvers were incorporated in the formulated twelve base blends: styrene-isoprene copolymer (SICP), styrene butadiene copolymer (SBCP), two ethylene-propylene copolymers (OCP)1 and (OCP)2, a non-dispersant polyalkyl-methacrylate (PMA), a dispersant polyalkyl-methacrylate (PMA-D) and two highly dispersant mixed copolymers (MIX)1 and (MIX)2. The two ethylene-propylene copolymers were provided from two different suppliers. Both of these OCP viscosity index improvers are from the non-dispersant type. The two mixed copolymers were provided from one supplier, each with different chemical structure and molecular weight characteristics. All these VI improvers were incorporated in the base blends at the treating rates recommended by their suppliers. Table 2.2.3 includes the most important characteristics for these eight VI improvers.

Kinematic viscosity for neat base blends (Vo) and for formulated multigrade blends (V) were measured at temperatures 40 and 100° C. Following viscometric properties were considered:

Table 2.2.1:   Base Blends: Constituents and Characteristics

Base Blends*	Base-Stocks, % wt			Kin. Vis., cSt.,		VI
	150 N	300 N	135 Brt.	40° C	100° C	
BO 1	40	60	—	50.22	7.14	99
BO 2	37	60	3	53.83	7.44	99
BO 3	35	60	5	55.45	7.68	99
BO 4	34	60	6	57.54	7.75	98
BO 5	32	60	8	58.82	7.88	98
BO 6	30	60	10	62.37	8.17	98
BO 7	27	60	13	66.21	8.49	97
BO 8	25	60	15	69.71	8.75	97
BO 9	22	60	18	74.33	9.15	97
BO 10	20	60	20	77.09	9.32	96
BO 11	18	60	21	78.98	9.48	96
BO 12	17	60	23	82.05	9.72	96

* A multipurpose additive is incorporated in all formulations at concentration 7.9 % wt.

Table 2.2.2: Base Blends: Chemical Structure Variables

	Chemical Groups, % wt					Nonhydron. % wt (S+N+O)	Mole %			
	SAT	TAR	MAR	DAR	PAR		C_P	C_A	C_N	S_m
BO 1	71	29	22.13	5.38	1.49	0.91	66.14	9.58	24.28	0.56
BO 2	69	31	22.32	6.81	1.87	1.12	65.97	9.65	24.34	0.65
BO 3	67.5	32.5	23.52	7.05	1.93	1.22	65.86	9.7	24.39	0.75
BO 4	66.2	33.8	24.44	7.15	2.21	1.26	65.81	9.72	24.4	0.78
BO 5	65.2	34.8	25.2	7.26	2.34	1.29	65.7	9.76	24.45	0.94
BO 6	64.58	53.42	25.51	7.51	2.4	1.43	65.59	9.81	24.49	1.09
BO 7	62.6	37.4	27.19	7.7	2.51	1.53	65.42	9.88	24.56	1.16
BO 8	61.71	38.29	27.69	7.9	2.7	1.61	65.31	9.93	24.6	1.22
BO 9	60	40	29.02	8.1	2.88	1.7	65.18	10.	24.67	1.31
BO 10	58.71	41.29	29.79	8.45	3.05	1.77	65.04	10.05	24.71	1.44
BO 11	55.5	44.5	32.04	9.1	3.36	1.87	64.96	10.08	24.74	1.54
BO 12	54.1	45.9	32.5	9.5	3.9	1.99	64.87	10.12	24.77	1.63
absolute change	−16.9	16.9	10.37	4.12	2.41	1.08	−1.27	0.54	0.49	1.07
percent change	−23.8	58.3	46.86	76.6	161.7	118.7	−1.92	5.640	2.02	188.9

Table 2.2.3: Viscosity Index Improvers Characteristics

Copolymer Chemical Structure	Symbol	Dispersing Activity	Co-Polymer Content, % wt	Dose, % wt
Styrene-isoprene copolymer	SICP	non-dispersant	solid	1.02
Styrene-butadiene copolymer	SBCP	non-dispersant	solid	1.3
Olefin copolymer	(OCP)1	non-dispersant	13	10.8
Olefin copolymer	(OCP)2	non-dispersant	11	12
Mixed copolymer	(MIX)1	highly dispersant	40	3.6
Mixed copolymer	(MIX)2	highly dispersant	45	3
Polyalkylmethacrylate	PMA	non-dispersant	26	5.5
Polyalkylmethacrylate	PMA-D	dispersant	26	5

- Kinematic viscosity of multigrade blend at 40 and 100° C (V40 & V100).

- Kinematic viscosity increase $(\Delta V) = V - Vo$ at 40 and 100° C (Δ V40 & Δ V100).

- Viscosity index (VI)

- Specific viscosity (Vsp) $= \dfrac{V - Vo}{Vo}$ at 40 and 100° C (Vsp40 & Vsp 100)

- Thickening tendency (Q) $= \dfrac{Vsp100}{Vsp40}$

A computer program was used to explore and derive the statistical relationships between the blends viscometric properties and the chemical structure variables for incorporated base-stocks (14 & 30).

2.2.4 Statistical Methodology

Regression analysis, as a statistical technique, is usually used to examine results and draw conclusions about the functional relationships existing between the independent variables and the dependent responses (14 & 36). Through regression analysis it is possible to get clear answers about the following three basic questions: Is it so?, To what extent is it so? and Why is it so?

Single linear regression analysis can be simply represented by the following equation:

$$Y = A + aX$$

Where,

Y = dependent response
X = independent variable
A = equation constant
a = regression constant

In this study Y, as dependent response, is one of the tested viscometric properties for formulated crankcase multigrade oils, while X, as independent variable, is one of the chemical structure variables for incorporated base-stocks.

As base-stocks contain multiple chemical structure variables, such single regression relationships are misleading. It is obvious that no single independent variable can alone explain the performance behaviour of a formulated blend. Therefore, it is more realistic to use multiple linear regression, which can be represented by the following equation:

$$Y = A + a_1 X_1 + a_2 X_2 \ldots + a_m X_m$$

where,

— $X_1, X_2 \ldots X_m$ are independent variables, i. e. base-stocks chemical structure variables.

— $a_1, a_2 \ldots a_m$ are regression constants.

Non-linear regression analysis can also be applied to improve the correlations as obtained via the linear relationships.

For each derived model the following statistical factors can be calculated and used to check its confidence and suitability:

— F: a statistical parameter which is taken to assess the whole model significance and to be compared with the statistical F-Distribution.

— t: a statistical parameter which is taken to assess the significance of each partial relation (i. e. with each of the independent variables) and to be compared with the statistical t-Distribution. Positive t value indicates direct proportionality between Y and X, while negative value shows inverse proportionality. Sign of t value is similar to that of F value. It is also possible to use t value for the entire model.

— R^2:   the model correlation coefficient.

Models with F and t values higher than those listed in the statistical tables for confidence of 99.5 % are the only models to be considered in this study. Any accepted model must also represent a logical physical meaning. References 14 & 30 include more data about the use of statistical methodology.

2.2.4.1 Chemical Variables

In this study the following single chemical structure variables are considered to derive the dependency relationships (i. e. statistical models):

Saturates (SAT), total aromatics (TAR), monoaromatics (MAR), diaromatics (DAR), polyaromatics (PAR), total non-hydrocarbons (S+N+O), sulphur percent mole (S_m), paraffinic carbon (C_P), naphthenic carbon (C_N) and aromatic carbon (C_A).

Absolute and percent changes in these single variables by going from oil BO1 to oil BO12 are listed in Table 2.2.2.

Some other combined variables were calculated, either by adding, multiplying or dividing of the basic single variables. These derived combined variables are included in the following:

67

Table 2.2.4: Combined Chemical Structure Variables

	$\dfrac{(DAR+PAR)}{(S+N+O)}$	$\dfrac{(SAT)}{(S+N+O)}$	$\dfrac{(TAR)}{(S+N+O)}$	$\dfrac{(DAR+PAR)}{(S)}$	$\dfrac{(TAR)}{(S)}$	$(TAR+S)$	$\dfrac{(S_m)}{(C_A)}$	$\dfrac{(C_A)}{(S_m)}$	$\dfrac{(C_A+S_m)}{(C_p/C_N)}$
BO 1	7.52	77.76	31.76	161.11	29.18	0.0587	17.03	27.63	38.17
BO 2	7.73	61.44	27.6	147.62	31.21	0.068	14.7	27.93	41.33
BO 3	7.35	55.28	26.62	135.42	32.74	0.077	12.93	28.22	37.42
BO 4	7.42	52.46	26.78	135.2	34.05	0.08	12.44	28.32	37.44
BO 5	7.42	50.39	26.89	116	35.1	0.096	10.411	28.07	32
BO 6	6.93	45.19	24.79	101.2	35.77	0.1115	8.969	29.2	28.31
BO 7	6.67	40.89	24.43	101.08	37.77	0.117	8.544	29.4	27.59
BO 8	6.57	38.26	23.74	98.18	38.68	0.1227	8.15	29.6	27.18
BO 9	6.47	35.34	23.56	95.24	40.42	0.1313	7.62	29.89	26.14
BO 10	6.5	33.17	23.33	89.76	41.75	0.143	6.99	30.24	25.00
BO 11	6.68	29.74	23.84	90.82	44.99	0.1525	6.56	30.5	25.43
BO 12	6.73	27.17	23.05	88.27	42.42	0.1606	6.23	30.76	25.77
absolute change	– 0.79	– 50.59	– 8.71	– 72.84	17.24	0.1019	– 10.8	3.13	– 12.3
percent change	– 10.51	– 65.05	– 27.42	– 45.21	59.08	173.59	– 63.42	11.33	– 32.49

68

(DAR+PAR)/(S+N+O),	(SAT)/(S+N+O),
(TAR)/(S+N+O),	(DAR+PAR)/(S),
(TAR)/(S),	(TAR+S),
(S_m/C_A), (C_A/S_m), and	(C_A+S_m) (C_P/C_N)

Previous publications were considered at selection of these combined variables (1−14, 22 & 25). The values of these variables, their absolute changes and percent changes are listed in Table 2.2.4. Combined variables are also used in exploring the dependency of oils viscometric properties on base-stocks chemical structures.

Stepwise regression procedure was applied to establish the multiple relationships, where trials were run by combining two or more of the single and/or combined variables in one equation, as far as such combinations have acceptable physical meaning.

2.2.5   Results and Discussion

2.2.5.1   Base-Stocks Chemical Structure Variables

As bright-stock (Brt 135) content in formulated base blends (BO1 − BO12) is increased in steps from zero up to 23 % wt, content of neutral base oil 150 N is decreased from 40 down to 17 % wt (Table 2.2.1). Kinematic viscosity at 40 and 100°C and VI for these base blends are also listed in Table 2.2.1, where, as expected by increasing bright-stock, viscosity at both 40 and 100°C are increased while viscosity index (VI) is decreased. By going from blend BO1 to blend BO12 the absolute increase in kinematic viscosity is found to be 31.83 cSt at 40°C and 2.58 cSt at 100°C, while VI showed a limited decrease (only 3 units).

The chemical structure variables in the base blends (BO1 − BO12) are also changed in steps, as listed in Table 2.2.2. It is possible to order the change in the chemical groups, either as absolute change or as percent change.

As absolute change:

SAT = TAR > MAR > DAR > PAR > (S+N+O)

As percent change:

PAR > (S+N+O) > DAR > TAR > MAR > SAT

Such high increase in the percent change for polyaromatics, non-hydrocarbons and diaromatics is directly related to the increase in bright-stock content.

With regard to the chemical structure variables, as percent mole, which are also listed in Table 2.2.2, it is possible to order the change in them according to the following:

.

As absolute change:

$$C_P > S_m > C_A > C_N$$

As percent change:

$$S_m > C_A > C_N > C_P$$

Combined chemical structure variables, as listed in Table 2.2.4, are also changed in steps by increasing bright-stock. Order of decrease in these variables is according to the following:

As absolute change:

(DAR + PAR)/S > (SAT)/(S+N+O) > (TAR/S) >
TAR/(S+N+O) > (DAR+PAR)/(S+N+O) > (TAR+S).

As percent change:

(TAR+S) > SAT/(S+N+O) > (TAR/S) >
(DAR+PAR)/S > TAR/(S+N+O) > (DAR+PAR)/(S+N+O)

The change in the combined chemical variables, as percent mole, is found to be according to the following order:

As absolute change:

$$(C_A + S_m) \, (C_P/C_N) > S_m/C_A > C_A/S_m$$

As percent change:

$$S_m/C_A > (C_A+S_m) \, (C_P/C_N) > C_A/S_m$$

2.2.5.2 Change in Viscometric Properties with increasing bright-stock content

VI improvers were incorporated in the base-blends (BO1 — BO12 — Table 2.2.1) at the treating rates recommended by their suppliers. The following viscometric properties were measured on the formulated blends:

— Kinematic viscosity at 40 & 100° C (V40 & V100).

— Kinematic viscosity increase (difference between kinematic visocosity of VI improver containing blend (V) and kinematic viscosity of neat blend (Vo) at 40 & 100° C (ΔV40 & ΔV100)).

— Viscosity index (V. I.)

70

— Specific viscosity at 40 & 100° C (Vsp 40 & Vsp 100). Where,

$$Vsp = \frac{V - Vo}{Vo}$$

— Thickening tendency (Q). Where,

$$Q = \frac{Vsp\ 100}{Vsp\ 40}$$

Table 2.2.5 includes these viscometric results for blends BO1 and BO12 with zero and 23 % wt of bright-stock respectively. While V40, Δ V40 & V100, Δ V100, Vsp 40 and Vsp 100 for all blends are increased with increasing bright-stock, their VI and Q value are decreased. For more illustration Fig. 2.2.1 includes the change in V100 and VI for the tested blends against increase in bright-stock. Accordingly, it is clear that oils viscometric properties are remarkably changed with increasing bright-stock.

Comparing the absolute changes in these viscometric properties, as listed in Table 2.2.6, with the changes in the chemical structure variables, as listed in Table 2.2.2, is also clearly supporting that the limited changes in chemical variables have resulted in great changes in oils viscometric properties. In other words, such comparison clearly shows the high dependency of oils viscometric properties on chemical structure variables of base-stock.

In general, these viscometric properties are indicating the following:

1. Increase of bright-stock content results in a very wide change in kinematic viscosity at 40°C. It is possible to explain such a change in V40 by the improvement of the blends solvation power due to the increase in their aromaticity as their contents of bright-stock are increased.

2. The wide increase in Δ V40 is in analogy to such a remarkable increase in V40.

3. Kinematic viscosity at 100°C for VI improver containing blends with zero % wt bright-stock (i. e. based on oil BO1) have not been acquired the requirement of SAE 50 viscosity (16.3 cSt at 100°C).
   According to Fig. 2.2.1, these blends have acquired the SAE 50 viscosity limit at the following bright-stock contents:

SICP	12.8 % wt	(Mix)1	20.5 % wt	
SBCP	19.2 % wt	(Mix)2	23 % wt	
(OCP)1	4.4 % wt	PMA	11 % wt	
(OCP)2	6.2 % wt	PMA-D	15 % wt	

4. Changes in Δ V100 are limited in comparison to Δ V40 changes.

5. Due to the increase in the oils solvation power the viscosity index for all blends is decreased.

6. Changes in Vsp 40, Vsp 100 and Q value can be considered limited in comparison to changes of other viscometric properties. As expected, Q values for all blends are decreased with increasing bright-stock. While both Vsp 40 and Vsp 100 for all blends are increased, those for SICP are decreased. It is difficult to explain these Vsp results for SICP without considering its Vsp at other temperatures.

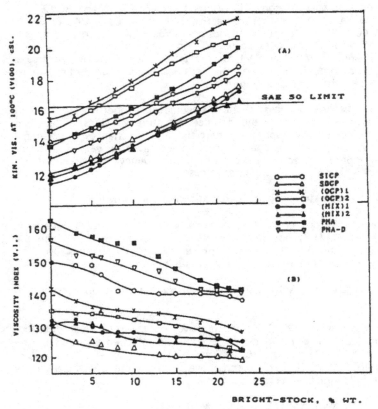

Figure 2.2.1: Change in kinematic viscosity at 100° C and viscosity index with increasing bright-stock content

Table 2.2.5: Viscometric properties for formulated multigrade crankcase oils with zero % wt and 23 % wt bright-stock

Oil		SICP	SBCP	(OCP)1	(OCP)2	(MIX)2	(MIX)2	PMA	PMA-D
BO 1	V40	96.06	89.22	115.07	112.77	80	86.35	87.48	83.65
(zero % wt	ΔV40	45.83	39	64.85	62.55	29.78	36.13	37.26	33.43
bright-stock)	V 100	14.04	12.03	15.52	14.8	11.45	11.91	13.86	13.05
	ΔV100	6.9	4.89	8.38	7.66	4.31	4.77	6.72	5.91
	V.I.	150	128	142	135	132	131	163	157
	Vsp 40	0.913	0.777	1.29	1.245	0.593	0.719	0.742	0.666
	Vsp 100	0.966	0.685	1.17	1.07	0.604	0.668	0.941	0.828
	Q	1.059	0.882	0.91	0.86	1.018	0.929	1.269	1.243
BO 12	V40	151.4	158	202	194	148.4	144	162.2	143.6
(23 % wt	ΔV40	69.35	75.95	119.95	111.95	66.35	61.95	80.15	61.55
bright-stock)	V100	18.6	17.36	21.6	20.38	17.08	16.46	19.78	18.16
	ΔV100	8.88	7.64	11.88	10.66	7.36	6.74	10.06	8.44
	V.I.	138	119	128	122	125	122	141	141
	Vsp 40	0.845	0.926	1.46	1.36	0.809	0.755	0.977	0.75
	Vsp 100	0.914	0.786	1.22	1.096	0.757	0.693	1.035	0.868
	Q	1.081	0.849	0.84	0.8	0.936	0.918	1.059	1.157

Table 2.2.6: Change in viscometric properties for formulated multigrade crankcase oils by increasing bright-stock content

VI Improver	V40	Δ V40	V100	Δ V100	VI	Vsp 40	Vsp 100	Q
SICP	55.35	23.52	4.56	1.98	− 12	− 0.067	− 0.05	− 0.022
SBCP	68.78	36.95	5.33	2.75	− 9	0.149	0.101	− 0.033
(OCP)1	86.93	55.1	6.08	3.5	− 14	0.171	0.648	− 0.072
(OCP)2	81.23	49.4	5.59	3	− 13	0.119	0.024	− 0.058
(MIX)1	68.4	36.57	5.63	3.05	− 7	0.216	0.154	− 0.082
(MIX)2	57.65	25.82	4.55	1.97	− 9	0.036	0.025	− 0.01
PMA	74.72	42.89	5.92	3.34	− 22	0.235	0.094	− 0.213
PMA-D	59.95	28.12	5.11	2.53	− 16	0.084	0.041	− 0.086

According to Table 2.2.6, it is possible to evaluate the response of the different VI improvers to bright-stock increasing. Such a response is considered in terms of the absolute changes in each viscometric property. The order of decrease in these responses is found according to the following:

high ← response of VI improver → low

V40	(OCP)1	>	(OCP)2	>	PMA	>	SBCP	>
ΔV40	(MIX)1	>	PMA-D	>	(MIX)2	>	SICP	

V100	(OCP)1	>	PMA	>	(MIX)1	$\geqslant$	(OCP)2	>
ΔV100	SBCP	>	PMA-D	>	(MIX)2	$\geqslant$	SICP	

V. I.	PMA	>	PMA-D	>	(OCP)1	>	(OCP)2	>
	SICP	>	SBCP	>	(MIX)2	>	(MIX)1	

As the differences in Vsp 40 and Vsp 100 are very close, the ordering of them in terms of VI improver responses is found not to be significant. With regard to Q values (cf. Table 2.2.5) the order of decrease in them for blends based on BO12 (with 23 % wt bright-stock) is found according to the following:

Q > 1	PMA-D	>	SICP	>	PMA,	(i.e. behave as VI improvers)
Q < 1	(MIX)1	>	(MIX)2	>	SBCP	$\geqslant$ (OCP)1 $\geqslant$
	(OCP)2, (i.e. behave as thickeners)					

In general these changes for the oils viscometric properties clearly prove the following:

1. The high dependency of oils viscometric properties on base-stocks chemical structures.

2. Response of tested VI improver samples towards change in base-stocks chemical structures is remarkably different.

2.2.5.3 Dependency of Viscometric Properties on Single Chemical Structure Variables

Single linear regression analysis (Y = A + aX) was applied to evaluate the degree of dependency for each of the measured viscometric properties on the chemical structure variables. It was on mind, at deriving the correlation between viscometric properties and single chemical structure variables, that no single chemical structure can exist alone or affect all viscometric properties with the same magnitude of power. Actually the main reasons to carry the single regression analysis are:

1. To define direction of proportionality (as direct or inverse) between viscometric properties and each of the chemical structure variables.

2. To establish magnitude of dependency of each viscometric property on each chemical structure variable.

Tables 2.2.7 – 2.2.14 include the F values for the derived statistical models between measured viscometric properties and each of the single and combined chemical structure variables.

As the F values for models of the single chemical variables are extremely high, it is clear the presence of high dependency of viscometric properties on most of the single chemical variables. The percent confidence in these models is also very high (over 99.5 %).

Exploring directions of proportionality and magnitude of dependency according to these models is discussed in the succeeding two sections.

2.2.5.3.1   Directions of Proportionality

Table 2.2.15 includes directions of proportionality between the different chemical structure variables and each of the measured viscometric properties. In spite of that, these directions of proportionality can be traced by referring to the trend of increasing or decreasing of each chemical variable comparing with each viscometric property, but applying the regression analysis technique has helped in defining them by referring to the sign of the F values for each derived model, as listed in Tables 2.2.7 – 2.2.14.

It can be seen, according to Table 2.2.15, that all single chemical variables, with the exception of SAT and C_p, are at direct proportionality to all viscometric properties. With the exception of VI and Q value which are at inverse proportionality. With regard to SAT and C_p, they are at direct proportionality to VI and Q and at inverse proportionality to the rest of the viscometric properties.

These observations concerning the direction of proportionality, agree with the previously published conclusions (19, 22 & 25) which have clarified that the viscosity of VI improver containing blend is increased as the structure of base-stock becomes more aromatic.

2.2.5.3.2   Magnitude of Dependency

Magnitude of dependency of viscometric properties on chemical structure variables can be expressed via considering the F values of the derived models. The high F value indicates a high degree of dependency and the opposite is also correct. Therefore, it was possible to order the decrease in magnitude of dependency in terms of the F values, as listed in Tables 2.2.7 – 2.2.14, for each incorporated VI improver. Table 2.2.16 includes the order of decrease in dependency of the different viscometric properties on the following single chemical groups: SAT, MAR, DAR, PAR and (S+N+O). In this table similar orders for magnitude of power are put together.

76

Table 2.2.7: F values for dependency relationships (SICP blends)

	V 40	ΔV 40	V 100	ΔV 100	VI	Vsp 40	Vsp 100	Q
Single								
SAT	— 399.5	— 174.5	— 289	— 145.3	— 32.3	L.C.(*)	L.C.	L.C.
MAR	499.4	252.3	435.1	222	31.8	L.C.	L.C.	L.C.
PAR	139.5	93.3	129.8	87.5	29.5	L.C.	L.C.	L.C.
DAR	70.1	49.7	63.8	41.4	24.7	L.C.	L.C.	L.C.
(S+N+O)	369.8	166.2	288.5	120.4	37.7	L.C.	L.C.	L.C.
C_P	— 499.4	— 308.4	— 848.7	— 219	39.2	L.C.	L.C.	L.C.
C_N	1342.2	302.8	817	208.4	38.4	L.C.	L.C.	L.C.
C_A	1244.9	269.4	879.6	210.5	36.5	L.C.	L.C.	L.C.
S_m	704.9	350.5	412.3	184.9	51.6	L.C.	L.C.	L.C.
Combined								
(TAR/S)	— 61	— 64.5	— 49.4	— 37.7	135.4	L.C.	L.C.	L.C.
TAR/(S+N+O)	— 30.2	— 25.4	— 27.4	— 19.5	98	L.C.	L.C.	L.C.
SAT/(S+N+O)	— 80.3	— 62.3	— 68.2	— 43.3	48.7	L.C.	L.C.	L.C.
(DAR+PAR)/S	56.4	74.1	47.2	43.9	94.4	L.C.	L.C.	L.C.
(DAR+PAR) (S+N+O)	37.2	38.9	35.1	31.1	25.1	L.C.	L.C.	L.C.
(TAR+S)	351	178	296	147	— 33	L.C.	L.C.	L.C.
C_A/S_m	74.4	73	59.5	42.5	117.6	L.C.	L.C.	L.C.
S_m/C_A	585.8	336.4	351.3	169.4	55.9	L.C.	L.C.	L.C.
(C_A+S_m) (C_P/C_N)	753.3	388	435.7	182.7	51.4	L.C.	L.C.	L.C.

(*) L.C.: low confidence

Table 2.2.8: F values for dependency relationships (SBCP blends)

	V 40	Δ V 40	V 100	Δ V 100	VI	Vsp 40	Vsp 100	Q
Single								
SAT	—	—	—	—	64.5	77.4	89.1	7
MAR	414.9	386.5	386.5	346.2	54	95.2	116.9	7.1
	561.2	589.9	525.9	518.95	53.2	50.4	59.1	5.8
DAR	157.6	149.2	150.7	140.2	66.2	34.12	34.1	6.4
	78.9	72.7	80.4	70.1	—	—	—	—
(S+N+O)	416.6	322.8	432.8	277.3	124.3	69.5	69.1	7.9
Cp	15.4	15	82.5	14.4	21.3	12.8	11.2	5.7
CN	2014.6	1271.2	2313.3	869.4	103.9	107.4	108.3	8.5
CA	2276.2	1166.7	2436.3	821	101.8	94.7	100.5	7.9
Sm	580.4	574.8	742.8	742.9	84.2	119.5	163.3	7.3
Combined								
TAR/S	52.9	52.7	52.7	55.3	111.3	58	44.9	9.5
TAR/S+N+O	29.8	27.4	27.4	26.4	106	19.4	16.1	8.2
SAT/S+N+O	78.9	71.9	82.5	68.7	239.5	42.8	36	9.3
DAR+PAR/S	46.1	48.7	49.5	52.7	21.8	75.1	71.7	8.7
DAR+PAR	—	—	—	—	—	—	—	—
S+N+O	32.8	32	33.7	31.3	35.4	29.5	24.8	8.7
TAR+S	428	398.4	416.8	357.7	65.1	78.4	90.5	7
C_A/S_m	65.7	64.3	70.9	68	165.3	60.3	50.3	9.9
S_m/C_A	468.1	466.6	585.1	585.7	88.1	120.5	160.6	7.4
(C_A+S_m) (C_P/C_N)	383.4	378.4	450.7	461.2	65.3	99.4	138.8	6.8

Table 2.2.9: F values for dependency relationships ((OCP)1 blends)

	V 40	ΔV 40	V 100	ΔV 100	VI	Vsp 40	Vsp 100	Q
Single								
SAT	− 461.8	− 445.3	− 292.1	− 214.6	− 127.8	− 35.4	− 20.8	17.2
MAR	651.7	696	452.3	364.5	79.6	65.5	27.3	− 14.5
PAR	162	155.8	144.3	92	− 129.7	38.5	14.6	− 18.2
DAR	80.8	76.8	66.3	53.7	− 278	28	9.8	− 23.1
(S+N+O)	386.1	305.7	321.4	203.7	− 186.7	42.6	18.2	− 15.7
Cp	− 1348.2	− 849.6	− 38.4	− 626.2	109.3	− 53.7	25.42	13.5
CN	1300.5	822.6	1754.7	671.3	− 103	54.2	52.6	− 13.5
CA	1622.4	914.1	1663.9	616.9	− 109.5	52.4	24.6	− 13.6
Sm	459.9	399.7	536.7	398	81.9	52.9	27.5	− 12.2
Combined								
TAR/S	− 49	− 58.6	− 46.6	− 56.7	48.9	55.5	− 20.5	L.C.
TAR/S+N+O	− 28.6	− 30	− 26.3	− 26.1	64	− 13.3	L.C.*	10.3
SAT/S+N+O	− 76.8	− 70.7	− 78.8	− 65.8	246.8	− 29.3	L.C.	15.6
DAR+PAR/S	42.7	42.1	55.7	62.4	21.8	32.3	45.6	5.3
DAR+PAR	30.8	29.2	39.6	41	16.6	17.8	23.9	
S+N+O								L.C.
TAR+S	474.6	456	300.2	219.7	− 127.3	55.5	21	− 17
CA/Sm	61.9	58.9	71.6	66.7	81.7	32.7	19.3	11.2
Sm/CA	379.5	331	455.9	354.5	81.8	52.2	27.6	12
(CA+Sm) (Cp/CN)	497.4	412.8	595.4	416.5	88.1	52.5	27	12.4

(*) L.C.: low confidence

Table 2.2.10: F values for dependency relationships ((OCP)2 blends)

	V 40	ΔV 40	V 100	ΔV 100	VI	Vsp 40	Vsp 100	Q
Single								
SAT	−478.7	483.8	203.5	−120.7	158.9	−48.8	158.9	20.7
MAR	695.5	827.1	226.9	133.7	−162.2	58	−162.2	22.1
PAR	149.9	139.2	139.2	66.1	−127	29.2	−127	17.7
DAR	83.5	80.8	80.8	55.8	−59.9	27.3	−59.9	14.1
(S+N+O)	458.6	383.2	418.5	218.4	−76	41.3	−76	13.5
Cp	−2371.7	−1564.3	−1593	−435.8	81.9	−51.3	81.9	13.9
CN	2123	1399.2	2387	499.6	−75.1	50.8	75.1	13.6
CA	2477.3	1361	1735.7	419.1	−79.7	47	−79.8	13.7
Sm	610.2	607.6	494.9	283.2	−80.3	57.5	5.1	14.3
Combined								
TAR/S	−54.2	−54.6	−54.6	−123.7	18.3	−3.3	18.3	5.1
TAR/S+N+O	−30.7	−29.3	−29.3	−48.4	−12.6	−8.2	12.6	3.4
SAT/S+N+O	−84.7	−81.8	−125.9	−128.4	−28.3	−7.2	28.3	7.4
DAR+PAR/S	46.3	48.2	67.5	91.5	16.7	49.2	14.6	5.5
DAR+PAR								
S+N+O	33.3	32.9	49.8	62.5	11.1	23	12	L.C.*
TAR+S	494.9	500	208.8	123.7	−157.5	49.1	3.3	−20.5
CA/Sm	68.7	69.2	114.5	151.6	22.2	45.9	11.6	6.2
Sm/CA	494.9	499.8	475.3	295.1	76	58.8	5.4	13.7
(CA+Sm) (Cp/CN)	669.7	640.2	613.4	326.7	79	56	5.2	13.8

(*) L.C.: low confidence

Table 2.2.11: F values for dependency relationships ((MIX)1 blends)

	V 40	ΔV 40	V 100	ΔV 100	VI	Vsp 40	Vsp 100	Q
Single								
SAT	504	526.8	390.4	259	64.9	123.2	65	L.C.*
MAR	66	769.8	532.9	369.3	63.7	133.6	77.7	L.C.
PAR	169.3	166.8	145.7	117.3	48.3	72.1	45.5	L.C.
DAR	86.1	84.1	78	63.3	44	56.2	30.9	−5
(S+N+O)	509.5	416.1	429.8	277	76	126.7	74.8	L.C.
Cp	2515.5	1396.8	2164.8	876.6	73	159.2	106.1	L.C.
CN	2081.1	1123	2020.2	728.2	70.7	152	98.8	L.C.
CA	2594.2	1157.7	2234.1	734	70.1	139.8	96.1	L.C.
Sm	917	1469.5	955.6	861.5	70.1	304.6	132.4	L.C.
Combined								
TAR/S	58	63.3	60	64.1	80.2	137.7	69.6	L.C.
TAR/S+N+O	31.2	29.6	30	28.8	38.9	31.6	23.7	L.C.
SAT/S+N+O	86.7	84.16	83.1	73	104.2	87.5	45.4	L.C.
DAR+PAR/S	49	54.4	52.7	63.4	52.6	117.1	98.6	L.C.
DAR+PAR	32.9	32	35.1	38.2	32	34.4	44.9	L.C.
S+N+O								
TAR+S	524.6	551.4	404.9	267.1	65.3	125.8	66.1	L.C.
CA/Sm	73.1	78	73.7	74.5	124.2	162.7	65.9	5
Sm/CA	697.4	1047	730.2	714	73.5	345.1	136.3	9.6
(CA+Sm) (Cp/CN)	2873.04	3887.7	2772	1258.7	66.6	213.8	117.7	L.C.

(*) L.C.: low confidence

81

Table 2.2.12: F values for dependency relationships ((MIX)2 blends)

	V 40	ΔV 40	V 100	ΔV 100	VI	Vsp 40	Vsp 100	Q
Single								
SAT	249.8	150.5	207.3	108.9	72.8	10.9	7	L.C.*
MAR	336.7	209.5	237.7	126.1	82.6	13.3	7.7	L.C.
PAR	110.5	75.2	98.3	61.4	51.9	7.7	5.2	L.C.
DAR	65.7	49.2	69.3	48.9	35.8	6.3	5.3	L.C.
(S+N+O)	343.5	199.5	350.6	163.1	81.6	12.6	8.7	L.C.
C_P	1259.9	478.8	928.1	281.8	107.8	15.8	10.2	5.8
C_N	1625.9	608.6	1294.2	333.7	107	16.9	10.7	6
CA	1698.9	517.5	1261.3	313.4	99.2	15.7	10.3	5.6
S_m	431	278.7	277.9	137.9	209.6	15.9	8	7.7
Combined								
TAR/S	56.5	58.2	55	47.7	129.6	18.4	8	9.7
TAR/S+N+O	32	30	36.5	34.2	26.4	8.9	8.3	L.C.
SAT/S+N+O	79	67.2	89	69.7	55.1	10.9	8.3	L.C.
DAR+PAR/S	51.5	61.1	45.4	42.3	179.4	34.8	8.8	22.3
DAR+PAR	41.3	51.6	42	47.8	38.8	42.3	16.4	6.8
S+N+O								
TAR+S	255.9	153.5	211.1	110.1	74.2	11	7	L.C.
CA/S_m	67.9	65.6	68	56	233	15.5	8	8
S_m/CA	373.7	258.5	252.1	131.1	233.8	16.2	8	8
(CA+S_m) (C_P/C_N)	482.7	299.8	315.2	150.3	196.7	16	8.3	7.5

(*) L.C.: low confidence

Table 2.2.13: F values for dependency relationships (PMA blends)

	V 40	ΔV 40	V 100	ΔV 100	VI	Vsp 40	Vsp 100	Q
Single								
SAT	– 350.6	– 252.9	– 557.1	585.9	151.8	– 98.4	– 72.7	– 60
MAR	436.9	312.2	627.3	679	– 161.4	113.1	75	– 67.5
PAR	158.9	138.7	190.3	212	– 84.8	72	69.9	– 43.1
DAR	75.5	65.5	93.2	90.5	– 64.2	41.6	37.3	– 31.9
(S+N+O)	249.8	154.7	475.9	317.8	– 187.3	62.6	43.2	– 50
Cp	– 148.1	– 123.8	– 200.4	– 228	89.5	– 68.5	– 79.6	– 40.9
CN	522.6	260.1	1426.6	614.6	– 409.2	89.7	49.4	– 72.3
CA	522.6	288.1	1909.5	712.2	– 436.6	92.7	50.4	– 73.2
Sm	255.7	161.4	565.8	475.1	– 139	65.6	60.2	– 44.2
Combined								
TAR/S	– 36.6	– 28.6	– 50.3	– 44.4	38.2	– 17.6	– 19.7	15.3
TAR/S+N+O	– 24.2	– 19.5	– 30	– 25.7	30.7	– 12.5	– 10.3	13.2
SAT/S+N+O	– 59.3	– 45.6	– 82.1	– 69.2	69.3	27	23.5	24.9
DAR+PAR/S	32.2	26.1	40.9	37.2	30.5	– 16.9	18.6	14.6
DAR+PAR	24.9	20.2	28.7	24.2	29.9	12.9	8.9	14.9
S+N+O	355.2	253.7	576.2	603	– 152.2	98	72.7	– 59.8
TAR+S								
CA/Sm	45.2	34.9	63.9	55.8	47.8	21.2	22.9	18.3
Sm/CA (Cp/CN)	217.2	140.9	452	370.9	127.7	60.1	57.5	41.2
(CA+Sm) (Cp/CN)	273.4	163	622	416.3	170.5	64.5	51.5	47.3

83

Table 2.2.14: F values for dependency relationships (PMA-D blends)

	V 40	ΔV 40	V 100	ΔV 100	VI	Vsp 40	Vsp 100	Q
Single								
SAT	– 322.2	– 228.7	– 354.7	– 294.6	74	– 22.5	26.2	13.4
MAR	410.5	301.7	447.7	417.9	66.5 –	24.7	31.5	13.6 –
PAR	130.5	101.1	137.4	121.3	51.2 –	16.5	20.1	10.1 –
DAR	75.9	64.9	79.4	68.6	62.2 –	15.6	14.1	12.1 –
(S+N+O)	470.4	332	528.4	372.8	138.2 –	27	25.9	17.3 –
Cp	– 2791.7	– 1158.6	– 3615.7	– 1650.5	146.5 –	– 31.3	32.9	17.9 –
Cn	5529.7	2058	6033.7	2030.5	154.8 –	33.4	33.2	19.3 –
CA	4441.3	1247	5249.8	1519.1	143.9 –	30.4	31	17.8 –
Sm	660	807	628.1	1052.1	117.4 –	35.6	43.5	18.3 –
Combined								
TAR/S	– 61.9	– 81.8	– 66.4	– 75.3	153.8 –	– 89	53.7	46.1
TAR/S+N+O	– 34.5	– 35.7	– 34.8	– 32.9	86.9 –	– 23.1	12.9	25.2 –
SAT/S+N+O	– 94.3	– 97.8	– 96.6	– 90.7	233.6 –	– 33.5	22.6	26.6 –
DAR+PAR/S	54.8	73	55.6	69	58.5	135.3	153.5	38.9
DAR+PAR	39.5	49.8	39.1	41.9	43.3	45.9	30.8	29.9
S+N+O								
TAR+S	332.2	235.2	366.9	304.7	74.9 –	22.7	26.5	13.5 –
CA/Sm	81.2	100	83.4	91.8	225.2	70.2	44	41.5
Sm/CA	557.5	583.3	664.4	891.88	124.4	37.8	45.7	19.2
(CA+Sm) (Cp/Cn)	183.8	735.6	971.2	1214.8	129.6	36	42.3	18.8

Table 2.2.15: Proportionality direction between viscometric properties and chemical structure variables

Chemical Variables		Proportionality Direction	
Single	Combined	Direct	Inverse
MAR	(S+N+O)	V 40	V. I.
DAR	(TAR+S)	Δ V 40	Q
PAR	S_m/C_A	V 100	
TAR	(C_A+S_m) (C_P/C_N)	Δ V 100	
C_N		Vsp 40	
C_A		Vsp 100	
S_m			
SAT	TAR/S	V. I.	V 40
C_P	SAT/(S+N+O)	Q	Δ V 40
	TAR/(S+N+O)		V 100
	DAR+PAR/(S+N+O)		Δ V 100
	DAR+PAR/S		Vsp 40
	C_A/S_m		Vsp 100

Table 2.2.16: Viscometric properties as affected by the single chemical structure variables

VI Improver	Viscometric Properties	Decrease in magnitude of power for single chemical groups
SICP	V 40 – Δ V 40 – V 100 – Δ V 100	
SBCP	V 40 – Δ V 40 – Δ V 100 – Vsp 40 Vsp 100 – Q	
(OCP)1	V 40 – Δ V 40 – Δ V 100 – Vsp 100	
(OCP)2	V 40 – Δ V 40 – Vsp 40	MAR > SAT > (S+N+O) > PAR > DAR
(MIX)1	Δ V 40	
PMA	V 40 – Δ V 40 – V 100 – Δ V 100 – Q	
PMA-D	Vsp 100	
(OCP)2	V 100 – Δ V 100 – Vsp 100	
(MIX)2	V 40 – Δ V 40 – V 100 – Δ V 100 Vsp 40 – Vsp 100 – V. I.	(S+N+O) > MAR > SAT > PAR > DAR
PMA	V. I.	
SBCP	V 100	
(OCP)1	V 100	
(MIX)1	V 100 – Δ V 100 – Vsp 40 – Vsp 100	MAR > (S+N+O) > SAT > PAR > DAR
PMA-D	V 100	
PMA-D	V 40 – Δ V 40 – V 100 – Vsp 40 – Q	(S+N+O) > MAR > SAT > DAR > PAR

According to Table 2.2.16, it is possible to observe and/or conclude the following:

1. Most of the viscometric properties for blends containing SICP, SBCP, (OCP)1, (OCP)2 (i. e. all hydrocarbon copolymers) and PMA are highly dependent on MAR. The order of decrease in magnitude of dependency on the five chemical groups is found according to the following:

   MAR > SAT > (S+N+O) > PAR > DAR

   This order clearly expresses the great role of monoaromatic compounds to affect the oils viscometric properties for formulated blends. On the contrary, this order shows the relative low magnitude for power for PAR and DAR in affecting these properties. This order also explains that the non-hydrocarbons, in spite of their presence at very low content, affect viscometric properties at power higher than those for PAR and DAR, which are presented at higher concentrations (cf. Table 2.2.2).

2. For most of the viscometric properties for blends containing the mixed copolymer (MIX)1 the following order for magnitude of dependency is found:

   MAR > (S+N+O) > SAT > PAR > DAR.

   In the case of the other mixed copolymer (MIX)2, the role of non-hydrocarbons in affecting its viscometric properties is found to be more powerful than is the case in (MIX)1. The order of decrease in power of chemical groups for this VI improvers is found according to the following:

   (S+N+O) > MAR > SAT > PAR > DAR.

3. With blends containing PMA-D, nearly the same order as that for (MIX)2 is found. The only difference is the exchange between DAR and PAR in their positions, i. e. the following order is found:

   (S+N+O) > MAR > SAT > DAR > PAR.

   The increasing role for the non-hydrocarbons with blends containing the following VI improvers: (MIX)1, (MIX)2 and PMA-D, can be explained by the presence of dispersant chemical groups (i. e. nitrogen containing groups), which are attached to the structure of these ester-based copolymers. It is obvious to mention that similar molecules are usually associated free, therefore they are showing good solvation power to groups similar to them. Accordingly, these three viscosity index improvers are greatly affected by the change in the non-hydrocarbon content (or type) of the base-stock chemical structures.

4. Most of the VI and Q value results have showed a great role for the influence of the non-hydrocarbons.

5. Most of the derived relations are supporting the low role of power for the diaromatics (DAR) and polyaromatics (PAR) in affecting the oils viscometric properties.

In general, the derived statistical models can be valuable in explaining the different behaviours for the viscosity index improvers under the influence of the change in base-stocks chemical structure.

With regard to the chemical structure variables, as presented by % mole, possibility to order dependency of viscometric properties on them did not show clear agreement, where each VI improver has acquired different order for its response to the change in these variables.

2.2.5.4 Dependency of Viscometric Properties on Combined Chemical Structure Variables

Single linear regression analysis $(Y = A + aX)$ was also applied to evaluate the degreee of dependency of viscometric properties on the calculated combined chemical variables, as listed in Table 2.2.4. The F values for the derived statistical models are also listed in Tables 2.2.7 — 2.2.14. According to these F values it is clear that most of the combined variables are in good correlation with the viscometric properties. In other words, measured viscometric properties showed a high dependency on the combined variables.

Directions of proportionality between the combined variables and viscometric properties are also listed in Table 2.2.15, where they are defined according to the sign of the F values for the derived statistical models (cf. Tables 2.2.7 — 2.2.14).

According to these F values, it is also possible to order the decrease in magnitude of dependency of viscometric properties according to the following:

(TAR+S) > SAT/(S+N+O) > TAR/S > (DAR+PAR)/S > (DAR+PAR)/(S+N+O) > TAR/(S+N+O).

Most of the measured viscometric properties are found to be strongly affected by the first three variables. (TAR+S), SAT/(S+N+O), (TAR/S). Table 2.2.17 includes the viscometric properties which are affected by each of these combined variables. Accordingly, it is clear that the combined variable (TAR+S) strongly affected most of the concerned viscometric properties. This variable is representing the sum of the two most important single variables (TAR & S) in the base-stock chemical structure. This variable (TAR+S) was previously postulated and used by Burn and Greig in their valuable study concerning effect of aromaticity on lubricating oil oxidation (3).

Table 2.2.17: Viscometric properties as effected by the combined chemical structure variables

VI Improver	Viscometric properties	Combined chemical variable
SICP	V 40 − Δ V 40 − V 100 − Δ V 100	
SBCP	V 40 − Δ V 40 − V 100 − Δ V 100 Vsp 40 − Vsp 100	
(OCP)1	V 40 − Δ V 40 − V 100 − Δ V 100 Vsp 40	
(OCP)2	V 40 − Δ V 40 − V 100 − Vsp 40 V. I. − Q	(TAR+S)
(MIX)1	V 40 − Δ V 40 − V 100 − Δ V 100	
(MIX)2	V 40 − Δ V 40 − V 100 − Δ V 100	
PMA	V 40 − Δ V 40 − V 100 − Δ V 100 Vsp 40 − Vsp 100 − V. I. − Q	
PMA-D	V 40 − Δ V 40 − V 100 − Δ V 100	
SICP	V. I.	
SBCP	Q	
(MIX)1	Vsp 40 − Vsp 100 − V. I.	(TAR/S)
(MIX)2	Vsp 40 − V. I.	
PMA-D	Vsp 40 − Vsp 100 − Q	
SBCP	V. I.	
(OCP)1	V. I.	
(OCP)2	Δ V 100 − Vsp 100	SAT/(S+N+O)
(MIX)2	Vsp 100	
PMA-D	V. I.	

The other two variables SAT/(S+N+O) and TAR/S were also previously success-fully used in explaining some of the oils performance properties (1, 4, 5 & 14). In general, it might be important to explain the oils viscometric properties to draw the attention on these three variables.

On the other hand, the order of decrease in dependency of oils viscometric properties on combined chemical variables, as based on % mole (cf. Table 2.2.4) is found to be according to the following:

$$(C_A+S_m) \ (C_P/C_N) > S_m/C_A > C_A/S_m$$

Most of the measured viscometric properties are also found to be strongly affected by these three variables. As listed in Table 2.2.18, the variable (C_A +S_m) (C_P/C_N) strongly affected most of the viscometric properties, while the other two variables affected less properties. These three combined variables were previously developed by Korcek and Jensen in their study concerning the relation between base oil composition and oxidation stability for the oils at high temperatures (5). It is the first time to use of them in explaining the viscometric properties of base-stocks.

2.2.5.5 Predictive Statistical Models

Most of the oils viscometric properties have shown high dependency on the two combined variables: (TAR+S) and (C_A +S_m) (C_P/C_N). Therefore, their derived statistical models were used for the prediction of oils viscometric proper-ties. These models were derived via single regression analysis (Y = A + aX), where

Y =   any of the measured viscometric properties
X =   any of the two combined variables, (TAR+S) or (C_A+S_m) (C_P/C_N)
A =   equation constant
a =   regression constant

Tables 2.2.19 and 2.2.20 include the results for A, a and F of the derived models via (TAR+S) and (C_A+S_m) (C_P/C_N) respectively.

Tests of these models with other base-stocks, either from local sources or im-ported, in the presence of different VI improvers, have proved their suitability for predicting of the oils viscometric properties at high confidence.

Figure 2.2.2 respresents calculated viscometric results via models of the variable (TAR+S) against observed results for the following viscometric properties:

Vsp 100 for PMA   Figure 2.2.2-A
Δ V 40 for (OCP)1   Figure 2.2.2-B
Δ V 100 for (MIX)1   Figure 2.2.2-C

90

Table 2.2.18: Viscometric properties as effected by the combined chemical variables (% mole)

VI Improver	Viscometric properties	Combined chemical variable
SICP	$V\ 40 - \Delta V\ 40 - V\ 100 - \Delta V\ 100$	
(OCP)1	$V\ 40 - \Delta V\ 40 - V\ 100 - \Delta V\ 100$ $Vsp\ 40 - V.I. - Q$	
(OCP)2	$V\ 40 - \Delta V\ 40 - V\ 100 - \Delta V\ 100$ $V.I. - Q$	
(MIX)1	$V\ 40 - \Delta V\ 340 - V\ 100 - \Delta V\ 100$	$(C_A + S_m)(C_P/C_N)$
(MIX)2	$V\ 40 - \Delta V\ 40 - V\ 100 - \Delta V\ 100$ $Vsp\ 100$	
PMA	$V\ 40 - \Delta V\ 40 - V\ 100 - \Delta V\ 100$ $Vsp\ 40 - V.I. - Q$	
PMA-D	$\Delta V\ 40 - V\ 100 - \Delta V\ 100$	
SBCP	$V\ 40 - \Delta V\ 30 - V\ 100 - \Delta V\ 100$ $Vsp\ 40 - Vsp\ 100$	
(OCP)1	$Vsp\ 100$	
(OCP)2	$Vsp\ 40$	S_m/C_A
(MIX)1	$Vsp\ 40 - Vsp\ 100$	
(MIX)2	$Vsp\ 40 - V.I.$	
PMA	$Vsp\ 100$	
PMA-D	$V\ 40 - Vsp\ 40$	
SICP	$V.I.$	
SBCP	$V.I. - Q$	
(OCP)2	$Vsp\ 100$	C_A/S_m
(MIX)1	$V.I.$	
(MIX)2	Q	
PMA-D	$Vsp\ 100 - V.I. - Q$	

Table 2.2.19: Predictive models via combined chemical variable (TAR+S)

VI Improvers	V 40	ΔV 40	V 100	ΔV 100	VI	Vsp 40	Vsp 100	Q
SiCP								
A	− 10.1769	− 1.5975	5.0486	2.5832	169.678	0.9504	0.9615	1.007
a	3.5423	1.5571	0.2973	0.1379	− 0.7211	− 0.0023	− 0.0011	0.002
SBCP								
A	− 40.5342	− 31.9548	1.9811	− 0.4843	140.342	0.4685	0.4578	0.928
a	4.3409	2.3557	0.3371	0.1777	− 0.482	0.101	0.0072	− 0.002
(OCP)1								
A	− 50.2117	− 41.6323	3.4277	0.9623	159.443	0.9161	0.9559	0.987
a	5.5213	3.5361	0.4029	0.2436	− 0.681	0.0121	0.0063	− 0.003
(OCP)2								
A	− 41.8952	− 33.3158	4.4862	2.0208	159.779	0.973	1.016	0.987
a	5.1956	3.2105	0.3569	0.1976	− 0.789	0.0091	0.0027	− 0.003
(MIX)1								
A	− 48.2602	− 39.6634	0.6979	− 1.7871	143.684	0.1887	0.263	1.085
a	4.3093	2.3239	0.3601	0.2019	− 0.427	0.0139	0.0112	− 0.003
(MIX)2								
A	− 25.8654	89.5552	3.4019	0.9365	149.572	0.5674	0.5959	1.026
a	3.7565	− 0.3591	0.29	0.1307	− 0.622	0.0047	0.0025	− 0.003
PMA								
A	− 54.6175	− 46.0381	2.9944	0.529	203.861	0.2621	0.7581	1.699
a	4.7059	2.7207	0.3661	0.2068	− 1.402	0.0153	0.0059	− 0.014
PMA-D								
A	− 29.2358	− 20.6564	3.4753	1.0099	181.766	0.4904	0.7175	1.379
a	3.8186	1.6564	0.3239	0.1646	− 0.945	0.0063	0.0035	− 0.379

Table 2.2.20: Predictive models via combined chemical variable $(C_A + S_m)$ (C_P/C_N)

VI Improvers	V 40	ΔV 40	V 100	ΔV 100	VI	Vsp 40	Vsp 100	O
SICP								
A	− 414.25	− 180.72	− 28.81	− 13.15	256.19	1.18	1.08	0.87
a	18.39	8.12	1.54	0.71	3.88	0.01	0.005	0.006
SBCP								
A	− 497.82	− 280.34	33.63	− 19.32	191.25	− 0.62	− 0.32	1.11
a	21.18	11.503	1.65	0.87	2.36	0.05	0.03	− 0.009
(OCP)1								
A	− 675.1	− 440.99	− 42.63	− 26.97	234.89	− 0.45	0.2	1.31
a	26.46	18.19	2.09	1.27	3.45	0.06	0.03	− 0.01
(OCP)2								
A	− 631.59	− 397.48	− 36.71	21.05	245.6	− 0.07	0.64	1.38
a	26.84	16.57	1.88	1.04	3.95	0.05	0.02	− 0.02
(MIX)1								
A	− 126.43	− 81.81	5.89	− 5.52	151.27	− 0.069	0.047	1.13
a	1.99	1.07	0.17	0.094	0.19	0.006	0.005	0.0014
(MIX)2								
A	− 455.53	− 221.71	− 29.72	− 14.06	223.63	− 0.02	0.29	1.36
a	19.52	9.26	1.5	0.68	3.33	0.026	0.014	− 0.014
PMA								
A	− 578.25	− 346.21	− 37.99	− 22.49	361.41	− 1.4	0.12	3.22
a	3.94	13.75	1.87	1.05	7.18	0.078	0.03	− 0.07
PMA-D								
A	− 466.05	− 231.94	− 33.57	− 17.91	292.25	− 0.28	0.29	2.06
a	19.84	9.58	1.68	0.88	4.99	0.034	0.019	− 0.03

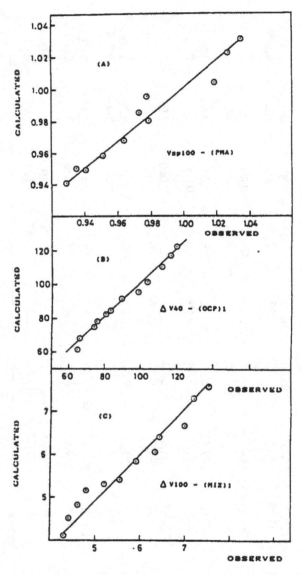

Figure 2.2.2: Calculated viscometric properties via (TAR+S) models against observed results

Similarly, Figure 2.2.3 represents calculated results via models of the variable (C_A+S_m) (C_P/C_N) against observed results for the following viscometric properties:

Δ V 100 for SICP	Figure 2.2.3-A
Δ V 40 for (OCP)2	Figure 2.2.3-B
Δ V 100 for PMA	Figure 2.2.3-C

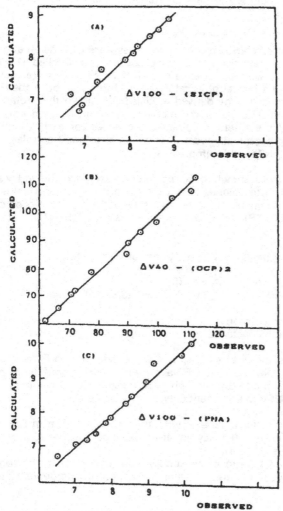

Figure 2.2.3: Calculated viscometric properties via (C_A+S_m) (C_P/C_N) models against observed results

It is clear, according to Figs. 2.2.2 & 2.2.3 the high degree of accuracy for derived models in predicting with oils viscometric properties.

2.2.5.6 Correlation via Multiple Models

Different trials were conducted to develop statistical models via multiple linear regression analysis:

$$(Y = A + a_1 X_1 + a_2 X_2 + \ldots a_m X_m)$$

Most of the single and combined chemical variables were considered at deriving of these models, where stepwise regression procedure was applied (14 & 36). Accepted models (i. e. with confidence higher than 99.5 %) were only considered if they satisfied a logical physical meaning. In general both the F and t statistical values for most of the derived models have shown their relative low confidence (lower than 97.5 %). Some models are found in good correlation with one or more of the tested viscometric properties for each VI improver but without achieving a general model, i. e. which gives good correlation with all viscometric properties for all VI improvers.

The only model that has shown good correlation to viscosity and viscosity increase results for all tested blends (i. e. with the eight VI improvers) is found to base on the two chemical variables: saturates (SAT) and ratio of di- and poly-aromatics (DAR+PAR) to total non-hydrocarbons (S+N+O), — i. e. (DAR+PAR)/(S+N+O).

The equation of this model is consequently the following one:

$$\text{Viscometric property} = A + a_1 (SAT) + a_2 (DAR+PAR)/(S+N+O)$$

where,
A   = equation constant
a_1 & a_2 = regression constants.

Table 2.2.21 includes the results of A, a_1 & a_2 in addition to F values for derived relationships, where their high F values indicate high confidence percent (higher than 99.5 %). In addition, the physical meaning for this model is quite significant as it includes two important chemical variables:

— Saturates (SAT), the single variable with the highest content in base-stocks, beside its well known efficiency in affecting most of the lubricants performance properties (1 — 14).
— (DAR+PAR)/(S+N+O), ratio of the undesirable aromatics to the total non-hydrocarbons, which are also forming undesirable compounds with high polarity.

Calculated viscometric properties via derived models are found to be very close to observed ones.

Table 2.2.21: Predictive relationships via the multiple model X_1: (SAT) and X_2: (DAR+PAR)/(S+N+O)

		A	a_1	a_2	F
SICP	V 40	375.71	− 2.881	− 10.296	496.39
	Δ V 40	170.27	− 1.202	− 5.427	163.22
	V 100	37.34	− 10.91	− 3.213	284.39
	Δ V 100	17.59	− 0.1128	− 0.393	92.91
SBCP	V 40	426.547	− 3.697	− 10.293	427.79
	Δ V 40	221.11	− 2.018	− 5.423	345.2
	V 100	39.34	− 0.284	− 0.838	454.28
	Δ V 100	18.597	− 0.1526	− 0.404	276.15
(OCP)1	V 40	540.2	− 4.809	− 11.61	410.19
	Δ V 40	334.76	− 3.13	− 6.74	329.65
	V 100	47.6	− 0.319	− 1.285	461.56
	Δ V 100	27.85	− 0.188	− 0.85	264.89
(OCP)2	V 40	517.67	− 4.42	− 12.36	619.42
	Δ V 40	311.83	− 2.743	− 7.489	594.99
	V 100	44.44	− 0.2596	− 1.4673	538.029
	Δ V 100	24.69	− 0.1278	− 1.034	276.437
(MIX)1	V 40	L.C.*	L.C.	L.C.	L.C.
	Δ V 40	209.396	− 2.007	− 5.117	616.16
	V 100	39.68	− 0.2988	− 0.951	511.73
	Δ V 100	20.32	− 0.1609	− 0.633	305.115
(MIX)2	V 40	388.106	− 2.92	− 12.82	378.99
	Δ V 40	182.67	− 1.241	− 7.95	257.75
	V 100	35.496	− 0.2216	− 1.043	268.194
	Δ V 100	15.75	− 0.0899	− 0.61	119.014
PMA	V 40	439.55	− 4.358	− 6.27	186.11
	Δ V 40	22.08	− 0.196	− 0.212	310.07
	V 100	41.83	− 0.328	− 0.645	431.04
	Δ V 100	L.C.	L.C.	L.C.	L.C.
PMA-D	V 40	388.83	− 3.043	− 11.93	598.5
	Δ V 40	183.39	− 1.367	− 7.062	565.4
	V 100	38.87	− 0.26	− 0.987	767.08
	Δ V 100	19.13	− 0.128	− 0.5538	669.94

(*) L.C.: low confidence

97

2.2.6 Summary

Multigrade crankcase oils were formulated to contain increasing percentage of bright-stock (up to 23 % wt). Different commercial VI improvers were used in these oils at the concentrations recommended by their suppliers. The changes in the oils viscometric properties by increasing bright-stock content were evaluated in terms of the chemical structure changes of these oils.

The different hydrocarbon groups in addition to the non-hydrocarbon elements were considered as the chemical structure variables, where they termed the single variables. In addition to the single variables, some other combined variables were calculated by adding, multiplying or dividing the single variables.

As single variables the following ones were considered: saturates (SAT), monoaromatics (MAR), diaromatics (DAR), polyaromatics (PAR), total non-hydrocarbon elements (S+N+O), paraffinic carbon (C_P), naphthenic carbon (C_N), aromatic carbon (C_P) and sulphur percent mole (S_m). The combined variables included: (TAR+S), (DAR+PAR)/(S+N+O), SAT/(S+N+O), (DAR+PAR)/S, TAR/(S+N+O), TAR/S, S_m/C_A, C_A/S_m and (C_A+S_m) (C_P/C_N).

The measured viscometric properties were the following ones: kinematic viscosity at 40 and $100°C$, kinematic viscosity increase (the difference between viscosity of VI improver containing blend and viscosity of neat blend at 40 & $100°C$), viscosity index, specific viscosity at 40 and $100°C$ and thickening tendency (Q).

Regression analysis, as a statistical technique, was applied to evaluate the degree of dependency of oils viscometric properties on base-stocks chemical structure variables.

The statistical models as derived via the single chemical variables have proved the following:

1. The high dependency of viscometric properties on monoaromatics content.

2. Decrease in magnitude of dependency for viscometric properties on base-stocks chemical structure variables is going at the following order:

    MAR > SAT > (S+N+O) > PAR > DAR.

3. In case of VI improvers with dispersant activity, decrease in magnitude of dependency is going at the following order:

    (S+N+O) > MAR > SAT > PAR > DAR.

4. Viscosity index and Q value are highly affected by the non-hydrocarbons.

With combined chemical structure variables, the gathered statistical results showed the dependency of viscometric properties on most of them. The best derived models are found via any of the following two combined variables: $(TAR+S)$ or (C_A+S_m) (C_P/C_N). Applying these best models for the prediction of the oils viscometric properties have shown their suitability and accuracy.

Correlations via multiple regression analysis have shown a high degree of confidence in a model derived via the combination between the two chemical variables (SAT) and $(DAR+PAR)/(S+N+O)$. This model also has a significant physical meaning.

2.2.7 Acknowledgement

The authors wish to express their appreciation to chemist Nagi S. Noaman and chemist Alaa A. Mohamed for their great effort in running the different requirements for this work.

2.2.8 References

(1) Murray, D.W., Clarke, C.T., MacAlpine, G.A., and Wright, P.G.: "The Effect of Basestock Composition on Lubricant Oxidation Performance", (1982), SAE Reprint 821236.

(2) Murray, D.W. et al.: "A New Concept of Lubricant Base Oil Quality", SP 15, Reprint of the Eleventh World Petroleum Congress, London (1983).

(3) Burn, A.J. and Greig, G.: "Optimum Aromaticity in Lubricating Oil Oxidation", J. Inst. Petr., 58 (1972) 564, 346—350.

(4) Cranton, G. and Noel, F.: "Aromatics and Lubricant Basestock Oxidation", J. Inst. Petr. (1974) 74—005, 1—19.

(5) Korcek, S. and Jensen, R.K.: "Relation between Base Oils Composition and Oxidation Stability at Increased Temperatures", ASLE Trans., 19 (1976) 2, 83—94.

(6) Hsu, S.M., Ku, C.S. and Line, R.S.: "Relationship between Lubricating Basestock Composition and the Effects of additives on Oxidation Stability", (1982) SAE Reprint 821327.

(7) Abou El Naga, H.H., and Abdel Ghany, M.A.: "Chemical Structure Bases for Oxidation Stability of Neutral Base Oils", ASLE Trans., 30 (1987) 2, 261—268.

(8) Abou El Naga, H.H., and Abdel Ghany, M.A.: "Engine Performance Correlation with Chemical Structure for Motor Oil Blends", Lubr. Eng., 42 (1986), 104—108.

(9) Rounds, F.G.: "Effect of Aromatic Hydrocarbons on Friction and Surface Coating Formation with Three Additives", ASLE Trans., 16 (1973) 2, 141—149.

(10) Rounds, F.G.: "Changes in Friction and Wear Performance Caused by Interactions Among Lubricant Additives", Esslingen (1986), P. 4.8—1 — 4.8—21, 5th International Colloquium; Additives for Lubricants and Operational Fluids, Techn. Akademie Esslingen, 14—16.01.86, West Germany.

(11) Watanabe, H., Fukushima, T. and Nose, Y.: "Demulsibility Characteristics of Industrial Lubricating Oils", Japan (1976), Proceedings of JSLE/ASLE International Lubrication Conference.

(12) Watanabe, H. and Kobayashi, C.: "Degradation of Turbine Oils", Japanese Turbine Lubrication Practices and Problems, Lubr. Eng., 34, (1978), 421—428.

(13) Biswajit Basu, Subhash Chand, Jayaprakash, K.C., Sirvastava, S.P. and Goel, P.K.: "Air-Entrainment Phenomenon in Mineral Lubricating Oils", ASLE Trans., 28 (1985) 3, 313—324.

(14) Abou El Naga, H.H., Bendary, S.A. and Salem, A.E.M.: "Industrial Oils Performance in Terms of Base Oils Chemical Structure", Esslingen (1988), P. 12.5—1 — 12.5—21, 6th International Colloquium; Industrial Lubricants: Properties, Application, Dispossal, Techn. Akademie Esslingen 12—14.01.88, West Germany.

(15) Brooks, F.C. and Hopkins, Vern: "Viscosity and Density Characteristics of Five Lubricant Base Stocks At Elevated Pressures and Temperatures", ASLE Trans., 20 (1977) 1, 25—35.

(16) Selby, T.W.: "The Non-Newtonian Teristics of Lubricating Oils", ASLE Trans., 1 (1958), 68—81.

(17) Schilling, A.: "Motor Oils and Engine Lubrication", Vol. 1, Chapter 2, Scientific Publication — U.K., (1968).

(18) Assef, P.A.: "Some Performance Characteristics of Hydrorefiend Lubricating Oils", Pet. Hydrocarbons, 5 (1971) 4, 168—178.

(19) Arlie, J.P., Denis, J. and Parc, G.: "Relations between the Structure and Viscometric Properties of Polymethacrylate Solutions in Lube Oils", J. Inst. Petrol., (1975), IP—75—006.

(20) Muller, H.G.: "Mechanism of Action of Viscosity Index Improvers", Tribol. Int., 6, (1978), 189—192.

(21) Bartz, W.J.: "Rheological Properties of Multigrade Engine Oils", Bombay (1987), L.7.1 — L.7.25, 5th LAWPSP Symposium, Indian Institute of Technology, 29.—31.01.87, India.

(22) Himmat, Singh and Gulati, I.B.: "Influence of Base Oil Refining on the Performance of Viscosity Index Improvers", Wear, 118, (1987), 33—56.

(23) Bartz, W.J.: "Investigations on the Rheological Behaviour of Multigrade Engine Oils", Bombay (1989), L.5.1 — L.5.31, 6th LAWPSP Symposium, Indian Institute of Technology, 19—21.01.89, India.

(24) Bartz, W.J. and Nemes, N.: "How Properties and Thickening Effect of Different Viscosity Index Improvers Used in Multigrade Engine Oils", Lubr. Eng., 33 (1977) 1.

(25) Abou El Naga, H.H. and Bendary, S.A.: "A study on Viscometric Properties for Multigrade Crankcase Oils", STLE (1989), to be published.

(26) Eckert, R.J.A. and Covey, D.F.: "Developments in the Field of Hydrogenated Diene Copolymers as Viscosity Index Improvers", Esslingen (1986), P.8.5—1 — 8.5—14, 5th International Colloquium; Additives for Lubricants and Operational Fluids, Techn. Akademie Esslingen, 14—16.01.86, West Germany.

(27) Spiess, G.T., Johnston, J.E. and VerStrate, G.: "Ethylene Propylene Copolymers as Lube Oil Viscosity Modifiers", Esslingen (1986), P.8.10—1 — 8.10—11, 5th International Colloquium; Additives for Lubricants and Operational Fluids, Techn. Akademie Esslingen, 14—16.01.86, West Germany.

(28) Nemes, N., Kovacs, M. and Kantor, I.: "Untersuchung der Wirkung von fließverbessernden Additiven", Esslingen (1986), P.8.4—1 — 8.4—13, 5th International Colloquium; Additives for Lubricants and Operational Fluids, Techn. Akademie Esslingen, 14—16.01.86, West Germany.

(29) Neudorfl, P.: "Der Einsatz von Polymethacrylaten in Schmierölen", Esslingen (1986) P.8.2—1 — 8.2—15, 5th International Colloquium; Additives for Lubricants and Operational Fluids, Techn. Akademie Esslingen, 14—16.01.86, West Germany.

(30) Abou El Naga, H.H., Abd El Aziem, W.M., Fathy Zanib and Ibrahim Saad, H.: "Some Styrene Alkylmethacrylate Copolymers as Viscosity Index Improvers", J. Chem. Tech. Biotechnol, 36 (1986), 583—592.

(31) IP Standards for Petroleum and its products, Parts 1 & 2, Institute of Petroleum, Applied Science Publishers, U.K., (1982).

(32) U.O.P.: Laboratory Test Methods for Petroleum and its Products, Universal Oil Products Co., Illinoise, USA (1959).

(33) Van Nes, K. and Van Westen, H.A.: "Aspects of the Constitution of Mineral Oils", Elsevier, New York (1951).

(34) Brands, G.: "The Chemical Structure of Petroleum Fractions. I. Analysis by Infrared Spectroscopy", Brennstoff Chem. 37 (1956) 263—167, "The Structural-Group Analysis of Petroleum Fractions", Erdöl, Kohle, 11 (1958), 700—702.

(35) Eckardt, H., Heil, G. and Rentrop, K.: "The Determination of Groups in Lubricating Oils with the Aid of Infrared Spectroscopy", Chem. Techn. (Berlin), 14 (1962), 305—309.

(36) Bernard, O. and Richard, W.M.: "Statistics in Research", The Iowe State University Press, USA, (1982).

Appendix 1

Chemical Structure Abbreviations Key-Words

Chemical Structure Groups:

(SAT) . . . . Saturates, % wt
(TAR) . . . . Total aromatics, % wt
(MAR) . . . . Monoaromatics, % wt
(DAR) . . . . Diaromatics, % wt
(PAR) . . . . Polyaromatics, % wt
(S) . . . . Sulphur, % wt
(O) . . . . Oxygen, % wt
(N) . . . . Nytrogen, % wt
(S+N+O) . . . . Total non-hydrocarbons, % wt

Chemical structure percent mole:

(C_p) . . . . Paraffinic carbon, % mole.
(C_N) . . . . Naphthenic carbon, % mole.
(C_A) . . . . Aromatic carbon, % mole.
(S_m) . . . . Sulphur, % mole.

2.3 Determination of Zinc and Calcium in Multigrade Crankcase Oils

H.H. Abou el Naga, M.M. Mohamed and M.F. el Meneir
Research Centre, MISR Petroleum Co., Cairo, Egypt

2.3.1 Abstract

Atomic absorption spectroscopy is used for the determination of zinc and calcium in multigrade crankcase oils.

It has been found that zinc and calcium concentrations are subject to matrix interferences from polymers incorporated in the oil formulations as VI improvers. Results indicate that the two most important factors are concentration and chemical structure type for the following VI improvers: styrene-isoprene, styrene-butadiene, polyalkylmethacrylate and ethylene-propylene.

Correlations between zinc or calcium concentrations and either the concentration or the type for each VI improver are established via regression analysis statistical technique. Acceptable models have been found with confidence higher than 99.5 %.

Key words:

Atomic absorption spectroscopy; zinc; calcium; multigrade crankcase oils; matrix interference; viscosity index improvers.

2.3.2 Introduction

Atomic absorption spectroscopy (AAS) has been rapidly developed and widely applied over the last three decades. Original papers on this technique were first published by Walsh and his co-workers (1 & 2). The main reasons for its wide applications are its simplicity, sensitivity, portability and accuracy. Theoretical principles and description of basic apparatus are mentioned elsewhere (1—5).

AAS is widely used in the petroleum industry, either for quality control purposes or for research activities. One of its versatile applications is the determination of zinc and calcium in new and used lubricating oils (3—6). The presence of these two metals in lubricating oil formulations is due to the incorporated detergent and antioxidant additives.

Applied methods for the determination of zinc and calcium via AAS basically rely on dilution of tested samples with kerosene followed by aspiration into the flame. It has been shown in case of zinc determination that such methods are subject to matrix interferences from other additives, especially copolymers incorporated as VI improvers in crankcase lubricating oil formulations (7 & 8).

Generally, the following copolymers are used in lubricant formulations as VI improvers: styrene-isoprene copolymer (SICP), styrene-butadiene copolymer (SBCP), ethylene-propylene copolymer (OCP) and polyalkylmethacrylate (PMA). References (9–14) include more data about the function and mechanism of action for these VI improvers.

The effects of VI improvers on measured zinc concentrations in unused crankcase oils have been studied (7 & 8) where only one concentration from each incorporated VI improver was considered and precision limits for these measurements were not established. On the other hand, the determination of calcium concentrations in these oils has not been considered before.

Therefore, the aims of the present study are as follows:

1. Confirmation of previously published results for zinc determination, using variable concentrations of different commercial VI improvers in multigrade crankcase oil formulations. These concentrations are selected to be within the applicable commercial doses for VI improvers in multigrade crankcase oils.
2. Establishing precision limits for zinc concentrations in these formulations.
3. Calculation of regression equations for each type of incorporated VI improver, so that those equations can be used in estimating the actual concentrations of zinc in multigrade crankcase oil formulations.
4. Similarly, the measurement of calcium concentrations in presence of different VI improvers.

2.3.3 Experimental

A basic lubricating oil formula (Table 2.3.1), i. e. without VI improvers, was used in all measurements. Components of this formula are: neutral base oil (300 N), bright-stock (Brt 135) and a multipurpose additive (i. e. detergent, dispersant, antiwear and antioxidant). Zinc and calcium contents in this additive are 1.46 and 1.65 % wt respectively.

VI improvers were incorporated in this basic blend at different concentrations, which are within the applicable commercial doses for VI improvers in multigrade crankcase oils (Table 2.3.2). Base oil content is reduced so as to keep the zinc and calcium concentrations at constant values in all formulations. The following commercial VI improvers were tested: SICP at concentrations from 0.6 to 1.2 % wt. SBCP at concentrations from 0.88 to 1.68 % wt, OCP at concentrations from 0.78 to 1.43 % wt and PMA at concentrations from 0.78 to 1.43 % wt. Table 2.3.3 includes the ingredients percentage in the formulated multigrade crankcase oils. Kinematic viscosities at 40 and 100°C, viscosity index, total base number and sulphated ash for those blends are also listed in Table 2.3.3.

Table 2.3.1:   Basic oil: Composition and properties

A-Composition	
Neutral base oil (300 N), % wt	60
Bright-Stock (Brt 135), % wt	32.1
Multipurpose additive, % wt	7.9
B-Properties:	
Kin.Vis. 40°C, cSt.	90.45
Kin.Vis. 100°C, cSt.	10.56
Viscosity index	99
Pour point, °C	− 9
T.B.N. mg KOH/g sample	6.45
Sulphated ash, % wt	0.82
Zn, % wt	0.1150
Ca. % wt	0.1325

Table 2.3.2:   Viscosity index improver characteristics

Polymer type	Symbol	Copolymer content, %wt	Copolymer content in crank-case oil formulations	
			as diluted form (% wt)	as solid form (% wt)
Styrene-isoprene copolymer	(SICP)	Solid	10–21	0.6–1.2
Styrene-butadiene copolymer	(SBCP)	Solid	11–21	0.88–1.68
Olefin copolymer	(OCP)	13	6–11	0.78–1.43
Polyalkyl-methacrylate	(PMA)	26	3.5–5	0.78–1.43

Table 2.3.3: VI improver blends: Composition and properties

Copolymer Type	V.I.I. (% wt)	Oil 300N (% wt)	Brt. 135 (% wt)	Package (% wt)	Kin.Vis., cSt. 40°C	100°C	V.I.	T.B.N., mgKOH/g	S. Ash (% wt)
SICP	0.60	59.48	32.02	7.9	100.05	13.99	142	6.49	0.81
	0.72	59.40	31.98	7.9	111.10	15.25	144	6.42	0.85
	0.84	59.32	31.94	7.9	128.76	16.22	130	6.43	0.80
	0.96	59.24	31.90	7.9	142.50	17.89	140	6.38	0.82
	1.04	59.19	31.87	7.9	155.32	19.29	142	6.37	0.84
	1.20	59.09	31.81	7.9	182.91	21.43	140	6.41	0.81
SBCP	0.88	59.29	31.93	7.9	119.72	14.61	124	6.39	0.82
	1.04	59.19	31.87	7.9	136.49	15.87	122	6.42	0.85
	1.20	59.09	31.81	7.9	148.51	17.22	126	6.35	0.81
	1.36	58.99	31.75	7.9	174.98	18.72	120	6.32	0.84
	1.52	58.89	31.69	7.9	180.61	19.53	124	6.41	0.84
	1.68	58.79	31.63	7.9	221.45	21.96	120	6.44	0.81
OCP	0.78	59.36	31.96	7.9	123.21	15.66	134	6.51	0.84
	0.91	59.27	31.92	7.9	135.17	16.68	133	6.50	0.82
	1.04	59.19	31.87	7.9	143.71	17.65	136	6.44	0.81
	1.17	59.10	31.83	7.9	159.11	18.57	131	6.49	0.82
	1.30	59.02	31.78	7.9	173.21	19.80	132	6.42	0.83
	1.43	58.94	31.73	7.9	189.24	21.17	133	6.44	0.82
PMA	0.78	59.36	31.96	7.9	111.17	15.70	150	6.38	0.85
	0.91	59.27	31.92	7.9	119.91	16.54	149	6.44	0.82
	1.04	59.19	31.87	7.9	132.65	17.72	148	6.46	0.84
	1.17	59.10	31.83	7.9	142.38	18.86	150	6.51	0.82
	1.30	59.02	31.78	7.9	153.11	19.71	148	6.43	0.82
	1.43	58.94	31.73	7.9	158.56	20.60	152	6.48	0.81

Neat blends with the same viscosity as those VI improvers containing formulations were prepared in order to study the effect of viscosity on measured zinc and calcium concentrations. Table. 2.3.4 includes ingredients percentage and properties for these neat blends.

Zinc and calcium concentrations were measured using a Perkin Elmer Model 460 atomic absorption spectrophotometer and according to operating conditions given in Table 2.3.5. A fraction of kerosene with boiling range 150–250°C was used as a solvent.

Table 2.3.4: Neat blends: Composition and properties

A: Composition	BO 1	BO 2	BO 3	BO 4	BO 5
Neutral base oil (300N), %wt	44.0	35.5	33.5	11.0	24.5
Bright-stock (Brt 135), %wt	56.0	64.5	66.5	69.0	75.5
Multipurpose additive, %wt	7.90	7.90	7.90	7.90	7.90
B: Properties					
Kin.Vis., cSt.					
40°C	134.88	149.32	158.44	173.53	195.95
100°C	13.69	14.96	15.5	16.23	17.5
Viscosity index	97	100	99	97	96
T.B.N., mg KOH/g sample	6.49	6.39	6.51	6.52	6.39
Sulphated ash, % wt	0.83	0.81	0.84	0.82	0.85
Zn, % wt	0.1150	0.1150	0.1150	0.1150	0.1150
Ca, % wt	0.1325	0.1325	0.1325	0.1325	0.1325

Table 2.3.5: Operating conditions for atomic absorption spectrophotometer

Instrument	Perkin Elmer Model 460
Wavelength, nm	
Zn	213,9
Ca	422.7
Lamp current, mA	15
Air flow, l/min	35
Acetylene flow, l/min	13
Slot width, mm	0.7
Response time, s	1
Burner head, in	4 (1 slot)

To estimate the precision limits of the method, one sample was tested at least 10 times under the same working conditions and by the same operator. The results obtained were substituted in the standard deviation equation. Duplicate results by the same operator should be considered suspect, if average of two results differ by more than the % repeatability limits.

To estimate the correction factors, a computer programme was used to explore and derive the anticipated statistical correlations between the percentage of reduction in zinc and calcium concentrations in the presence of each type of incorporated VI improver.

2.3.4 Results and Discussion

2.3.4.1 Method precision limits

Calculated precision limits for zinc and calcium are found to be: ± 2.62 and ± 3.00 ppm from obtained concentrations respectively. Therefore, duplicate results by the same operator should be considered suspect, if the average differs by more than 0.23 %. These results indicate that the method has high precision and good sensitivity.

2.3.4.2 Effect of VI Improvers on Measured Zinc and Calcium Concentrations

Table 2.3.6 includes the measured zinc and calcium concentrations in the presence of variable concentrations of tested VI improvers. Percent reduction in measured concentrations from the actual ones, i. e., as incorporated in the basic formula, were calculated. Fig. 2.3.1 illustrates the percent reduction in AAS analytical data for zinc and calcium against the solid content (% wt) for each of the four tested VI improvers.

According to these results, it is possible to make the following observations:

1. Increase of VI improved concentrations have resulted in a significant reduction in both zinc and calcium content analysed.
2. VI improver chemical type has a significant effect on both the zinc and calcium measured concentrations. Percent reduction at the 1 % wt concentration as solid VI improver and at maximum used concentration from each of the incorporated VI improver are listed in Table 2.3.7.
3. It is possible to rank the percent reduction in zinc and calcium concentration in terms of VI improver chemical type according to the following:

$$\text{high} < \frac{\% \text{ Reduction}}{\text{SICP} > \text{SBCP} > \text{OCP} > \text{PMA}} < \text{low}$$

Table 2.3.6: Reduction in measured zinc and calcium concentrations with increasing VI improver concentrations

Copolymer Type	Diluted Copolymer % wt	Solid Copolymer % wt	Kin.Vis. 100°C cSt	Conc. of Zn, ppm Cal.	Conc. of Zn, ppm Meas.	% Red.	Conc. of Ca, ppm Cal.	Conc. of Ca, ppm Meas.	% Red.
SICP	10	0.60	13.99	1150	1111	3.39	1325	1300	1.89
	12	0.72	15.25	1150	1086	5.57	1325	1269	4.23
	14	0.84	16.52	1150	1052	8.52	1325	1252	5.51
	16	0.96	17.89	1150	1014	11.83	1325	1222	7.77
	18	1.04	19.29	1150	966	16.00	1325	1197	9.66
	20	1.20	21.43	1150	939	18.35	1325	1187	10.49
SBCP	11	0.88	14.61	1150	1108	3.65	1325	1300	1.89
	13	1.04	15.87	1150	1087	5.48	1325	1270	4.15
	15	1.20	17.22	1150	1053	8.43	1325	1254	5.36
	17	1.36	18.72	1150	1021	11.22	1325	1225	7.55
	19	1.52	19.53	1150	969	15.74	1325	1198	9.58
	21	1.68	21.96	1150	942	18.09	1325	1188	10.42
OCP	6	0.78	15.66	1150	1125	2.17	1325	1309	1.21
	7	0.91	16.68	1150	1111	3.39	1325	1288	2.79
	8	1.04	17.65	1150	1090	5.22	1325	1281	3.32
	9	1.17	18.57	1150	1067	7.22	1325	1256	5.21
	10	1.30	19.80	1150	1040	9.57	1325	1245	6.04
	11	1.43	21.17	1150	1020	11.3	1325	1232	7.17
PMA	3	0.78	15.70	1150	1136	1.22	1325	1316	0.68
	3.6	0.91	16.54	1150	1129	1.83	1325	1304	1.58
	4	1.04	17.72	1150	1117	2.87	1325	1297	2.11
	4.5	1.17	18.86	1150	1113	3.22	1325	1285	3.02
	5	1.30	19.71	1150	1087	5.48	1325	1276	3.70
	5.5	1.43	20.60	1150	1076	6.43	1325	1272	4.15

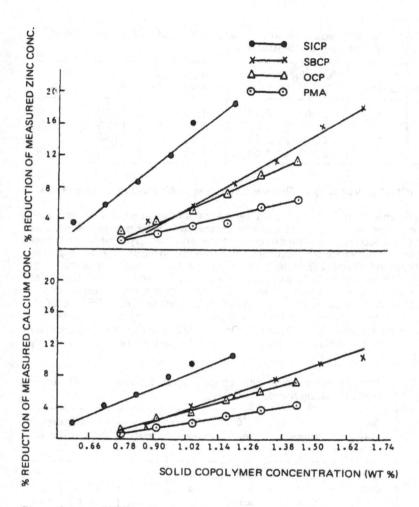

Figure 2.3.1: Effect of copolymer concentration measured zinc and calcium concentrations

Table 2.3.7: Percent reduction in zinc and calcium at 1 % wt and maximum VI improver concentrations

VI Improver	1 % wt Conc.	% Reduction Zn	Ca	max. conc. % wt	% Reduction Zn	Ca
SICP	1	13.90	8.68	1.20	18.35	10.49
SBCP	1	4.69	3.06	1.68	18.09	10.42
OCP	1	4.30	3.03	1.43	11.30	7.17
PMA	1	2.35	1.85	1.43	6.43	4.15

2.3.4.3 Effect of Viscosity on Percent Reduction of Measured Concentrations

It was very important to find a clear answer for the following question: can such reduction in measured zinc and calcium concentrations be due to the increase in the blend viscosity or to the presence of the different types of VI improver at variable concentrations? Therefore, neat blends without any VI improver were formulated (Table 2.3.4) to give nearly the same viscosities as those VI improvers containing formulations (Table 2.3.3). Zinc and calcium were also measured in these neat blends. Table 2.3.8 includes the results obtained for zinc concentration. As the SICP concentration is increased from 0.6 to 0.96 % wt (as solid copolymer weight) percent, reduction in measured zinc concentration is increased from 3.39 to 11.83 %. On the other hand, increasing viscosity for neat blends by the same magnitude increased percent reduction by only 0.96 to 1.74 %.

Table 2.3.8: Percent reduction in zinc concentration for neat and VI improvers containing blends

Copolymer Type	Solid polymer (% wt)	Kin.Vis.,cSt.100°C Polymer	Neat	Zn(ppm) Polymer	Neat	red.in Zn conc.(%) Polymer	Neat
SICP	0.60	13.99	13.69	1111	1139	3.390	0.96
	0.66	14.52	14.96	1094	1138	4.870	1.04
	0.72	15.25	15.50	1086	1136	5.570	1.22
	0.84	16.52	16.23	1052	1133	8.520	1.48
	0.96	17.89	17.50	1014	1130	11.83	1.74
SBCP	1.20	17.22	17.50	1053	1130	8.430	1.74
OCP	1.04	17.65	17.50	1090	1130	5.220	1.74
PMA	1.04	17.72	17.50	1117	1130	2.870	1.74

Figure 2.3.2 illustrated the change in percent reduction of measured zinc concentration for neat blends and SICP-containing formulations.

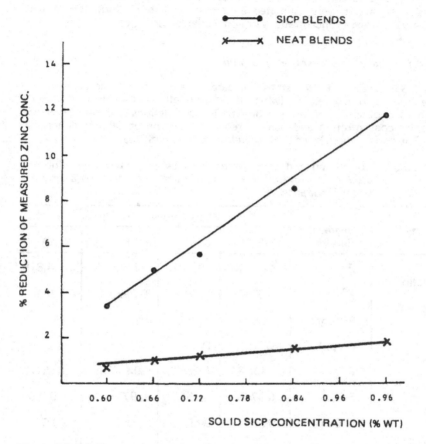

Figure 2.3.2: Percent reduction in zinc concentrations for neat and SICP containing blends

These results clearly show that viscosity increase has a limited effect on reducing measured zinc concentration, and the reduction in concentration is due mainly to the presence of the VI improvers. Other measurements in presence of SBCP, OCP and PMA have also confirmed these SICP results.

Measurements of calcium concentrations also confirm that neat blend viscosity has a limited effect on the percent reduction of the measured calcium concentration.

2.3.4.4 Estimating Correction Factors

To determine zinc and calcium concentrations in multigrade oil formulations, regression equations were estimated by either considering formulations with variable VI improver concentrations or with variable viscosities.

2.3.4.4.1 Via VI Improver Concentration

Table 2.3.9 includes: F (the statistical parameter which is taken to assess the whole model significance); R^2 (which is the correlation coefficient); confidence percent and direction of proportionality for correlations between VI improvers solid concentration and percent reduction in zinc or calcium concentrations. The derived models for these correlations are as follows:

Table 2.3.9: Correlation and proportionality between reduction in zinc and calcium concentrations and VI improver concentrations via single linear analysis

Property	Regression Results	VI improvers			
		SICP	SBCP	OCP	PMA
Reduction in zinc conc., % wt	F	426.713	805.191	1262.573	554.815
	R^2	0.9777	0.9769	0.9929	0.9840
	% Conf.	99.5	99.5	99.5	99.5
	Prop. *	D	D	D	D
Reduction in calcium conc., % wt	F	380.395	175.930	393.774	132.777
	R^2	0.9769	0.9513	0.9777	0.9365
	% Conf.	99.5	99.5	99.5	99.5
	Prop. *	D	D	D	D

* D = Direct proportionality

Percent reduction in zinc concentration:

In SICP formulations = $- 12.9412 + 26.5756$ (conc. % wt)
In SBCP formulations = $- 13.7367 + 18.9705$ (conc. % wt)
In OCP formulations = $- 9.2920 + 14.4231$ (conc. % wt)
In PMA formulations = $- 5.4281 + 8.2937$ (conc. % wt)

Percent reduction in calcium concentration:

In SICP formulations = – 6.8646 + 15.2669 (conc. % wt)
In SBCP formulations = – 7.3502 + 11.0534 (conc. % wt)
In OCP formulations = – 5.4322 + 8.9690 (conc. % wt)
In PMA formulations = – 3.3392 + 5.4690 (conc. % wt)

2.3.4.4.2   Via Blend Viscosity

Table 2.3.10 includes F & R^2 values, confidence % and direction of proportionality for correlations between blend viscosity and percent reduction in zinc or calcium concentration. The derived models for these correlations are as follows:

Table 2.3.10: Correlation results and proportionality between reduction in zinc and calcium concentrations and viscosity via single linear regression analysis.

Property	Regression Results	V.I. Improvers			
		SICP	SBCP	OCP	PMA
Reduction in Zinc conc., % wt	F	702.406	327.134	1467.325	1401.06
	R^2	0.9873	0.9732	0.9939	0.9396
	% conf.	99.5	99.5	99.5	99.5
	Prop.*	D	D	D	D
Reduction in Calcium conc., % wt	F	267.332	239.499	448.885	1467.679
	R^2	0.9674	0.9638	0.9803	0.9939
	% conf.	99.5	99.5	99.5	99.5
	Prop.*	D	D	D	D

* D = Direct proportionality

Percent reduction in zinc concentration:

In SICP formulations = – 27.3514 + 2.1845 (visc., cSt at 100°C)
In SBCP formulations = – 28.5886 + 2.1708 (visc., cSt at 100°C)
In OCP formulations = – 25.8449 + 1.7707 (visc., cSt at 100°C)
In PMA formulations = – 15.6308 + 1.0505 (visc., cSt at 100°C)

Percent reduction in calcium concentration:

In SICP formulations = − 14.4186 + 1.2119 (visc., cSt at 100°C)
In SBCP formulations = − 15.8972 + 1.2477 (visc., cSt at 100°C)
In OCP formulations = − 15.5983 + 1.0909 (visc., cSt at 100°C)
In PMA formulations = − 10.1314 + 0.6972 (visc., cSt at 100°C)

The testing of these models with several multigrade crankcase oil samples has proved their suitability for prediction of actual zinc and calcium concentrations.

2.3.4.5 Explanation of Results

Previous workers (8 & 15) have explained the phenomena for the reduction in measured metal concentration in presence of VI improvers as follows: the breakup point of a polymer-thickened fluid exiting from a capillary is affected and proportional to the polymer concentration. Upon breakup, the polymer-containing viscoelastic fluid forms droplets connected to other droplets by thread-like structures, thus producing droplets that are effectively larger. More dilute solutions can result in shorter threads and smaller sized particles, which affects the capability of the spectrometer to completely analyse metals contained in the oil.

2.3.5 Conclusions

This study shows that VI improvers can affect the determination of zinc and calcium concentration in unused multigrade crankcase oils formulations using atomic absorption spectroscopy. The differences in zinc and calcium concentration are found up to 18.5 % and 11 % respectively. Furthermore, it has been demonstrated that the VI improver concentration as well as its chemical structure type greatly affects these determinations.

Regression equations are calculated to assist in the estimation of the actual content of zinc and calcium in presence of the following VI improvers: SICP, SBCP, OCP and PMA. It is essential in applying these equations to know the type of the incorporated VI improvers. Developed models can be run either via the polymer concentration or the blend viscosity at 100°C.

2.3.6 References

(1)   Walsh, A.: Spectrochim Acta, 7 (1955), p. 108.
(2)   Russell, B.J.; Shelton, J.P.; Walsh, A.: Spectrochim Acta, 8 (1956), p. 317.
(3)   Sychra, V.; Lang, I.; Sebar, G.: Analysis of Petroleum Products by Atomic Absorption Spectroscopy and Related Techniques, Prog. Anal. At. Spectrosc., 4 (1981), p. 341.
(4)   Abou El Naga, H.H.; Nader, M.: The Application of Atomic Absorption Spectroscopy for Measuring Metals in Petroleum Products, 8th Algiers (1972, paper no. 81 (C-1) p. 1, 8th Arab Petroleum Congress, 28.05. – 3.06.1972.
(5)   Barnett, W.B.; Kahn, H.L.; Peterson, G.E.: Rapid Determination of Several Element in a Single Lubricating Oil Sample by Atomic Absorption Spectroscopy. At. Absorption Newslett., 10, 1971, p. 100.
(6)   Means, E.A.; Ratcliff, D.: Determination of Wear Metals in Lubricating Oils by Atomic Absorption Spectroscopy, At. Absorpt. Newsl., 4 (1965), p. 174.
(7)   Lukasiewicz, R.J.; Buell, B.F.: Anal. Chem., 47 (1975), p. 1673.
(8)   Oliphant, T.L.; Terry, J.R.; Klaus, E.E.: The Effect of Viscosity Index Improvers on the Determination of Zinc Using Atomic Absorption Spectroscopy, SAE reprint, No. 860548 (1986).
(9)   Stewart, W.T.; Start, F.A.: Lubricating Oil Additives, in Vol 7 of K. A. Kobe and J. J. Mcketta Advances in Petroleum Chemistry and Refining, John Wiley and Sons Inc., New York (1963).
(10)   Schilling, A.: Motor Oils and Engine Lubrication, Scientific Publications, Vol. 1, 2nd edn., Chapter 2, United Kingdom (1968).
(11)   Selby, T.W.: The Non-Newtonian Characteristics of Lubricating Oils, ASLE Trans., 1 (1958), p. 68.
(12)   Arlie, J.P.; Denis, J.; Parc, G.: Relations Between the Structure and Viscometric Properties of Polymethacrylate Solutions in Lube Oils, J. Inst. Petrol. IP 75-000 (1975).
(13)   Eckert, R.J.A.; Covey, D.F.: Developments in the Field of Hydrogenated Diene Copolymers as Viscosity Index Improvers. Esslingen (1986), p. 8.5.1.–8.5.14, 5th International Colloquium, Additives for Lubricants and Operational Fluids, West Germany, Tech. Akad. Esslingen, West Germany, 14. – 16.01.1986.
(14)   Spiess, G.T.; Johnston, J.E.; Verstrate, G.: Ethylene Propylene Copolymers as Lube Viscosity Modifiers, Esslingen (1986), p. 8.10.1–8.10.11, 5th International Colloquium, Additives for Lubricants and Operational Fluids, Tech. Akad. Esslingen, West Germany, 14.–16.01.1986.
(15)   Goldin, M.; Yerushalmi, J., Pfeffer, R.; Shinnar, R.: Fluid Mech., 38 part 4, 1969, page 689.

2.4 The Characterisation of Synthetic Lubricant Formulations by Field Desorption Mass Spectrometry

K. Rollins, M. Taylor, J.H. Scrivens and A. Robertson,
ICI Wilton Materials Research Centre, Wilton Middlesborough, Great Britain
H. Major, VG Analytical Ltd., Whythenshane, Great Britain

Field desorption-mass spectrometry (FD-MS) has been used to characterise a wide variety of compound types which are commonly used in synthetic lubricant formulations. The technique's ability to deal with compounds which vary widely in their polarity and thermal stability characteristics make it a particularly powerful tool in this application area.

The analysis of complex mixtures in multicomponent matrices is a recurring requirement for the industrial mass spectrometry laboratory. This often dictates that the full analysis of a particularly complex sample may require a strategy involving gas chromatography-mass spectrometry (GC-MS) and perhaps other spectroscopies (IR,NMR). Typical of this type of problem is the analysis of lubricant formulations.

These formulations typically consist of combinations of a mineral or synthetic oil base, a polyolefin and a synthetic ester. An oil/additives package of this type, by its nature, contains compounds of widely differing volatility and polarity and as such provides a considerable challenge to the mass spectroscopist. An initial approach in the study of these materials was undertaken using GC-MS and direct insertion probe (DIP) methodology in combination with electron impact and chemical ionisation.

As Figure 2.4.1 clearly demonstrates, the extraction of useful data regarding unknown compounds from such an overwhelming background in GC-MS mode is extremely difficult.

An additional problem in GC/MS and DIP analysis is that the volatility and thermal stability of unknown components present in the mixture may prevent their passage through a gas chromatograph or evaporation from a solid probe surface.

At this point, FD-MS was evaluated as a potential technique in the characterisation of mixtures of this type. The use of FD-MS in the characterisation of complex mixtures has been described by a number of groups (1—3) where the adavantages of the technique in terms of its ability to deal with samples of widely varying polarity under very "soft" ionisation conditions without the attendant suppression effects experienced in other ionisation methods far outweigh its disadvantages (experimentally difficult, relatively poor sensitivity and lack of expertise).

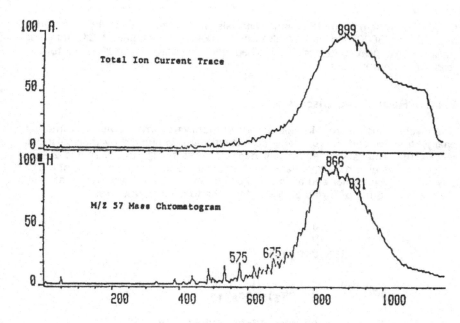

Figure 2.4.1: GC/MS Analysis of Lubricant Formulation

2.4.1 Experimental

All field desorption/ionisation mass spectra were recorded using a VG Analytical
ZAB2—SE mass spectrometer of reserve geometry equipped with the standard
VG combined FD/FAB source. 13 m FD emitters were prepared using the high
temperature indene activation procedure of Rabrenovic et al (4) using an FD
emitter activation unit supplied by Linden ChroMasSpec (Bremen, W. Germany).
Lubricants were initially screened using the field ionisation (FI) technique which
is experimentally somewhat easier to deal with when considering large numbers
of samples; any samples meriting further investigation were then subjected to
field desorption (FD) analysis. For FD analysis, a solution of the lubricant in
hexane was deposited onto the emitter using the standard syringe technique (5).
After introduction into the FD source, the emitter current was then programmed, typically, between 0 and 40 mA or until the signal intensity disappeared.
For FI experiments, the FD probe was left in position and an aliquot of the
lubricant in hexane solution introduced into the FD source using the standard
VG FI probe. The FI probe current was then programmed from 0 to 1 A
(equating to a temperature ramp of ambient to 400°C); the surface of the FD
emitter being replenished during the interscan period "flashing" the emitter
current to its maximum value. In both cases, data was acquired into the VG

Opus data system in multi-channel analysis (MCA) mode over the mass range m/z 100 to 3000 and at a resolution of 1000. The acquired MCA data was subsequently peak detected, centroided and converted to stick data format ready for interpretation.

2.4.2 Results and Discussion

The mechanisms involved in field desorption/ionisation involve very low internal energy transfer and the resultant mass spectra are dominated by molecular ion species (6). This is very effectively demonstrated in Figures 2.4.2a and 2.4.3b which are the electron impact and field desorption spectra for the ester of a C_{18} unsaturated acid dimer and a C_8 alkyl chain length alcohol which is commonly used in these types of formulations and has the following structure:

$$
\begin{array}{c}
\text{O} \\
\parallel \\
(CH_2)_7COC_8H_{17} \\
| \qquad\qquad\qquad \text{O} \\
CH \qquad\qquad \parallel \\
\diagup \quad CH \diagdown \quad CH(CH_2)_7COC_8H_{17} \\
CH \qquad\qquad\qquad | \\
\parallel \qquad\qquad\qquad CH-CH_2-CH=CH-(CH_2)_4-CH_3 \\
CH \diagdown \quad CH \diagup \\
| \\
(CH_2)_5CH_3
\end{array}
$$

Elemental Composition: $C_{52}H_{98}O_4$

Molecular Weight: 786

As can be seen, the protonated ion at m/z 787 is effectively the only species seen in the FD spectrum whilst the FI spectrum contains much more detailed, complementary, structural information. The dominance of molecular ion species in the FD spectrum is of course of major significance when considering the possibilities for FD–MS–MS work.

Field ionisation, by the very fact that it does involve a heating step in presenting the analyte molecules to the emitter, has a requirement for a degree of volatility in the samples to be examined. This effect is well demonstrated by a comparison of Figures 2.4.3a and 2.4.3b which are the field ionisation and desorption mass spectra respectively of lubricant formulations (A).

118

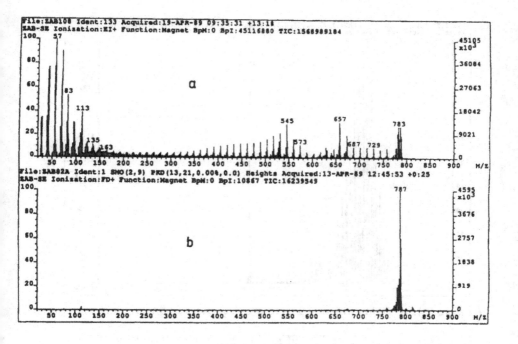

Figure 2.4.2a: Electron Impact Spectrum of 2-Ethyl Hexyl Dimer Acid Ester

Figure 2.4.2b: Field Desorption Spectrum of 2-Ethyl Hexyl Dimer Acid Ester

This formulation has two major components; a synthetic oil base of molecular weight distribution m/z 300 − 600 and the same C_{18} unsaturated acid dimer /C_8 alcohol ester shown above. Qualitatively there is an increase in relative response on the protonated molecular ion of the ester at m/z 787 if field desorption is used to study this mixture rather than field ionisation. This emphasises the sample volatility is still a consideration when utilising FD/FI techniques and both methods are utilised in our laboratory − FI as the initial screening technique and FD as the definitive characterisation method for samples which merit further study.

119

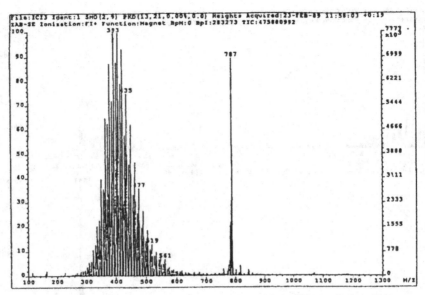

Figure 2.4.3a: Field Ionisation Spectrum — Lubricant Formulation (A)

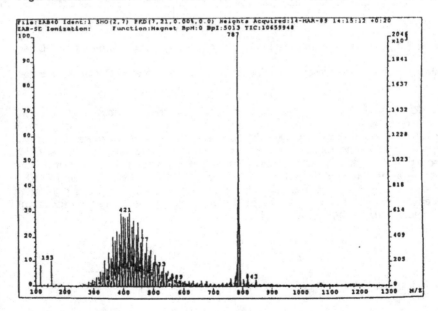

Figure 2.4.3b: Field Desorption Spectrum—Lubricant Formulation (A)

120

Three more examples of the type of problem suitable for analysis by FD—MS are presented here as an illustration of the general utility of the technique. The first example, presented in Figure 2.4.4, concerns the analysis of oils based upon polyisobutylenes of structure

$$\left[-CH_2 - \overset{\overset{\displaystyle CH_3}{|}}{\underset{\underset{\displaystyle CH_3}{|}}{C}} - \right]_n$$

which are widely used in the formulation of 2-stroke oils. The characteristic bimodal hydrocarbon distribution at 56n atomic mass units provides extremely useful molecular weight distribution information which in turn is very diagnostic in terms of oil performance.

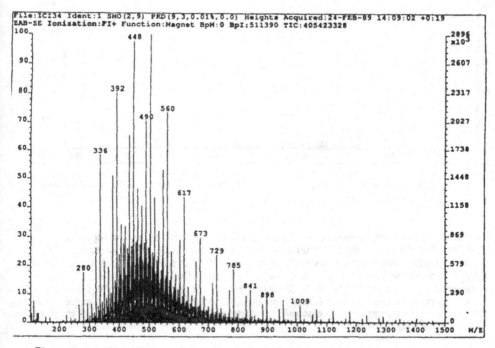

Figure 2.4.4: Field Desorption Spectrum—Polyisobutylene Oil

Figure 2.4.5 shows a different type of formulation frequently encountered which involves the combination of a synthetic oil base with a polyalphaolefin additive. The oil base shows the exptected hydrocarbon distribution in the m/z 300 to 550 area while the olefin component may be recognised by the 140 amu separation ($C_{10}H_{20}$) between m/z 423 to 563.

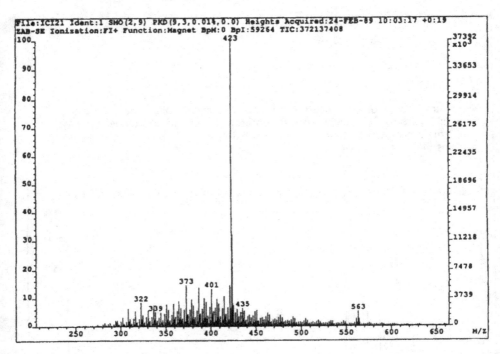

Figure 2.4.5: Field Desorption Spectrum—Synthetic Oil/Polyalphaolefin Lubricant

The final example presented in Figure 2.4.6 shows what might be regarded as a fully formulated blend containing all three potential additive types. The hydrocarbon distribution extends between m/z 250 to 650, the presence of a polyalphadecene component can again be identified by the series of peaks extending from m/z 423 to 140 atomic mass unit intervals and the final ester component is seen at m/z 639 which corresponds to the protonated molecular ion for the trimethylol propane/C_{11} carboxylic acid tri-ester utilised in this particular application.

122

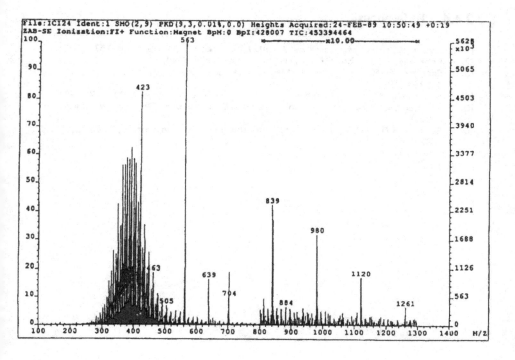

Figure 2.4.6:   Field Desorption Spectrum-Fully Formulated Lubricant

2.4.3 Conclusions

Field desorption/ionisation mass spectrometry provides an elegant approach to the characterisation of a wide range of compound types used in lubricant formulations. The technique's ability to deal with compounds of widely differing polarity and thermal characteristics in a single experiment make it an attractive tool in the analysis of complex mixtures. The addition of a tandem MS/MS instrument of BEBE geometry (the VG analytical ZAB-T) in late 1989 will be used to extend our present studies to the use of FD-MS-MS in the characterisation of sysnthetic lubricant formulations. The instrument will be equipped with a multichannel plate array detector which very effectively overcomes the traditional problem of low ion currents generated during field desorption ionisation thus giving access to the extra dimension of MS-MS information which this combination will permit.

2.4.4 References

(1) Ryska, M.; Kuras, M.; Mostecky, J.: Int. J. Mass Spec. Ion Physics, **16**, 257 (1975).
(2) Forehand, J.B.; Kuhn, W.F.: Anal. Chem., **42**, 1839 (1970).
(3) Schulten, H.-R.; Beckey, H.D.; Meuzelaar, H.L.C.; Boerboom, A.J.H.: Anal. Chem., **45**, 191 (1973).
(4) Rabrenovic, M.; Ast, T.; Kramer, V.: Int. J. Mass Spec. Ion Physics, **37**, 297 (1981).
(5) FD Sample Supply Unit Manual, Linden-ChroMasSpec (Bremen, W. Germany) (1989).
(6) Beckey, H.D.: Principles of Field Ionisation and Field Desorption, Pergamon Press (1977).

2.5 Tailor Making Polyalphaolefins

R.L. Shubkin and M.E. Kerkemeyer, Ethyl Corporation, Baton Rouge, USA
D.K. Walters and J.V. Bullen, Ethyl Petroleum Additives Ltd., Bracknell, Great Britain

Tailor Making PAO's

The production of polyalphaolefins (PAO's), an important class of synthetic functional fluids, can be better controlled to form materials with given sets of desired properties than can the production of analogous petroleum-based fluids.

Until now, commercial PAO's have been based almost exclusively on 1-decene as the starting material. Furthermore, they have been sold on the basis of the viscosity at $100°C$. Work at Ethyl Corporation has shown the need to focus attention on other physical properties for various alternative applications. Important properties may or may not include low temperature viscosity, pour point, volatility, high temperature stability, flash and fire point, and oxidative stability.

This paper will show how the physical properties of the PAO product can be controlled by the judicious choice of starting olefin, reaction catalyst and co-catalyst, reaction temperature, reaction time, and other pertinent reaction parameters. It will also make comparisons between PAO's and hydrocracked mineral oils, often referred to as XHVI (Extra High VI) fluids.

2.5.1 Introduction

The use of polyalphaolefins (PAO's) as high performance functional fluids has received growing acceptance in recent years. Each of the current commercial products has the widest range of physical properties possible from the known technology, and research demonstrating performance advantages of these products in a variety of applications, particularly automotive applications, has been reviewed (1). The objective of the research presented here is to broaden the scope of *potential* applications for PAO's by the development of technology that will allow for the custom design of new products to meet specific end-use requirements. Such new products may not have the same broad range of physical properties as the current commercial products, but they may be superior in meeting one or more specific application requirements. This aim may be accomplished by evolving a more complete understanding of the relationships between reaction parameters and end-product properties.

The process variables that have been found to have a significant impact on the physical properties of the PAO product are:

— Catalyst
— Co-catalyst
— Temperature
— Time
— Olefin chain length
— Position of double bond

2.5.2   Background

A variety of catalytic methods are known for the polymerization of linear α-olefins to low molecular weight oligomers. These include Friedel-Crafts or cationic (2), Ziegler or anionic (3), and free-radical (4) types of catalysts. In general, the fluids obtained from these catalysts contain a wide range of molecular weights, ranging from dimers through high oligomers. The compositions and/or internal structures often result in relatively poor viscosity/temperature characteristics. In 1968, Brennan at Mobil Oil showed that oligomerization of α-olefins could be carried out in a reproducible fashion with boron trifluoride in the presence of a protic co-catalyst. We have demonstrated that, when hydrogenated, the primarily trimeric products from the BF_3-catalyzed oligomerization of $C_6 - C_{14}$ α-olefins have unexpectedly good temperature/viscosity characteristics (6) (7). We later showed that the unique properties could be attributed to the presence of excess branching in the molecular skeleton (8).

The advantages that the current commercial PAO's offer over the very best in mineral oil base-stocks in terms of physical property performance over broad ranges of operating temperatures is impressive. Table 2.5.1 compares Ethyl's Hitec ® 164 with five topgrade 4 cSt fluids. The PAO is significantly superior in both low and high temperature characteristics. Tables 2.5.2 and 2.5.3 show similar comparisons for 6 and 8 cSt fluids. A comparison of gas chromatograms provides a graphic understanding of why these superior characteristics are inherent in PAO's and are not achievable in even the most highly refined mineral oils. Figure 2.5.1 shows traces of British Petroleum BP LHC (a highly refined mineral oil from base-stocks rich in wax) and Ethyl Hitec® oil 164, a PAO with approximately the same viscosity (4.0 cSt) at 100°C. Figure 2.5.2 makes a similar comparison between Shell XHVI and Hitec® 166. The highly refined mineral oil fluids contain a broad range of molecules, the structures and molecular weights of which are controlled in only a limited fashion by extensive reforming and distillation. In this paper we intend to show that the production of PAO's, on the other hand, offers significant advantages and flexibility in obtaining desirable properties. This is because PAO's are synthetic materials whose molecular size and structure, and hence performance characteristics, can be controlled in a predictable and understandable manner. Furthermore, PAO's are manufactured from linear α-olefins rather than from a very limited world supply of high wax petroleum crudes.

126

Table 2.5.1: Comparison of 4.0 cSt PAO with mineral oils

	Ethyl Hitec 164	Petro Can MCT-5	Chevron RLOP 100N	Exxon 100NLP FN1365A	Sun HPO-100	BP LHC
Viscosity at 100° C, cSt	4.06	3.81	4.06	4.02	4.06	3.75
Viscosity at 40° C, cSt	17.4	18.6	20.2	20.1	20.4	16.2
Viscosity at − 40° C, cSt	2490	solid	solid	solid	solid	solid
Pour Point, °C	↓ 65	− 15	− 12	− 15	− 12	− 27
Flash Point (COC), °C	224	200	212	197	206	206
NOACK volatility, %	12.9	37.2	30.0	29.5	29.2	22.2

Table 2.4.2: Comparison of 6.0 cSt PAO with mineral oils

	Ethyl Hitec 166	Petro Can 160 HT	Chevron RLOP 240N	Exxon 200SN	Sun HPO-170	Shell XHVI
Viscosity at 100° C, cSt	5.91	5.77	6.96	6.31	5.57	5.49
Viscosity at 40° C, cSt	31.4	33.1	47.4	40.8	32.1	25.9
Viscosity at −40° C, cSt	7877	solid	solid	solid	solid	solid
Pour Point, °C	− 63	− 15	− 12	− 6	− 12	− 9
Flash Point (COC), °C	235	220	235	212	224	226
NOACK volatility, %	7.5	16.6	10.3	18.8	15.2	14.3

Table 2.5.3: Comparison of 8.0 cSt PAO with mineral oils

	Ethyl Hitec 168	Exxon 325SN	Sun HPO-325N
Viscosity at 100° C, cSt	7.78	8.30	8.20
Viscosity at 40° C, cSt	46.7	63.7	58.0
Viscosity at −40° C, cSt	18305	solid	solid
Pour Point, °C	− 60	− 12	− 12
Flash Point (COC), °C	254	236	250
NOACK volatility, %	3.5	7.2	5.1

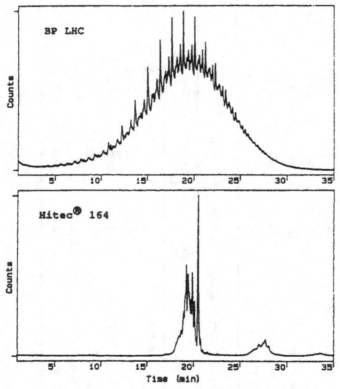

Figure 2.5.1: Comparison by GC of Hitec® 164 and BP LHC

128

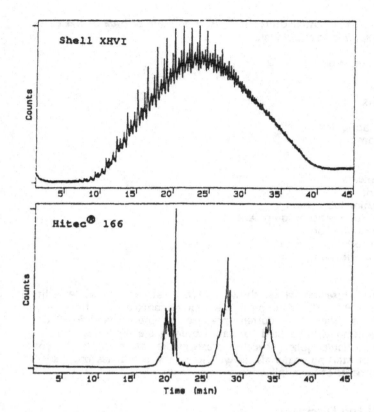

Figure 2.5.2: Comparison by GC of Hitec® 166 and Shell XHVI

Today's commercial PAO's are made from 1-decene. The low viscosity products (2 – 10 cSt at 100° C) are made using BF_3 catalysts. Higher viscosity PAO's are made using catalysts such as trialkylaluminium or alkylaluminium halide in conjunction with a halogen or halide source (9). These products are excellent for a variety of applications, especially in the automotive areas which account for the bulk of the current usage (10). Other applications, however, may place more or less importance on specific physical property requirements. For instance, two entirely different end-uses may require a fluid of approximately the same viscosity at a given temperature, but one may require a lower pour point while the other a higher flash point.

Some of the areas of potential utility for PAO's that are *outside of the auto-motive crankcase application field* include:

Automatic transmission fluids
Brake fluids
Circulating oils
Compressor oils
Cutting fluids
Electrical insulating oils
Fuel additive package diluents
Gear oils
Grease bases
Heat transfer media
Hydraulic fluids
Industrial machine lubrication
Lubrication of food contact equipment
Marine engine lubrication
Personal care products
Refrigeration lubrication
Turbine oils

This paper reports research which shows that a great deal more flexibility exists in designing PAO fluids to meet the requirements of specific end-use applications than is reflected in current practice. As the demand for PAO's grows, more attention will be paid to manufacturing the most cost-effective products to do particular jobs. This will mean greater attention to the cost and availability of starting materials as well as to the actual (as opposed to idealized) requirements of the application.

2.5.3 Results and Discussion

Polyalphaolefin fluids are manufactured in a two-step process. The first is the oligomerization step in which the starting linear α-olefin is contacted with a catalyst in such a manner that the olefin oligomerizes to a variety of low molecular weight products, each of which has a molecular weight that is a unit multiple of the starting material.

$$x C_n H_{2n} \rightarrow C_{2n} H_{4n} + C_{3n} H_{6n} + C_{4n} H_{8n} + C_{5n} H_{10n} + \text{etc.}$$

Fig. 2.5.3 is a gas chromatography trace of a typical oligomeriziation product from 1-decene. It shows quite clearly that the composition of the product consists of dimers, trimers, tetramers, etc. The relative abundance of these oligomers plays a major role in determining the physical properties of the final product. Part of the research presented in this paper is aimed at understanding the role of reaction variables in determining the relative composition, and hence the resultant physical properties, of the mixture of oligomers produced.

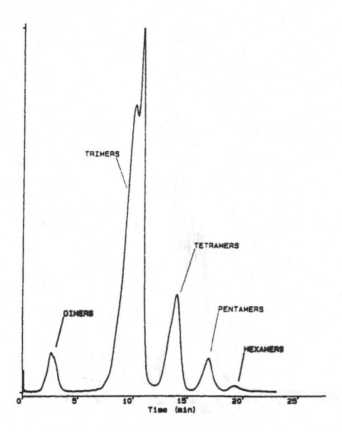

Figure 2.5.3: Gas chromatogram of a typical oligomer

Fig. 2.5.4 is a high resolution GC trace of just the trimeric product from 1-decene. It shows that this single oligomer fraction is composed of a very large number of isomers. Because physical characteristics are dependent upon molecular structure as well as carbon number, the influence of reaction parameters on structure plays an important role in determining product properties.

The second step in the manufacture of PAO is hydrogenation. This is necessary to obtain a chemically inert product, especially towards oxidation. Hydrogenation has a very minimal effect on the product performance properties.

$$C_{xn}H_{2xn} + H_2 \rightarrow C_{xn}H_{2xn+2}$$

Our studies on the effects of oligomerization variables on the composition, structure and physical properties of PAO products are presented here on a variable-by-variable basis.

131

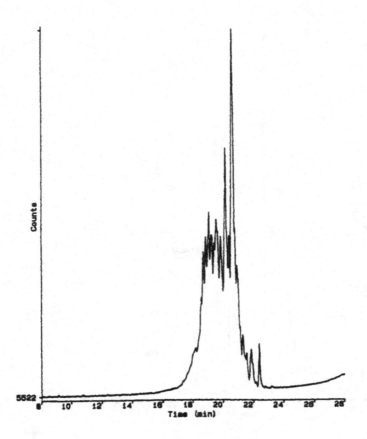

Figure 2.5.4:   High resolution gas chromatogram of decene trimer

Catalyst

The choice of catalyst is critical for the successful manufacture of a functionally useful hydrocarbon from linear α-olefins. Table 2.5.4 indicates the differences in results using five different catalyst systems. Oligomerization reactions were carried out at 25°C using 1-decene as the starting olefin. The exception was the run with di-*tert*-butyl peroxide. In that case a reaction temperature of 150 − 160°C was required. The reactions were continued until the monomer was approximately 95 % consumed. The products were washed of catalyst, stripped of unreacted monomer and dimer, and hydrogenated.

Table 2.5.4:   The effect of catalyst type on product properties

Catalyst[1]	Avg. M.W.	Pour Point (°C)	$V_{40°C}$ (cSt)	$V_{100°C}$ (cSt)	VI
$AlCl_3$	779	− 45	258	27.8	150
$(t\text{-}Bu)_2 O_2$[2]	−	− 45	124	16.0	149
$Et_3 Al/TiCl_4$	729	− 51	78	12.2	165
$Et_3 Al_2 Cl_3/TiCl_4$	575[3]	− 44	46	7.8	149
$BF_3 \cdot H_2 O$[4]	445[3]	− 60	23	4.6	130

1. Reaction conditions:
   Temp.   = 25 − 30° C
   Olefin   = 1-decene
   Product is dimer-free and hydrogenated
2. Reaction temp. = 150 − 160° C
3. Calculated from GC data
4. BF_3 (g) bubbled through reaction

Through the series $AlCl_3$, $(C_4 H_9)_2 O_2$, $Et_3 Al/TiCl_4$, $Et_3 Al_2 Cl_3/TiCl_4$, and $BF_3 \cdot H_2 O$ there is a marked trend towards a lower degree of oligomerization. The $BF_3 \cdot H_2 O$ catalyst system, however, is unique in its ability to peak the oligomer at the trimer. It is a combination of this characteristic, the skeletal branching induced by this catalyst, and the consequent wide range of physical properties obtained when this catalyst is used in conjunction with 1-decene that led to the adoption of BF_3/decene as the basis of manufacture by every major producer of low viscosity PAO's. The remainder of the research presented here will deal only with BF_3 catalyst systems.

Co-catalyst

Boron trifluoride itself is incapable of catalyzing the oligomerization of linear α-olefins. It must be used in conjunction with a protic co-catalyst. Water, alcohols and carboxylic acids have been used successfully. Furthermore, it has been found that a molar excess of the BF_3 over the co-catalyst must be present for the reaction to proceed. This may be accomplished by bubbling excess BF_3 gas through the reaction mixture or, more practically on a commercial scale, running the reaction under a slight BF_3 pressure (3, 4). For the remainder of this paper, $BF_3 \cdot ROH$ will be understood to mean that the BF_3 was used in conjunction with a protic co-catalyst and that excess BF_3 was present.

A series of three reactions was carried out under identical conditions using three different co-catalysts. The composition of the reaction mass was monitored over time by gas chromatography (GC). The results are presented on Figs. 2.5.4 – 2.5.7. With co-catalyst A, a maximum content of 81 % dimer was reached very rapidly. After that, the dimer appears to be undergoing a secondary reaction with itself to form tetramer. With B, the initial reaction is even faster than with A, but there is much less secondary reaction. The maximum observed dimer composition is 75 %. Finally, co-catalyst C yields a much lower maximum dimer composition of 49 % and very little secondary reaction. A direct comparison of the percent dimer in the oligomer mixture as a function of time for the three different co-catalysts is given in Fig. 2.5.8.

Before concluding that the differences in behaviour observed above are due solely to the chemical nature of the co-catalyst, one must look at the effect of co-catalyst concentration. Fig. 2.5.9 compares the formation of dimer with co-catalyst A at a somewhat lower temperature and at two different concentrations. The "high" concentration is 10x that of the "low" concentration. The behaviour at the "low" concentration is similar to that observed for co-catalyst C at the "high" concentration.

The function of the co-catalyst is obviously complex. It reacts with the BF_3 to form a species capable of protonating the olefinic double bond:

$$BF_3 + ROH \rightleftharpoons H^+ (ROBF_3)^-$$

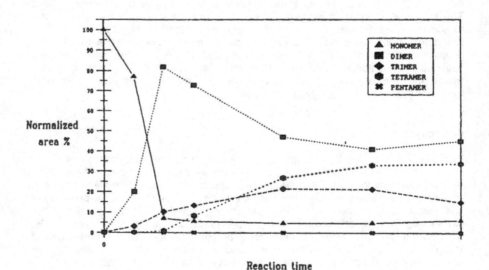

Figure 2.5.5: Oligomerization with BF_3-A

134

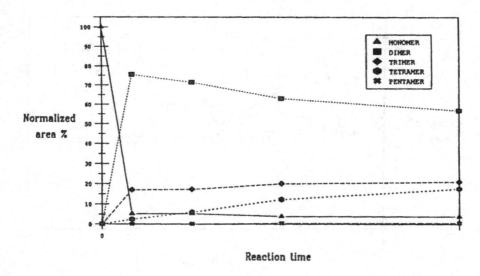

Figure 2.5.6: Oligomerization with BF$_3$-B

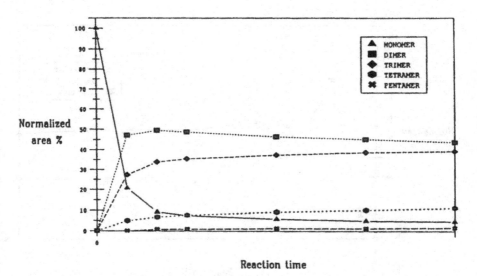

Figure 2.5.7: Oligomerization with BF$_3$-C

135

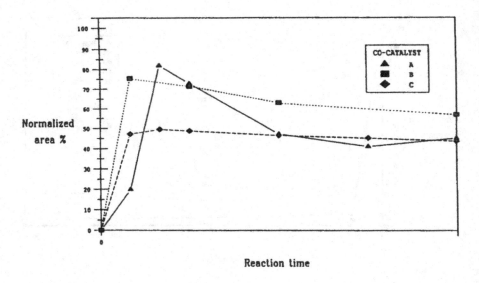

Figure 2.5.8: Effect of co-catalyst on dimer formation

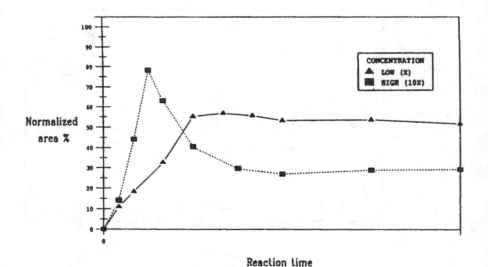

Figure 2.5.9: Effect of co-catalyst concentration

136

The necessity for an excess of BF_3 has not been explained, and the relationships between the catalyst species, the reaction medium, and the olefinic reactants have not been very well understood. Empirically, however, we have shown that the choice of both the co-catalyst and the concentration at which it is used plays an important role in determining the composition of the product PAO.

Temperature

Temperature has a marked effect on the product composition from the BF_3-catalyzed oligomerization reaction (5). Figs. 2.5.10 − 2.5.11 trace the progress of a BF_3 · A-catalyzed oligomerization reaction at two different temperatures. At the lower temperature the production of dimer and trimer parallel each other until the monomer is nearly consumed. After that, the dimer content drops rapidly as the tetramer builds up while the trimer decreases slowly. There is little change in the relative composition after an extended reaction time. At the highest temperature the build-up of dimer far outpaces that of trimer with a maximum dimer content of over 80 % being reached. A similar consumption of dimer to form tetramer is then observed. Fig. 2.5.12 compares the percent concentration of trimer as a function of time at six different reaction temperatures with all other variables held constant.

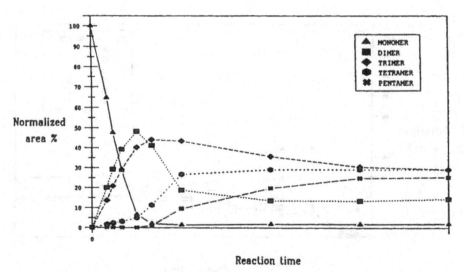

Figure 2.5.10: Oligomerization reaction at low temperature

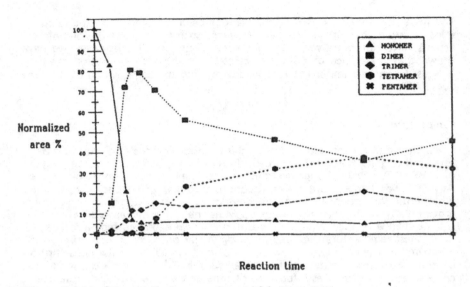

Figure 2.5.11: Oligomerization reaction at high temperature

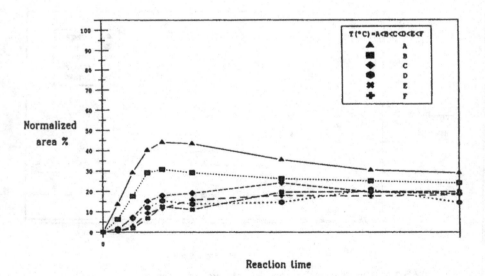

Figure 2.5.12: Trimer formation at various temperatures

138

Examination by ^{1}H NMR (Table 2.5.5) of both the unreacted olefin and the dimer products from reactions similar to those described above reveals that the unreacted olefin undergoes both positional and skeletal isomerization which is increased with increasing temperature. Continued reaction shows an increase in trisubstituted product relative to linear internal olefin as the reaction time is extended. The increase in branched monomer relative to the decrease in total monomer indicates that the linear internal is undergoing both oligomerization and skeletal isomerization. The highly branched trisubstituted olefin does not undergo further oligomerization. This is consistent with the leveling out of the product distribution. By way of comparison, unreacted monomer distilled from a reaction at lower temperature shows some linear internal olefin (22.7 %) but no trisubstituted.

Table 2.5.5: The composition of unreacted monomer recovered from oligomerization reactions at different temperatures

Reaction Temperature (°C)	Unreacted Monomer (Area %)	Monomer Composition Mole % by ^{1}H NMR				Dimer Product Avg.No.Branches per Molecule (^{1}H NMR)
		Vinyl	Internal	Trisubstituted	Vinylidene	
low	44.5	77.3	22.7	N.D.	< 0.1	1.9
high	10.2	3.7	51.2	44.2	0.9	—
high	6.9	4.1	33.3	58.1	4.4	2.6
higher	11.7	4.8	40.4	51.6	3.3	2.9

In addition, Table 2.5.5 shows that the degree of branching in the dimer that is formed is also temperature dependent. At low temperature the mixture of dimeric isomers contains an average of 3.9 methyl groups per molecule. This is an average of 1.9 branches in the skeleton. At high temperature the ratio of methly groups increases to 4.6, and at sill slightly higher temperature it is 4.9 (or 2.9 branches per molecule).

As indicated above, higher reaction temperatures may increase the degree of branching in the oligomer product. One of the results is an increase in the low temperature viscosity of the product. The viscosities of saturated, dimer-free oligomers prepared with $BF_3 \cdot C$ at three different temperatures are given in Table 2.5.6.

A different perspective on the effect of oligomerization temperature is given in Table 2.5.7. A mixture of 1-olefins was oligomerized using $BF_3 \cdot A$ catalyst at six different temperatures. The reactions were carried to 80 − 90 % conversion of the monomer. Interestingly, the time required to reach the desired level of conversion went up with each increase in temperature. Unreacted monomer was distilled off, and the product (including the dimer) was hydrogenated. As expected, the ratio of dimer to higher oligomers shows a strong correlation to reaction temperature. Similarly, the viscosities, pour points and volatilities all show a smooth progression of values that correlate to the reaction temperature. The volatilities reported in the table were determined by a procedure developed at ethyl that correlates very well with the standard NOACK procedure.

Table 2.5.6: The effect of reaction temperature on product properties with BF$_3$ · C

Reaction Temperature (°C)	V$_{100°C}$ (cSt)	V$_{40°C}$ (cSt)	V$_{-40°C}$ (cSt)	VI	Pour Point (°C)
low	4.41	20.4	3548	128	− 69
moderate	4.39	20.7	3702	124	− 69
high	4.48	21.5	4218	122	− 69

Table 2.5.7: The effect of reaction temperature on product properties with BF$_3$ · A

Reaction[1] Temperature (°C)	Ratio Dimer/ Trimer+	V$_{40°C}$ (cSt)	V$_{100°C}$ (cSt)	VI	Pour Point (°C)	Volatility[2] (Wt % Loss)
A	0.68	31.26	6.36	161	− 21	3.4
B	1.11	27.19	5.71	158	− 21	4.7
C	1.61	24.82	5.37	159	− 21	5.2
D	1.92	24.86	5.37	158	− 21	5.8
E	2.64	23.16	5.06	149	− 23	7.9
F	2.98	22.84	4.89	143	− 29	8.3

1. Temperature (°C): A < B < C < D < E < F
2. Volatility = In-house test to simulate Noack volatility

Time

Figs. 2.5.5 − 2.5.10 indicate a changing composition of the PAO relative to the time reaction. As already indicated, both the composition and the isomeric distribution of the product change. This phenomenon presents both an

140

operational problem and a useful opportunity for tailor making PAO's with special physical properties. Two strategies are possible. First, the oligomerization reaction can be terminated prior to complete consumption of the starting monomer. Second, continued reaction after the monomer is essentially consumed results in secondary reactions of the oligomers with each other. The rate of these reactions is much slower than those involving monomer.

Table 2.5.8 shows the effect of early termination on the properties of the dimer-free product. The data show that the products obtained at lower monomer conversion have lower viscosities and pour points. Even before significant secondary reaction takes place, the oligomer distribution is a function of reaction time. The implication is that the oligomerization process is sequential in the early stages of the reaction. In other words, the formation of tetramer and higher oligomers result from the reaction of monomer with trimer as long as there is an excess of monomer. Later, tetramer appears to be formed from the reaction of two molecules of dimer.

Table 2.5.8: The effect of early termination of the oligomerization reaction on product properties

Temperature (°C)	Monomer Conversion (Wt %)	$V_{40°C}$ (cSt)	$V_{100°C}$ (cSt)	VI	$V_{-40°C}$ (cSt)	Pour Point (°C)
low	95.2	22.93	4.61	130	3748	− 59
low	72.7	20.27	4.28	131	2873	− 62
moderate	97.2	24.71	4.78	126	5007	− 57
moderate	79.9	18.38	3.98	125	2599	− 62

Olefin chain length

Perhaps the most obvious variable available to the researcher attempting to tailor-make a PAO is the chain length of the starting olefin. Table 2.5.9 lists the properties of saturated PAO's prepared under identical conditions. As expected, the viscosities, pour point, flash and fire points all increase with increasing chain length of the monomer. Conversely, volatility decreases.

The data on Table 2.5.9 also indicate another important variable available to the PAO producer. Particularly with the heavier starting olefins, the choice of whether or not to remove some or all of the dimer provides a new dimension in flexibility. Removal of dimer leads to lower volatilities, but to higher low temperature viscosities, pour points and flash points.

Table 2.5.9: The effect of olefin chain length on the product properties

Starting Olefin	Dimer (Wt %)	V$_{100}$°C (cSt)	V$_{40}$°C (cSt)	V$_{-18}$°C (cSt)	V$_{-40}$°C (cSt)	Pour Point (°C)	Flash Point (°C)	Noack Vol. (%)
Octene	1.2	2.77	11.2	195	1320	+ 65	190	55.7
Decene	0.3	4.10	18.7	409	3031	+ 65	228	11.5
Dodecene	15.7	4.94	23.3	534	–	– 48	235	11.4
	2.8	5.70	27.8	703	–	– 45	256	3.5
Tetradecene	40.9	5.34	24.3	519	Solid	– 24	230	8.5
	11.1	6.91	36.2	966	Solid	– 21	250	3.7
	3.0	7.59	41.3	1150	Solid	– 18	272	2.3

As mentioned earlier, PAO's produced from 1-decene have the broadest spectrum of important physical properties over the widest temperature range. For this reason decene-based PAO's have become universally adopted as the standard for high performance by the highly demanding automotive industry. For other applications, however, the optimum choice of starting olefin depends on a number of factors. The first consideration is the actual physical property requirements. Often there is a conflict between the requirements at low and high temperatures, but sometimes the importance of one property overshadows that of the others. After a realistic assessment is made of the requirements to be placed on the product PAO, then the candidates for starting olefin can be chosen. The next choice in the selection process is availability and cost. The cost of a particular olefin stream is often related to availability, and this may be dependent on unrelated market conditions. Another factor is the cost of production. The use of a light starting olefin may entail the need to separate and dispose of a light (dimer) fraction while the employment of a heavier stream may not.

Position of double bond

The work presented so far was carried out exclusively with 1-olefins. We reported earlier that oligomer products from $BF_3 \cdot ROH$-catalyzed reactions of 1-olefins at moderate temperatures have more branching in the carbon skeleton than would be predicted by conventional cationic catalyst mechanisms, and that this was the reason for the excellent low temperature viscometrics (8). We speculated at that time that a skeletal rearrangement was occurring at the dimer stage. Onopchenko et al., at Gulf confirmed our findings but postulated that the rearrangement occurs at the monomer stage (11). Driscoll et al., at Suntech proposed that the excess branching arises from positional isomerization of the double bond followed by oligomerization of internal olefins (12). In order to understand better the role of positional isomerization of the monomer in determining the physical properties of the product, we undertook a series of experiments to determine the effect of using internal olefins in the starting olefin mix (13).

A series of oligomerization reactions was carried out using $BF_3 \cdot C$ as the catalyst system. All reactions were at the same temperature. The starting 1-olefin contained 0, 10, 20 and 50 %, respectively, of a mixture of 2-olefin and 3-olefin. In order to separate the effect of isomer content from oligomer composition, the pure trimer fraction was isolated from each reaction product by distillation and hydrogenated. Increased levels of internal olefins in the starting mixture resulted in slower reactions and increased levels of dimer in the product. At 0 % internals the yield of dimers was 3.0 %, but at 50 % it was 27.7 %. When 100 % internals were used the reaction was very slow, and the product was nearly all dimer. The hydrogenated trimer fractions are compared in Table 2.5.10.

Table 2.5.10: The effect of double bond position on product properties

Starting Alpha (wt %)	Olefin Internal (wt %)	Pour Point (°C)	$V_{-40°C}$ (cSt)	$V_{40°C}$ (cSt)	$V_{100°C}$ (cSt)	Me/Mole (NMR)
100	0	− 54	2330	15.5	3.64	4.4
90	10	− 57	2140	15.5	3.62	4.5
80	20	− 66	2240	17.2	3.67	4.4
50	50	← 69	2980	17.1	3.76	4.3

The most startling result from the series of experiments just described is the profound effect that the presence of internal olefins in the starting olefin has on the low temperature properties of the trimer product. As the percent internals increase from 0 % to 50 %, the pour point decreases from − 54° C to ← 69° C, and the − 40° C viscosity increases from the 2140 − 2330 cSt range to 2980 cSt. Proton NMR shows that the degree of branching remains approximately the same, indicating that the pour point and viscosity effects are a result of the various branch chain lengths rather than the number of branches. Another implication of this data is that the observed branching is not a result of internals reacting with internals.

2.5.4 Conclusion

Polyalphaolefins are a versatile class of high performance functional fluids. The physical properties of these fluids may be adjusted by a variety of techniques in order to meet the requirements of specific end-use applications. While the current decene-based products represent an excellent choice of fluids with wide operating ranges for automotive crankcase use, they may not always meet all of the requirements for other uses, or meet them in the most cost-effective manner. We have shown that a thorough knowledge of the effects of the oligomerization reaction variables on the product properties is an important element in the effort to "tailer make" PAO's for the marketplace.

2.5.5 References

(1) Shubkin, R.L.: Synthetic Lubricants. In "Alpha Olefins Applications Handbook", Lappin, G.R. and Sauer, J.D.; eds.; Marcel Dekker, New York, 1989; Chapter 13, pp 353–373.

(2) Sullivan Jr., F.W.; Vorhees, V.; Neely, A.W.; Shankland, R.V.: Ind. Eng. Chem., 23: 604 (1931).

(3) Southern, D.; Milne, C.B.; Moseley, J.C.; Beynon, K.I.; Evans, T.G.: British Patent 873,064 (1961), to Shell Research.

(4) Garwood, W.E.: U.S. Patent 2,937,129 (1960), to Socony Mobil.

(5) Brennan, J.A.: U.S. Patent 3,383,291 (1968), to Mobil Oil.

(6) Shubkin, R.L.: U.S. Patent 3,763,244 (1973), to Ethyl Corp.

(7) Shubkin, R.L.: U.S. Patent 3,780,128 (1973), to Ethyl Corp.

(8) Shubkin, R.L.; Baylerian, M.S.; Maler, A.R.: "Olefin Oligomers: Structure and Mechanism of Formation", Synthetic Lubricant Symposium, 178th National Meeting, American Chemical Society, Washington, D.C., Sept. 9–14, 1979; also, Ind. Eng. Chem. Prod. Res. Dev., 19: 15–19 (1980).

(9) Loveless, F.C.: U.S. Patent 4,469,910 (1984), to Uniroyal. Loveless, F.C.; Merijanian, A.V.; Smudin, D.J.; and Nudenberg, W.: U.S. Patent 4,594,469 (1986), to Uniroyal.

(10) Blackwell, J.W.; Bullen, J.V.; Shubkin, R.L.: "Current and Future Polyalphaolefins", Conference on Synthetic Lubricants, The Hungarian Hydrocarbon Institute Section of the Hungarian Chemical Society, Sopron, Hungary, Sept. 12–14, 1989.

(11) Onopchenko, A.; Cupples, B.L.; Dresge, A.N.: "Ind. Eng. Chem. Prod. Res. Dev.", 22: 182–191 (1983).

(12) Driscoll, G.L.; Linkletter, S.J.G.: "Synthesis of Synthetic Hydrocarbon via Alpha Olefins", Air Force Wright Aeronautical Laboratories report designation AFWAL-TR-85-4066, May 1985.

(13) Part of this work was previously reported. See reference 10.

3. Additives and Mechanism of Effectiveness

3.1 Engine Oil Additives: A General Overview

C. Kajdas
Technical University at Radom, Poland

Additives are incorporated in engine oils to improve existing desirable properties of base stocks and to impart them new specific properties required by modern engines. The additives include antioxidants, detergents, dispersants, corrosion and rust inhibitors, viscosity index improvers, pour point depressants, foam inhibitors and tribological agents. This paper gives a general description of these additives and provides references for detailed information related to both their application and action mechanisms. Tribological additives including friction modifiers and load-carrying agents are discussed in more detail. An interaction of the additives between themselves and with base oils are also taken into account.

3.1.1 Introduction

Engine oils form the most important group of liquid lubricants. They amount to over 40 percent of all the lubricating oils. These oils are made from base oils —— mostly petroleum base stocks —— and a set of additives.

Existing properties of the mineral base oils which primarily relate to the physical properties are not sufficient to provide an adequate lubricant performance for modern engines. Thus, they have to incorporate about 10 to 25 percent of additives. The additives give the base oils desirable characteristics which they lack —— this relates mostly to chemical properties —— and improve existing properties.

It is widely recognized that the performance of finished engine oils is sensitive to both properties of base oils and additives used. This is associated with a specific interaction of the additives among themselves and with the base stock.

The aim of this paper is to describe all the groups of additives used to formulate modern engine oils and provide the general information on interaction of the additives between themselves and with base oils. Some more specific information is connected with tribological additives.

3.1.2 Engine Oil Properties

3.1.2.1 Properties of Base Oils

Primarily they include physical properties such as:

— viscosity and flash point which are determined by the boiling range

— pour point and viscosity change with temperature that depend on both the crude oil type and refining processing.

The typical chemical properties of base oils

—— are oxidation resistance and chemical activity

—— depend on the crude oil type used and refining processes. The refining processes control the oxidation stability of base oils significantly.

Usually, the raw vacuum cuts from most crude oils contain compounds which provide undersirable characteristics for finished engine oils. These compounds are removed (solvent extraction) or reconstituted (selective hydrocracking, hydrotreating). The undesirable compounds negatively influence pour point, viscosity index, oxidation stability, carbon- and sludgeforming tendency etc.

The processes used to improve the characteristics of the raw vacuum cuts encompass:

— Solvent deasphalting

— Solvent extraction or selective hydrocracking or hydrotreating

— Dewaxing

— Finishing (hydrogenation or clay treatment).

3.1.2.2 Properties Imparted by Additives

These are mostly associated with the chemical properties of engine oils. They include such general performance aspects as:

— Oxidation resistance

— Corrosion/rust resistance

— Deposit formation

— Friction and wear

150

For example, for the SF engine oils these properties relate to:

- Oxidation and bearing corrosion under high temperature operation (CRC L-38 or Petter W1 test)

- Engine rust under short-trip, low temperature operation (Oldsmobile IID test)

- Deposits, oxidation and wear under high temperature operation (Oldsmobile IID test or Petter W1 and Ford Cortina tests)

- Overhead cam wear and sludge and varnish under stop-and-go operation (Ford V-D test or Daimler Benz OM-616 and Fiat 600 D tests).

For the CD engine oils these properties are mostly associated with oxidation and bearing corrosion under high temperature operation (CRC L-38 test) and diesel piston deposits formation. The latter include:

- High-temperature diesel piston deposits with high-sulfur fuel.

These tests stress certain performance aspects of the engine oils imparted or improves by adequate additives and have been correlated with field experience.

3.1.3 Types and General Characteristics of Additives for Engine Oils

They include most of the known lubricating oil additives:

- Oxidation inhibitors (antioxidants) and metal deactivators which primarily increase oxidation stability of engine oils and, thus, reduce varnish and sludge formation; this provides increased life of the oils.

- Detergents that reduce or prevent formation of deposits at high temperatures, thus, provide cleanliness of lubricated engine surface.

- Dispersants which retard or prevent sludge formation due to keeping insoluble oxidation and combustion products in suspension.

- Corrosion inhibitors aiming at protecting alloy bearing and other metal surfaces against chemical attack expressed in terms of corrosion action of formed acids and peroxides.

- Rust inhibitors protecting ferrous metal surfaces from rusting in presence of moisture.

- Viscosity index improvers which primarily reduce the rate of change of viscosity with temperature change.

— Depressants providing for the oil lower pour point.

— Foam inhibitors that prevent the oil from stable foam formation.

— Tribological agents controlling friction and wear; they reduce wear in steel on steel mating elements and increase film strength and load-carrying capacity.

Additives can carry out their task of enhancing or imparting new engine oil properties in three ways (1):

(i)   Protection of engine surfaces
(ii) Modification of oil properties
(iii) Protection of the base stock.

The first group of the additives include tribological boundary agents, corrosion and rust inhibitors, detergents and dispersants. The second group — oil modifiers — encompasses viscosity index improvers, pour point depressants and swell agents. The last group (base stock protectors) include antioxidants, metal deactivators and foam inhibitors.
Adequate combination of these additives with a given base stock provides unique performance features which are required to satisfy the lubrication needs of modern engines, service conditions and the oil change intervals.

Since modern engine oils, especially multiviscosity ones, are very complex mixtures of base oils and additives, it is necessary to bear in mind that some components may either enhance or interfere with the function of another component. Therefore, the finished engine oil should be formulated to achieve maximum synergistic effects and minimum antagonistic ones.

Additives used to formulate engine oils have to possess certain general properties to allow them to be effectively incorporated in base stocks. These general properties include:

— Solubility in the base stock

— Chemical stability

— Compatibility

— Non-toxicity

— Low volatility

— Flexibility

— Controlled activity.

Other properties, e.g. color and odor are also of importance.

Additive package formulation demands consideration of individual additive performance as well as interaction and competition compatibility among the additives. Often the solution to an engine oil formulation problem lies not in changing the relative concentration of an additive but in changing the additives themselves; molecular weight, molecular weight distribution, purity, reaction conditions, and many other variables influence these parameters. Small amounts of chemicals are added during additive interaction performance; these "additive additives" are considered highly proprietary by additive manufacturers (2).

3.1.4 Description of the Engine Oil Additives

3.1.4.1 General Information

An engine oil additive is defined as a material designed to enhance or to impart the performance properties of the base stock. Usually they are materials that have been chemically synthesized to supply the desired performance features. Individual additives are used at concentration levels ranging from several parts per million to greater than 10 volume percent (1).

Different additives can assist each other, resulting in a synergistic effect or they can lead to antagonistic effects. Typical commercial additives perform more that one function, and all the engine oil functions usually can be represented by only a few separate additives. All engine oil additives used in higher concentration contain hydrocarbon groups which are required to make them soluble or dipersible in base stocks. High molecular weight hydrocarbon groups are mostly derived from olefin polymers. The size, specific structure, and location of the hydrocarbon groups profoundly affect the ways in which the additives function (3).

The development and commercialization of new additives is a long, complex and expensive process involving many technical disciplines.

According to (4) the sequence of events generally includes:

- the definition of a need by the marketing personnel or by liason representatives to the engine manufacturers;
- the review of the need by the technical staff in an attempt to arrive at a fundamental understanding of the chemistry and physics involved;
- a hypothesis of the kinds of molecular structures likely to correct the problem;
- synthesis in the laboratory new materials fitting the hypothesis;
- bench and laboratory engine testing.

Development which follows these steps is shown in Table 3.1.1.

Table 3.1.1:   Steps in Additive Development (4)

A   1. Definition of Need
     2. Proposal of Chemical Structure
     3. Laboratory Preparation
     4. Bench Testing
     5. Laboratory Engine Testing
     6. Decision — Go

B.   1. Optimization and Laboratory Process Development
     2. Preliminary Economic Evaluation
     3. Formulation for Application
     4. Storage and Compatibility Testing
     5. Expanded Laboratory Engine Testing
     6. Toxicity Testing
     7. Premanufacturing Notification to EPR
     8. Decision — Go

C.   1. Pilot Plant Preparation
     2. Manufacturing Process Design Preparation
     3. Advanced Economic Evaluation
     4. Formulation for Application
     5. Field Testing
     6. Decision — Go

D.   1. Plant Construction or Adaptation
     2. Initial Manufacture
     3. Confirmational Engine Testing
     4. Final Economic Evaluation
     5. Limited Marketing

3.1.4.2   Additives Responsible for Formation of Deposits

Engine oil oxidation products form varnish and sludge. Thus, the oxidation process of the oil must be retarded. This can be achieved by using oxidation inhibitors. To control buildup of varnish and sludge detergents have to be applied. Furthermore, dispersants are responsible for keeping the formed sludge from agglomerating and depositing in the engine. Since the oxidation process is catalyzed by some metals, an application of metal deactivators provides further reduction of the deposit formation. Therefore, oxidation inhibitors, metal deactivators, detergents and dispersants are primarily responsible for keeping the engine mating elements clean.

154

3.1.4.2.1   Oxidation Inhibitors

Oxidation is the major process of engine oil deterioration. The mechanism of this process involves free radical reactions which are catalyzed by metals and accelerated by heat. In engine oils composed of hydrocarbons, free radicals react with oxygen to form peroxy free radicals and hydroperoxides. The latter undergo further reactions to form alcohols, ketones, aldehydes, carboxylic acids, and other oxygen containing compounds.

These products usually have molecular weights close to the base stock and remain in solution. As the oxidation process proceeds, the oxygenated compounds polymerize to form viscous materials which, at a particular point, become oil-insoluble.

Consequently, the oxidation process generates viscous soluble materials which thicken the oil and insoluble materials forming deposits.

Two factors are of importance

(i)   Free radicals are formed faster than they are used and the rate of oxidation increases;

(ii)   Nitrogen oxides formed in the combustion process are oxidizing agents and can result in unique oxidation products, which include nitrate esters and other nitrocompounds (5).

Some of the oxygenated compounds are active, polar materials, e.g. acids, that accelerate corrosion and rust.

To retard and reduce the oxidation product formation from hydrocarbons oxidation inhibitors are used. They work by reducing organic peroxides. The reduction of organic peroxides terminates the oxidation process (oxidation chain) and thus, minimizes the formation of:

— Acids
— Resins/Polymers
— Varnish
— Sludge.

The formation of oxidation products is complicated by the nature or the base stock, the presence and influence of the additives used, and the environment. All these factors can change the specific compounds formed and their rates of formation (5).

Petroleum based oils may contain some natural inhibitors. Their nature and amount depends on the crude oil type and the mode and degree of refining. However, the great majority of oxidation inhibitors is provided by synthetic materials. These materials encompass:

— Hindered phenols

— Zinc dialkyldithiophosphates

— Metal dithiocarbamates

— Aromatic amines

— Sulfurized fats and hydrocarbons

— Metal phenol sulfides

— Phospho-sulfurized fats and olefins

— Metal salicylates and many other compounds.

In many engine oils zinc dialkyldithiophosphates are the only antioxidants used, while in other oils they are supplement by other types of antioxidants, such as amines, hindered phenols, sulfides, etc. (6).

Sometimes metal deactivators, i.e. additives which react with metal ions and surfaces to reduct their catalytic activity, are also incorporated into oxidation inhibitors. Taking this into account, one can say that that antioxidants may function by one of the following three mechanisms:

— Free radical inhibition

— Peroxide decomposition

— Metal deactivation.

Hindered phenols are effective free radical (or radical scavenging) antioxidants as they react with free radicals to form nonfree-radical compounds. Some sulfur containing antioxidants decompose peroxides into stable compounds. Metal deactivators/passivators inhibit catalytic activity of metal ions and surfaces.

The action mechanism of zinc dialkyldithiophosphates (ZnDDP) seems to be very complex since it may involve all of the three mechanisms. Laboratory investigations of individual ZnDDP compounds have shown that they act as radical scavenging oxidation inhibitors (7, 8), peroxide decomposers (9, 10), and metal passivators (11). Further investigations (12, 13), combining results of laboratory and engine fleet testing with a commercial, synthetic engine oil containing a combination of ZnDDP and ashless oxidation inhibitors, have shown that in the early stages of engine operation:

(i)   antioxidant agents are consumed by their reactions with combustion-derived free radicals in the piston-cylinder area

(ii)   ZnDDP species are consumed at a higher rate than ashless antioxidant additives.

These findings lead to a conclusion that in a new engine oil the free radical scavenging reactions of ZnDDP are important. As an engine oil ages it will accumulate an ever increasing amount of products from combustion-derived free radical reactions and subsequent reactions in the oil sump; thus, in an aged oil the other mechanisms of ZnDDP consumption, such as peroxide decomposition, will become increasingly important (6). Additional work on antioxidant behavior of pure neutral and basic zinc dialkyldithiophosphates, dialkyldithiophosphoric acid, tetraalkylthioperoxydithiophosphate (disulfide), and of neutral zinc dialkyldithiophosphate in combination with a hindered phenol antioxidant was performed (6). This work included studies of the effect of hydroperoxides and hydrocarbon oxidation products on the radical scavenging activity of the above compounds. Among other things, it was found that only neutral ZnDDP and dialkyldithiphosphoric acid as such are radical scavenging antioxidants. Basic ZnDDP scavenge peroxy radicals only after interactions with oxidation products and disulfide does not scavenge peroxy radicals at all.

Detailed information on a model for oxidation of engine oils in internal combustion engines and the assessment of high temperature antioxidant capabilities are provided in (14, 15).

3.1.4.2.2   Heavy Duty (HD) Additives

HD additives encompass detergents and dispersants. Both additives provide a cleaning function. The purpose of these additives in engine oils is:

— to keep oil—insoluble combustion products in suspension and

— to prevent resinous and asphalt-like oxidation products from agglomeration into solid particles.

The combustion products mostly include carbon-like deposits formed by pyrolysis of deteriorated oil products which accumulate in the ring-belt area. These products include soot and coke-like materials and can amount to 10 percent in the case of diesel oil.

It is reasonable to include to HD additives also alkaline agents which neutralize and buffer the offending acids. This group of additives prevent:

— deposits to be formed on metal surfaces

— sludge deposition in the engine

— corrosive wear by the neutralization of acidic combustion products.

3.1.4.2.2.1   Detergents

Usually all detergents contain:

— Polar groups, e.g. sulfonate, carboxylic;
— Aliphatic, cycloaliphatic or alkylaromatic radicals;
— One or several metal ions or amino-groups.

The most important agents providing detergency —— which is a surface phenomenon of cleaning surface deposits —— include:

— Sulfonates

— Phenates

— Sulfurized phenates

— Salicylates

— Thiophosphates.

Detergents work by lifting deposits from the surfaces to which they adhere. Extensive literature (16–22) associated with various kinds of detergent additives describes in detail their chemistry, manufacturing, application and action mechanism. Problems related to the engine oil formulation are also discussed (23).

3.1.4.2.2.2   Dispersants

Dispersants are nonmetallic materials characterized by a nitrogen or oxygen containing polar group attached to a high molecular weight hydrocarbon chain which solubilizes the additive in hydrocarbon base stocks. The most important agents providing dispersancy —— which is a bulk lubricant phenomenon of keeping contaminants suspended in the oil — include:

— Copolymers (polymethacrylates, styrenemaleinic ester copolymers, etc.);

— Substituted succinamides;

— Polyamine succinamides;

158

— Polyhydroxy succinic esters;

— Polybutene hydroxy benzyl polyamine.

Dispersants have a strong affinity for dirt particles and surround themselves with oil soluble molecules which keep the sludge from agglomerating and depositing in the engine.

Dispersants and detergents each perform both dispersancy and detergency functions and differ in their relative ability to function in the bulk of lubricant or at the engine surface.

Usually, many papers associated with dipersants (16,18, 21–23) also relate to the detergent additives.

3.1.4.2.2.3   Alkali Agents

These additives are formed by incorporating calcium or magnesium carbonates into sulfonate or phenate soaps, in a dispersed form in which a tiny core of the metal carbonate is solubilized by the attached soap (16, 17). By the over-basing technology the soap molecules are able to incorporate 10 to 20 times the stoichiometric soap equivalent of alkaline earth metal; overbased sulfonates containing 50 percent active ingredient may have a total base number as high as 500 mg KOH/g (4). Micellar structure of alkaline agents was recently studied in (20).

3.1.4.3 Additives Modifying Oil Properties

The major additives of this group include viscosity index inprovers and pour point depressants. They have been used by the lubricating oil industry not only when the nature of the engine oil makes it impractical to obtain a product with the needed high viscosity index and low pour point by the refining process but also in applications where extremely wide temperature variations are involved. Usually, engine oils with high viscosity index and low pour point, i.e. the oils which function efficiently, are obtained by a combination of solvent refining procedures and application of these additives. Broad literature (22, 24 – 31) discusses all specific features relating to these additives.

3.1.4.3.1   Viscosity Index Improvers

They are oil soluble polymers in the 50,000 to 1,000,000 molecular weight range such as:

— Polyisobutylenes

— Polymethacrylates

— Ethylene/Propylene copolymers

— Polyacrylates

— Styrene/Maleic ester copolymers

— Hydrogenated Styrene/Butadiene copolymers.

Dispersant viscosity index improvers incorporate polar groups as do dispersants.

Viscosity index (VI) improvers reduce the rate of change of viscosity with temperature, i.e. they cause minimal increase in engine oil viscosity at low temperature but considerable increase at high temperature. This is due to the fact that the polymer molecule assumes a compact curled form in a cold base stock —— which is poorly solvent —— and an uncurled large surface area in a hot base stock that is better solvent. The uncurled form thickens the oil.

In this context it has to be remembered that:

— The addition of the viscosity index improver additive alters the flow be-havior of the base stock; dynamic viscosity of the formulated oil changes with the rate share;

— The sensitivity of the viscosity index improver additives towards mechanical stresses increases with increasing molecular weight of the additives;

— Shear stresses, as they occur, e.g. between piston and cylinder walls in the engine, lead to irreversible breakdown of the polymer molecules into smaller fragments; this results in a drop of viscosity.

Effect of viscosity index improvers as function of the viscosity index of the base oils will be discussed during the conference presentation.

3.1.4.3.2   Pour Point Depressants

Normal paraffin hydrocarbons tend to form waxy crystals at moderately low temperature. Usually, they are not present in engine oil base stocks. Other components such as isoparaffins, alkylnaphthenes, alkylaromatics and alkyl-naphthene-aromatics which are present in the petroleum engine base stocks show much less tendency to crystallize.

Pour point depressants are organic compounds which lower pour point of the oil by retarding the formation of full-size wax crystals by coating or co-crystal-lization with the wax. The pour point depressants have no effect on the precipi-tation temperature (cloud point), the amount and the crystal lattice of the separated wax. They only change the external shape and size of the crystals. Spherical crystals are formed instead of needles and thin platelets. Such change diminishes the ability of the wax crystals to overlap and interlock to form large conglomerates of wax which would impede the flow of the oil.

The majority of depressant additives include polymerization and condensation products. Some of them act the same time as viscosity index improvers. The main products used for this application encompass:

— Alkyl methacrylate polymers and copolymers

— Alpha-olefin polymers and copolymers

— Vinyl carboxylate-dialkyl fumarate copolymers.

The molecular weight range of polymers effective as pour point depressants is generally below that of polymers used as VI improvers and is usually in the range of 5,000 to 100,000.

Wax alkylated naphthalene and long chain alkyl-phenols and phthalic-acid dialkylaryl esters have also been used as depressants.

In paraffinic oils depressants show a better effect than in napththenic oils. This item will be discussed in detail during the conference presentation.

3.1.4.4 Corrosion Inhibitors

The function of an oxidation inhibitor is to minimize the formation of organic peroxides, acids and other oxygenated materials which deteriorate engine oils. Thus, it also acts as a corrosion inhibitor. Consequently one can say that corrosion inhibitors just enhance the function of antioxidants.

These additives protect bearing and other metal surfaces from corrosion, which also applies to the decomposition of nonferrous alloys. For example, engine copper-lead bearings are structured so that small pockets of lead occur dispersed throughout the continuous copper phase and the surface. Oxidative corrosion of the bearing occurs in two steps (4):

— In the first step the lead at the surface of the bearing is oxidized by the peroxides formed in the fuel and oil;

— In the second step the oxidation products, such as organic acids, dissolve away the lead oxide renewing the surface for reoxidation.

161

High end-point and unleaded gasolines aggravate copper-lead bearing corrosion, suggesting that the fuel hydrocarbons are the preliminary instrument of corrosion, modern engines have bearings provided with a very light tin overlay on the surface.

Corrosion inhibitors form an adsorbed protective film on metal surfaces which prevents contact between corrosive agents:

- acids
- peroxides
- others

and base metal. The adsorbed protective film stops also the catalytic effect of metals on oxidation.

The film formed by corrosion inhibitors must adhere tightly to bearing surface to avoid it be removed by dispersants or detergents and expose the underlying metal surface to attack by acidic components in the engine oil.

Corrosion inhibitors include:

- Metal dithiophosphates
- Sulfonates
- Metal dithiocarbonates
- Sulfurized terpenes
- Sulfurized olefins
- Many other compounds.

Since metal corrosion and oil oxidation inhibitors are closely related to one another they are sometimes discussed together (32).

3.1.4.5 Other Additives But Tribological Ones

3.1.4.5.1 Rust Inhibitors

The term is used to designate materials which protect ferrous metals against rust. Mostly it relates to the formation of hydrated iron oxide. Rust inhibitors prevent water from penetrating the protective oil film. This is achieved by application of polar molecules which are adsorbed preferentially on the metal surface and serve as barrier against water. To be effective, the additive molecules have to adsorb tightly on the iron surface and form a very stable film.

Rust inhibitors used in engine oils include polar compounds such as sulfonates, amine phosphates, esters, ethers, and derivatives of dibasic acids. Calcium and magnesium sulfonate used as detergents also provide antirust characteristics. Additionally, the overbased detergents neutralize the acids which catalyze rusting.

162

The effectiveness of rust inhibitors is controlled by the alkyl chain length of the additives. This relates to the fact that decrease in the size of the alkyl groups increases the tendency of the additive molecules to come out of solution and adhere on the iron surface.

3.1.4.5.2 Foam Inhibitors

Strong foaming affects the lubricating properties of engine oils and decreases their oxidation stability due to the intensive mixing with air. Strong foaming which relates to the splashing action of the crankcase and connecting rods can also lead to the oil transport in circulation systems.

The most universally used foam inhibitors are liquid silicones, especially polydimethylsiloxanes. In order to achieve a maximum effect, the silicones must be insoluble in the oil. They have to be finely dispersed in order to be sufficiently stable and must have a lower surface tension than the oil. In order to obtain stable dispersions of silicones, the droplet size must not exceed $10 \, \mu m$.

They function by attacking the oil film surrounding each bubble and thereby reducing interfacial tension so that the film breaks. Consequently, the small bubbles liberated combine to form large ones which float to the surface.

3.1.4.6 Tribological Additives

3.1.4.6.1 Introduction

In the regime of fluid film lubrication there is no contact between solids. The thickness of the film that supports the load is governed by the lubricant viscosity. However, when the severity of operating conditions increases (high load, low speed, high surface roughness), a point is eventually reached where the load can not longer be carried completely by the fluid film. Asperities of the solids have to share with the fluid film in load support. The lubrication fluid film regime shifts to mixed film and then to complete boundary lubrication. The situation is presented in Fig. 3.1.1. The contact of the solids involves wear, increased friction and welding of asperities. To reduce friction and wear and prevent damage of the mating surfaces tribological additives were developed.

Tribological additives present an extremely important group of chemicals incorporated into engine oils. They are also called boundary additives. The tribological additives encompass organic, metal-organic (organometallics), and inorganic compounds. They may function as friction modifying (FM), antiwear (AW), and extreme pressure (EP) additives. EP additives are also called load-carrying agents. Performance of these additives depends on chemical structure of the additives and the composition of the base stock used.

163

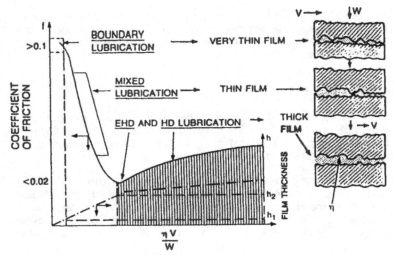

Figure 3.1.1 : Lubrication Regimes

3.1.4.6.2 Friction Modifying Additives

Friction modifying additives can be described as chemicals that allow to reduce coefficient of friction and achieve smooth sliding or to increase coefficient of friction and achieve no sliding. Usually, they increase oil film strength and thereby keep metal surface apart and prevent oil film breakdown. FM additives that reduce coeeficient of friction conserve energy. They are mostly applied in engine oils and automotive engine drive-train gear oils. These additives provide 3-4 percent improvement of fuel economy in automotive vehicles. Generally, they are used when smooth sliding with no vibration and minimum coefficient of friction is needed.

Fig. 3.1.2 presents the effect of FM additive on gasoline engine friction losses in a motored engine. This figure clearly shows that at high-crankcase temperature, low viscosity of the lubricant increases engine friction losses and the FM additive reduces these losses effectively. However, at low-crankcase temperature the FM has little effect; it is due to higher lubricant viscosity that reduces boundary lubrication. Screening tests described in (34) can provide guidance in developing fuel-efficient oils. Problems connected with fuel economy improvement by modification of oils have been broadly discussed in many papers (35 − 48). It is also to note that engine power loss can be reduced through improved filtration of motor oil (49 − 50).

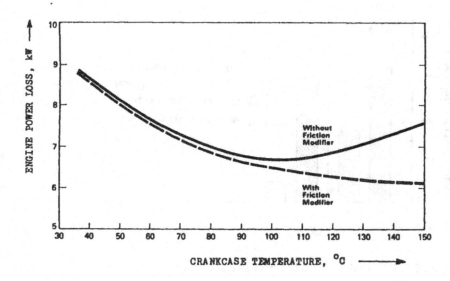

Figure 3.1.2:   Effects of Friction Modifying Additive in Engine Oil (33)

Fig. 3.1.3 (48) summarizes the results of two experiments on friction reduction. The friction reduction by a FM additive is equal to about a 3 percent fuel saving. A 6-cylinder engine was used for these tests (51). The engine was operated without combustion; it was driven by a balancing engine. The temperature of the CC quality engine oil was controlled through the cooling system of the engine. The friction was measured by means of a weight balance (51) with an accuracy of ±0.25 percent. The FM additive was added to the lubricant in the engine. Experiment A demonstrates that the original friction status of the engine was improved using the FM additive. In experiment B, air cleaner fine tests dust ACFTD was used as a contaminant (Fig. 3.1.3, Bar 3). It was also found that adding other types of very effective friction modifiers, such as PTFE or molybdenum disulfide, had no additional influence on the friction. These additional tests were carried out using the same 6-cylinder engine a year later. These results also show a fairly good repeatability of the measurements. Further decreasing of friction can be achieved by controlling the particle filtration (Fig. 3.1.3, Bar 4).

FM additives reducing friction coefficient include many compounds containing oxygen and nitrogen, molybdenum, copper and others. These additives increase oil film strength mostly by physical adsorption and thereby reduce friction. The film strength is connected with the length of the alkyl chain of FM additive molecules. Thus, as the FM additives long-chain compounds such as fatty acids, fatty alcohols, fatty esters or fatty amines and amides are used. FM additives also encompass sulfur containing compounds, e.g. sulfurized fats, oil soluble molybdenum sulfur compounds, molybdenum disulfide and graphite.

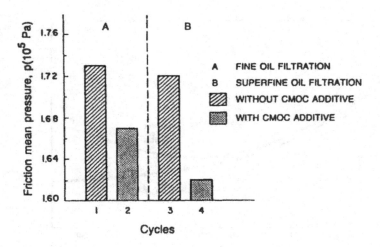

Figure 3.1.3:   Friction of a 6-Cylinder Diesel Engine (47)

FM additives first of all adsorb on the metal surface and remain in place to separate mating surfaces. Generally, they are effective when the loading is not too excessive, i. e. where penetration of the oil film by surface asperities is not significant.

It was shown (52) that under boundary lubrication conditions, the friction coefficient is reduced by increasing the additive concentration and approaches a minimum friction coefficient at higher concentrations. The following order of increasing minimum friction coefficient was obtained for various long chain additives (52), as compared with stearic acid:

C_{18} acids: stearic $<$ elaidic $<$ oleic

terminal groups on C_{18} chain-

$-COOH$ $<$ $-SH$ $<$ $-OH$ $<$ $-Br$

$-COOH$ $<$ $-NH_2$ $<$ CN $<$ $-CONH_2$

$-COOH$ $<$ $-COOCH_2 DH_3$ $<$ $-COOCH_3$

According to (52) the ranking order within each group of single compounds was explained by the chain dispersion interactions; it means that the greater the chain dispersion interaction the lower the minimum friction coefficient which relates to the friction coefficient at monolayer coverage.

166

The thickness and hence the effectiveness of the adsorbed film of FM additives that reduce coefficient of friction is a function of several variables (53):

— Polar group- the stronger the polarity, the greater the thickness and tenacity of the adsorbed film.

— Chain length- the longer the chain, the thicker the adsorbed film.

— Configuration- slender molecules allow closer packing than bushy or squat ones; this affects packing density.

— Base oil- both the size of the base molecule and the presence of polar competitors can influence film formation.

— Concentration- the higher the concentration the higher the FM additive effect, up to a point; then it could become uneconomical or tending to produce results opposite those desired.

— Metallurgy- the higher the heat of adsorption on the particular metal, the stronger the film; the oxide coating on the metal surface is the key.

— Temperature- too high a temperature might provide enough energy to desorb the adsorbed molecules and thus weaken the film.

A huge number of FM additives is disclosed in the literature, notably the patents around the world, and a large part of it is actually in use.

3.1.4.6.3   Antiwear Additives

A group of tribological additives that are effective in the mixed lubrication region, where penetration of the lubricant film by surface asperities is intermittent. In localized metallic contacts on the rubbing surfaces these additives chemisorb and react with metal to form a surface compound which is deformed by plastic flow to allow a new distribution of load. AW additives encompass several classes of chemicals. Perhaps the most important and most effective AW additive for controlling or eliminating wear in the valve train system is zinc dialkyldithiophosphate. Other important AW additives are organic phosphorus compounds tricresyl phosphate, dilauryl phosphate, didodecyl phosphate and compounds containing suflur, sulfurized terpenes or combinations of sulfur and oxygen sulfurized fats and fat derivatives. Since some FM additives may reduce wear, AW additives involve also compounds containing oxygen.

The general function of AW additives may be depicted in Fig. 3.1.3. They are responsible for the formation of very complex intermediate compounds, especially from chemicals containing sulfur, oxygen, and phosphorus, e.g. zinc dialkyldithophosphates. Fig. 3.1.5 shows that decomposition reactions of

ZnDDP yield both products that form deposited antiwear layer and new compounds. The latter, in turn, undergo further reactions which may produce antiwear and, to some extent, antiseizure reaction layer with the surface. However, for many compounds chemisorption is the main mechanism of their antiwear action. Chemisorption occurs when the adsorbed compound reacts chemically with the metal surface without removal of this surface. Chemisorption may be treated as the bridge between antifriction and antiwear functions of some substances for example, fatty acids. In the case of fatty acids the physically adsorbed film can reduce friction and may be converted to a metal soap which is anchored in the surface. After the chemisorption reaction the film has much higher resistance to shear than the adsorbed film, thus, apart of maintaining its antifriction property it also provides an antiwear property.

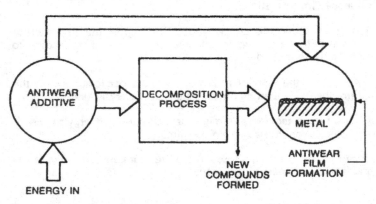

Figure 3.1.4:   General Action Mechanism of Antiwear Additives

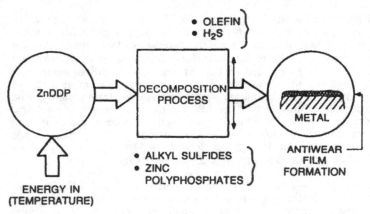

Figure 3.1.5 :   General Antiwear Action Mechanism of ZnDDP

168

Similarly, chemical reaction may be treated as the bridge between antiwear and antiseizure functions of some substances, for example, disulfides. Disulfides in the chemisorbed form function as antiwear additive —— it is connected with the mercaptide layer formation —— but in the form of iron sulfide the action is associated with both antiwear and antiseizure functions.

Good examples of tribological additives combining both friction and wear reducing functions are oil soluble molybdenum-sulfur compounds. Compounds of this group, e.g. molybdenum dithiophosphates and molybdenum dithiocarbamates are widely used as additives for reducing friction and wear and increasing load-carrying capacity. In addition, they are well known as excellent antioxidants. Friction and wear characteristics of these compounds are described in many publications (54 — 57).

3.1.4.6.4 *Extreme Pressure Additives*

These are a group of tribological additives that prevent seizure and welding between metal surfaces working under severe operating conditions. Usually, they control damage when the number of metallic contacts increases and seizure takes place. Increasing load-carrying capacity of these additives may be associated with a decrease or an increase of the wear. It is connected with the fact that EP additives usually are effective only by chemical reactions. Thus, their use involves possible corrosion problems. As these additives are not of specific importance for engine oils they will not be discussed here in detail.

3.1.5   Interactions of Engine Oil Components

Modern motor oil formulations contain may kinds of additives to achieve the desired oxidation stability, viscosity temperature, detergency-dispersancy and corrosion prevention properties. Table 3.1.2 gives an example of additive formulation for SF/CC 10W-30 passenger car oil. The antiwear property is usually provided by ZnDDP which also supply antioxidant and corrosion inhibitor functions.

While the properties of each additive are well characterized, the interactions of the additives among themselves and with the polar molecular compounds in the base stocks (58) are not always understood. It was shown (59) that interactions of the additives and lubricating base stocks are complex and the mechanisms of interactions at low temperature appear to be different from high temperature catalyzed conditions. However, the interaction studies (59) in oxidation stability among ZnDDP, calcium sulfonate, and succinimide in paraffins, high sulfur and low sulfur light neutral base oils provide very interesting information:

- Base oil has a significant effect on oxidation stabilities as a result of the interactions.
- The sulfonate appears to be a prooxidant but the effects may be minimized under optimum conditions.
- Succinimide, while synergistic in binary system, appears to be antagonistic with sulfonate. Sulfur in base oils may be advantageous in providing some synergistic effects.
- Within the complex interaction network, there is a concentration dependence. Often an optimum combination of various additives occurs in terms of oxidation stability. Thus, new additive components and formulations need careful concentration study.
- Different oxidation levels were observed for different base stocks under the same additive combinations. While some interactions and the interaction responses sometimes are significantly different.
- No sufficient data base exists at present to develop a generalized theory to define the optimum.

Table 3.1.2: SF/CC Additive Treatment for Passenger Car Service (4)

Additive	% wt. in Oil
DISPERSANT Polymer Amine	4.0
RUST INHIBITOR—ALKALINE AGENT High Base Sulfonate	1.0
ANTIOXIDANT—ANTIWEAR ZnDDP — A ZnDDP — B	0.8 0.5
PRIMARY DETERGENT Low Base Sulfonate	0.5
ANTIOXIDANT Sulfurized Hydrocarbon	0.5
FRICTION REDUCER Sulfurized Fat	0.7
VISCOSITY INDEX MODIFIER Ethylene-Propylene Copolymer	10.0
	18.0

Other additives in engine oils affect the ZnDDP performance. Usually, highly polar agents can actively compete with this additive for the friction surfaces. Four-ball studies described in (60) have clearly demonstrated that other additives in the blend can have large effects on ZnDDP antiwear performance. Detrimental effects on wear were obtained with primary alkyl amine friction modifiers, metallic dithiocarbamate oxidation inhibitors and a basic barium sulfonate detergent rust inhibitor. On the other hand, detergent-dispersants, viscosity index improvers, and oxidation inhibitors were found to have litte or no effect. Corresponding data (60) for all the second additives are summarized in Fig. 3.1.6. The bars in Fig. 3.1.6 represent the range of results observed for a given type of additives.

ZnDDP CONCENTRATION REQUIRED TO REDUCE WSD TO 0.30 mm

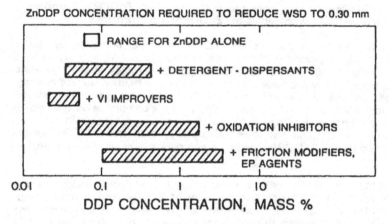

DDP CONCENTRATION, MASS %

Figure 3.1.6: Change in ZnDDP Antiwear Performance Caused by Presence of Other Additives (60)

Since adverse interactions of ZnDDP with other additives in an engine oil is one factor contributing to the high phosphorus content of some engine oils, ways to reduce this adverse interactions — on an example of ZnDDP-amine interaction — was investigated (61). It is associated with the well-known fact that phosphorus poisons exhaust gas emission catalyst. For commercial ZnDDP — three alkyl and one aryl zinc dithiophosphates — were evaluated. Antiwear performance was measured in four-ball wear tests. The amine selected for this study was lauryl amine friction modifier. The amine concentrations were varied independently over a range from 0 to 5 mass percent. The most important results described in (61) might be summarized as follows:

— For a given zinc dithiophosphate (ZnDP) concentration, there was a critical amine concentration at which there was a rapid loss of the ZnDP antiwear benefit with increasing amine concentrations.

171

— This critical amine concentration was found to be independent of both load and temperature for ranges studied.

— Organic acids reduced the detrimental effect of the amine on essentially a mole for a mole basis.

Other work (62) investigated the coadsorption of ZnDDP and sulfonate or succinimide onto iron. Recently, the relationship between the relative magnitude of the changes in zinc diisobutyldithiophosphate adsorption resulted from the presence of various classes of additives typically used in engine oils has been determined. These results were correlated with those described in (60, 61). It was found that the reducing effect of the presence of sulfonates, succinimides and dibenzyldisulfide on the magnitudes and rates of ZnDDP adsorption is also in a good correlation with decreasing ZnDDP AW performance.

To account for the interaction among ZnDDP and other additives, especially for amines, it was suspected (62) that the second additive effectively cancels the effect of the ZnDDP by forming some type of complex in the bulk fluid that makes both additives unavailable for forming an effective surface coating. Recently (64) complex formation from ZnDDP and amines was proven by isolation and identification of solid crystals. Chelate type complexes were found for short-chain diamines and bridge type was isolated from long-chain diamine. Complex from monoamine has lower adsorptivity than free ZnDDP; thus the power adsorptivity of ZnDDP/monoamine complex makes its antiwear property worse.

Wear data obtained in (65) with samples of engine oil from service indicate that the products of the deterioration reactions of ZnDDP are in a fully formulated engine oil in normal service ineffective as AW agents. It is very interesting that an addition of a mixed hydrocarbon/polyolester base stock resulted in increased wear at low concentrations of ZnDDP. This effect is mitigated by the addition of overbased detergent. On the other hand, a hindered phenol such as 2,6-di-tert-butyl-4-methylphenol can not effectively protect ZnDDP from degradation as an AW additive. It was suggested (65) that the increased wear may conceivably be due to a corrosion process involving reaction of the steel surface with hydrogen sulfide which is generated from ZnDDP upon heating in the presence of esters.

The effect of adding of ZnDDP mineral oil solution to a mixed polyol ester/n-hexadecane base stock on wear and tribochemical reactions of ZnDDP were studied recently (66). It was found that in this system the characteristic feature of ZnDDP tribochemical reactions is formation of disulfide and lack of surface sulfur products in the initial period of the wear test. Lack of sulfur products on the rubbing surfaces coincides with sharp increase in disulfide formation and with the highest wear rate. On the base of these observations it seems that the higher wear rates at lower ZnDDP concentrations in the studied oil system are connected with corrosion wear due to changes of the ester.

To understand the mechanism of this interaction further tests and analytical work have been made (67).

Some more detailed results relating to the interactions of engine oil components especially from the point of view of their action mechanism, will be provided and discussed during presentation of this paper.

3.1.6 References

(1) SAE Handbook: Fuels and Lubricants. Warrendale, Pa.: Society of Automotive Engineers 1987, P. 23.01–23.17.
(2) O'Brien, J.A.: Lubricating Oil Additives. In: CRS Handbook of Lubrication. Vol. 2. Boca Raton, Florida: CRS Press Inc. 1986, P. 301–315.
(3) Watson, R.W.: The Role of Alkyl Groups in Petroleum Additives. Tokyo 1975. Proc. JSLE-ASLE International Conference. Jap. Soc. of Lubrication Engineers. 1975.
(4) Watson, R.W.; McDonnel, Jr., T.F.: Additives — the Right Stuff for Automotive Engine Oils. In: Fuels and Lubricants Primer for Automotive Engineers. SP-671. Warrendale, Pa.: SAE 1986, P. 17–28.
(5) Coates, J.P.; Setti, L.C.: Infrared Spectroscopic Methods for the Study of Lubricant Oxidation Products. ASLE Trans. 29 (1986) 394–401.
(6) Korcek, S.; Mahoney, L.R.; Johnson, M.D.; Siegl, O.W.: Mechanism of Antioxidant Decay in Gasoline Engines: Investigations of ZnDDP Additives. SAE Technical Paper Series Paper No. 810014. Warrendale, Pa.: SAE 1981.
(7) Howard, J.A.; Ohkatsu, Y.; Chenier, J.H.B.; Ingold, K.U.: Metal Complexes as Antioxidants, I. The Reaction of ZnDDP and Related Compounds with Peroxy Radicals. Can. J. Chem. 51 (1973) 1543–1553.
(8) Howard, J.A.: Inhibition of Hydrocarbon Autooxidation by Some Sulfur Containing Transition Metal Complexes. In: Frontiers of Free Radical Chemistry. New York: Academic Press 1980, P. 237–282.
(9) Bridgewater, A.J.; Dever, J.R.; Sexton, M.D.: Mechanisms of Antioxidant Action. Part 2. Reactions of ZnDDP and Related Compounds with Hydroperoxides. J. Chem. Soc. Perkin II (1980) 1006–1016.
(10) Kennerly, G.W.; Patterson, W.L.: Kinetic Studies of Petroleum Antioxidants. Ind. Eng. Chem. 48 (1956) 1917–1924.
(11) Klaus, E.E.; Cho, L.; Dong, H.: Adaptation of the Penn State Microoxidation Test for Evaluation of Automotive Lubricants. SAE Paper 801362 presented at 1980 Fuels and Lubricants Meeting, Baltimore, 1980.
(12) Willermet, P.; Mahoney, L.R.; Bishop, C.M.: Lubricant Degradation and Wear II. Antioxidant Capacity and IR-Spectra in Systems Containing ZnDDP. ASLE Trans. 23 (1980) 217–224.
(13) Mahoney, L.R.; Otto, K.; Korcek, S.; Johnson, M.D.: The Effect of Fuel Combustion Products on Antioxidant Consumption in a Synthetic Oil. Ind. Eng. Chem., Prod. Res. Dev. 19 (1980) 11–15.
(14) Johnson, M.D.; Korcek, S.; Zinbo, M.: Inhibition of Oxidation by ZnDDP and Ashless Antioxidants in the Presence of Hydroperoxides at 160°C — Part I. SAE Paper 831684 presented at 1983 Fuels and Lubricants Meeting, San Francisco, 1983.

173

(15) Korcek, S.; Johnson, M.D.; Jensen, R.K.; Zinbo, M.: Assessment of High Temperature Antioxidant Capabilities of Engine Oils and Additives. Ostfildern 1986. P. 3.2-1-3.2-10. Proc. Additives for Lubricants and Operational Fluids. Technische Akademie Esslingen. 14.–16.01.86. Ostfildern.

(16) Bray, W.B.; Dickey, C.R.; Voormes, V.: Dispersions of Insoluble Carbonates in Oil. Ind. Eng. Chem., Prod. Res. Dev. 14 (1975) No. 4.

(17) Gergel, W.C.: Detergents – What Are They. Tokyo 1975. Proc. JSLE-ASLE International Lubrication Conference. Jap. Soc. of Lubrication Engineers. 1975. Tokyo.

(18) Raddatz, J.; Bartz, W.J.: Detergent-/Dispersantadditive – Herstellung, Wirkungsweise und Anwendung. Ostfildern 1986. P. 9.1-1 – 9.1-20. Proc. Additives for Lubricants and Operational Fluids. Technische Akademie Esslingen. 14.–16.01.86. Ostfildern.

(19) Heldeweg, R.F.: Salicylates as Detergent Additives for Engine Lubricants. Ostfildern 1986. P. 9.2-1 – 9.2-14, ditto.

(20) Tricaud, C.; Hipeaux, J.C.; Lemerle, J.: Micellar Structure of Alkaline Earth Alkylarylsulfonate Detergents. Ostfildern 1986. P. 9.3-1 – 9.3-7, ditto.

(21) Zalai, A.: Effect of Detergent-Dispersant Additives. Ostfildern 1986. P. 9.4-1 – 9.4-8, ditto.

(22) Klamann, D.: Schmierstoffe und verwandte Produkte. Weinheim, Deerfield Beach; Basel: Verlag Chemie, 1982.

(23) McGeehan, J.A.; Rynbrandt, J.D.; Hansel, T.J.: Effect of Oil Formulations in Minimizing Viscosity Increase and Sludge Due to Diesel Engine Soot. Ostfildern 1986. P. 9.5-1 – 9.5-13. Proc. Additives for Lubricants and Operational Fluids. Technische Akademie Esslingen. 14.–16.01.86. Ostfildern.

(24) Schoedel, U.: Viskositaets-Index-Verbesserer: Typen und Wirkungsweise. Ostfildern 1986. P. 8.1-1 – 8.1-12, ditto.

(25) Neudörfl, P.: Der Einsatz von Polymethacrylaten in Schmierölen aus heutiger Sicht. Ostfildern 1986. P. 8.2-1 – 8.2-15, ditto.

(26) Neven, C.; Huby, F.: Solutions Properties of Polymethacrylate VI-Improvers. Ostfildern 1986. P. 8.3-1 – 9.3-14, ditto.

(27) Nemes, N.; Kovacs. M.: Untersuchung der Wirkung von fließverbessernden Additiven. Ostfildern 1986. P. 8.4-1 – 8.4-13, ditto.

(28) Eckert, R.J.A.; Covey, D.F.: Development in the Field of Hydrogenated Diene Copolymers as VI Improvers. Ostfildern 1986. P. 8.5-1 – 8.5-13, ditto.

(29) Denis, J.: Pour Point Depressants in Lubricating Oils. Ostfildern 1986. P. 8.6-1 – 8.6-12, ditto.

(30) Spiess, G.T.; Johnson, J.E.; VerStrate, G.: Ethylene Propylene Copolymers as Lube Oil Viscosity Modifiers. Ostfildern 1986. P. 8.1-1 – 8.10-11, ditto.

(31) Marsden, K.: Literature Review of OCP Viscosity Modifiers. Ostfildern 1986. P. 8.11-1 – 8.11–10, ditto.

(32) Hamblin, P.C.; Kristen, U.: Areview: Ashless Antioxidants and Corrosion Inhibitors. Their Use in Lubricating Oils. Ostfildern 1986. P. 7.3-1 – 7.3-27, ditto.

(33) Passut, C.A.; Kollman, R.E.: Laboratory Techniques for Evaluation of Engine Oil Effects on Fuel Economy. SAE Paper 78061 (1978).

(34) Chamberlin, W.B.; Saunders, J.D.: Screening Tests Used for Developing Fuel Efficient Oils. ASLE Paper 79-AM-2C-2 (1979).

(35) Badioli, F.L.; Cassiani-Ingoni, A.A.; Pusateri, G.: Friction Power Loss of Mineral and Synthetic Lubricants in a Running Engine. SAE Paper 780376 (1978).

(36) McGeeham, J.A.: A Literature Review of the Effect of Piston and Rig Friction and Lubricating Oil Viscosity on Fuel Economy. SAE Paper 780673 (1978).

(37) O'Conner, B.M.; Graham, R.; Glover, I.: European Experience with Fuel Economy Gear Oils. SAE Paper 790746 (1979).

(38)   Haviland, M.L.; Goodwin, M.C.: Fuel Economy Improvements with Friction-Modified Engine Oils in Environmental Protection Agency Road Tests. SAE Paper 790945 (1979).

(39)   Papay, A.G.; Rifkin, E.B.; Shubkin, R.L.; Dawson, R.B.; Jackisch, P.F.: Advanced Fuel Economy Engine Oils. SAE Paper 790947 (1979).

(40)   Bartz, W.J. Kraftstoffeinsparnung durch Reibungsminderung bei Motoren- und Getriebeölen. MTZ 41 (1980), 1, 7-12.

(41)   Bartz, W.J.: Mögliche Energieeinsparung durch tribologische Maßnahmen. Erdöl und Kohle 33 (1980), 78-87.

(42)   Bartz, W.J.: Improving Fuel Economy by Friction Reducing Engine and Gear Oils. Proc. AGELFI European Automotive Symp. (1980), 313—342.

(43)   Bartz, W.J.: Some Considerations Regarding Fuel Economy Improvement by Engine and Gear Oils. ASLE Paper 81-LC-4B-2 (1981).

(44)   Damrath, J.G.; Papay, A.G.: Fuel Economy Factors in Lubricants. SAE Paper 821226 (1982).

(45)   Bush, K.P.; Robert, D.C.; Villis, T.: European Test Methods for Fuel Economy Oils. SAE Paper 831741 (1983).

(46)   Irvin, R.F.; Fernandez, I.: Energy Conserving Oil-Techniques and Technology. SAE Paper 830164 (1983).

(47)   Kajdas, C.; Fodor, J.: Action Mechanism of Load-Carrying Performance of a New Lubricant Additive. Tribol. Trans. 31 (1988) 476—80.

(48)   Kuo, L.L.K.; Chang, S.T.; Hsiek, S.K.; Chu, C.Y.; Tung, C-Y.: Fuel Economy Engine Oils via Friction Modifiers. Lubrication Engineering 45 (1989) 81—90.

(49)   Fodor, J.: Improving the Economy of IC Engines by Controlling the Contaminants. Proc. World Filtration Congress III, Philadelphia, 1982. P. 707—710.

(50)   Fodor, J.; Ling, F.F.: Friction Reduction in an IC Engine Through Improved Filtration and a New Lubricant Additive. Lubr. Eng. 41 (1985) 614—618.

(51)   Pasztor, E.; Levai, L.: Report on the Inner Friction Measurement of a 6-Cylinder Diesel Engine. Techn. Univ. of Budapest. Report No. 497015 (1983).

(52)   Jahanmir, S.; Beltzer, M.: Effect of Additive Molecular Structure on Friction Coefficient and Adsorption. Journal of Tribology 108 (1986) 109—116.

(53)   Papay, A.G.: Friction Reducers for Engine and Gear Oils — A Review of the State of the Art. Ostfildern 1986. P. 5.1-1 — 5.1-10. Proc. Additives for Lubricants and Operational Fluids. Technische Akademie Esslingen 14.—16.01.86. Ostfildern.

(54)   Braithwaite, E.R.; Greene, A.B.: A Critical Analysis of the Performance of Molybdenum Compounds in Motor Vehicles. Wear 46 (1978) 405—432.

(55)   Mitchel,, P.C.H.: Oil-Soluble Mo-S Compounds as Lubricant Additives. Wear 100 (1984) 281—300.

(56)   Yamamoto, Y.; Gondo, S.: Frictional Characteristics of Molybdenum Dithiophosphates. Wear 112 (1986) 79—87.

(57)   Yamamoto, Y.; Gondo, S.: Friction and Wear Characteristics of Molybdenum Dithicoarbamate and Molybdenum Dithiophosphate. Tribology Trans. 32 (1989) 251—257.

(58)   Hsu, S.M.; Ku, C.S.; Lin, R.S.: Relationship Between Lubricating Basestock Composition and the Effects of Additives on Oxidation Stability. SAE SP 526 (1982).

(59)   Hsu, S.M.; Lin, R.S.: Interactions of Additives and Lubricating Base Oils. SAE Paper 831683 (1983).

(60)   Rounds, F.G.: Additive Interactions and Their Effect on the Performance of a ZnDDP. ASLE Trans. 21 (1978) 91—101.

(61)   Rounds, F.G.: Some Effects of Amines on ZnDDP AW Performance as Measured in 4-Ball Wear Tests. ASLE Trans. 24 (1981) 431—440.

175

(62) Inoue, K.; Watanabe, H.: Interactions of Engine Oil Additives. ASLE Trans. 26 (1983) 189—199.

(63) Plaza, S.: The Effect of Other Lubricating Oil Additives on the Adsorption of ZnDDP on Fe and Fe_2O_3 Powders. ASLE Trans. 30 (1987) 241—247.

(64) Shiomi, M.; Tomizawa, H.; Kuribayashi, T.; Tokashiki, M.: Interaction Between ZnDDP and Amine. Ostfildern 1986. P. 3.7-1 — 3.7-10. Proc. Additives for Lubricants and Operational Fluids. Technische Akademie Esslingen 14.—16.01.86. Ostfildern.

(65) Willermet, P.A.; Mahoney, L.R.; Bishop, C.M.: Lubricant Degradation and Wear III. ASLE Trans. 23 (1980) 225—231.

(66) Plaza, S.: Effect of Ester/Mineral Oil Compositions on Tribochemical Reactions of a ZnDDP. Ostfildern 1988. P. 4.5.1-8. Proc. Industrial Lubricants. Technische Akademie Esslingen 12.—14.01.88. Ostfildern.

(67) Plaza, S.; Kajdas, C.: Tribochemical Reactions of ZnDDP in Ester — Mineral Oil Solution Under Boundary Lubrication Conditions. Paper to be submitted for publication.

3.2 Effects of NO_x on Liquid Phase Oxidation and Inhibition at Elevated Temperatures

S. Korcek and M.D. Johnson
Ford Motor Company, Dearborn, USA

Abstract

Nitrogen oxides, NO_x, play a role in degradation of lubricating oils in internal combustion engines. It is well established that NO_x reactions contribute to sludge and varnish formation (1-9) and to oil oxidation and thickening (10, 11, 12). Therefore, evaluation of engine oils in engine test, such as ASTM Sequences VE (13) and IIIE (14), is performed under conditions producing high levels of NO_x or high levels of blowby in order to accelerate degradation and increase test severity.

Nitrogen oxides, formed during combustion, come into contact and react with oil components in the piston/cylinder areas and enter the crankcase environment as a part of blowby gases which leak past the rings. The blowby gases travel through the engine and come into contact with the oil in different locations depending upon the type of ventilation system.

Despite the fact that a causal relationship between NO_x and engine oil degradation, insolubles accumulation, and sludge and varnish formation has been established, the mechanisms by which these effects occur have not been fully elucidated.

It is generally accepted that deposit formation in engines is initiated by formation of precursors resulting from interactions of nitrogen dioxide, NO_2, with unsaturated fuel derived components of blowby (1, 5). Interactions of NO_x with additive systems and base oil components have not, however, been investigated in detail as a source of lubricant degradation and deposit formation. Therefore, this work was aimed at developing an understanding of such interactions using a model system. Hexadecane was utilized as a model lubricant and 2,6-di-tert-butyl-4-methylphenol (MPH) as a radical trapping antioxidant. This model system was oxidized in the presence of NO_2. The effects of NO_2 on oxidation processes and the fate of NO_x in these systems have been investigated. The significance of results obtained in this study with respect to mechanisms of oil oxodation in internal combustion engines is also discussed.

3.2.1 Blowby Composition

The principal NO_x components in blowby are NO and NO_2 which co-exist in a temperature dependent equilibrium, equation 1. NO predominates at high temperature while a shift towards NO_2 occurs at lower temperatures.

$$2NO + O_2 \rightleftharpoons 2NO_2 \qquad\qquad (1)$$

Thus, the principal component in exhaust is NO due to the high temperatures and low oxygen levels. At the lower temperatures and higher oxygen contents found in the crankcase the fraction of NO_x from blowby which is NO_2 increases.

The content and composition of nitrogen oxides in crankcase gases have been investigated previously (14, 3). Concentrations of NO and NO_2 in crankcase and exhaust were measured as a function of operating variables. A higher ratio of NO_2 to NO was found in blowby than in the exhaust. NO_x levels in blowby were, however, lower than what was predicted from the CO_2 content of the blowby. An attempt was made to explain the sources of NO_2 in crankcase gases. It was concluded that NO_2 concentrations in the crankcase cannot be estimated from NO and NO_2 concentrations in the exhaust.

Measurement of NO_x in crankcase gases underestimates the amount of NO_x to which the oil was exposed since some NO_x is consumed by reactions with the oil. Nitrogen contents of about 7000 ppm have been reported (8) in used oils after 288 hours ov VE testing. At a blowby rate of 2.0 cfm this corresponds to approximately 30 ppm NO_x reacting with and remaining in the oil. Since some compounds formed from NO_x reactions with the oil are thermally unstable and may release NO_x as a result of their decomposition, the nitrogen content of the oil may also underestimate the extent of NO_x reactions which are occurring.

Despite these uncertainties, analysis of blowby gives an indication of the amount of NO_x that comes in contact with the oil. The composition of blowby gases typically fall in the ranges reported in Table 3.2.1. This data is derived from the literatures (3, 14) and from our recent measurements.

Table 3.2.1: BLOWBY COMPOSITION

COMBUSTION PRODUCTS	15 – 20 %
AIR + FUEL	80 – 85 %
O_2	16 – 17 %
NO_x	8 – 500 ppm
NO_2	0.3 – 155 ppm

Results from measurements of NO_x in blowby in the Sequence VE test are listed in Table 3.2.2-

Table 3.2.2:   2.3 L — SEQUENCE VE *

	Speed	Load	Oil Temp.	NO_2	NO
	rpm	bhp	°F	ppm	ppm
Stage 1	2500	33.5	155	110 – 160	275 – 410
Stage 2	2500	33.5	210	100 – 150	320 – 350
Stage 3	750	1	115	40 – 105	60 – 85

* No EGR, modified ventilation, increased blowby rate

3.2.2   Experimental

The experiments described in this paper were conducted in a batch reactor at 160°C using apparatus and techniques described previously (17, 18). Hexadecane, 40 mL, alone or in combination with an antioxidant was oxidized by passing 200 mL/min of oxidizing gas mixture through the liquid. Prior to oxidation, the reactor was flushed with argon and the sample to be tested was placed in the reactor and purged with argon until it had reached test temperature. At that time the gas flow was switched from argon to the oxidizing gas mixture which consisted of 20 % O_2, 0 – 768 ppm NO_2, and the balance nitrogen.

Tests were conducted in two general ways. In one case, samples of reaction mixture were taken for analysis as a function of test time and in the other the content of NO_x in the gases exiting the reactor was measured. These two types of experiments when run under otherwise identical conditions are not equivalent since, in the one case, the volume of liquid in the reactor changes and results have been shown to be affected by volume of liquid in the reactor. The flow rate of oxidizing gas mixture also affects the results obtained in these experiments. Unlike oxidation experiments with pure oxygen where excess O_2 is used in order to keep the concentration in solution constant, the total amount of NO_2 available for reaction in these experiments, where low levels of NO_2 are introduced, is determined by the flow rate. Therefore, when comparisons are made, it is important to compare the same types of experiments performed under the same conditions.

Nitrogen oxides were analysed by chemiluminescence using a Monitor Labs Model 8430 oxides of nitrogen analyser and a Model 8750 Thermocon NO_2 to NO converter. The distance between the reactor and the NO_x analyser has been kept at a minimum; nevertheless, some uncertainty in determination of NO_x remains since additional reactions could occur between the reactor and the analyser. Further, even through NO_x plots have been adjusted for the delay time before the analyser begins to respond to NO_x, there is some uncertainty regarding the accuracy of initial points since some additional time is required for the sampling lines to be fully purged and stable readings obtained. The figures are labeled according to the following convention:

$$NO_x \text{ retained} = NO_x \text{ in} - NO_x \text{ out}$$
$$NO_2 \text{ consumed} = NO_2 \text{ in} - NO_2 \text{ out}$$
$$NO \text{ formed} = NO \text{ out} - NO \text{ in}$$

Hydroperoxides were determined iodometrically, hexadecane oxidation products by gas chromatography, and 2,6-di-tert-butyl-4-methylphenol using HPLC or GC.

3.2.3 Reactions of NO and NO_2

A review of the literature describing reactions of NO and NO_2 with hydrocarbons and their reaction intermediates (Appendix I) reveals that:

- NO_2 reacts as a free radical and abstracts hydrogen from hydrocarbons with formation of an alkyl radical (R·) and nitrous acid (HNO_2),

- HONO decomposes to give hydroxyl radical (·OH) and NO,

- both NO and NO_2 react with R· and alkylperoxyl radicals (RO_2·) with formation of nitro (RNO_2) and nitroso (RNO) compounds, nitrites (RONO), nitrates ($RONO_2$), and peroxynitrates ($ROONO_2$),

- NO_2 could also react with alkoxy radicals (RO·) and hydroperoxides (ROOH) with formation of $RONO_2$,

- RONO decomposes at elevated temperatures to form RO· and NO,

- NO can be converted into NO_2 by reaction with RO_2·,

- NO and NO_2 can react with water in the gas phase to form nitrous acid (HNO_2),

- upon heating, HNO_2 can decompose to form nitric acid (HNO_3).

As follows from the review of possible NO and NO_2 reactions, the concentration of NO and NO_2 in the oxidizing gas at the exit of the reactor will depend on availability of various oxidation intermediates in the liquid hydrocarbon and on the presence of other blowby components such as water in the oxidizing gas. In this work we have investigated oxidation in the absence of water.

180

3.2.4 Effects of Uninhibited Hexadecane Oxidation

The kinetics and mechanisms of hexadecane autoxidation at elevated temperatures have been extensively investigated (17 — 19). A simplified reaction scheme describing this free radical chain process includes initiation, propagation and termination reactions.

Initiation of oxidation occurs due to homolytic decomposition of hydroperoxides (ROOH) leading to formation of alkoxy (RO·) and hydroxyl (·OH) free radicals (Reaction 2). These free radicals are very reactive and immediately abstract hydrogen from hexadecane (RH) to produce alkyl radicals (R·) (Reactions 3 und 4). Alkyl radicals undergo very fast oxygen addition (Reaction 5) to give peroxy radicals (RO₂·).

$$
\begin{array}{lll}
\text{ROOH} & \longrightarrow & \text{RO·} + \text{·OH} & (2) \\
\text{RO·} + \text{RH} & \longrightarrow & \text{ROH} + \text{R·} & (3) \\
\text{·OH} + \text{RH} & \longrightarrow & \text{H}_2\text{O} + \text{R·} & (4) \\
\text{R·} + \text{O}_2 & \longrightarrow & \text{RO}_2\text{·} & (5)
\end{array}
$$

Propagation of oxidation occurs by a chain reaction process. Hydrogen abstraction from RH by RO₂· leads to formation of ROOH and R· (Reaction 6). This reaction, followed by reaction 5 which regenerates another RO₂·, is responsible for accumulation of hydroperoxides in the oxidizing system. Termination of oxidation occurs due to self reaction RO₂· (Reaction 7).

$$
\begin{array}{lll}
\text{RO}_2\text{·} + \text{RH} & \longrightarrow & \text{ROOH} + \text{R·} & (6) \\
\text{RO}_2\text{·} + \text{RO}_2\text{·} & \longrightarrow & \text{non radical products} & (7)
\end{array}
$$

Based on this reaction scheme, oxidation results can be conveniently displayed by plotting the square root of hydroperoxide concentration versus time. If the only source of initiation is homolytic decomposition of hydroperoxides (Reaction 2) a linear relationship is observed. This is illustrated in Figure 3.2.1 where extent of oxidation is expressed by the concentration of hexadecyl monofunctional hydroperoxides. With O_2 only (20 % O_2 + 80 % N_2), a linear plot with zero intercept is obtained. When NO_2 is added to the oxidizing gas mixture the initial rate of oxidation is significantly accelerated but the subsequent rate is apparently unaffected. In the initial stages NO_2 contributes to initiation probably due to free radical formation, as depicted in equations 8 and 9,

$$
\begin{array}{lll}
\text{RH} + \text{NO}_2 & \longrightarrow & \text{R·} + \text{HONO} & (8) \\
\text{HONO} & \longrightarrow & \text{HO·} + \text{NO} & (9)
\end{array}
$$

which is followed by reactions 4 and 5. Thus, per one molecule of NO_2 reacting with RH, two RO₂· radicals and one molecule of NO should be formed.

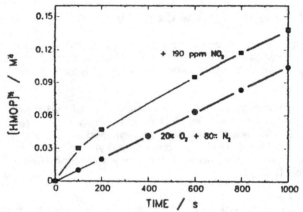

Figure 3.2.1: Uninhibited Hexadecane Oxidation

In later states, when the level of hydroperoxides and extent of oxidation begin to increase, the contribution of NO_2 to the rate of initiation apparently becomes negligible. This is either due to its contribution becoming small relative to initiation from hydroperoxide decomposition or due to depletion of NO_2 by other reactions. This same behaviour was observed with higher concentrations of NO_2 as well. The level of increased oxidation obtained during that initial period was comparable up to 792 ppm NO_2. Subsequent oxidation at higher reaction times was nearly independent of NO_2 concentration.

The extent of NO_2 reaction during the oxidation is shown in Fig. 3.2.2. Initially, the concentration of NO_2 in the exiting gas increases until it reaches its maximum value at about 200 s and then it gradually decreases. Correspondingly, the amount of NO_2 consumed (taken as the difference between the amounts of NO_2 entering and exiting the reactor) is first decreasing, reaches a minimum, and then increases. Since, in our system, about 100 s is needed to reach reaction equilibrium it can not be conclusively distinguished wether the observed decrease in NO_2 consumed is real. In spite of expectations based on reactions 8 and 9, formation of NO is not observed. Moreover, the small amount of NO initially present in the oxidizing gas is consumed during the initial stages of oxidation. This suggests that NO reacts with species produced upon oxidation.

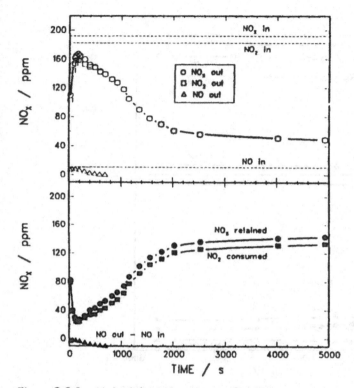

Figure 3.2.2: Uninhibited Hexadecane Oxidation

To compare NO_x reactions in the absence of oxygen (Fig. 3.2.3) and oxidation products, hexadecane (RH) was exposed to NO_2 under the same conditions, however, without O_2 present in the gas. The results of these experiments (Fig. 3.2.3) show that NO is being formed under these conditions and that the NO_2 consumed in the initial stages is greater than that in oxidation experiments. In later stages, however, NO_2 consumed does not increase but remains nearly constant.

The values of the rates of NO_2 consumption and NO formation in our oxidizing system are proportional to the NO_2 consumed and NO formed as determined from the analyses of oxidizing gas before and after the reactor (for method of calculation of rates see Appendix II).

183

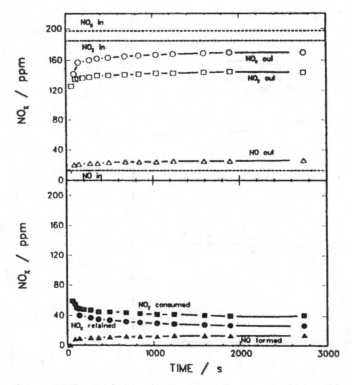

Figure 3.2.3: Reaction of NO$_2$ with Hexadecane in the Absence of O$_2$

The rate of NO$_2$ consumption under oxygen free conditions, after equilibrium is established in the system , is approximately equal to three times the rate of NO formation (Fig.3.2.3), i.e.,

$$\left(-\frac{d[NO_2]}{dt} \right)_{noO_2} \approx 3 \left(\frac{d[NO]}{dt} \right)_{noO_2}$$

while under oxidizing conditions (Fig. 3.2.2) the rate of NO$_2$ consumption at 200 s is about half of the observed under non-oxidizing conditions and at 600 s these rates are about equal, i.e.,

184

$$\left(-\frac{d[NO_2]}{dt} \right) \approx \frac{1}{2} \left(-\frac{d[NO_2]}{dt} \right)_{noO_2} \quad \text{at 200 s}$$

$$\left(-\frac{d[NO_2]}{dt} \right) \approx \left(-\frac{d[NO_2]}{dt} \right)_{noO_2} \quad \text{at 600 s}$$

The results obtained under non-oxidizing conditions are consistent with the reaction sequence consisting of reactions 8, 9 and 4 followed by reactions 10 and/or 11

$$R\cdot \ + NO_2 \longrightarrow RONO \tag{10}$$
$$R\cdot \ + NO_2 \longrightarrow RNO_2 \tag{11}$$

which lead to molecular products, alkyl nitrites (RONO) and/or nitroalkanes (RNO_2). Thus, under non-oxidizing conditions, formation of one molecule of NO is accompanied by consumption of three molecules of NO_2. Under oxidizing conditions, however, $R\cdot$ will react with O_2 to give $RO_2\cdot$ (Reaction 5). Consequently, one molecule of NO_2 should react to produce one molecule of NO and up to two $RO_2\cdot$ radicals. However, formation of NO was not detected under oxidizing conditions. This is probably due to reaction between $RO_2\cdot$ and NO (Reaction 12)

$$RO_2\cdot \ + NO \longrightarrow RO\cdot \ + NO_2 \tag{12}$$

which does not change the number of free radicals formed but consumes NO and regenerates NO_2. Of course, there is also a possibility that NO is oxidized to NO_2 by hydroperoxides, ROOH, formed during the oxidation. This could explain the absence of NO in the exiting gas and the apparent lower rate of NO_2 consumption in the early stages of oxidation experiments (200 s) as compared to experiments under non-oxidizing conditions. If reaction 5 would completely replace reactions 10 and/or 11 and reaction 12 or reaction with hydroperoxides would regenerate NO_2, the apparent rate of NO_2 consumption should be equal to zero. However, the observed rate of NO_2 consumption is about 1.5 times that assumed for NO formation based on non-oxidizing conditions. Thus, it seems that in the early stages some NO_2 is still consumed by reactions 10 and 11 or by similar reactions with $RO_2\cdot$ and in later stages probably also with other oxidation products.

3.2.5 Effects on Inhibited Hexadecane Oxidation

In order to determine NO_2 effects on inhibited oxidation, 2,6-di-tert-butyl-4-methlyphenol (MPH) has been used as a model radical trapping antioxidant. Results of experiments with different levels of NO_2 and MPH are shown in Figs. 3.2.4 – 3.2.8 and Table 3.2.3.

In the presence of MPH (PhOH) oxidation is inhibited by reactions 13 and 14 which are much faster and replace propagation reactions 6 and 7 and thus substantially reduce the steady-state concentration of RO•

$$PhOH + RO_2• \longrightarrow PhO• + ROOH \tag{13}$$
$$PhO• + RO_2• \longrightarrow O=PhOOR \tag{14}$$

Oxidation is inhibited until all MPH is consumed. The rate of MPH consumption during inhibition is greatly increased by addition of NO_2. The increase depends on both the concentration of NO_2 in the oxidizing gas and the initial concentration of MPH in hexadecane (Fig. 3.2.4 and 3.2.5). The increase in inhibition time, τ. At high concentration of MPH (5.85 mH) a complete consumption of MPH is observed at about half of the induction time while at lower MPH concentration (ca. 1 mH), MPH is consumed just prior to the induction time (compare τ and t_{MPH} in Table 3.2.3). At high concentration of MPH formation of an antioxidant intermediate is observed. This intermediate reaches a maximum concentration when all MPH is consumed and then its concentration gradually decreases until the end of the induction time (Fig. 3.2.5).

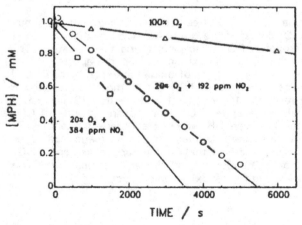

Figure 3.2.4: Depletion of MPH during Inhibited Oxidation

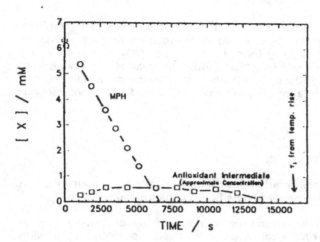

Figure 3.2.5: Inhibited Oxidation with 384 ppm NO$_2$

Table 3.2.3: Hexadecane Oxidation Inihibited by MPH

MPH, mM	0.98	1.05	1.02	5.86
NO$_2$, ppm	0	192	384	384
O$_2$ %	100	20	20	20
−d[MPH]/dt, M/s	2.9×10^{-8}	18.6×10^{-8}	27.7×10^{-8}	94×10^{-8}
d[NO]/dt, M/s				
initial	−	12×10^{-8}	21×10^{-8}	60×10^{-8}
first stage	−	10×10^{-8}	26×10^{-8}	94×10^{-8}
−d[NO$_2$]/dt, M/s				
initial	−	25×10^{-8}	46×10^{-8}	68×10^{-8}
first stage	−	22×10^{-8}	48×10^{-8}	94×10^{-8}
τ[a], s		6700	3750	16000
t_{MPH}[b], s		5500	3500	7000

a) Inhibition time from an increase in reaction temperature.
b) Time at which MPH is consumed.

The extent of NO_x reactions (Figs. 3.2.6 – 3.2.8) suggests that inhibition by MPH in the presence of NO_2 is a two stage process during which NO_2 is being consumed and NO is being formed. The rates of NO_2 consumption and NO formation depend on the initial concentration of MPH and the concentration of NO_2 in the oxidizing gas. Under all experimental conditions, the apparent rate of NO generation is lower than the rate of NO_2 consumption, i.e.,

$$\frac{d[NO]}{dt} < -\frac{d[NO_2]}{dt}$$

however, at high MPH concentration these rates are nearly equal during the first stage of inhibition (approximately first half of inhibition time).

In the initial stages of inhibition the apparent rates of NO formation are always smaller than the rates of MPH consumption. At higher concentrations of MPH or NO_2 (Fig. 3.2.6, Table 3.2.3) the rate of NO formation initially is about 2/3 of the rate of MPH consumption, then it gradually increases and later during the first stage (FS) it becomes equal to the rate of MPH consumption, i.e.,

$$\left(\frac{d[NO]}{dt}\right)_{FS} \approx \left(-\frac{d[MPH]}{dt}\right)_{FS}$$

The above observations suggest that under our experimental conditions NO_2 is effectively trapped by MPH and converted to NO while preventing reactions of NO_2 with the oxidizing substrate.

Brunton et. al (16) previously proposed a mechanism for reactions of NO_2 with MPH at room temperature in the absence of oxygen. The initial step in this mechanism is reaction of NO_2 with the phenolic hydrogen of MPH, reaction 15, followed by NO_2 addition to the resulting phenoxy radical (Reaction 16).

PhOH	+	NO_2 $\longrightarrow$	PhO$\cdot$ + HONO	(15)
PhO$\cdot$	+	NO_2 $\longrightarrow$	$O=PhNO_2$	(16)
$O=PhNO_2$	+	NO_2 $\longrightarrow$	$O=Ph(NO_2)_2$	(17)
$O=Ph(NO_2)_2$	+	NO_2 $\longrightarrow$	$O=Ph(NO_2)_2(ONO)$	(18)

These reactions are similar to reactions with $RO_2\cdot$ described above (Reactions 13 and 14). However, unlike reactions with $RO_2\cdot$, Brunton observed that further additions of NO_2 to the aromatic ring are possible (Reactions 17 and 18). These additions effectively increase the stoichiometric factor for NO_2 trapping by MPH to 4. Our results indicate that this mechanism (or its modifications) are operative also under our experimental conditions.

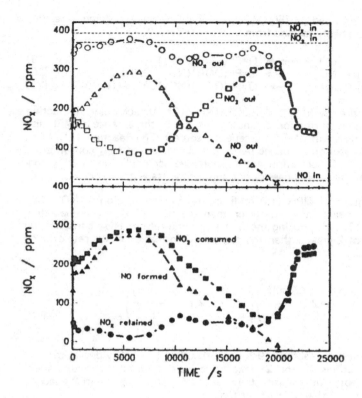

Figure 3.2.6: Inhibition by 5.9 mH MPH — high NO$_2$

At high MPH and NO$_2$ concentrations (Fig. 3.2.6) reaction 15 should be fast. It should be followed immediately by formation of RO$_2\cdot$ via reactions 9, 4 and 5 and then either by addition of RO$_2\cdot$ to PhO$\cdot$ (reaction 14) or by abstraction reaction 13. In the first case the rate of MPH consumption should be equal to the rates of NO$_2$ consumption and NO formation. In the second case the rate of MPH consumption should be twice the rate of NO formation. The observed rates (Table 3.2.3) suggest that initially reactions 13, 14 and 15 probably occur simultaneously, while after ~ 4000 seconds reaction 14 follows reaction 15 and reaction 13 can be neglected.

189

After all MPH is consumed ($\sim$ 7500 s) NO_2 probably reacts according to reactions 19 and 20. NO is probably formed from

$$O=PhOOOR \quad\quad + \; NO_2 \longrightarrow \quad O=Ph(OOR)(NO_2) \tag{19}$$
$$O=Ph(OOR)(NO_2) + \; NO_2 \longrightarrow \; O=Ph(OOR)(NO_2)(ONO) \tag{20}$$
$$O=Ph(OOR)(NO_2)(ONO) \longrightarrow \; O=Ph(OOR)(NO_2)O\cdot \; + \; NO \tag{21}$$

decomposition of the resulting $O=Ph(OOR)(NO_2)(ONO)$ according to reaction 21 which also leads to formation of another $RO_2\cdot$. In the absence of MPH and $PhO\cdot$, $RO_2\cdot$ may react with NO to regenerate some NO_2 (Reaction 12). As an overall result, this reaction sequence leads to a higher rate of NO_2 consumption than is the rate of NO formation and to increased accumulation of NO_2 containing products in the liquid phase during the second stage of inhibition.

At low concentration of MPH (1.0 mM) and high concentration of NO_2 (384 ppm) (Fig. 3.2.7) reaction 14 is faster than reaction 13 and it immediately follows reaction 15. Also, during the first stage, the rate of NO_2 consumption is about a factor of 2 greater than the rate of NO formation or the rate of MPH consumption, i.e.,

$$\left(-\frac{d[NO_2]}{dt}\right)_{FS} \approx 2\left(\frac{d[NO]}{dt}\right)_{FS} \approx 2\left(-\frac{d[MPH]}{dt}\right)_{FS}$$

This could indicate that reaction 19 follows reaction 18 already during the first stage and that reactions 20 and 21 take place mainly during the second stage. Based on this proposed mechanism, the rate of NO_2 consumption in the second stage (SS) should be one half of that for the first stage, i.e.,

$$\left(-\frac{d[NO_2]}{dt}\right)_{SS} \approx \frac{1}{2}\left(-\frac{d[NO_2]}{dt}\right)_{FS}$$

and inhibition time corresponding to the second stage should be shorter than that corresponding to the first stage. Results in Figure 3.2.7 confirm these predictions.

At low MPH concentration (1.0 mM) and low NO_2 concentration (192 ppm) (Fig. 3.2.8) the two stages of inhibition are less distinguished, nevertheless, they are present. The general mechanisms proposed above seem to be applicable also in this case. The rates of reactions involving NO_2 and MPH are lower due to lower concentration of both rectants. This causes the first stage to be less enhanced and significant overlap with secondary reactions occurs.

In all inhibited experiments at the end of induction time, when chain oxidation begins, a drastic increase in the rate of NO_2 consumption and ending of NO formation are observed. This is in agreement with observations made in the noninhibited system.

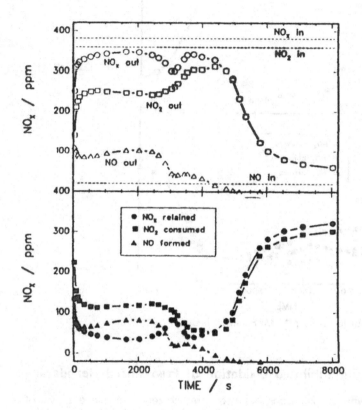

Figure 3.2.7: Inhibition by 1 mM MPH — high NO_2

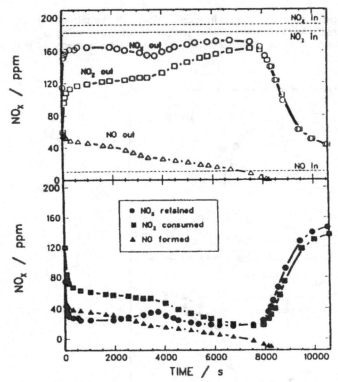

Figure 3.2.8: Inhibition by 1 mM MPH – low NO_2

3.2.6 Effects on Inhibited Oxidation of Preoxidized Hexadecane

Experiments in which hexadecane has been preoxidized to a selected level of hydroperoxide concentration before adding NO_2 in the oxidizing gas and MPH in hexadecane were conducted with the aim of determining concurrent effects of NO_2 and hydroperoxides on inhibited oxidation. Results of these experiments (Figures 3.2.9 and 3.2.10 and Table 3.2.4) confirmed that NO_2, even in the presence of hydroperoxides, increases the rate of MPH consumption and shortens inhibition time. This increase in the rate approximately corresponds to the rate of MPH consumption in the non-preoxidized system at given concentration of MPH. The rate of NO_2 consumption, however, is independent of MPH concentration and corresponds to consumption of ~50 % of NO_2. No information of NO is observed under these conditions. NO is probably oxidized to NO_2

192

either by RO• or hydroperoxides. If this is the case, then the observed rate of NO_2 consumption is the difference between the rates for NO_2 trapping and NO_2 formation, i.e.,

$$\left(-\frac{d[NO_2]}{dt}\right)_{consumption} \approx \left(-\frac{d[NO_2]}{dt}\right)_{trapping} - \left(\frac{d[NO_2]}{dt}\right)_{formation}$$

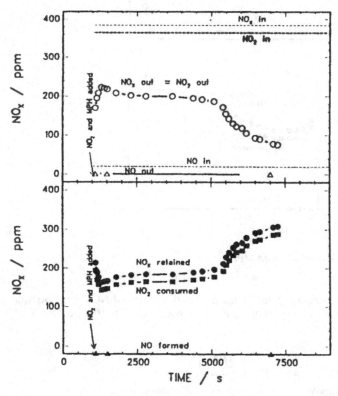

Figure 3.2.9: Preoxidized − 5.9 mM MPH

193

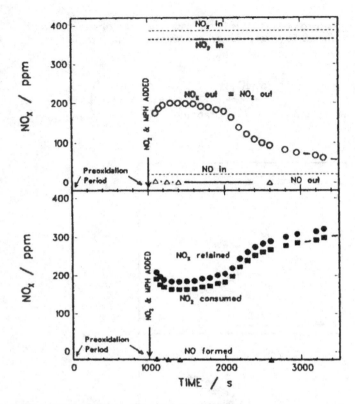

Figure 3.2.10: Preoxidized — 1 mM MPH

Since the rate of NO_2 consumption is equal to half the rate of NO_2 introduction in the reactor, the rate of NO_2 trapping must be equal to twice the rate of NO_2 formation,

$$\left(-\frac{d[NO_2]}{dt}\right)_{trapping} \approx 2 \left(\frac{d[NO_2]}{dt}\right)_{formation}$$

Thus, for every two molecules of NO_2 trapped one molecule of NO_2 is regenerated. Under such conditions, concentrations of NO_2 in oxidizing gas at the exit from the reactor can not be lower than half of the original concentration as long as all NO_2 entering the reactor is trapped by MPH or ROOH. This suggests that the rates of NO_2 consumption observed in our experiments both at low and high

194

MPH concentrations are the maximum rates possible at given concentrations of NO_2 in the oxidizing gas. Consequently, the differences in the observed rate of MPH consumption can be explained only by differences in mechanisms and stoichiometric factors for NO_2 and $RO_2\cdot$ trapping by MPH at different MPH concentrations and possibly also by including in mechanistic considerations hydrogen abstraction from ROOH by NO_2. To conclusively interpret our experimental observations will require more work.

Table 3.2.4: NO_2 and MPH Consumption in Preoxidized HD

MPH, mM	1	1	5.9
NO_2, ppm	0	384	384
O_2, %	20	20	20
[ROOH]/mM	18	18	18
$-d[MPH]/dt$, M/s	45×10^{-8}	71×10^{-8}	159×10^{-8}
$\Delta(-d[MPH]/dt)$		26×10^{-8}	114×10^{-8}
$d[NO_2]/dt$, M/s		65×10^{-8}	66×10^{-8}
τ [a], s	3700	2100	4800
t_{MPH} [b], s	3250	2300	4500

a) Inhibition time from an increase in reaction temperature.
b) Time at which MPH is consumed.

3.2.7 Summary

Investigations of the effects of NO_2 on oxidation of hydrocarbons (hexadecane) in the presence and absence of a radical trapping antioxidant (2,6-di-tert-butyl-4-methylphenol) were carried out at $160°C$ and ~ 110 kPa total pressure. Concentrations of NO_2 and oxygen in the oxidation gas were selected to be similar to those existing in an engine. For comparison some experiments were conducted in the absence of oxygen or in the presence of hydroperoxide oxidation products.

Results of investigations in an uninhibited system showed that:

• in the absence of oxygen, NO_2 is partially converted to NO,

- under oxidizing conditions, formation of NO is not observed, instead its interactions with oxidation products $(RO_2 \cdot)$ are believed to lead to regeneration of NO_2,

- NO_2 initiates oxidation of pure hydrocarbons, however, in the presence of oxidation products, NO_2 is preferentially consumed by non-initiating interactions with oxidation intermediates.

During oxidation inhibited by a radical trapping hindered phenol antioxidant:

- NO_2 preferentially reacts with antioxidant; this reaction accelerates antioxidant consumption and shortens inhibition time,

- the rate of NO_2 trapping is dependent on concentration of antioxidant in hydrocarbon and concentration of NO_2 in the oxidizing gas,

- consumption of NO_2 is accompanied by formation of NO,

- stoichiometric factor for NO_2 trapping by MPH can be as high as 4.

During inhibition of oxidation of hydrocarbon containing hydroperoxides:

- NO_2 contributes to initiation which otherwise takes place only by homolytic decomposition of hydroperoxides,

- only half of NO_2 could be consumed since for every two molecules of NO_2 reacted one molecule is being regenerated from oxidation of NO probably by hydroperoxides.

An attempt has been made to explain and rationalize experimental observations in terms of reactions previously described. Combining mechanisms of oxidation, nitration and inhibition, however, results in extremely complex sequences which would be difficult and time consuming to prove.

Results obtained in this study suggest that the presence of NO_2 in blowby gas contributes to accelerated consumption of antioxidants in engine oil and thus shortens the useful life of the oil in an engine. Observed differences in reaction of NO_2 and NO and conversion of NO_2 to NO and back to NO_2 during inhibited and uninhibited oxidation suggest that the concentration of these two gases in blowby will differ not only due to differences in operational conditions of the engine but also depending on the degradation state of the engine oil. In the early stages of use, when antioxidants are still present in the engine oil, NO_2 and NO concentration in blowby should be higher than in stages following a complete consumption of antioxidant.

3.2.8 References

(1) Spindt, R.S.; Wolfe, C.L.; Stevens, D.R.: "Nitrogen Oxides, Combustion, and Engine Deposits", SAE Trans., 64, 797-811 (1956).

(2) Dimitroff, E.; Quillian, Jr., R.D.: "Low Temperature Engine Sludge — What? — Where? — How?", SAE Technical Paper 650255 (1965).

(3) Dimitroff, E.; Moffitt, J.V.; Quillian, Jr., R.D.: "Why, What, How: Engine Varnish", Transactions of the ASME (1969) 91, 406—416.

(4) Geyer, J.: "The Mechanism of Deposit Formation and Control in Gasoline Engines", Am. Chem. Soc. Preprints, Div. Petrol. Chem., ACS, 14(4(, A15—A23 (1969).

(5) Vineyard, B.D.; Coran, A.Y.: "Gasoline Engine Deposition: I. Blowby Collection and the Identification of Deposit Precursors", Am. Chem. Soc. Preprints, Div. Petrol. Chem., ACS, 14(4), A25—A32 (1969).

(6) Coran, A.Y.; Vineyard, B.D.: "Gasoline Engine Deposition: II. Sludge Binder", Am. Chem. Soc. Preprints, Div. Petrol. Chem., ACS, 14(4), A35—A44 (1969).

(7) Zeelenberg, A.P.; Wortel, J.M.: "More Information on Oil and Engine from Sludge Analysis", SAE Technical Paper 770643 (1977).

(8) Hanson, J.B.; Harris, S.W.; West, C.T.: "Factors Influencing Lubricant Performance in the Sequence VE Test", SAE Technical Paper 881581 (1988).

(9) Nakamura, K., et al.: "Effect of Ventilation and Lubricants on Sludge Formation in Passenger Car. Gasoline Engines", SAE Technical Paper 881577 (1988).

(10) Kreuz, K.L.: "Gasoline Engine Chemistry", Lubrication, Vol. 55, No. 6 (1969).

(11) Kuhn, R.R.: "The Development of a Lube Oil Oxidation and Thickening Bench Test to Simulate the ASTM Sequence IIIC Test". Preprints, Div. Petro. Chem., ACS, 18(4), 694 (1973).

(12) Bardy, D.C.; Assef, P.A.: "Motor Oil Thickening — A CLR Engine Test Procedure Which Correlates with Field Service", SAE Technical Paper 700508 (1970).

(13) Nahumck, W.M.; Hyndman, C.W.; Cryvoff, S.A.: "Development of the PV-2 Engine Deposit and Wear Test — An ASTM Task Force Progress Report", SAE Technical Paper 872123 (1987).

(14) Smolenski, D.J.; Bergin, S.P.: "Development of the ASTM Sequence IIIE Engine Oil Oxidation and Wear Test", SAE Technical Paper 881576 (1988).

(15) Spearot, J.A.; Gallopoulos, N.E.: "Concentrations of Nitrogen Oxides in Crankcase Gases", SAE Technical Paper 760563 (1976).

(16) Brunton, G.; Cruse, H.W.; Riches, K.M.; Whittle, A.: "On the Mechanism of Nitration of 4-methyl-2,6-ditertiarybuthylphenol by Nitrogen Dioxide in the Liquid Phase", Tetrahedron Letters, No. 12, 1093 (1979).

(17) Jensen, R.K.; Korcek, S.; Mahoney, L.R.; Zinbo, M.: J. Am. Chem. Soc. 101, 7574 (1979).

(18) Hamilton, E.J.; Korcek, S.; Mahoney, L.R.; Zinbo, M.: Int. J. Chem. Kinet. 12, 577 (1980).

(19) Jensen, R.K.; Kircek, S.; Mahoney, L.R.; Zinbo, M.: J. Am. Chem. Soc. 103, 1742 (1981).

(20) Titov, A.I.: Tetrahedron 19, 557 (1963).

(21) Albright, L.F.: "Nitration" in Encyclopedia of Chemical Technology, 15, 841 (1981).

(22) Adachi, H.; Basco, N.: Int. J. Chem. Kinet. 14, 1242 (1982).

(23) Atkinson, R.; Aschman, S.M.; Carter, P.L.; Winer, A.M.; Pitts, J.N.: Phys. Chem. 86, 4563 (1982).

(24) Pryor, W.A.; Castle, L.; Church, D.F.: J. Am. Chem. Soc. 107, 211 (1985).

(25) March, J.: Advanced Organic Chemistry: Reactions, Mechanisms, and Structure, McGraw-Hill Book Company, New York, p 61 (1968).
(26) Cotton, F.A.; Wilkinson, G.: Advanced Inorganic Chemistry, John Wiley and Sons Inc., p 349 (1966).
(27) Yost, D.M.; Russell, Jr., H.: Systematic Inorganic Chemistry, Prentice Hall Inc., p 60 (1944).

Appendix I
Reactions of NO and NO₂

Reaction	Number*	Reference
$2NO + O_2 \rightleftharpoons 2NO_2$	(1)	(1)
$RH + NO_2 \longrightarrow R\cdot + HNO_2$	(8)	(20)
$HONO \longrightarrow \cdot OH + NO$	(9)	(20)
$R\cdot + NO_2 \longrightarrow RNO_2$	(10)	(20)
$R\cdot + ONO \longrightarrow RONO$	(11)	(20)
$R\cdot + NO \longrightarrow RNO$		(21)
$RO\cdot + NO_2 \longrightarrow RONO_2$		(21)
$RO_2\cdot + NO \longrightarrow RO\cdot + NO_2$	(12)	(22)
$RO_2\cdot + NO \longrightarrow [ROONO] \longrightarrow RONO_2$		(23)
$RO_2\cdot + NO_2 \rightleftharpoons ROONO_2$		(23)
$ROOH + NO_2/N_2O_4 \longrightarrow \text{products} \longrightarrow RONO_2$		(24)
$RONO \longrightarrow RO\cdot + NO$		(21)
$ROONO_2 \longrightarrow RO\cdot + NO_3\cdot$		(20)
$R_2CHNO \rightleftharpoons R_2C=NOH$		(25)
$NO + NO_2 + H_2O \longrightarrow 2HNO_2$		(26)
$3HNO_2 \longrightarrow HNO_3 + 2NO$		(27)

*Reaction numbers used in this paper.

198

Appendix II

Calculation of NO or NO₂ Rates of Formation or Consumption

The rate of NO formation or NO_2 consumption can be obtained using the following equations:

$$\frac{d[NO]}{dt} = 3.41 \times 10^{-9} \, [NO]_{formed}$$

$$-\frac{d[NO_2]}{dt} = 3.41 \times 10^{-9} \, [NO_2]_{consumed}$$

where $[NO]_{formed}$ and $[NO_2]_{consumed}$ are in ppm. The equations are valid for the total gas flow rate through the reactor of 200 mL/min and a volume of hexadecane in the reactor equal to 40 mL. The proportionality constant (3.41×10^{-9}) corresponds to moles of NO formed or NO_2 consumed per liter of solution per second for each ppm of NO or NO_2 added to or removed from the gas stream as it passes through the reactor.

3.3 Application of a New Concept to Detergency

J.M. Georges, J.L. Loubet, N.Alberola and G.Meille, Ecole Centrale de Lyon, France
H. Bourgognon, P. Hoornaert and G. Chapelet, Centre de Recherche Elf Solaize,
Saint-Symphorien d'Ozon, France

Abstract

The purpose of this paper is to describe a new concept in the detergency phenomenon for engines. It is well known that oil detergency is associated with additive adsorption neutralisation of acid formed and insoluble encapsulation. But bibliography is poor relative to the mechanical properties of the oxidized oil film created during the coking process.

Basic experiments taking a pure base oil and an overbased additived oil show clearly two different mechanical behaviours in the burned oil film. The base oil creates a ductile film very adherent to the substrate. Conversely, the additived film can be very brittle and shows clearly cohesive cracks followed by adhesive film-substrate cracks.

A physico-mechanical model is proposed to explain the role played by the internal stress developed in the film during its gel state. This analysis leads to a very simple thermo-oxidation laboratory test, where mechanical behavior is evaluated by peeling and is correlated with engine detergency.

3.3.1 Introduction

The term "detergent", as Shilling (1) indicates, is used in the engine field to bring out the notion of engine cleanliness and especially the hotter parts of the engine, pistons, piston-rings, and liners, or lubricant pipes. It needs to be pointed out here that detergent oils are not appropriate for cleaning a previously dirty engine. They aim at avoiding deposit build-up.

Those neutral detergents generally used in lubricants are of hydrocarbon soluble alkaline or alkaline earth metal soap types. The anion structure is of the sulphonate phenate phosphonate carboxylate types. The detergents may contain an excess of base, for example, in the form of carbonate to neutralize the acids produced by oxidation of the lubricants.

In the literature several detergent additives modes of action have been presented, along with their corresponding tests (2):

(i) the molecules of detergent cover the polar products resulting from oxidation of the lubricant and solubilize them. The polar products, in this way, can neither condense nor give insoluble resins or varnishes (3) (4),

(ii) these molecules cover the metal surfaces of the parts by adsorption. The film thus formed reduces adhesion of deposits (5) (6),

(iii) the adsorption of the detergents onto the small particles of soot, resulting from combustion, controls their flocculation (7) (8),

(iiii) the basicity brought by overbased and by some neutral detergents neutralizes the acid compounds, which have the feature, upon recombining, of accelerating the formation of insolubles (9) (10).

Up to the present day, though, amongst these modes of action, the role played by additives on deposits mechanical properties, e.g. their fragile or ductile behaviour and their adherence, has been underestimated. Microscopical observation of deposits obtained on a diesel engine piston, however, shows two types of deposits. The first type are brown in colour and mainly correspond to the coking of the lubricant. The other type, very black in colour, result from compaction of carbon soot. Cracking of the deposits is easily observable, when they are of a certain thickness. Their capacity to crack is, moreover, linked to the cleanliness of the engine parts. (Fig. 3.3.1).

Figure 3.3.1: Optical microscopic observations of deposits on the top ring groove of diesel engine piston

(a) brown varnishes of few micrometers thick cover the surface. Cohesive cracks are observable in the deposit (arrow).

(b) brown varnishes are also observable in some place (index A), are removed in other place (index B), are covered with carbon black soots (index C).

We present two basic experiments in this article, along with interpretations of results, which enable the importance of these effects to be appreciated.

3.3.2 A Coking Experiment

3.3.2.1 Principle of the Experiment

We are interested in deposits left by evaporation of a drop of lubricant placed onto a hot slide. A parallelepiped shaped strip of glass (25 x 25 x 1 mm) is placed, then, on the upper surface of a steel cylinder (height 35 mm, inside diameter 20 mm, outside diameter 45 mm) (Fig.3.3.2) (11).

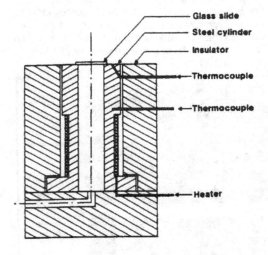

Figure 3.3.2: Schematic drawing of the coking equipment (Test ELF/ECL)

In this experiment, the cylinder is heated to a regulated temperature of 285°C. A constant, verified volume of lubricant (0.05 ml) is placed in the centre of the glass strip corresponding the centre of the cylinder bore. The drop spreads due to the wetting phenomenon. As the thermal gradient existing between the centre of the slide and its edges controls the wetting process, the diameter of the drop of liquid stays constant and equal to 20 mm. We have, then, a disc of liquid with a thickness close to 150 μm. This set-up constitutes the basic equipment for the ELF/ECL micro-coking procedure (11).

3.3.2.2 Results

Two lubricants are considered here: a pure base 200 Neutral (A) and a formulated oil containing the 200 N base + 15 % calcium dialkyl benzene sulfonate overbased with calcium carbonate (B).

Fig. 3.3.3 shows the evolution, as a function of the isothermal hold time, of the remaining fraction of lubricant found by weighing. Comparison of these thermo-oxidation kinetics demonstrates that lubricant B leaves more deposits than lubricant A. After cooling at the different stages of thermo-oxidation the state of the drops was noted. This was performed by examination under an optical microscope and by touching them with a very fine glass needle placed at the type of a micro-manipulator. For heating times of less than 20 minutes, drop of lubricant A is in liquid state. After that time it takes the form of a solid characterized by a brownish deposit, strongly adherent to the glass. Oil "B" demonstrates quite a different behaviour. For heating times of less than 10 minutes (t) it remains in liquid state, then changes into a gel state (t_G) with a viscoelastic behaviour, and finally adopts a solid state for heating times exceeding 50 minutes (t_s). These differences in mechanical behaviour can be detected (20). Simple wiping with a dry cloth moves the drop of liquid, but not the gelified or solid drop.

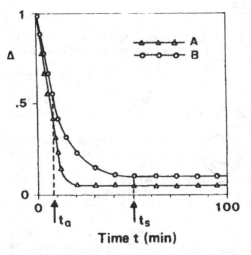

Figure 3.3.3: Evolution of the remaining weight of lubricant versus thermo-oxidation time for a given temperature (285°C). Δ is the ratio of the weight of deposit at a given time to the initial weight of the lubricant drop

A peeling test has also been performed on the deposits using a strip of adhesive tape. We observe that for the lubricant A, the strip removes some quantity of the film for a cured treatment time of $t < 20$ min, because the film is in a liquid state. But for a time $t > 20$ min, the film is highly adherent and cannot be removed from the substrate. The situation is more complex for films formed by treatments of lubricant B. In the liquid state $t < t_G$, the strip removed some quantity of lubricant and leaves nevertheless some on the glass slide (Fig. 3.3.4). At a critical time $t_G \approx 10$ min the film state is a gel and the experiment shows that the the adhesion glass-film is better than film-tape. At time $t > t_S$ (50 min) the film is solid and can be removed easily. For $t_G < t < t_S$ we observe an intermediate situation between (b) and (d) (Fig. 3.3.4) * .

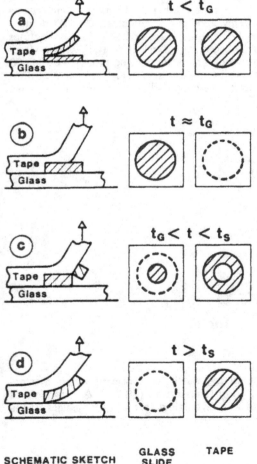

$t < t_G$

$t \approx t_G$

$t_G < t < t_S$

$t > t_S$

SCHEMATIC SKETCH

GLASS SLIDE

TAPE

Figure 3.3.4:
A peeling test is realized with a tape on a cured film of over-based lubricant B
(a) If the film after a time treatment of $t < t_G$ is liquid, half of the film wets tape, half wets glass slide
(b) At time t_G gel occurs, the tape cannot remove the fim
(c) At time $t_G < t < t_S$ the central region is not cracked and the film adheres to the glass
(d) At time t_S (50 min) cracks occur in all regions of the film and the tape can remove the film

* The time t_G and t_S varies with the temperature and the drop quantity.

The transfer from the gel state to the solid state is more rapid at the edges of the drop, where it was hotter than in the centre. Rings of film can consequently be stripped off as shown in Fig. 3.3.4.

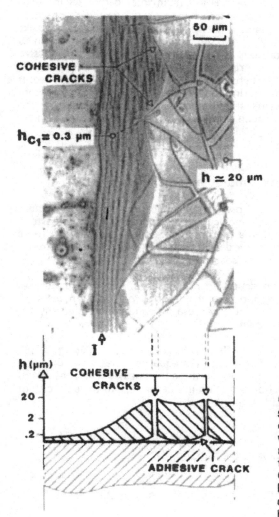

COHESIVE CRACKS

$h_{C_1} \simeq 0.3\ \mu m$

$h \simeq 20\ \mu m$

50 µm

h (µm)

COHESIVE CRACKS

20
2
.2

ADHESIVE CRACK

I

Figure 3.3.5:
Optical microscopical observations and cut sketch of the solidified wetting wedge. Optical interference (I) gives an estimation of the film thickness. Cohesive cracks begin for $h_{C_1} = 0.3\ \mu m$. Adhesive cracks are very visible for $h = 2\ \mu m$

205

Microscopic examination of a film (B) in the solid state shows the presence of cracks (Fig. 3.3.5). These cracks are of two types: the first one are cohesive cracks which occur in the thick part of the deposit. The second type are film-substrate adhesive cracks. Film thickness is a fundamental parameter in crack appearance. Film (B) with a thickness of less than h_{C1} = 0.3 μm does not crack. When the thickness exceeds this value a network of rectilinear cohesive cracks develops with a spacing of about 50 to 400 μm between cracks. With a thickness greater than h_{C2} = 20 μm spontaneous stripping of the film occurs: an adhesive crack has spread. The film is easily stripped off with the peeling test, when its thickness is h_{C1}. The driving force behind these phenomena, as we shall see, is the internal energy which appears during the evaporation process. When the internal energy reaches a critical value, the film breaks first and then bents it. The peeling test reveals the presence of such a non-adherent solid state of the film.

3.3.3 Gel Formation and its Consequences

The formulated lubricant contains those elements necessary for gel formation. The solvent is formed by the small molecules corresponding to the basestock light fractions. On the other hand, the overbased detergents as well as the more or less polycondensed products of oxidation which give rise to the resins or colloids can provide a gel network by association (2) (12).

Consider a small cut in the film deposited on the glass. This one undergoes evaporation on its upper surface, and thus the fraction per unit volume Φ of the gelifying product increases over time because we can assume that only the light part of the lubricant is evaporated.

We can make a sketch of the cut in the film in the gel state $(t = t_G)$ (Fig. 3.3.6). At the time t_G, we have a piling of solid plates of equal thickness (17). Evaporation of the solvent brings about a reduction in these thicknesses. The cohesion of the film leads to a bidimensional deformation in tension, ϵ, and the elasticity of the gel to the presence of the residual tensile stress:

$$\sigma = E(\Phi) . \epsilon \qquad (1)$$

where $E(\Phi)$ is the Young's modulus of the film.

Therefore per unit volume of film an internal elastic deformation energy U_e is created which is:

$$U_e = \frac{1}{2} \sigma \epsilon = \frac{1}{2} E(\Phi) . \epsilon^2 \qquad (2)$$

The viscoelastic properties of such a product vary very strongly with the volume fraction. It is known that when $\Phi < \Phi_G$ (with $\Phi_G \approx 0.05$) there is no network formation and the solution behaves as a liquid. For $\Phi > \Phi_G$ we reach the gel state, which presents elastic behaviour with Young's modulus $E(\Phi)$ and a viscosity $\eta(\Phi)$.

The Young's modulus E strongly increases with Φ according to a power law (13).

$$E\ (\Phi)\ a\ (\Phi - \Phi_G)^m \text{ with } m \approx 3 \text{ to } 4 \tag{3}$$

This law can be explained by percolation theories (14) (15) (16).

Simultaneously the viscosity increases according to a law of the type:

$$\eta(\Phi)\ a \exp \left[\frac{2.5\ \Phi}{1 - 1.6\ \Phi} \right] \tag{4}$$

For the high value of Φ, the medium acquires the solid state because the viscosity value is very high ($\eta > 10^8$ Pa.s^{-1}).

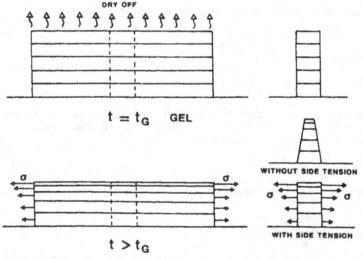

Figure 3.3.6: A model of the film in gel state during the evaporation process. The gel slab is treated as a combination of thin layers. The lowest is mechanically fixed to the substrate but the top layer is free to contract

3.3.4 Cracking and Overbased Detergent

We prove that the evaporation phenomenon is the dominant factor in the film cracking process and that colloidal particles of the overbased additive can be the gelifying product. Basic amorphous calcium carbonate is used. It is spherical colloidal particles (diameter 9.5 nm ± 0.5 nm) covered with calcium sulfonate alkyl benzene molecules.

This product is dissolved in pure toluene at a volumic concentration Φ of 0.2. A drop of this solution is placed on a glass slide. It is left to evaporate at ambient temperature ($\theta = 25°C$). Examination under an optical microscope shows the same type of cracking on the dry deposit as that seen previously on the oil (B) at high temperature (Fig. 3.3.5). The presence of cohesive cracks, which appear for a certain thickness of deposit can be noted. The thickness is easily determined using the colour of the optical interference fringes ($h_{c1} \approx 0.3 \mu m$). The deposit peeled off for greater thicknesses.

For this experiment, where a solvent and a colloid are present, the gel occurs due to association between the calcium carbonate spheres. We can express the critical concentration upon gelifying as Φ_G, with a characteristic value of the order of $\Phi_G = 0.05$ (12). We express the extreme concentration of the solid state as Φ_S, which corresponds to maximum piling up of the spheres. The close packed sphere density is $\Phi_S = 0.74$.

The maximum admissible tension deformation ϵ can be expressed as :

$$\epsilon = \frac{V}{3\,V} \approx \frac{1}{3} \left[\frac{V_G - V_S}{V_G} \right] \tag{5}$$

where V_G is the volume under consideration in the gel state corresponding to V_S in the solid state after evaporation of the solvent. As the colloid does not evaporate, its volume stays constant, this gives:

$$V_G\,\Phi_G = V_S\,\Phi_S \tag{6}$$

Then taking relations (5) and (6), we have:

$$\epsilon = \frac{1}{3} \left[1 - \frac{\Phi_G}{\Phi_S} \right] \approx 0.3 \tag{7}$$

This deformation corresponds well to the one found when, under microscopic examination, the inital length of a platelet prior to cracking and the length of the cracked, liberated platelet, are estimated.

208

For a steady test, a long crack in an elastic system is considered. Non-dimensional analysis shows that the mode I stress intensity factor at the crack front K_1 is independent of the crack length and can be expressed as (18) (19):

$$K_1 \approx \sqrt{EG} = \Omega\,(\Sigma)\,\sigma\sqrt{h} \tag{8}$$

where σ is the film stress G is the strain energy release rate, h is the film thickness, Σ is the Young's modulus ratio of the film to the substrate i.e. E/E_S, $\Omega\,(\Sigma)$ is a dimensionless quantity here $\Omega\,(\Sigma) \approx 2$ (18).

Taking into account equations (1), (7), (8), we have:

$$\epsilon = 0.5\sqrt{\frac{G}{E.h}} \approx 0.3 \tag{9}$$

We can evaluate G for $h = h_{c1} = 0.3\ \mu m$, where $E = 10^8$ Pa (20), then the average residual stress is $\sigma \approx 3.10^7$ Pa and $G \approx 10$ J/m^2 which is an acceptable value for such a material (21) (22).

3.3.5 Correlation between the ELF/ECL Coking Test and a Renault 30 TD Engine Test (11)

An experimental procedure (temperature, duration, measurement parameters) has been developed on the basis of 14 experimental diesel lubricants. These lubricants have about the same phosphorus contents. They have been perfectly characterized in terms of tests on Renault 30 turbo diesel 2086 cc engines (piston rating). Operating conditions for this test are as fallows : no oil-change, a 50-hour run, 4.200 rpm, 150 m.da.N load, temperatures: oil 130°C, water 88°C, air 45°C. Best correlation is obtained with the following experimental conditions: first carbon groove rating of a Renault 30 TD engine, and for the coking test a temperature of 330°C.

Results show that at this temperature, t_G and t_S vary with the nature of the lubricant, though $t_S - t_G$ stay almost constant. Therefore we used here t_G as the measurement parameter of the coking test (Fig. 3.3.7). The linear regression coefficient of 0.83 gives a satisfactory correlation.

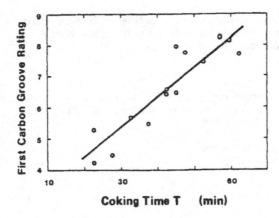

Figure 3.3.7: Correlation between ELF/ECL coking test and Renault 30 TD engine Test

3.3.6 Conclusions

The behaviour of lubricant films has been observed upon thin thermo-oxidation film. Addition of an overbased detergent brings about the formation of cohesive and adhesive cracks in the film. These are due to the presence of residual stress which builds up within the film upon evaporation. Consequently a new coking test is proposed.

3.3.7 References

(1)   Shilling, A.: "Automative Engine lubrication", Vol 2, Scientific Publication, London (1972).
(2)   Hsu, S.M.: A.S.L.E., Lubrication Engineering, Vol 37, 12, pp. 722—731 (1981).
(3)   Bell, G.H.; Groszeck, A.H.: ACS Meeting D89-D102 (1965).
(4)   Fowkes, F.M.: "Solvent Property of Surfactant Solutions". Chapter 3. Marcel Dekker Inc., New York, pp. 65—115 (1967).
(5)   Jager, G.; Knispel, B.: Schmierungstechnik 8, pp. 1—27 (1977).
(6)   Tamura, K.; Tse, J.T.; Adamson A.W.: J. Japan Petr. Inst. 27, (5), pp. 385—391 (1984).
(7)   Courtel, R.; Bernelin, B.; Labre, J.: Revue de I.F.P., Vol 5, pp. 447—486 (1955).
(8)   Glavaty, O.L.; Marchenko. A.T.; Kravchook, G.G.; Glavaty, E.V.: Acta. Chem. Hung., 116 (4), pp. 367—375 (1984).
(9)   Hosonuma, K.; Tamura, K.: J. Japan Petr. Inst., Vol 27, No. 2 (1984).
(10)   Salino, P.; Volpi, P.: Annali di chimica, Vol 77, pp. 145—146 (1987).
(11)   Alberola, N.; Vassel, A.; Bourgognon, H.; Rodes, C.: CEC Symposium, Paris (1989).

(12) Georges, J.M.; Loubet, J.L.; Tonck, A.: "New Materials Approaches to Tribology: Theory and Applications", L. Pope, L.L. Fehrenbacher, W.O. Winer, Eds. M.R.S., Vol 131, pp. 67)78, Boston (1989).
(13) Bauthier-Manuel, B.; Guyon, E.: J. Phys. Letter (France), 41, p. 503 (1980).
(14) de Gennes, P.G.: "Scaling concept in polymer physics". Cornell University Press, Ithaca (1979).
(15) Stauffer, D.: J. Chem. Soc. Faraday Trans., 72, p. 1354 (1972).
(16) de Gennes, P.G.: J. Phys. Letter (France), 37, L.1. (1976).
(17) Tanaka, T.: Phys. Rev. Lett. 40, pp. 820–823 (1978).
(18) Gille, G.: Thin Solid films, 110, p. 219 (1984).
(19) Hu, M.S.; Thouless, M.D.; Evans, A.G.: Acta Metall. 36, p. 1333 (1988).
(20) Loubet, J.L.; Tonck, A.; Georges, J.M.: Measurements of the elastic properties of detergent film with indentation test (to be published).
(21) Ashby, M.F.; Jones, D.H.: "Engineering materials: an introduction to their properties and applications", Int. Series, Vol 34, Pergamon Press (1985).
(22) Kendall, K.: Powder Metallurgy, Vol 31, 1, pp. 28–31 (1988).

211

3.4 Overbased Lubricant Detergents — A Comparative Study of Conventional Technology and a New Class of Product

S.P. O'Connor, BP Chemicals, Hull, Great Britain
J. Crawford, Adibis, Redhill, Great Britain
C. Cane, Adibis, Hull, Great Britain

Abstract

Overbased detergents are commonly used in automotive and marine lubricants. Their main functions are to neutralise potentially corrosive acids and to contribute to engine cleanliness. The major types of overbased detergents that are produced commercially are phenates, sulphonates and salicylates. This paper will describe the chemistry of these additive classes and review their properties and performance in bench and engine tests. The paper will also describe a new class of overbased detergents. Novel chemistry has been used to produce new types of highly overbased phenate, sulphonate and salicylate. These new products exhibit significant improvements in properties and performance over conventional detergents. Evidence from bench and engine tests will be presented to demonstrate the benefits of the new detergent types.

3.4.1 Introduction

Detergent additives are widely used in the formulation of lubricating oils for automotive and marine applications. The main functions of these oil-soluble detergents are:

1. To neutralise the acidic by-products of the combustion process and thereby reduce corrosive wear.

2. To neutralise the acidic products of lubricant oxidation.

3. To act as a "detergent" and help to keep pistons and other high temperature surfaces clean of deposits.

The role of neutralising agent is particularly important in marine applications where the burning of high sulphur diesel fuel leads to the formation of significant amounts of sulphuric acid in the combustion chamber. In certain marine engines, lubricants containing high concentrations of detergent additive are fed directly into the combustion chamber to neutralise acids at source.

3.4.2 Review of Detergent Types

3.4.2.1 General

Lubricant detergent additives are composed of 2 main structural features:

1. A metal surfactant salt or soap
2. A store of inorganic base stabilised by soap

Some lubricant detergents only comprise the soap component and these are generally known as neutral detergents. Detergents with a store of inorganic base are known as overbased or basic detergents as these materials formally contain more moles of base than moles of acid.

A number of different surfactant types have been used to produce lubricant detergents. Common examples include:
Sulphonates
Alkylphenates
Sulphurised alkyl phenates
Carboxylates
Salicylates
Phosphonates
Phosphinates

Some of these will be discussed in more detail.

The metals most commonly employed are:
Calcium
Magnesium
Barium

Other metal types that have been used include:
Sodium
Potassium
Zinc

The production of neutral detergents involves a simple acid/base reaction (see Fig. 3.4.1).

ACID/BASE REACTION

$$2\ R\text{-}H\ +\ M(OH)_2\ \longrightarrow\ R\text{-}M\text{-}R\ +\ 2H_2O$$

| ORGANIC ACID | INORGANIC BASE | NEUTRAL DETERGENT (SOAP) | WATER |

Figure 3.4.1: Neutral Detergent Production

In hydrocarbon solution the neutral detergent molecules act with typical ionic surfactant behaviour and tend to form ordered micellar arrangements (see Fig. 3.4.2).

Overbased detergents can be made using the acid/base reaction but overbasing is achieved by employing a chemical equivalent excess of inorganic base (see Fig. 3.4.3).

A more common process, however, is the carbonation technique which converts excess inorganic base to metal carbonate. This inorganic carbonate is stabilised by metallic soap to form an oil soluble overbased detergent (see Fig. 3.4.4).

The excess basicity is usually considered to be accommodated in the micellar core of the neutral detergent molecules (see Fig. 3.4.5).

IONIC AMPHIPHILE

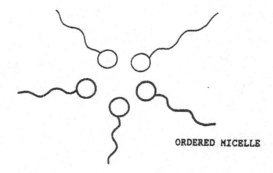

ORDERED MICELLE

Figure 3.4.2:   Neutral Detergent Molecular Arrangements

OVERBASING

$$2 \text{ R-H } + \text{ nM(OH)}_2 \longrightarrow \text{R-M-R (n-1) M(OH)}_2 + 2\text{H}_2\text{O}$$

| ORGANIC ACID | INORGANIC BASE | OVERBASED DETERGENT | WATER |

Figure 3.4.3:   Overbased Detergent Production

214

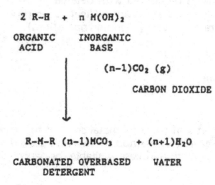

$$2 \ R-H \ + \ n \ M(OH)_2$$

ORGANIC   INORGANIC
 ACID   BASE

$$(n-1)CO_2 \ (g)$$

CARBON DIOXIDE

$$R-M-R \ (n-1)MCO_3 \quad + \ (n+1)H_2O$$

CARBONATED OVERBASED   WATER
    DETERGENT

Figure 3.4.4: Carbonated Overbased Detergent Production

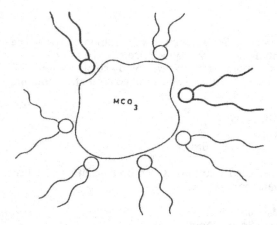

Figure 3.4.5: Structure of Carbonated Overbased Detergents

This constitutes the store of base useful in the neutralisation of potentially harmful acidic components in the combustion chamber and the crankcase.

The extent to which a lubricant detergent is overbased can be conveniently determined by either ASTM D2896 (AV) or ASTM D664 (TBN). TBN(Total Base Number) and AV (Alkalinity Value) measure the basic equivalent of the detergent additive in milligrams of KOH per gram of detergent. The TBN value (or AV) and the % metal w/w give measure of the concentration of base present in the detergent additive.

Another key analytical parameter for lubricant detergents is additive viscosity (e.g. viscosity at 100°C), which should always be quoted with detergent TBN. Clearly, additives possessing excessively high viscosities are difficult or impossible to handle during and after manufacture (e.g. during blending operations).

Other parameters for detergents are often used and include:

* Other microanalytical data (e.g. % sulphur)
* Metal ratio or basicity index (BI)
* % neutral detergent (% soap)

Metal ratio (or basicity index) is a simple ratio of the total metal content to the metal contained in the neutral soap. However, this method of expressing overbasing does not give a measure of the concentration of base available for neutralisation of harmful acids.

The most common types of lubricant detergent will now be reviewed.

3.4.2.2 Sulphonates

Neutral and overbased sulphonates are derived from organic sulphonic acids. The earliest sulphonic acids were obtained as by-products from white oil manufacture. Removal of the aromatic components of mineral oil by sulphonation with sulphuric acid led to the production of white oil and a mixed aromatic sulphonic acid stream.

Mineral Oil + Sulphuric → White + Sulphonic
            Acid   Oil   Acid
                                      Mixture

The sulphonic acid mixtrue was then neutralised and the resulting salt was separated into oil soluble and water soluble fractions.

Sulphonic Acid + MOH → Mahogany + Green
   Mixture   Acid Soap Acid
                                       Soap

           M=K, Na   oil   water
                        soluble soluble

The mahogany acid derived sulphonates are also known as neutral sulphonates or petroleum sulphonates.

However, the increased use of hydrogenation for the production of white oils has led to a reduction in the availability of the mahogany sulphonic acids. Alternative sulphonation feedstocks have been developed and these include:

216

* polydodecylbenzenes — obtained as "bottoms" form the alkylation of benzene to give dodecylbenzene, an intermediate in the production of household detergents.

* synthetic alkylbenzenes specifically produced for use in the lubricant additive application (see Fig. 3.4.6).

Typical olefins used for this route include

* branched chain olefins (average chain length >C18)
* linear α—olefins (average chain length >C18)

A large volume of work has been carried out on processes to neutral and especially overbased sulphonates. The optimum process conditions for a particular target sulphonate depend on the structure of starting sulphonic acid, the metal salt employed and the nature of the promotor used.

Structural studies (1—4) on overbased sulphonates confirm a micellar structure with the size of the core (the store of excess base) dependent on TBN (and % metal content).

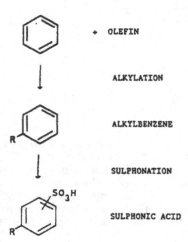

Figure 3.4.6:   Synthetic Sulphonic Acid Production

3.4.2.3 Phenates

The acidic organic substrate used for the production of metal phenates are alkylphenols and sulphurised alkyl phenols (SAP's) (see Fig. 3.4.7).

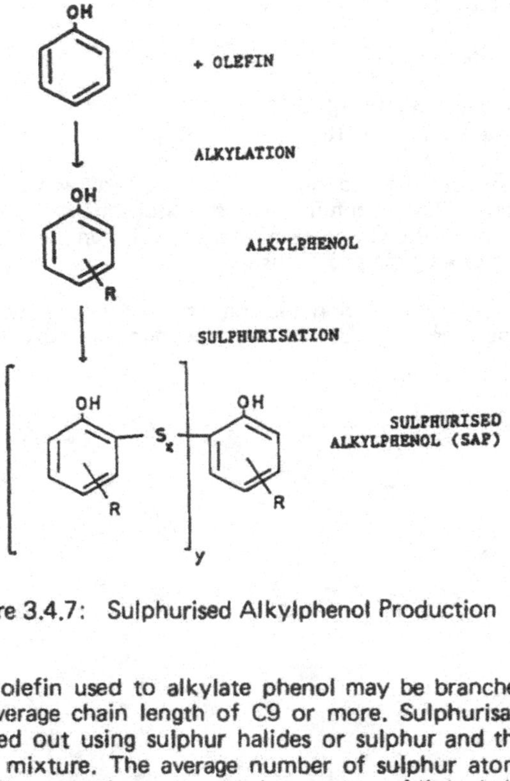

Figure 3.4.7: Sulphurised Alkylphenol Production

The olefin used to alkylate phenol may be branched or linear and is generally of average chain length of C9 or more. Sulphurisation of alkyl phenol can be carried out using sulphur halides or sulphur and this process yields a complex SAP mixture. The average number of sulphur atoms per sulphur bridge (x) is usually more than one and the number of linked phenol rings (y) is also usually greater than one. This substrate is then reacted with the appropriate quantity of metal base (and optinally carbon dioxide) to produce the target neutral or overbased phenate.

Many variations on the process to overbased phenates have been reported (5) and usually involve carbonation to give metal carbonate overbasing.

"Mixed detergents" have been produced in which a mixture of substrates are used in reaction with the metal base e.g. mixed phenate sulphonate overbased detergents.

218

3.4.2.4 Salicylates

The substrates used for the production of metal salicylate detergents are alkyl-salicylic acids. These are derived from alkyl phenols using the Kolbe-Schmidt reaction (see Fig. 3.4.8).

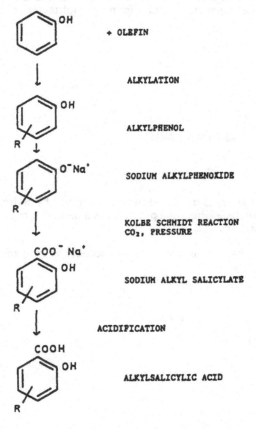

Figure 3.4.8:   Alkylsalicylic Acid Production

The resulting alkali metal salicylate is usually converted to an alkaline earth salt by metal exchange (e.g. with alkaline earth halide salt). As with phenates the olefin used to alkylate phenol may be branched or linear (usually C9).

The micellar behaviour of some neutral metal salicylates has been studied (6,7).

3.4.2.5 Others

3.4.2.5.1 Naphthenates

Naphthenate based detergents are derived from naphthenic acid substrates. Naphthenic acids occur in varying quantities in crude oil (0—2.5%) and are extracted during refining. A typical chemical structure found in naphthenic acid is shown in Figure 3.4.9.

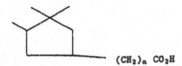

$(CH_2)_n \; CO_2H$

Figure 3.4.9: Typical Structure in Naphthenic Acid

3.4.2.5.2 Phosphonates

A common substrate for phosphonates is produced by the reaction of polyiso-butene with phosphorous pentasulphide where the product has a general formula approximating to that in Figure 3.4.10.

The metal employed is usually barium as there are oil solubility problems associated with the calcium salts.

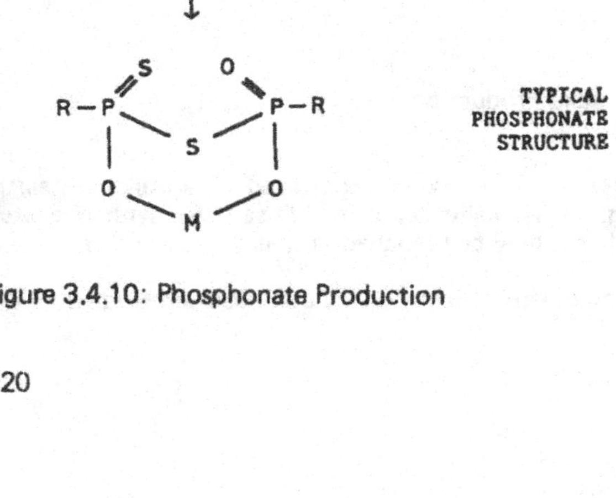

POLYISOBUTENE (PIB) = R

$+ \; P_4S_{10}$

TYPICAL
PHOSPHONATE
STRUCTURE

Figure 3.4.10: Phosphonate Production

220

3.4.3 A New Class of Overbased Detergent

3.4.3.1 Introduction

A new class of overbased detergent has recently been developed. This novel range of products offers significant advantages over conventional overbased detergents. In particular the new products possess:

(i) Higher basic strength and greater neutralising power than conventional detergents, primarily phenates and salicylates.

(ii) Added performance advantages over conventional products.

The new product range includes:
New high TBN phenates
New high TBN salicylates
New high TBN mixed phenate/sulphonates
New high TBN sulphonates

Certain of the products are available on a commercial scale.

Evidence to support these claims is presented below.

Table 3.4.1: Comparison of Conventional Detergents

Property	Phenates	Sulphonates	Salicylates	Phosphonates
Approx. Commercial TBN Range (ASTM D2896)	0–300	0–500	0–300	0–80
Hydrolytic Stability	Good	Moderate	Good	Moderate
Oxidation Stability	Very good	Poor	Very good	Good
Thermal Stability	Excellent	Excellent	Excellent	Moderate
Detergency	Good	Good	Excellent	Good
Rust Inhibition	Low	Good	Low	Good
Anti–Oxidant Effect	Very good	None	Very good	Good

3.4.3.2 Properties and Performance

3.4.3.2.1 Basic Strength

The new detergent product range provides the formulator with a flexible series of options. See Tables 3.4.2 and 3.4.3.

Table 3.4.2: New Overbased Detergents

Parameter	Range
Metal Type	Ca, Mg
TBN(ASTM D2896)	50–500
%Metal	1.5–18.0
Basicity Index	1–10
Viscosity cSt at 100°C	50–700

Table 3.4.3: New Overbased Detergents

Typical Products

Type	% Ca w/w	TBN (ASTM D2896)	V100
New Phenate	14.3	400	300
New Salicylate	14.2	400	250
New Sulphonate	15.7	425	100
Conventional Phenate	9.2	258	300
Conventional Salicylate	10.0	280	160
Conventional Sulphonate	15.3	400	70

222

The new detergent technology allows the production of overbased phenates and salicylates of higher basic strength than has previously been readily available.

Consequently, the new phenates and salicylates have a significantly greater neutralising power per unit weight than the conventional products and therefore treatment level in the finished lubricant can be reduced. This permits a greater oil content in the lubricant thereby contributing to oil film reinforcement.

Furthermore, the new products possess viscosity properties consistent with large-scale handling and blending operations.

3.4.3.2.2 *Viscosity-Temperature Behaviour*

The general move towards more severe operating conditions in marine engines has led to an increase in the significance of the viscosity index (VI) of marine oils. The oil must possess a sufficiently high VI to provide an adequate oil film at the higher operating temperatures to prevent scuffing and minimise wear.

Additive components of the new detergent class have been found to make a more significant contribution to lubricant viscosity index than components of the conventional type. This property in demonstrated in Table 3.4.4.

Table 3.4.4:   Contribution to Lubricant VI — Comparative Data

Additive Components	Δ % VI
New Phenate vs Conventional Phenate	+ 26
New Sulphonate vs Conventional Sulphonate	
New Salicylate vs Conventional Salicylate	

*   a typical MCL treatment levels blended into bright stock

3.4.3.2.3   *Friction Reduction*

An important function of automotive and marine lubricants is to reduce engine friction. This can lead to reduced wear and improved fuel consumption.

The new detergent additive have been found to possess markedly improved friction modification properties over detergents of the conventional type.

223

This friction reduction property has been demonstrated in 2 types of rig test which model the contact between piston ring and cylinder bore. These are (i) the pin-on-disc technique and (ii) a technique employing a reciprocating test rig, the Cameron—Plint TE77 friction and wear machine.

Table 3.4.5 presents pin-on-disc machine results on marine lubricants formulated with new and conventional detergent types.

A marked reduction in friction is observed for the lubricants containing the new components especially for the new phenate and sulphonate. Cameron-Plint machine results also indicate improved friction performance for the new detergents, with the improvement over conventional products increasing with increasing temperature (see Fig. 3.4.11).

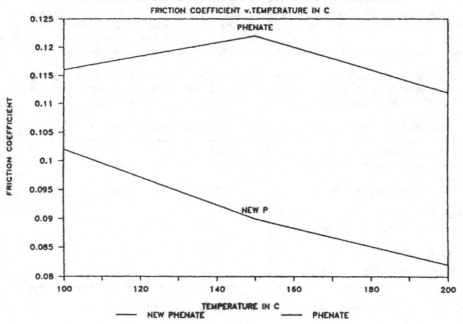

Figure 3.4.11

Table 3.4.5:   Pin-on-Disc Friction Reduction Results

Component	Friction Co-Efficient	%Decrease in Friction Co-Efficient
New Phenate	0.09	
		36
Conventional Phenate	0.14	
New Sulphonate	0.08	
		33
Conventional Sulphonate	0.12	
New Salicylate	0.076	
		11
Conventional Salicylate	0.085	

Components blended into full marine formulations conditions — slow sliding speed, 100°C.

3.4.3.2.4   *Anti-oxidant Performance*

The new detergent components were evaluated in the Rotary Bomb oxidation Test (RBOT, ASTM D2272) which determines anti-oxidant performance in the presence of water. This refelcts the wet environment in which a marine lubricant has to operate in the field.

Table 3.4.6:   Rotary Bomb Oxidation Test Results (ASTM D2272)

Component	Time to 15 PSI Pressure Drop in O_2 (min)
Base Oil	33, 29
New Phenate	139, 137
Conventional Phenate	63, 61
Conventional Sulphonate A	34, 32
Conventional Sulphonate B	19, 32

Test oils blended to 70 TBN in base oil.

225

The results (see Table 3.4.6) show the superior anti-oxidant performance of the new phenate over conventional phenate. It can be seen that overbased sulphonates perform only as well as or marginally worse than base oil in this test.

3.4.3.2.5 Other Bench Tests

Representatives of the new detergent class have been evaluated in a series of standard bench tests to establish comparative performance data against conventional products (see Table 3.4.7).

The new phenate and new sulphonate detergents showed improved or equal performance in all tests over conventional products. In particular compatibility of the new components with conventional detergents of different type was much improved over that for 2 conventional detergents of different type:
compatibility (new phenate/conventional
sulphonate) > >
compatibility (conventional phenate/
conventional sulphonate)

Table 3.4.7: Comparison of new Detergents with Conventional Products in Bench Tests

	New Phenate of Conventional Phenate	New Sulphonate of Conventional Sulphonate
1. Demulsibility (ASTM D1401)	improved	—
2. Centrifuge Test Sludge (AF2C5)	improved	—
TBN Loss %	improved	—
3. Rust Protection (ASTM D665B)	improved	—
4. Panel Coker (330°) Deposits	improved	improved
5. Wear Protection FZG	equal	—
6. Compatibility with conventional Phenate/Sulphonate	much improved	much improved

226

3.4.3.2.6 Reciprocating Rig Wear Test

This test was designed to simulate ring/cylinder contact under typical marine engine operating conditions. The test employed high specimen temperatures (250°) in the presence of sulphuric acid to simulate the corrosive environment caused by the combustion of residual fuel oils.

Table 3.4.8: Reciprocating Rig Wear Test Results
Marine Formulations Containing new and Conventional Phenate

H_2SO_4 Added (% W/W)	%Base Removed		Total Wear (Microns)		Friction Co-Efficient	
	New P	Conv.P	New P	Conv. P	New P	Conv.P
0.0	0	0	0	0	0.050	0.101
2.0	56	61	0	0	0.059	0.101
2.5	70.6	77	0	18.2	0.073	0.106
3.0	84.7	91.9	14.5	68.9	0.084	0.108

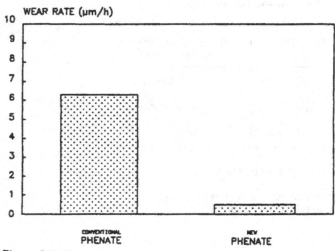

CAMERON–PLINT MACHINE TEST
MARINE FORMULATIONS

Figure 3.4.12

The improved performance of the new overbased phenate over the conventional phenate type can be observed in the results table (Table 3.4.8). The new phenate is more effective at forming a boundary film which acts as a barrier to corrosive attack at the surface and reduces the rate of wear (see Fig. 3.4.12).

3.4.3.2.7 Engine Tests

A representative of the new detergent class, a highly overbased phenate (400 TBN), has been evaluated in a number of bench engine tests. These include a Petter AVB Marine Test and the Bolnes Engine Test, specifically designed to evaluate marine formulations.

The Petter AVB Marine Test employs a radio-active ring wear measurement technique and candidate oils are evaluated alternately between reference oils over the life of a cylinder linear. Marine cylinder lubricants (MCLs) formulated with the new overbased phenate were found to exhibit equivalent performance to a conventional MCL at equal TBN. Evaluation of the new phenate in a MCL formulated to 50 TBN led to marginally poorer performance than the 70 TBN reference oil (see Table 3.4.9).

The Bolnes Engine Test uses a 3 cylinder marine 2-stroke engine with each cylinder separately fed allowing concurrent evaluation of 3 different oils. Marine formulations (70 TBN) containing the new phenate were found to perform at an equivalent standard to a commercial MCL (70 TBN) in this test, both in terms of total ring weight loss and piston rating.

Table 3.4.9: Petter AVB Marine Test Results

Candidate Oil	Oil TBN	Performance Relative to Reference Oil
A	70	equivalent
A	70	equivalent
C	50	poorer

Candidate Oils A—C contain new phenate

Reference oil = commercial marine cylinder lubricant
(70 TBN, V100 = 18 cSt)

Table 3.4.10: Comparison of new and Conventional Detergents

Property	Conv. Phenate	New Phenate	Conv. Sulphonate	New Sulphonate	Conv. Salicylate	New Salicylate
Typical TBN (ASTM D2896)	250	400	400	430	280	400
Hydrolytic Stability	good	very good	moderate		good	
Oxidation Stability	very good	excellent	poor		very good	
Thermal Stability	excellent	excellent	excellent	excellent	excellent	excellent
Detergency	good	very good	good	excellent	very good	
Rust Inhibition	low	moderate	good		low	
Anti-oxidant Effect	very good	excellent	none	excellent	very good	
Friction Effect	moderate	excellent	moderate	excellent	good	excellent

3.4.4 Conclusion

A new class of detergent additive has been developed which offers significant improvements in properties and performance over conventional products. New high TBN phenates and salicylates are now available with excellent performance credentials. Benefits include improved anti-oxidant, friction reduction and wear control performance. Furthermore, the chemistry associated with this new technology is sufficiently flexible to allow the production of "designer detergents" to meet the requirements of particular applications.

3.4.5 Acknowledgement

The authors wish to acknowledge the contribution of Dr. A. Moore (BP Research Sunbury) in designing the bench friction and wear tests and providing data.

3.4.6 References

(1) Marsh, J.F.: Colloidal Lubricant Additives, Chemistry and Industry (1987), 470–473.
(2) Tricaud, C.; Hipeaux, J.C.; Lemerle, J.: Micellar Structure of Alkaline Earth Metal Alkylarylsulphonate Detergents. Ostfildern 1986. P 9.3-1 – 9.3-7. Additives for Lubricants and Operational Fluids, 5th International Colloquium, Technische Akademie Esslingen 14/1/86 – 16/1/86. Ostfildern.
(3) Glavati, O.L.; Fialkovskii, R.V.; Marchenko, A.I.; Premyslov, V.Kh.; Alekseev, O.L.: Stabilisation of Colloidal CaCO, Dispersions in Hydrocarbons containing Anionic Furfactants Kolloid Zh (1970), 42, 26–30.
(4) Markovic, I.; Ottewill, R.H.; Cebula, D.J.; Field, I.; Marsh, J.F.: Small Angle Neutron Scattering Studies on Non-Aqueous Dispersion of Calcium Carbonate Colloid and Polymer Sci. (1984), 262, 648–656.
(5) Morin, S.V.; Pavlova, T.V.: Methods of Preparation of High Alkalinity (Overbased) Additives of the Alkylphenol Type (Review of Patents). Khim. Tekhnol. Topl. Masel (1978), 3, 61–63.
(6) Inoue, K.; Watanabe, H.: Micelle Formation in Detergent-Dispersant Additives in Non-aqueous Solutions, J. Japan. Petrol Inst. (1981), 24(2), 92–100.
(7) Inoue, J.: Carbon-13 Nuclear Magnetic Resonance Study of Reversed Micellar Structure of Calcium Dodecylsalicylates in Chloroform. J. Japan, Petrol Inst. (1982), 25(5), 335–339.

3.5 Synthesis of Additives Based on Olefin-Maleic Anhydride Reactions

G. Deak, L. Bartha and J. Proder
Veszprem University of Chemical Engineering, Hungary

Alkenyl succinic anhydrides and olefin-maleic anhydride copolymers were synthesized from straight and branched olefins which were reacted with alcohols and amines. Some of these compounds showed rust preventing, pour point depressing or emulsifiying effects in lubricants. Relationships were established between the efficiency and structural properties of the additives.

3.5.1 Introduction

Many industrial additives, including several of those used in petroleum products are produced by reacting an olefin — usually alpha olefin — and maleic anhydride. In this reaction either alkenyl succinic anhydride of maleic anhydride — olefin copolymers are produced depending upon the reaction parameters. The potential use of the reaction products depends not only on this difference, but also on the characteristics of the olefin raw material. According to data in the literature the carbon number of olefin molecules used in such reactions is within the range 2 to 60, and the structure of the olefin molecules can be different. Most of the suggested hydrocarbons are, however, alpha olefins. In the present paper the reaction of maleic anhydride and various olefins — polyethylene and atactic polypropylene degradation products, crack gasoline oligomers — were investigated and the potential use of the produced anhydrides or copolymers and some of their derivatives as additives for lubricants are evaluated and correlations among the performance and olefin type are tried to be established.

3.5.2 Materials

Since the effects of the olefin structure on the performance of the anhydrides or copolymers were to be investigated, olefins in a wide molecular mass range having different CH_3/CH_2 ratio were selected for the experiments (Table 3.5.1).

Raw materials were mixtures of hydrocarbons with the exception of alpha octadecene. Alpha olefins (AO-X) were commercial products (Fluka, Germany). Mixtures PE-X and APP-X were prepared by thermal degradation of polyethylene and atactic polypropylene, resp. that was followed by distillation into narrower fractions. Samples KO-X were prepared from a 32 — 112° C cat. crack gasoline by oligomerization and subsequent fractionation.

Table 3.5.1: Olefin raw materials

Property	Alpha olefins[1]			Polyethylene degradates			
	AO-14-16	AO-16-18	AO-18	PE-1	PE-2	PE-3	PE-4
Boiling range, °C	14–16[2]	16–18[2]	18[2]	120–180	180–210	240–300	300–350
CH_2/-CH_3 ratio[3]	8.6	8.2	8.9	3.1	4.8	8.0	9.0
Olefin distribution[4]							
trans-	5	8	–	2	4	10	9
vinyl-	76	63	97	58	52	39	28
vinylidene-	19	29	–	7	6	5	4
Σ alpha olefins	95	91	97	65	58	44	32

Property	Atactic polypropylene degradates				Oligomers from crack gasoline			
	APP-1	APP-2	APP-3	APP-4	KO-1	KO-2	KO-3	KO-4
Boiling rate, °C	120–182	182–260	240–280	270–340	120–180	170–205	200–225	220+
CH_2/-CH_3 ratio[3]	1.1	1.2	1.7	1.5	1.3	1.2	1.1	1.0
Olefin distribution[4]								
trans-	5	9	16	8	77	71	70	74
vinyl-	–	–	–	–	–	–	–	–
vinylidene-	58	68	76	77	–	–	–	–
Σ alpha olefins	58	68	76	77	<1	<1	<1	<1

(1) Fluka, Germany
(2) Carbon number range
(3) Determined as in (4)
(4) Determined as in (5)

232

While alpha olefin samples (AO-X) contained almost no internal double bonds, oligomers made from gasoline contained only an insignificant amount of alpha olefins. About 50 — 70 % of the double bonds in the polymer degradation products were terminal double bonds. Those in products obtained from polyethylene were in vinyl and those obtained from atactic polypropylene were in vinylidene groups.

Unfortunately the boiling range and average molecular mass range in different olefin types were not identical and especially in the case of alpha olefins the ranges studied were rather limited. This latter olefin type has been widely investigated. In the present work more emphasis was put on studying other olefin types — polyolefin degradation products and olefins obtained from crack gasoline.

Other materials used in the experiments were of technical grade.

3.5.3 Methods

Analyses of raw materials, intermediate and final products and performance testing of the latter were carried out according to standard test methods or other methods described in the literature.

Olefin — maleic anhydride adducts and copolymers were prepared by processes known from the literature (1, 2).

The addition of olefins and maleic anhydride was carried out under inert atmosphere within the 160 — 200° C temperature range (depending on the type of olefin being alpha olefins easiest to be reacted) using a 1 : 1 olefin : maleic anhydride molar ratio and a reaction time of 2—6 hours. Products were purified by filtering and distillation.

Olefin — maleic anhydride copolymers were prepared by a radical initiated catalytic reaction in xylene solvent at a temperature within the 100 — 160° C range, using a 1 : 1 molar ratio and a reaction time of 5 — 12 hours. Products were purified by distillation.

3.5.4 Olefin-Maleic Anhydride Adducts

Thermal reaction of olefins and maleic anhydride resulted mostly in alkenyl succinic anhydrides as it is illustrated in Table 2.5.2 and Figs. 3.5.1 — 3.5.3. The average molecular mass and the molecular mass distribution of the reaction products show that one olefin molecule reacted with one maleic anhydride molecule (Figs. 3.5.1 — 3.5.2). This is also indicated by the IR spectrum of the products (Fig. 3.5.3).

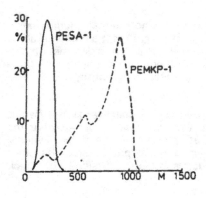

Figure 3.5.1:
Molecular mass distribution of the reaction products of a polyethylene degradate (PE-1) and maleic anhydride

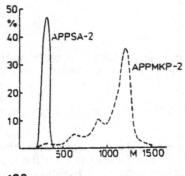

Figure 3.5.2:
Molecular mass distribution of the reaction products of an atactic polypropylene degradate (APP-2) and maleic anhydride

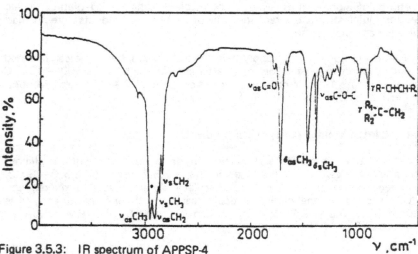

Figure 3.5.3: IR spectrum of APPSP-4

Table 3.5.2: Alkenyl succinic anhydrides

Property	AOSA-18	PESA-1	APPSA-2	KOSA-2
Raw material	AO-18	PE-1	APP-2	KO-2
$\overline{M}_n$*	364	205	299	294
$\overline{M}_w/\overline{M}_n$*	1.01	1.02	1.05	1.01
Yield, %	89	91	82	66

* by GPC

The ease of the occurrence of the reaction is, however, greatly dependent upon the type of the olefin. This can be judged by the yields of the anhydrides. While there are no significant differences between the reactivity of olefins having vinyl or vinylidene double bonds, those containing trans double bonds are rather difficult to be reacted with maleic anhydride.

3.5.5 Olefin-Maleic Anhydride Copolymers

The reaction of olefins and maleic anhydride resulted in the formation of copolymers if it was carried out in the presence of a peroxide catalyst. This is clearly demonstrated by data in Table 3.5.3 and by the molecular mass distribution of the products (Figs. 3.5.1 — 3.5.2). It is also apparent that alpha olefins are easiest to be reacted, degradation products of polyethylene coming next and those of atactic polypropylene are still harder to be reacted with maleic anhydride. The copolymers contain 4 — 5 olefin-maleic anhydride units.

Table 3.5.3: Olefin-maleic anhydride copolymers

Property	AOMKP-18	PEMKP-1	APPMKP-2	KOMKP-2
Raw material	AO-18	PE-1	APP-2	KO-2
$\overline{M}_n$	1760	814	928	350—350
$\overline{M}_w/\overline{M}_n$	1.80	1.66	1.40	1.08
B*	5.0	4.0	3.2	1.4
Yield, %	98	99	81	83

* Average number of olefin-maleic anhydride blocks in copolymer molecules

There is, however, one exception. In the case of olefin mixtures obtained from crack gasoline there is only a slight copolymerization, the main reaction also in the presence of peroxide catalysts being that of adduct formation. This is of course not really surprising since these olefin mixtures contain almost no alpha olefins and their molecules are highly branched ones.

3.5.6 Olefin-Maleic Anhydride Reaction Products as Raw Material for Additives in Lubricants

Products of olefin-maleic anhydride reactions were evaluated as raw materials for additives in lubricants.

3.5.6.1 Emulsifiers

Alkenyl succinic acids are highly polar compounds and their reaction with alcohols of basic compounds results in surfactants of many varieties.

In Fig. 3.5.4 results obtained with triethanolamine salts of some alkenyl succinic acid are presented. Emulsions contained 44 % oil, 50 % water and 6 % emulsifier. Best results were obtained with triethanolamine salts prepared from polyethylene degradation products and emulsifying effect was poorest with derivatives of atactic polypropylene, while that of crack gasoline was in between.

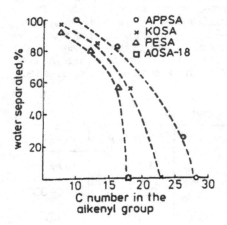

Figure 3.5.4: Emulsion stabilizing effect of alkenly succinic anhydrides.
(Amount of water separated after 24 hours from an emulsion containing 44 % oil, 50 % water and 6 % alkenyl succinic anhydride vs. the carbon number in the alkenyl group.)

236

Good emulsion stabilizing effect was reached only if the average molecular mass of the olefin raw material was high, i. e. if the carbon number of the alkenyl group was high: 18, 23 and 27 in the case of polyethylene degradates, crack gasoline derivatives and polypropylene degradates, resp.

These values are higher than those in usual surfactants. Fig. 3.5.4 also contains the value obtained with the triethanolamine salt of C_{18} alkenyl succinic acid (AOSA-18). It fits well into the curve of polyethylene derivatives.

3.5.6.2 Rust Prevention

Alkenyl succinic anhydrides were evaluated themselves as rust preventing agents. In Fig. 3.5.5 some of the results are shown that were obtained according to the test method ASTM D 665-60. Products of thermal reactions always proved to be better than those obtained in the presence of a peroxide, the only real exception being that of crack gasoline derivatives. Since they contained mainly anhydrides irrespective of their manufacturing process, this was to be expected. On the other hand the higher the average molecular mass of the olefins of a certain type the better the rust preventive effect (see the series of APP derivatives). It was also found that more complex olefin chain structures resulted in better rust preventive effect of the anhydride.

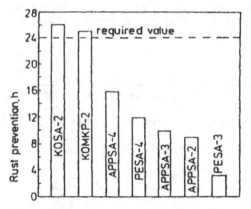

Figure 3.5.5: Rust prevention of olefin-maleic anhydride reaction products (ASTM D 665)

3.5.6.3 Pour Point Depressants

Copolymers of alpha olefins and maleic anhydride and some of their derivatives proved to be good pour point depressants and flow improvers of petroleum products (3). Since it was clear from the literature that linear olefins were good raw materials for this purpose, present investigations were restricted for studying polyethylene degradation products as potential raw materials for such additives. For the sake of comparison some individual alpha olefins were also investigated and results were compared to the performance of a commercial polymethacrylate additive, REONIT M (Nitrokemia, Hungary).

Experiments showed that polyethylene degradation products were suitable raw materials for flow improvers but they also showed that the performance of the additives were highly dependent on

— the molecular mass of the olefin mixture;
— the molecular mass distribution of the olefin mixture;
— the olefin : maleic anhydride molecular ratio in making the copolymer;
— the molecular mass of the copolymer;
— the form of the copolymer used (e. g. as a mono or diester, the alcohol used for esterification);
— the type of the base oil the additive is to be used in.

The evaluation of these effects are still under way. The following examples are meant only to illustrate some characteristics of pour point depressants made from polyethylene degradation products.

In Table 3.5.4 some data are given on copolymer monoesters found effective as pour point depressants in lube oils. They were selected so as that their average molecular mass and acid number be not too different and they were esterified with a mixture of C_{14-20} primary alcohols.

Table 3.5.4:   Pour point depressants

Property	AOKPE-14-16	AOKPE-16-18	PEMKPE-4
Copolymer	AOMKP-14-16	AOMKP-16-18	PEMKP-4
$\bar{M}_n$	1260	1380	1490
acid number, mg KOH/g	350	342	361
Alcohol	C_{14-20}	C_{14-20}	C_{14-20}
Solubility in oil	complete	complete	complete

Their effect on the pour point of two base oils are shown in Figs. 3.5.6 and 3.5.7, where the pour point depressing effect to the above mentioned commercial polyalkylmethacrylate (REONIT M) is also given.

238

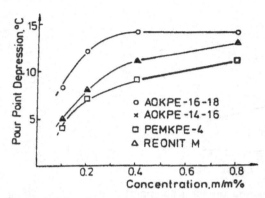

Figure 3.5.6: Pour point depressing effect of esters of olefin-maleic acid copolymers I
(Base oil: $VK_{100° C}$ = 5.1 mm^2/s; VI_E = 101; pour point = 16° C)

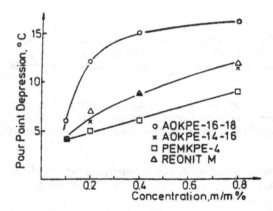

Figure 3.5.7: Pour point depressing effect of esters of olefin-maleic acid copolymers II
(Base oil: $VK_{100° C}$ = 9.6 mm^2/s; VI_E = 99; pour point = -18°C)

Since the process of making the olefin-maleic anhydride copolymers and their esters were identical, only the olefin raw materials were different, results show that olefin has great effect on the performance of the additives, hence there is scope for selecting better raw materials even of polyethylene degradation products. Results also show that these additives have comparable effects to those obtained with polymethacrylates. If the shear stability of these additives are compared to that of polyalkylmethacrylates they prove to be more advantageous (Fig. 3.5.8).

239

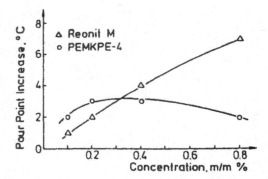

Figure 3.5.8: The effect of shearing on the pour point of an oil containing pour point depressant (Base oil: $VK_{100°C} = 9.6$ mm^2/s; $VI_E = 99$; pour point = $- 19°$ C; shearing by ultrasound TG-250 Schoeller) at $60°$ C for 7 minutes; intensity: 70 %)

3.5.7 Conclusions

Commercial alpha olefins, olefins obtained by thermal degradation of polyethylene and atactic polypropylene and olefins prepared by oligomerizing the olefin content of a crack gasoline were evaluated as raw materials for additives for lubricants.

While it was possible to prepare both alkenyl-succinic anhydrides and copolymers from the alpha olefins and polyolefin degradation products, reaction of maleic anhydride and crack gasoline oligomers resulted always in the formation of adducts.

Triethanolamine salts of alkenyl succinic acids had emulsifying effects if the alkenyl chain was long enough, i. e. C_{18} in the case of the polyethylene derivatives and still higher with the other olefin types.

The rust preventive effect of alkenyl succinic acids was highly dependent upon the structure of the alkenyl group. Larger and more branched groups ensure better performance. Thus olefins prepared from crack gasoline proved to be the best raw material for this purpose.

Monoesters of maleic anhydride-olefin copolymers prepared with polyethylene degradation products had a pour point depressing effect comparable to that obtained with derivatives of commercial alpha olefins or with a commercial polymethacrylate. The shear stability of these esters is, however, better than that of polymethacrylates.

240

3.5.8 References

(1) Kurada, K.; Voshimi, K.; Baba, T. (Mitsubishi Chemical Industries Co., Ltd): Low temperature Fluidity improver Can. CA 1,208,423, 1986.

(2) Beck, H.; Frassek, K.-H.; Holtvoigt, W.; Mukerjee, A. (AKZO GmbH): Olefin-Maleinsäureenhydrid-Copolimerisate. Gebr. Offen 27 48 367, 1979.

(3) Rossi, A. (EXXON Research and Engineering Co.): Additif pouvant abaisser le point de goutte des huiles lubrifiantes. Fr. Demande 75 31686.

(4) Bélafi-Réthy, K.; Bor, Gy.: A C-H vegyérték-rezgés infravörös abszorpciós sávjainak mennyiségi vizsgálata a szénhidrogének metil- és metiléncsoportjainál. MÁFKI Közlemények 6 (1965) 27—34.

(5) Bélafi-Réthy, K.; Bor, Gy.: Olefin szénhidrogének csoportelemzése IR spektrometriás módszerrel. MÁFKI Közlemények 8. (1967) 9—15.

3.6 Effects of the Structure on Various Performances of Polyisobutenylsuccinimides

K. Endo and K. Inoue, Nippon Oil Company Ltd., Yokohama, Japan

Abstract

In this paper, the various types of performance of polyisobutenylsuccinimides such as thermal stability, oxidation stability, dispersancy, solubility and wear prevention, were investigated in connection with their structure (molecular weights of polyisobutenyl groups, number of imino groups and treatment by various acids). These types of performance were evaluated using the panel coaking test, hot tube test, thin film oxygen uptake test, NO_x bubbling test and four-ball test.

3.6.1 Introduction

Succinimides play a major role in today's engine oils. They prevent deposition of sludge formed by deterioration of engine oils. They, therefore, are required as essential additives for high performance oils, such as API SG grade oils. in which higher sludge dispersancy properties are needed than in previous engine oil formulations. Their structure affects such types of performance as oxidation stability and thermal stability of engine oils.

There are some studies on the mechanisms for the action of poylisobutenyl-succinimides. Forbes and Neustadter reported[1] that polyisobutenylsuccini-mides did not form micelles in lubricating oil basestock. They estimated that the additive stabilized the sludge suspension by a steric mechanism.

Inoue and Watanabe reported[2][3] that high molecular weight succinimides formed micelles (the number of aggregations was 2–10); however, low molecular weight succinimides did not form micelles in nonpolar media. Concerning the interaction between succinimides and the other additives, they also reported [4] that there was strong interaction between succinimides and zinc dithiophos-phates, and there was interaction between succinimides and sulfonates, though it was not so strong, by using the vapor pressure osmometer. Shiomi et al. studied[5] the complex formation between zinc dithiophosphates and low molecular weight amines as a model of succinimides and reported that the ratio of complex formation of succinimides versus zinc dithiophosphates was 1 : 2 for most primary amines and was 1 : 1 for secondary amines of primary amines that have a branched chain.

In the present paper, the molecular effects of the structure of polyisobutenyl-succinimides on their performance are dicussed by means of laboratory bench tests.

242

3.6.2 Experimental

3.6.2.1 Materials

As polyisobutenylsuccinimides, both commercial and synthesized samples were used. A series of polyisobutenylsuccinimides with varying polyisobutene molecular weights (610 – 2350), number of polyisobutenyl groups (mono- and bistype) and polar amine head groups was synthesized. These additives were prepared by reacting the proper polyisobutenylmaleicanhydride with the suitable polyamine in xylene as a solvent.

The acid treatment of polyisobutenylsuccinimides was made with boric acid, sulfuric acid phosphoric acid and 2-ethylhexanoic acid. The acid treatment was carried out at 80°C for boric acid and at 0°C – 6°C for others in hexane.

3.6.2.2 Apparatus and Procedure

The thermal stability was evaluated by the panel coking test [6] and hot tube test manufactured by Komatsu Engineering Co. Evaluation was carried out by the lacquer level of the used test tube. A schematic figure of the apparatus is shown in Figure 3.6.1. In evaluating the formulated engine oil, we use the more severe conditions in general. In the present study, the panel coking test was performed at 280°C for 3 hours; however, ordinarily this test is carried out the temperature range of 300 to 320°C for 3–24 hours in evaluating SF or SG engine oils. In the hot tube test, the ordinary temperature is 280 – 300°C.

The oxidation stability was evaluated by the Thin Film Oxygen Uptake Test (TFOUT)[7].

The detergency and dispersancy were evaluated by the NO_x bubbling test. The test method was as follows. No gas, O_2 gas and N_2 gas were bubbled into the sample oils containing the degradation promoter and water for 5 hours. Evaluation was carried out be measuring the pentane insolubles in the tested oils (ASTM D893–75).

The antiwear performance was determined by the four-ball test (ASTM D2783).

Details of the test conditions are given in Table 3.6.1.

Unless otherwise noted, SAE 10 basestock was used as a diluent for most of the experiments.

Table 3.6.1: Test Conditions

Panel coking test	Panel temperature:	280°C
	Oil temperature:	100°C
	Test duration:	3 hours
	Test cycle:	15 sec (splash)/45 sec (stop)
Hot tube test	Tube temperature:	280°C
	Sample volume:	5 ml/16h
	Test duration	3 hours
TFOUT	Temperature:	120°C
NO_x bubbling test	Temperature:	140°C
	NO:	5 ml/min
	O_2:	50ml/min
	N_2:	111 ml/min
	Promoter	Caprylic acid 4 mass%
		Tetrahydronaphthalene 1 mass%
	Sample volume:	95 g
	Test duration:	5 hours
Four-ball test	Temperature:	80°C
	Load:	30 kg
	Rotation speed:	1200 rpm
	Test duration:	30 min

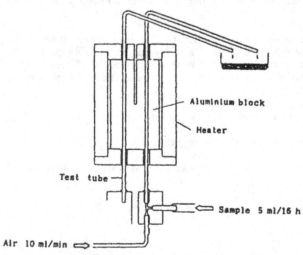

Figure 3.6.1: Schematic Figure of Apparatus of Hot Tube Test

244

3.6.3   Results and Discussions

3.6.3.1 Thermal Stability

Figure 3.6.2 shows the relation between the amount of carbon deposit versus the concentrations of polysobutenylsuccinimides in the panel coking test. Irrespective of the acid treatment, the amount of deposits increased with increasing concentrations of polyisobutenylsuccinimides. Especially, polyisobutenylsuccinimides treated by boric acid were the most effective in preventing coking deposition, more than untreated polyisobutenylsuccinimides.

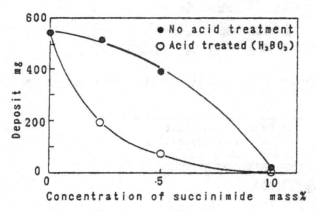

Figure 3.6.2:   The results of Panel Coking Test

Figure 3.6.3 shows the effects of the molecular weights of polyisobutenyl-succinimides on thermal stability in the panel coking test and hot tube test. It appeared that the smaller the molecular weight of the succinimides, the better the thermal stability: however, there are not any clear tendencies. No distinction could be seen from the difference in the number of nitrogen, i.e. NH groups, and polyisobutenyl groups (mono or bis) in either the panel coking test or the hot tube test.

As shown in Figure 3.6.3 the acid treatment of polyisobutenylsuccinimides was found to be very effective. In Figure 3.6.4 polyisobutenylsuccinimides treated by boric acid and sulfuric acid showed better results in both the panel coking test and hot tube test in comparison with untreated polyisobutenyl-succinimides.
Especially in the polyisobutenylsuccinimides treated with boric acid on all its active nitrogen, there were almost no deposits in either the panel coking test or the hot tube test. However, marked effects could not be seen from polyiso-butenylsuccinimides treated with 2—ethylhexanoic acid and phosphoric acid.

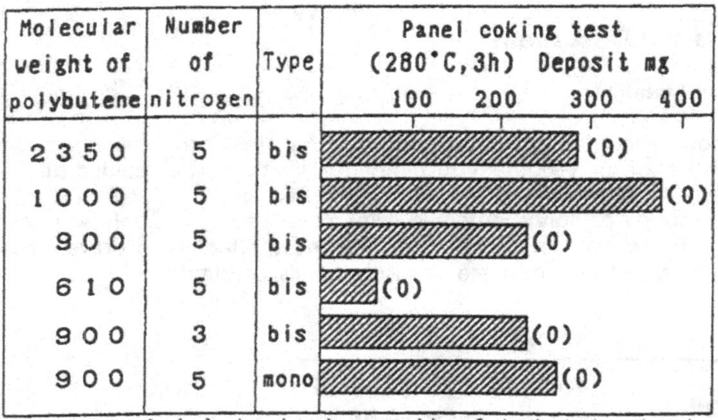
Molecular weight of polybutene
Number of nitrogen
Type
Panel coking test (280°C, 3h) Deposit mg
100 200 300 400
2350 5 bis (0)
1000 5 bis (0)
900 5 bis (0)
610 5 bis (0)
900 3 bis (0)
900 5 mono (0)

*Inside of ( ) is test result of hot tube test.

Figure 3.6.3: Effect of the Structure on the Thermal Stability of Succinimides

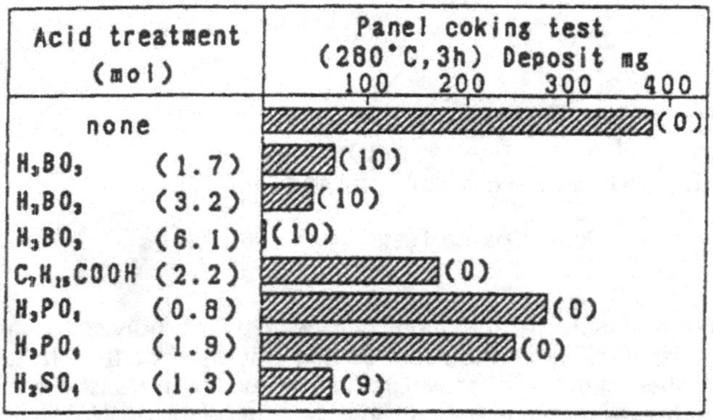
Acid treatment (mol)
Panel coking test (280°C, 3h) Deposit mg
100 200 300 400
none (0)
H₃BO₃ (1.7) (10)
H₃BO₃ (3.2) (10)
H₃BO₃ (6.1) (10)
C₇H₁₅COOH (2.2) (0)
H₃PO₄ (0.8) (0)
H₃PO₄ (1.9) (0)
H₂SO₄ (1.3) (9)

*Inside of ( ) is test result of hot tube test.

(polyisobutene's m.w = 1000)

(nitrogen = 5, bis-type)

Figure 3.6.4: Effect of the Acid Treatment on the Thermal Stability of Succinimides

3.6.3.2 Oxidation Stability

Figure 3.6.5 shows the relationship between the induction time in TFOUT and the concentration of polyisobutenylsuccinimides. The molecular weight of polyisobutenyl group of succinimides used here was 1000. They were bis-type and had 5 nitrogens for each molecule.

The induction time increased with an increase in the concentration of polyiso-butenylsuccinimides. With the exception of 2-ethylhexanoid acid, the induction time was prolonged by the acid treatment.

Kimura and Ishida reported[8] that polyisobutenylsuccinimides have the anti-oxidant performance since they have the active hydrogen, which acts as a radical trapper in the ethylene group adjacent to the imino groups. Thus the reason that the acid treated polyisobutenylsuccinimides have the antioxidant performance could be considered to be as follows.

1) the increase in the ability as a radical trapper by acid treatment

2) the addition of performance as a hydroperoxide decomposer by acid treatment.

To clarify the mechanisms mentioned above, we determined the radical trapping ability of succinimide. 1,1-Diphenyl-2-picrylhydrazyl (DPPH) was used as a model radical.
The test conditions are as below.

Test temperature: 114°C
Concentration: Succinimide = 0.10 mmol/l
DPPH = 0.10 mmol/l
Solvent: Monochlorobenzene
Detecting method: Adsorption of visible spectrum (532 nm)

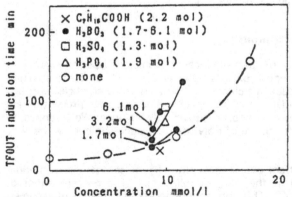

Figure 3.6.5: Effect of Concentration of Succinimide and Acid Treatment on TFOUT Induction Time

Figure 3.6.6 shows the rate of trapping of DPPH by succinimides. The greater the decrease in the retention ratio of DPPH as a stable free radical, the higher the ability of the additives as a radical trapper. In Figure 3.6.6, it can be seen clearly that polyisobutenylsuccinimides treated with boric acid increased the retention ratio of DPPH in proportion to the amount of acid treatment. This fact showed that the hydrogen release rate had been delayed and, therefore, the radical trapping ability of polyisobutenylsuccinimides had been reduced by boric acid treatment. It was expected from these results obtained here that the improvement of antioxidant performance of acid treated polyisobutenyl-succinimides as shown in Figure 3.6.5 was not caused by the above-mentioned hypothesis 1). And it has been known that boric acid and sulfuric acid have the ability of hydroperoxide decomposition [9][10]. So it might be obtained by added acids, probably because of the ability of hydroperoxide decomposition of these acids.

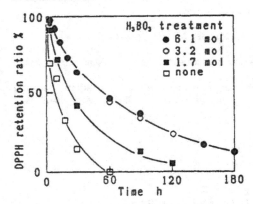

Figure 3.6.6:   Reaction of Succinimide and DPPH

3.6.3.3 Sludge Preventing Performance

The most important type of performance of ashless dispersants such as succinimides is to prevent deposition of sludge caused by NO_x, precombustion fuel and its oxidation products which are incorporated in the crankcase through blow-by. It has been considered that the mechanisms of sludge prevention are deactivation of sludge precursor and/or dispersion of sludge. We intended to evaluate these tapes of performance of polyisobutenylsuccinimides by the NO_x bubbling test.

Figure 3.6.7 shows the effect of the structure of polyisobutenylsuccinimides on the pentane insolubles in the NO_x bubbling test. The concentrations of succinimides were 12 mmol/l. This corresponds to about 5 mass% of commer-cially available bis-type succinimide. The molecular weight of polyisobutenyl group is about 1000. The increase in the molecular weights of polyisobutenyl

248

group decreases in pentane insolubles, showing that the sludge preventing effect had been improved. Figure 3.6.7 also shows the effect of mono- or bis-type succinimides on the sludge preventing effect. As seen in Figure 3.6.7 bis-type succinimide showed better sludge prevention effect than mono-type at the range of low molecular weights of polyisobutenyl groups.

The number of nitrogen was found to be independent of the sludge preventing effect. The change of the number of nitrogen, that is, the number of active polar groups, did not affect the results. The results show that the strength of the interaction between succinimides and insolubles was not always in proportion to the number of active nitrogen in the succinimides.

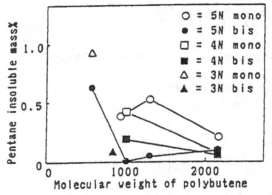

Figure 3.6.7: Sludge Preventing Effect of Polyisobutenylsuccinimides

Figure 3.6.8 shows the results of the NO_x bubbling test of the acid treated succinimides. The polyisobutenylsuccinimides used were bis-type and had a number of nitrogen of 5. The molecular weight of polyisobutenyl group was 1000. It can be seen that the acid treatment with boric acid and phosphoric acid lowered the sludge preventing performance. This may be due to the lowering of the ability of solubilization and/or dispersancy of succinimides by masking to imino groups.

The NO_x bubbling test was also carried out in the presence of overbased Ca sulfonates. The purpose of this test was to put stress on the evaluation of dispersancy by introducing carboxylate, which is considered to be a major component in the actual engine sludge in the test oils. As shown in Figure 3.6.9 though there was 0.3 % of insolubles for untreated polyisobutenylsuccinimides, the boric acid treatment reduced the insolubles to less than half. It can be seen that boric acid treatment was effective in preventing sludge in this condition. Table 3.6.2 shows the infra-red absorbance of tested oils to indicate the reason for this difference dependent on the presence of detergent. It can be seen that degradation was reduced by the presence of detergent, especially at 1550 cm^{-1} which indicates the formation of nitro compound.

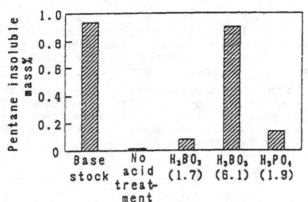

Figure 3.6.8: Effect of Acid Treatment on Sludge Preventing Performance

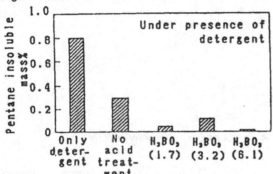

Figure 3.6.9: Effect of Acid Treatment on Sludge Preventing Performance under Presence of Detergent

Table 3.6.2: IR Adsorption of Tested Oils in NO_x Bubbling Test

Succinimide	Yes	Yes	Yes	Yes
Acid treatment	No	No	H_3BO_3 (1.7 mol)	H_3BO_3 (3.2. mol)
Detergent	No	Yes	Yes	Yes
IR adsorption (0.1 mm) 1550 cm^{-1} 1630 cm^{-1} 1720 cm^{-1}	0.309 0.372 0.438	0.093 0.263 0.339	0.106 0.260 0.297	0.078 0.342 0.284

250

3.6.3.4 Antiwear Performance

Figure 3.6.10 shows the results of the four-ball wear test. Polyisobutenylsucci-nimides treated with boric acid and 2-ethylhexanoic acid were effective to reduce wear in comparison with untreated polyisobutenylsuccinimides. How-ever, phosphoric acid and sulfuric acid were not effective. Improvement of anti-wear performance by boric acid and 2-ethylhexanoic acid treatment may be due to the improvement of adsorption ability by the formation of salt. Lowering by sulfuric acid and phosphoric acid may be due to corrosive wear.

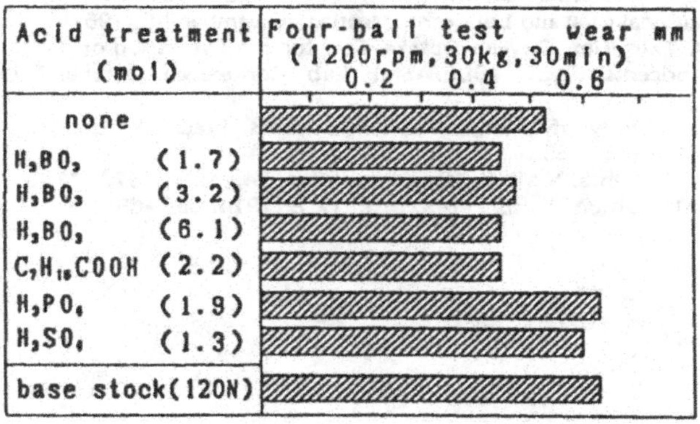

Acid treatment (mol)	Four-ball test wear mm (1200rpm,30kg,30min) 0.2 0.4 0.6
none	
H_3BO_3 (1.7)	
H_3BO_3 (3.2)	
H_3BO_3 (6.1)	
$C_7H_{15}COOH$ (2.2)	
H_3PO_4 (1.9)	
H_2SO_4 (1.3)	
base stock(120N)	

(polyisobutene's m.w = 1000)
(nitrogen = 5 , bis-type)

Figure 3.6.10: Effect of the Acid Treatment on the Antiwear Performance

3.6.4   Conclusion

1. Bis-type succinimides have higher sludge preventing ability than mono-type.

2. The higher the molecular weight of polyisobutenylsuccinimide is, the higher the performance of sludge preventing becomes.

3. The thermal stability and oxidation stability of polyisobutenylsuccinimides were improved by treatment with acids.

4. Sludge preventing ability of polyisobutenylsuccinimides was lowered by the acid treatment.

5. Acid treatment of succinimides with boric acid and 2-ethylhexanoic acid improves the antiwear performance.

251

3.6.5 References

(1) Forbes, E.S.; Neustadter, E.L.: The mechanism of action of polyisobutenyl succi-nimide lubricating oil additives. Tribology, **5** (1972) 72—77.

(2) Inoue, K.; Watanabe, H.: Micelle Formation of Detergent-Dispersant Additives in Nonaqueous Solution. J. Jap. Petrol. Inst., **24**, 2 (1981) 92—99.

(3) Inoue, K.; Watanabe, H.: Interactions of Engine Oil Additives. ASLE Trans., **26**, 2 (1983) 189—199.

(4) Inoue, K.; Watanabe, H.: Interactions between Engine Oil Additives. J. Jap. Petrol. Inst., **24**, 2 (1981) 101—107.

(5) Shiomi, M.; Tokashiki, M.: Prepr. Annual Mtg. Jap. Trib. Inst., **28**, (1982) 69—73.

(6) Rounds, F.G.: SAE National Fuel and Lubricants Meeting, November 7-8 (1957).

(7) Ku, C.; Hsu, S.M.: A Thin Film Oxygen Uptake Test for the Evaluation of Auto-motive Crankcase Lubricants. Prepr. ASLE/ASME Lub. Conference, October 5-7 (1982).

(8) Kimura, S.; Ishida, N.: Study of Ditergent as Antioxidants. Prepr. Annual Mtg. Jap. Petrol. Inst., **13**, (1970) 50—55.

(9) Sakaguchi, H.; Kamiya, Y.; Ohta, N.: Bull. Japan Petrol Inst. 14, No. 1 (1972) 71.

(10) Ohkatu, S.: Oxidation Inhibition. J. Jap. Petrol. Inst., **17**, 8 (1974) 686—693.

3.7 Resistance of Ashless Dispersant Additives to Oxidation and Thermal Decomposition

L. Bartha and J. Hancsok, Veszprem University of Chemical Engineering, Hungary
E. Bobest, Komarom Petroleum Refinery, Komarom, Hungary

Ashless dispersants of the alkenyl-succinic anhydride derivatives type were prepared. The resistance of these compounds to oxidation and thermal decomposition were investigated by thermoanalytical methods. Results were compared to those obtained in the panel coking test of these additives at 300° C in a multifunctional package. It was found that data obtained by DSC can be used for rating the stability of polyisobutenyl-succinimides against thermooxidation.

3.7.1 Introduction

In the past two decades the alkenyl-succinic anhydride based ashless dispersant additives were successfully used in formulation of motor oils. In the last few years one of the main fields of this additive type has been the optimization of the structure (1, 2).

Structure modifications were generally carried out by empirical methods because there were no exact correlations between the additive characteristics and the performance test results. In this work analytical methods which can be used for screening additives with different properties are of great importance. In this paper results are presented in which the application of a suitable thermo-analytical method is combined with some conventional laboratory test methods. Relationship between the structure of alkenyl-succinimides and their derivatives and stability to oxidation and thermal decomposition have been investigated. Hence the structure of the apolar side-chain and the polar group were changed.

3.7.2 Methods

The polyolefins, the alkenyl-succinic anhydrides and the final products were identified and evaluated by standard test methods or by methods described in the literature (3, 4, 5). The thermoanalytical measurements were carried out by a dynamic method (6) on a Differential Scanning Calorimeter (Perkin Elmer DSC-2C), the mass of the sample being 2 − 5 mg, heating rate 5°C/min, the oxygen flow rate through ofen 40 cm^3/min.

Three laboratory tests were used that separately characterize individual detergent-dispersant (DD) efficiencies and complement one another. In a previous paper it was shown that results obtained by these methods can be correlated to results of engine test. The methods (4) used were as follows:

a) The dispersion stabilizing efficiency of the DD additives was studied by a method based on centrifugation. Detergent index (DI) in a range of 0—100 % was evaluated.

b) The washing efficiency (M) was measured by the modified Zaslavskii's methods. The efficiency was evaluated within 0—125 mm interval.

c) The deposit formation blocking efficiency at high temperature (300°C) was determined by the modified Jolie's panel coking method. After a 9 hours test period the deposit on the panel was determined in mg and the visual colour merit number (0—10) was estimated.

The ashless DD additives were synthesized by laboratory processes published before (7, 8).

3.7.3 Materials

The properties of polyolefins for the production of ashless dispersants are given in Table 3.7.1. The molecular mass of the polymers was limited by the solubility in oils of the final products. Alkenyl-succinic anhydrides were prepared at 200°C at a 1 : 1 maleic-anhydride polyolefin molar ratio, without any catalyst (Table 3.7.2). From these intermediates alkenyl-succinimides and derivatives were prepared. The active ingredient of the samples was over 50 m/m% and known in the literature als follows in Table 3.7.3. Because of stepwise dehydration the composition was changed from amine salts — to amides and finally to imides in the process of acylation of tetraethylene pentamine. The effects on this change in the composition on the properties and DD efficiency of the additive were studied by two intermediates of a polyisobutenyl-bis-succinimide (PIBBSI/1 and PIBBSI/2).

Table 3.7.1: Properties of polyolefins

Properties	PE	PP	PIB
Olefin content, m/m%	45	50	98
M_n(1)	250	260	524
M_w	268	870	1454
$\alpha = \dfrac{M_n}{M_w}$	1.07	3.34	2.77
M_{max}	255	600	900

(1) Determined by gel-permeation-chromatography

254

Table 3.7.2: Properties of alkenyl-succinicanhydrides

Properties		PIBBA	PPBA	PEBA
Molecular mass[1]	M_n	690	340	374
	M_w	1540	980	538
	M_{max}	1010	660	450
$\alpha = \dfrac{M_n}{M_w}$		2.2	2.9	1.4
Acid number, mg KOH/g		38.7	43.6	53.1
$VK_{100° C}$, mm^2/s		39.6	38.3	37.2

(1) Determined by gel-permeation-chromatography

Table 3.7.3: Properties of ashless dispersants

Ashless Dispersant		Structure	IR Spectrum C=O (cm⁻¹)
PIBMSI	N-(tetraethylenetetramino)poly-butenyl-succinimide	R - CH - C(=O) - N-(CH₂-CH₂-NH)ₙ-H / CH₂-C(=O) n=4	1710, 1760
PIBBSI	N,N'-(tetraethylentriamino)-bis-polybutenyl-succinimide	R - CH - C(=O) - N(CH₂-CH₂-NH)₃-CH₂-CH₂-N - C(=O)-CH₂ / CH₂-C(=O) ... C-CH~R	1710, 1760, 1650
PIBMBSI	PIBMSI:PIBBSI Ratio 1:3	mixture	1710, 1760, 1650
PIBE	Pentaerithritol-polybutenyl-succinate	R - CH - C(=O) - O - CH₂ - C - CH₂ - OH / CH₂ - C(=O) - O - CH₂ ... CH₂ - OH	1735
PPMSI	N-(tetraethylene-tetramino)poly-propenyl-succinimide	R - CH - C(=O) - N-(CH₂-CH₂-(NH))ₙ-H / CH₂-C(=O)	
PEMSI	N-(tetraethylene-tetramino)poly-ethylene-succinimide	R - CH - C(=O) - N-(CH₂-CH₂-NH)ₙ-H / CH - C(=O)	1710, 1760
PBAE	Reaction product of polybutenyl-succinic-anhydride and tetra-ethylene pentamin and pentaerithritol	R - CH - C(=O) - O - CH₂ - C - CH₂ - OH / HO - CH₂ ... HO - CH₂ / CH₂ - C(=O) - NH-(CH₂-CH₂-NH)ₙ-H	1710, 1735

3.7.4 Results and Discussion

Analytical and DD properties of ashless dispersants are summarized in Table 3.7.4. The wide range of the TBN/N values (25.5 — 16.0) points to great differences in the ratios of amino-, imino- and imide groups in the additives.

There was a clear relationship between the DSC values and the structure of the additives. In Fig. 3.7.1 typical TG and DSC curves made by the dynamic method are shown for the additive PIBMBSI. Apparent breaking points were only on the DSC curve. It means that significant exotherm processes took place without greater changes in the mass of the sample. The most important breaking points of the DSC curve shown in Fig. 3.7.1 are given in Table 3.7.4 for all additives. In Fig. 3.7.3 it can be observed that temperatures of the first exotherm decomposition (between T_0 and T_1) increased in the order of mono-, mono-bis-, and bis-succinimides. In the case of the mono-bis-succinimide mixture the first exotherm decomposition process took place in a wider temperature range. The cause of differences in resistance of additives to thermal and oxidation decomposition shown by DSC curves is presumably the difference in the relative concentration of dispersants containing heat-sensitive amino- and imino groups. This tendency is consistent with practical experiences of motor oil formulations.

Similar increase of stability was observed in the series of bis-succinimides. Intermediates with greater TBN/N value and relatively less succinimide rings had lower stability.

The results suggest that DSC values could be used suitably for rating the thermal and oxidation stability of different succinimides and their intermediates both in development and quality control.

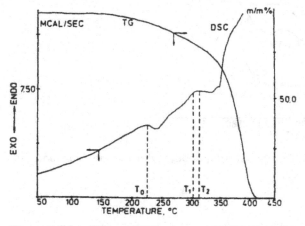

Figure 3.7.1: TG and DSC curves of PIBMBSI additive

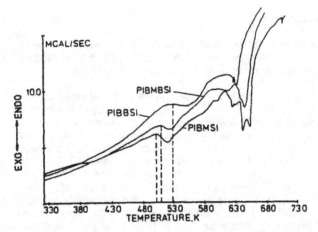

Figure 3.7.2: DSC curves of polybutenyl-succinimides

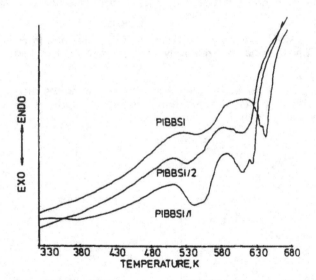

Figure 3.7.3: DSC curves of PIBBSI polyisobutenyl-bis-succinimide and its intermediates

258

Table 3.7.4: Analytical and DD properties of ashless dispersants (diluting oil content 40 – 60 m/m%)

Properties	Ashless dispersant								
	PIBMSI	PIBMBSI	PIBBSI/1[3]	PIBBSI/2[3]	PIBBSI	PIBPE	PPMSI	PEMSI	PIBAE
Nitrogen cont.(N), m/m%	3.07	1.72	1.30	1.23	1.19	–	1.60	1.50	0.65
Total Basic Number[1] (TBN), mg KOH/g	78.3	32.2	22.1	20.1	19.0	–	38.9	37.0	11.2
TBN/N	25.5	18.7	17.0	16.3	16.0	–	24.3	24.7	17.2
DSC values — Temp. of the 1. decomp., °C	188	219	239	241	244	207	165	195	209
Heat of 1. decomp., cal/g	–431	–163	–171	–119	–119	–413	–450	–410	–190
End temp. of the 1. decomp., °C	289	313	314	308	311	286	280	287	298
End temp. of the 2. decomp., °C	302	316	319	320	335	288	282	307	302
DD properties[2] — DI, % (max. 100)	90	89	89	87	87	27	82	9	96
W, mm (max. 125)	59	54	44	54	50	11	72	19	88

(1) MSZ 60,0111 Standard method
(2) Additive concentration 3 m/m% in SAE base oil
(3) Intermediates of PIBBSI production

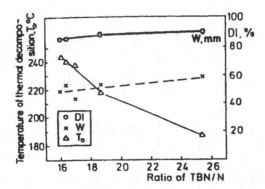

Figure 3.7.4: Temperature of decomposition and DD efficiency of different polyisobutenyl-succinimides

The mono-succinimide with polypropylene side-chain showed significantly lower stability then the polyethylene- and polyisobuthylene-succinimides.

The stability of the reaction products of polyisobutenyl-succinic anhydride and pentaerythrithol and pentaerytrithol-tetraethylenepentamine mixture was between the stability of mono-, and bis-succinimide dispersants. Comparing the potential DD efficiency of the additives (Table 3.7.4) shows that polyethylene-succinimide (PEMSI), and polyisobutenil pentaerythrithol-ester have significantly lower efficiency. With other species there were no great differences. The highest efficiency was achieved by the polyisobutenyl-succinic-anhydride-pentaerythrithol-tetraethylenepentamine reaction product. This was caused presumably by the great polar group in this dispersant which provides higher ability of adsorption to sludge particles in motor oils.

DD efficiency of polyisobutenyl-succinimides and their thermal and oxidation stability changed contrary to one another. From the point of stability mono-succinimide was the most effective dispersant.

Thermoanalytical properties of the dispersants and the results of their panel coking test are given in Table 3.7.5. Experiments were carried out in combination with commercial additives generally used for formulation of motor oils. Washing efficiency of the additive package in motor oils was measured before and after the coking test. In the case of oils containing polyisobutenyl-succinimides and polyisobutenyl-succinic-ester-amide test panels did not show significant differences of the deposit blocking efficiency.

Because of thermal and oxidation effects the highest decrease of washing efficiency (23 %) was observed with the composition containing mono-succinimide (PIBMSI). Washing efficiency of the ester-amide type dispersant (PIBEA) was practically unchanged. It seems likely that decomposition products of the additive also have dispersant efficiency.

Table 3.7.5: Results of the panel coking test of the ashless dispersant containing motor oils

Additive*	Rating of the panels after 9 hours		Washing efficiency, W (mm) (max. 125)		
	deposit (mg)	colour merit number (max. 10)	fresh oils	after coking test	change in the W (%)
PIBMSI*	16.0	7.5	104	80	− 23
PIBMBSI	14.2	7.5	101	93	− 8
PIBBSI	12.4	7.5	90	85	− 6
PIBPE	60.5	7.0	91	85	− 7
PPMSI	40.1	5.0	95	60	− 37
PEMSI	89.9	2.5	29	19	− 34
PIBAE	10.6	7.5	107	105	+ 2

* Composition of the test oils

  1.7 m/m% ashless dispersant
  2.1 m/m% overbased calcium sulphonate
  1.2 m/m% zinc-dialkyl-dithiophosphate
  0.5 m/m% polymethacrylate

The results back up the supposition that DSC values could be used for rating the resistance to oxidation and thermal decomposition only dispersants containing compounds of the same type. In addition they make it possible to reduce the number of the preliminary laboratory test thus making the screening of additives more efficient.

3.7.5 References

(1)   Raddatz, J.; Bartz, W.J.: Detergent-Dispersant additive Herstellung und Anwendung. 5th Int. Coll. Additives for Lubricants and Operational Fluids. 1986. Band II. 9.1-1 — 9.1-20.
(2)   US Pat. No. 4.548.724.
(3)   Hlavay, J.; Bartha, L.; Bartha, A.; Vigh, Gy.: Integrated Analytical Scheme for the Development of an atactiv polypropylene-based Olefin-polymer Producing Technology. J. of. Chromatography. 1982. 241. 121—128.
(4)   Bartha, L.; Deak, Gy.; Kantor, I.; Pechy, L.: Method for the Determination of the Optimum Composition of Detergent-Dispersant Engine Oil Additives. Hung. J. of Ind. Chem. Veszprem, 1979. 7. 359—366.
(5)   Bartha, L.: Investigation of Boundary Layers Arising trough Reactions of Engine Oil Additives with Aluminium, Conf. on Additives. 1983. Siofok. 6.1-6.17.
(6)   Lovasz, Cs.: Thermische Stabilität von Schmierstoffen. Erdöl und Kohle. Bd. 30. 5. 1977.
(7)   Hung. Pat. No. 170.349.
(8)   Grünekker, A.: Öllösliche, aschefreie Detergens-Dispergens-Zubereitung und Verfahren zu ihrer Herstellung. Ger. Offen. Pat. No. 2.417.868.

3.8 Anti-Wear Actions of Additives in Solid Dispersion

M.F. Morizur and O. Teysset
Institut Francais du Pétrole, Rueil-Malmaison, France

Abstract

We compare the action mechanism of different "non-soluble" additives: a dispersion of potassium triborate and overbased detergents formed by calcium carbonate or borate aggregates dispersed by a detergent. These additives act by producing a deposit issued from their mineral phase, on the friction surface.

With the potassium triborate, a solid film of borate, bound to the metal, is only detected on a negatively-polarized surface or in conditions of severe wear when friction surfaces have been activated by abrasion. On the contrary, with the overbased detergent, a solid film is always detected on the friction surfaces, even without activation by abrasion. This additive is decomposed in the contact zone by the effect of the elevated pressure.

This study shows that the method of placement in suspension of the inorganic phase by the detergents is important. It offers the advantage of forming deposits only in the friction zone, instantly and independently of the type and state of the friction surface.

3.8.1 Introduction

The authors investigate the action mechanism of additives based on inorganic solids in suspension. Anti-wear and extreme pressure additives act in boundary conditions by forming a solid film on the friction surfaces.

Standard thiophosphorus anti-wear additives act by their thermal decomposition products. Their anti-wear and extreme pressure effectiveness is related to the high reactivity of these decomposition products in the friction zone, and hence to the low thermal stability of the additives. This property is not always compatible with the increasing thermal stability requirements of engine and gearbox lubricants (1).

Additives based on solids in suspension such as potassium triborate and hyperbasic detergents display anti-wear and extreme pressure properties that are independent of this thermal stability.

During friction, the inorganic phase in suspension is deposited on the surface and forms a protective film (2) (3).

Different inorganic phases are tested here: potassium triborate, calcium carbonate, and mixtures of calcium borate and carbonate. They were placed in suspension by an ashless surfactant for potassium triborate, and by a detergent (sulfonate and sulfophenate) for the other substances.

The main problems that arise concerning the action mechanism of these additives are the following:

— decomposition mode of suspensions in the friction zone and formation mode of the boundary film,

— presence or absence of a chemical bond between the film and the substrate to explain the adhesion of the film in friction,

— relationship between the type of film and its tribological properties.

In this study, we shall examine the decomposition of the additives in oil, by heating and under an electric field. These two activation modes co-exist in the friction zone, in addition to the pressure increases and the presence of highly reactive abraded surfaces in contact (4).

In the second part, we shall examine the properties of these additives in the presence of a metallic surface (adsorption and reactivity).

We shall conclude by testing the additives in different tribological conditions. We shall characterize the films formed by their friction behaviour and their composition.

In this study, we shall compare the behaviour of the additives in solid dispersions with that of zinc dithiophosphates, which are conventional soluble anti-wear additives.

3.8.2 Experimental Condition

3.8.2.1 Lubricant

The lubricants used in this study consist of a base oil containing an anti-wear additive to be tested.

The additives tested and the lubricant composition are given in Tables 3.8.1 and 3.8.2. Sulfophenates are obtained by co-synthesis of sulfonates and phenates. In borate detergents, part of the calcium carbonate has been replaced by calcium borate. With the exception of potassium triborate, the additives used were synthesized at IFP. In this specific case, we use the commercial mixture, which is a dispersion of particles of triborate $K_2O(B_2O_3)_3$.

Table 3.8.1

Additives	TBN*	% (MASSIC)			B/Ca (MOLAR)
		B	Ca	S	
Neutral sulfonate	23.2	0	2.3	2.4	
Overbased sulfonate	419	0	16.1	1.9	
Overbased borated sulfonate (0.4)	414	1.6	15.1	1.8	0.4
Overbased borated sulfonate (1.3)	369	4.72	13.8	1.0	1.3
Overbased borated sulfophenate	329	1.63	12.7	1.9	0.4
Overbased borated sulfophenate	304	3.96	11.9	2.05	1.3

	ZN	S	P		
Isobutylpentyl ZN DTP	10.1	20.1	9.5		

	B	K	S	N	Ca
$K_2O(B_2O_3)_3$	7.6	8.9		0.22	0.09

* TBN: Total Base Number

Table 3.8.2

LUBRICANTS	% MASSIC		
	Additives	B	Ca
B* + neutral sulfonate	7.9		0.18
B + overbased sulfonate	10		1.6
B + overbased borated sulfonate (0.4)	13.1	0.21	2
B + overbased borated sulfonate (1.3)	13.8	0.65	1.9
B + overbased borated sulfophenate (0.4)	12.8	0.21	1.62
B + overbased borated sulfophenate (1.3)	16.4	0.75	1.95
B + $K_2O(B_2O_3)_3$	8.6	0.65	
B + Zn DTP	1	Zn 0.1	

*B: base oil → pure paraffin oil for adsorption tests 130 N for other experiments

3.8.2.2 Friction Tests

The friction materials are carbonitride steels and cast irons A and S (Table 3.8.3). Carbonitriding treatment is representative of the treatments used on automotive gear wheel parts. Cast iron A is a lamellar graphite cast iron representative of the cylinder metallurgy. Cast iron S is a ring cast iron, with modular graphite.

The friction conditions in the ring/linear zone and in the transmission components are given in Table 3.8.4. Our test conditions on the tribometer approach them partly.

Table 3.8.3

	Poisson Coefficient	Young Modulus	Vickers Hardness
35NCD4 Carbonitrided	0.3	21 000	800
Cast iron S	0.25	11 400	250
Cast iron A	0.25	12 000	250

Table 3.8.4

	Motor		Wear-Test ↓	Fatigue Test ↓
	Liner/Ring	Trans-mission	Alternating Tribometer	Three balls/cone
Hertz pressure (GPa)	0.01 → 0.8	0.5 → 1.5	2 (crossed cylinders) 0.85 and 0.7 (cylinder on plane)	3
Sliding rate (5)	100	0 → 30	100	~ 0
Sliding speed (m/s)	6 → 10	0.3 → 0.5	0.4	/
Temperature (°C)	250°C max.	120 → 130	120	120

3.8.3 Results and Discussion

3.8.3.1 Additive decomposition mode in homogeneous phase

In the first part, we examined the degree of decomposition of the additives subjected to heating at 180° C or to an electric field of 4000 V/cm, by the decrease in absorbence of the characteristic infrared peaks (Table 3.8.5).

These temperatures and electric fields are approximately the same as those found in the friction zone (4).

The results obtained confirm the sensitivity of ZnDTP to the temperature and to the electric field. This sensitivity varies according to the type of organic radical of the molecule.

Table 3.8.5

Additives	ISO C4C5 Zn DTP			Neopentyl Zn DTP		Overbased Sulfonate		Overbased Borated Sulfonate			Potassium Triborate
Bound	P-O-C	P=S	P=S	P-O-C	P=S	S-O	CO_3^{2-} ($CaCO_3$)	S-O	B-O	C-O ($CaCO_3$)	B-O
Wave number (cm^{-1})	1006.8	661	644	1003	667.4	1049.3	864.1	1045.4	1020	871.8	1072
Heating treatment (180° C; 8 h)	33	35	27	21	26	0	0	0	0	0	0
Treatment in an electrical field (4000 V/cm; 8 h)	3	2	5	Total decomposition		< 10* 0	0	< 0 0	0 10	0 0	56 − 15**

*   polarization of electrodes

**   increase of the absorbence (15 %)

In the same conditions, the hyperbasic sulfonates display practically no decomposition. It is well known that the sulfonate function is highly stable, but we expected a break of the micellar arrangement leading to a drop in the $CaCO_3$ content of the oil.

Potassium triborate is thermally stable. A sharp decrease in the borate content is observed in the neighbourhood of the anode, and a slight increase in the neighbourhood of the cathode, which is explained by a migration of the particles dispersed in the lubricant under the effect of the electric field (Fig. 3.8.1). Thus, electrophoresis of the triborate particles is possible between the surfaces polarized by friction (5). The borate contained in the borate detergent does not show the same behaviour.

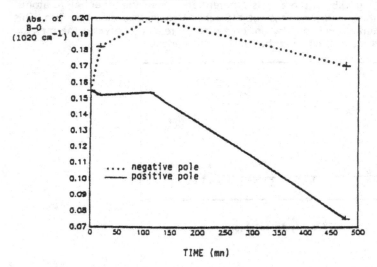

Figure 3.8.1: Migration of KB in an electrical field (4000 V/cm)

3.8.3.2 Behaviour of Additives in the Presence of Metallic Surfaces

Two types of test were conducted:

— characterization of the adsorbed phase on a nickel powder in equilibrium with the lubricant (120° C),
— ex situ characterization of the film formed on metal plates by immersion at temperature (180° C) or under electric fields (4000 V/cm).

3.8.3.2.1   Adsorption Isotherms

For the first type of test, we do not have a direct analytical method for the adsorbed layers in contact with the lubricant. Accordingly, a determination method on the remnant of the quantities adsorbed is employed. In this method, the lubricant hetero-atoms are determined by plasma spectroscopy for boron, and by X-ray fluorescence for the heavier elements, before and after contact with the lubricant. On the isotherms obtained, the values are divided by the relative proportions of each element in the additive molecule before adsorption. Thus, if the isotherms concerning an additive are not merged, the relative proportions in the adsorbed phase and in the oil are different.

For ZnDTP (Fig. 3.8.2), the zinc is differentiated from the other hetero-atoms. At 120° C, these additives react at the surface to yield a zinc-rich film. This zinc enrichment is explained by an ion exchange reaction between the metallic ion of the dithiophosphate and that of the adsorbent (6).

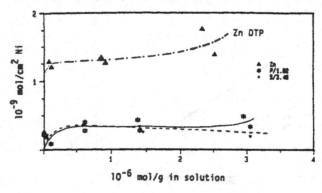

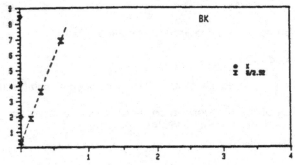

Figure 3.8.2:   Adsorption isotherm at 120° C

In the same concentration ranges, the isotherm levels are not reached for potassium triborate (Fig. 3.8.2). This reflects a great affinity of the additive for the surface.

With sulfonates, the proportions of sulfur and calcium are identical at the surface and in the oil. Hence the micellar arrangement appears to be the same in the adsorbed phase and in the oil (Fig. 3.8.3). The adsorption of these additives can only be physical and micellar.

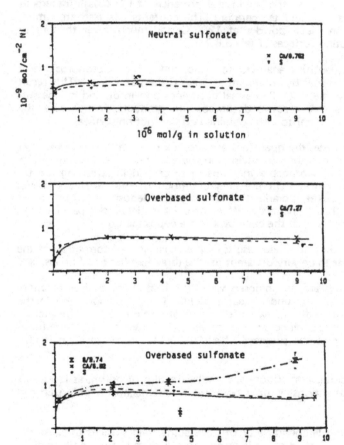

Figure 3.8.3: Adsorption isotherm at 120° C

3.8.3.2.2 Surface Reactivity

The surfaces treated by immersion in the lubricant are analysed by a Castaing microprobe and GDS (Glow Discharge Spectroscopy).

If no additive decomposition is detected in the oil, the quantities analysed on the metallic surfaces approach the detection limit (Table 3.8.6).

This applies to the samples treated in the presence of hyperbasic detergent by polarization or at elevated temperature. The contents analysed are similar to those obtained on metal powder at 120° C, and much lower than those measured on the friction surfaces (Table 3.8.7).

With ZnDTP, the quantities analysed for zinc, sulfur and phosphorus are at least hundred times higher by treatment at 180° C than at 120° C. This result can be explained by the ZnDTP thermal decomposition in oil and by the surface reactions of its very reactive decomposition products. In this case, the contents analysed are similar to these obtained on the friction surfaces.

With potassium triborate, the quantities analysed ex situ, on the surfaces, after treatment, are lower than those contained in the adsorbed phase in equilibrium with the lubricant. It is probable that rinsing succeeded in removing the triborate particles concentrated in the neighbourhood of the metallic surfaces. By contrast, the negatively polarized surface, a boron deposit of 0,3 μm is observed (Fig. 3.8.4). Thus, under the effect of electric fields, the particles migrate in the neighbourhood of the cathode and are deposited on it.

In conclusion, according to these results, the mechanisms of formation of the boundary film appear to be very different for the three families of additives.

The sulfonate micelles are decomposed very little if at all by surface reaction, at elevated temperature or under electric fields. Their decomposition in the contact zone, leading to the formation of a film of calcium salt on the friction surfaces, can only be explained by the mechanical stresses to which the lubricant is subjected, or to the presence of surface activated by abrasion in this zone.

In the presence of potassium triborate, an adhesive film of boron, as it appears in severe friction conditions, is only observed on the negatively polarized surfaces.

The effect of elevated temperature or of the electric field suffices to cause the ZnDTP to decompose and to make it react at the surface to form a film similar to that formed in friction.

Table 3.8.6

Treatment	Additive	Superficial concentrations (% atom)		
		S	Ca	Fe
Heating treatment (180° C, 8 h)	Overbased sulfonate	0.2	0.1	99.6
	Overbased borated sulfonate (1.3)	0.3	0.3	99.4
		S	K	Fe
	BK	0.3	0.1	99.5
		Zn	S	P
	iC$_4$-C$_5$ DTPZn	1.9	2.1	3
		S	Ca	Fe
Polarization (2000 V) without heating	Overbased sulfonate	+ 0.2 − 0.2	0 0.5	99.8 99.3
	Overbased borated sulfonate (1.3)	+ 0.2 − 0.6	0.2 0.2	99.6 99.2
		S	K	Fe
	BK	+ 0.1 − 0.1	0.1 0.2	99.8 99.7

273

Table 3.8.7

Test Conditions	Additive	Mobile Sample				Stationary sample			
		superficial				Superficial Concentrations (% Atom.)			
Carbonit./carbonit. P_{Hz} = 2GPA	overbased sulfonate	Ca 1.0	S 0.2	F 98.8		Ca 4.5	S 0.2	Fe 95.3	
	overbased borated sulfonate (0.4)	Ca 3.3	S 0.5	F 96.2		Ca 2.4	S 0.4	Fe 97.3	
	overbased borated sulfonate (1.3)	Ca 1.6	S 0.6	F 97.8		Ca 4.6	S 0.2	Fe 95.2	
	overbased borated sulfophenate (0.4)	Ca 1.8	S 0.6	F 97.6		Ca 8.3	S 0.7	Fe 91.1	
	overbased borated sulfophenate (1.3)	Ca 3	S 0.5	F 96.5		Ca 3.1	S 0.3	Fe 96.6	
	potassium triborate	K 0.2		F 99.8		K 0.4		Fe 99.7	
	Zn DTP	Zn 3.5	S 3.6	P 2.7	Fe 90.2	Zn 2.3	S 1.8	P 2.4	Fe 93.4
Carbonit./carbonit. P_{Hz} = 0.85 GPA	overbased sulfonate					Ca 2.3	S 0.3	Fe 97.5	
	overbased borated sulfonate (1.3)					Ca 2.5	S 0.2	Fe 97.4	
	overbased borated sulfophenate (0.4)					Ca 6.3	S 1.1	Fe 92.6	
	potassium triborate					K 0.2		Fe 99.8	
Cast iron/cast iron P_{Hz} = 0.7 GPA	overbased sulfonate	Ca 1.8	S 0.2	Fe 98.8		Ca 6.3	S 0.2	Fe 93.5	
	overbased borated sulfonate (0.4)	Ca 4.9	S 0.1	Fe 95.0		Ca 9.0	S 0.2	Fe 90.8	
	overbased borated sulfonate (1.3)	Ca 4.3	S 0.1	Fe 95.6		Ca 4.3	S 0.1	Fe 95.6	
	overbased borated sulfophenate (0.4)	Ca 2.9	S 0.2	Fe 96.9		Ca 7.4	S 0.2	Fe 92.4	
	overbased borated sulfophenate (1.3)	Ca 4.4	S 0.2	Fe 95.4		Ca 4.3	S 0.2	Fe 95.5	
	potassium triborate	K 0.7		Fe 99.3		K 1.8		Fe 98.2	
	Zn DTP	Zn 4.5	S 4.5	P 5.7	Fe 85.2	Zn 3.9	S 3.2	P 4.6	Fe 88.3

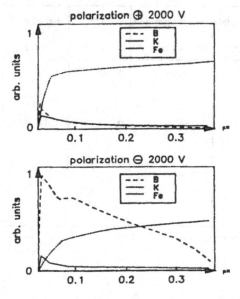

treatment : immersion in lubricant
containing potassium triborate

polarization ⊕ 2000 V

polarization ⊖ 2000 V

Figure 3.8.4:   GDL analysis of cast iron sample

3.8.3.3   Friction Tests

We shall first present the analytical results of the friction surfaces after the test, and shall then try to relate the data concerning the type of boundary film and its formation mode to the friction and wear results.

3.8.3.3.1   Rolling Tests

The cones resulting from the tests are analyzed by XPS (X-ray photon electron spectroscopy) and nuclear analysis (Tables 3.8.8 and 3.8.9).

In the presence of borate detergent, the friction surface is covered by a borate and carbonate film rich in Ca^{2+} ions, a few hundred angströms thick. The surface boron content, measured by nuclear analysis, lies between 1 and 2 %. It increased with the boron content of the oil.

Table 3.8.8

Additives	Film Thickness* (A)	Superficial Concentrations** (%)		B/Ca (arb. unit)	Fatigue Test Life (h)
		B(10 m)	OS(0.1 m)***		
Overbased borated sulfophenate B/Ca = 2.5	640	1.91	15 → 25	0.262	1.78
Overbased borated sulfonate B/Ca	640	1.39	20 → 30	0.17	1.34
Potassium triborate	− 0	0.04	6 → 13		1.0
Base oil					0.72

*   obtained by Rutherford backscattering analysis on Fe
**   obtained by nuclear reaction
*** analysed thickness

In the presence of potassium triborate, only a few traces of boron and potassium are detected in similar quantities outside the friction trace and on the friction trace.

For the thick film formed by the borate hyperbasic detergent, the XPS results show that the boron exhibits two chemical states in the boundary film (both being attributed to the borate):

— the first (189.3 eV), also present outside the friction trace, is hence attributed to the initial state of the borate in the oil,

— the second (191.3 eV), which is more oxidized, appears to indicate that the boron undergoes a chemical transformation during the formation of the boundary film.

In the presence of potassium triborate, a single chemical state of the boron is detected at 190 eV.

In rolling, the oil film exists between the friction surfaces, there is no mechanical abrasion of these surfaces, and the decomposition of the detergent micelles can only be explained by the pressure exerted on the lubricant in the contact zone. The inorganic borate and carbonate phase, and not the detergent, is then deposited on the surface. In the same conditions, potassium borate does not adhere to the surface.

276

Table 3.8.9

ADDITIVES	FRICTION CONDITIONS	BINDING ENERGIES (eV) / ATOMIC CONCENTRATIONS (%)		
		BORON (1s)	**CALCIUM (2p)**	**POTASSIUM (2p)**
Potassium triborate	Rolling PH$_z$=3 GPa, 120 °C carbonit. ***	depth → (Å): 0 | 15 | 300 | 600 * |190/1.5 /0 | 189.9/4.3 /0 | 2 | 2 ** |190.3/1.0 /0 | 190.1/3.4 /0 | 11 | 8.3		depth →: 0 | 15 | 300 | 600 * |292.5/2.9 /0 | 292.9/8.3 /0 | 293.3/1.6 | 293.3/1.2 ** |292.5/0.4 /0 | 293.3/1.8 /0 | 293.8/3 | 294.9/1.9
	Sliding PH$_z$=0.85 GPa, 120 °C carbonit.	depth →: 0 | 50 * |191/12.4 /0 | 4.6 ** |193/0.8 /0 | 14		depth →: 0 | 50 * |293/6.2 /0 | 293.3/1.4 ** |294.5/0.4 /0 | 294.5/3.6
	Sliding PH$_z$=0.7 GPa, cast iron	depth →: 0 | 100 | 400 | 750 | 2200 * |190.7/12 | 9 | 9 | 8.3 | 5 ** |192.6/0 | 5.4 | 3 | 1.5 | 0		depth →: 0 | 100 | 400 | 750 | 2200 * |293/8.1 /0 | 2.4 | 2.8 | 2.0 | 0.7 ** |294.8/0.4 /0 | 2.3 | 2.4 | 1.9 | 0.6
Overbased sulfonate	Sliding PH$_z$=0.85 GPa, 120 °C carbonit.		depth →: 0 | 50 | 300 | 600 * |347.8/7.2 /0 | 10.6 | 8 | 6.2 ** |349.1/0.1 /0 | 5.7 | 9.5 | 9.6	

Overbased borated sulfonate B/Ca=2.5	Rolling PH$_z$=3 GPa, 120 °C carbonit. ***	depth →: 0 | 15 | 65 | 200
		* |189.4/0.8 | 2.2 | 3.2 | 5
		  |191.3/3.3 | 4.4 | 6.5 | 14.1
		** |189.6/3.0 | 4.4 | 3.5 | 6.9
		  |190.6/0.9 | 2.5 | 1.8 | 4.3
			depth →: 0 | 15 | 65 | 200
			* |346.8/5.1
			/0 | 346.9/7
			/0 | 347/9 | 346.8/14.8
			** |346.8/1.2
			/0 | 346.8/1.5
			/0 | 347/2 | 347/3.4
			/0	

Overbased borated sulfonate B/Ca=1.3	Sliding PH$_z$=0.85 GPa, 120 °C carbonit.	depth →: 0 | 50 | 300 | 600 * |190.8/2.0 | 0.3 | 0.3 | 0.3 ** |193/0.3 | 3.0 | 3.8 | 3.4	depth →: 0 | 50 | 300 | 600 * |347.6/5.4 /0 | 1.0 | 0.7 | 0.7 ** |349.2/0 | 9.2 | 11.0 | 10.6	

* inside wear-path
** outside wear-path
*** analysis conditions don't allow to compare the potassium and boron contents in the film

According to the results of the three balls/cone tests, the formation of borate-rich surface film favours the fatigue strength of the materials (Table 3.8.8).

It should be noted that potassium borate exerts undeniable anti-fatigue action, although it does not form a solid film on the friction surfaces in these conditions.

3.8.3.3.2   Type of Surface Film Formed in Slip

The analyses performed with a Castaing microprobe, GDS and XPS, in the different friction conditions tested in pur slip, show that the type of boundary film is substantially the same for a given lubricant.

3.8.3.3.2.1 Chemical state of elements at the surface (Table 3.8.9)

For ZnDTP, XPS shows the same binding energies of zinc, sulfur and phosphorus in the initial additive and in the boundary film. By contrast, for additives in a solid dispersion, part of the boron, calcium or potassium goes into a more oxidized chemical state on the boundary film. The chemical transformation already identified in pure rolling appears to be more pronounced here, because, the chemical displacements are more intense and also affect the cations. This result was not identified for the carbonate carbon. This chemical transformation can be explained by:

— hydrolysis of the inorganic phase in contact with the surface metallic (7),

— electron transfer between the mechanically-activated surface and the deposit (8),

— modification of the structure of the inorganic phase under the effect of the pressure, tending to create a phase displaying better cohesion: for boron, for example, an increase in the coordination number is possible, enabling this atom to pass from the plane trigonal structure to the more rigid tetrahedral structure (9) (10).

3.8.3.3.2.2 Boundary Films Formed by Hyperbasic Detergents
(Tables 3.8.7, 3.8.10, Figure 3.8.5)

In the presence of hyperbasic detergents, the sulfonate is not detected in the boundary film. For the sulfophenate, a slightly sulfur content is observed, and is attributed to the surface reaction of the more labile sulfur atoms of the phenate (Table 3.8.7).

Table 3.8.10

Molar Ratio →	B/Ca	B/C(carb.)	B/Ca	B/C(carb.)	B/S
in the lubricant	0.4	0.5*/0.46**	1.3	4.03* (2.4**)	13.8
in the boundary film	0.3	1.25	0.8	5.9	95.6
		sulfonate 1		sulfonate 2	

Two hypotheses for the calcium borate in the lubricant

* $Ca(BO_2)_2$; (CaO, B_2O_3)
** $Ca_2 B_6 O_{11}$

Table 3.8.11

Additives	Arb. Unit*	
	Ca	B
Overbased sulfonate	55.1	
Overbased borated sulfonate B/Ca = 1.3	91.8	196.1
Overbased borated sulfophenate B/Ca 0.4	42.2	11.7
Overbased borated sulfophenate B/Ca = 1.3	60	129.7
Potassium triborate		194.2

* quantity in the boundary film evaluated by the area under the peak of each element

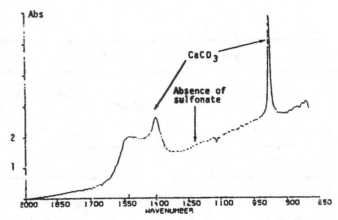

Figure 3.8.5: Infra-red spectra of the boundary film. Lubricant: overbased sulfonate

XPS and infrared spectroscopy of the surface enable us to identify the carbonate at the surface (Fig. 3.8.5). However, according to the XPS results obtained in the presence of non-borated hyperbasic detergents, it is probable that the boundary film contains another calcium salt than the carbonate. This calcium salt is not identified.

With borate additives, GDS analysis shows that the higher the boron content in the oil, the thicker the boundary film and the richer it is in boron and calcium (Table 3.8.11). These results are not confirmed by microprobe. In this case, in fact, a systematic error is introduced due to the ignoring of the widely varying boron contents in the different additives employed. The XPS results show that the B/Ca ratio of the lubricant is preserved at the surface for the very low boron contents. For high contents, this ratio is lower in the boundary film. By contrast, the borate/carbonate ratio of the film is always much higher than of the lubricant calculated by making different assumptions on the stoichiometry of calcium borate in the micelles (Tabele 3.8.11). This confirms the assumption of the decomposition of a part of the calcium carbonate during the formation of the boundary film.

3.8.3.3.2.3 Boundary Film Formed in the Presence of Potassium Triborate

In the slip conditions investigated, a solid film of potassium borate about 0.2 μm thick is formed at the surface. It is probable that the adhesion of the borate is made possible here by the abrasion of the friction surfaces. For this additive, a friction surface shows similar behaviour to a negatively-polarized surface. The quantities of boron estimated by GDS are identical to those formed in the presence of borated hyperbasic detergents (B/Ca = 1.3) (Table 3.8.11). Its

280

thickness distribution is different. It is more localized at the surface for potassium borate (Fig. 3.8.6). The B/K ratio is approximately preserved in the boundary film.

3.8.3.3.2.4 Thickness of Films Formed on Antagonistic Specimens

Given the depth distribution of the elements analysed, we shall treat similarly the surface concentrations measured by the microprobe and the thickness of the boundary film. The analyses performed on the antagonistic specimens enable us to identify the following tendencies: the film formed in the presence of ZnDTP or overbased detergent with high boron content appears to be thicker on the mobile sample, whereas, in the presence of non-borated hyperbasic detergents, the reverse process is observed. These results can be interpreted as follows (Table 3.8.7).

The decomposition products of ZnDTP react at the surface to form a boundary film. This tribochemical reactions are favoured if the surface is under greater mechanical stress. For the tribometer used, this applies to the surface of the mobile sample compared with that of the fixed sample.

For hyperbasic detergents, the rates of film formation on the two specimens opposed, by mechanical compaction, are probably identical, but the abrasion rate is greater on the mobile sample, hence the lower concentrations on these specimens. As the boron content increases in the micelles, another process occurs, and, in fact, the adhesion of the borate film is favoured on an abraded surface, hence the higher proportion of the boundary film in the mobile specimen.

3.8.3.3.3   Results of Mechanical Tests with Pure Slip

In pure slip, the tendencies identified on the reciprocating tribometer in the different testing conditions are as follows (Table 3.8.12).

3.8.3.3.3.1 Friction Properties

The borate and phosphate films display high friction factors. On the films consisting of calcium carbonate and calcium salt resulting from the carbonate, the friction factor is lower, but it is unstable under high pressures. Thus, the introduction of borate into the carbonate micelles has the effect of increasing the friction factor and stabilizing it. This increase, for high boron contents, is greater for sulfophenates than for sulfonates.

Table 3.8.12

Friction Conditions	Additives	Friction Factor	Wear (µm)	
			Mobile Sample	Stationary Sample
Crossed cylinders P_{Hz} = 2 GPa; 120° C Carbonit./carbonit.	Overbased sulfonate	0.070 0.072 0.068	410* 390 410	400* 420 400
	overbased borated sulfonate (B/Ca = 0.4)	0.069 0.066	420 40	400 390
	Overbased borated Sulfonate (B/Ca = 1.3)	0.086 0.087 0.088	510 530 520	500 520
	Overbased borated sulfophenate (B/Ca = 0.4)	0.075 0.074	410 410	370
	Overbased borated sulfophenate (B/Ca = 1.3)	0.096 0.090 0.095	530 600 510	480 560 490
	Potassium triborate	0.104 0.100 0.102	670 680 640	670 600 570
Cylinder/plane P_{Hz} = 0.85 PGa; 120° C carbonit./carbonit.	Overbased sulfonate	0.139 0.196 0.136		
	Overbased borated sulfonate (B/Ca = 1.3)	0.149 0.147		
	Overbased borated sulfophenate (B/Ca = 0.4)	0.156 0.153		
	Potassium triborate	0.167 0.161 0.163		

Table 3.8.12 (continued)

Friction Conditions	Additives	Friction Factor	Wear (μm)	
			Mobile Sample	Stationary Sample
Cylinder/plane P_{Hz} = 0.7 GPa; 120° C cast iron/cast iron	Overbased sulfonate	0.081 0.091 0.088	4.6* 3.5 5.6	2.5** 4
	Overbased borated sulfonate (B/Ca = 0.4)	0.107 0.100	3.8 3	1.5 1.2
	Overbased borated Sulfonate (B/Ca = 1.3)	0.107 0.103 0.100	6.4 8.9 6.7	4.0 9.5 6.0
	Overbased borated sulfophenate (B/Ca = 1.3)	0.107 0.103 0.100	6.4 8.9 6.7	4.0 9.5 6.0
	Overbased borated sulfophenate (B/Ca = 0.4)	0.099 0.101 0.101	3.3 3.0 3.8	3.5 1.5 2.0
	Overbased borated sulfophenate (B/Ca = 1.3)	0.119 0.119	4.6 6.0	2.0 3.5
	Potassium triborate	0.116 0.124 0.125	5.9 6.1 4.6	4.0 2.4 1.0

*   width of the wear-path
**   depth of the wear-path

3.8.3.3.3.2 Formation Rate of the Boundary Film

The change in the boundary film during the test is monitored by measuring the potential difference at the terminals of the friction specimens. With the additives in a solid dispersion, the film is formed instantly. With ZnDTP, the rate of formation is lower.

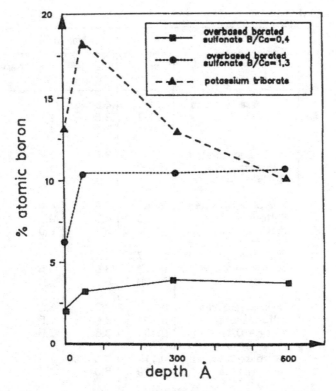

Figure 3.8.6:   Depth profile of boron in the boundary film

3.8.3.3.3.3 Wear

The tests under 700 MPa with carbonitrided steel are not severe enough to test the lubricants from the standpoint of their anti-wear capacity. In the other cases, it is in the presence of low-borate or non-borated hyperbasic detergents that the wear is lowest. The classification of the other additives depends on the testing conditions:

— carbonitrided steel; 200 MPa,
ob det = bor. ob det (0.4) $<$ ZnDTP $<$ bor. ob det (1.3) $<$ KB,

— cast iron, 700 MPa,
ob det = bor. ob det (0.4) $<$ KB $<$ bor. ob det (1.3) $<$ ZnDTP.

increasing wear

284

3.8.3.3.4 Interpretation

These tests were conducted with two iron-based metallurgies with different surface chemical compositions and different mechanical properties. The differences in behaviour of the lubricants containing ZnDTP can be explained by the difference in reactivity with respect to the two metallurgies. In particular, the high reactivity of ZnDTP with cast iron can lead to corrosive wear in the conditions investigated.

For the other additives, due to their low reactivity, the surface chemical compositions have little effect, whereas the difference in hardness can play an important role.

Thus, for the borate products, the boundary film is more effectively covering and more stable on a low-hardness surface subjected to severe abrasion.

According to the examination of the surfaces by the optical microscope, the limit of these additives appears to be the difficulty of forming an adhesive film that covers the friction surfaces. This explains their mediocre anti-wear effectiveness in these tests.

The weak point of overbased detergents under high loads comes from the instability of the boundary film, which is explained by the low shear strength and by the plastic properties of this film, which in our testing conditions (low slip speed), are offset by a rapid rate of formation. These properties are improved by introducing small amounts of borate in the micelles.

3.8.4 Conclusions

Additives in inorganic suspensions display a basically different action mechanism from conventional phosphor-sulfur additives. The latter can form a boundary film thanks to their thermal decomposition products by reaction with the metallic substrate, whereas hyperbasic detergents do not decompose and are only deposited on the surface under the effect of pressure. Potassium triborate only forms an adhesive film on a negatively-polarized or mechanically-abraded surface.

This study shows that the method of placement in suspension of the inorganic phase by the detergents is important. It offers the advantage of forming deposits only in the friction zone, instantly and idependently of the type and state of the friction surface. Yet, the composition of the inorganic phase remains to be optimized. This optimization entails the investigation of the rheological behaviour of thin films.

For these additives in a solid dispersion, no chemical bond was identified between the components of the boundary film and the metallic substrate. The boundary film consists exclusively of the elements of the inorganic phase but part of these elements is oxidized during the formation of this film.

Acknowledgement

The authors wish to thank Mr. J. C. Hipeaux, G. Parc for providing the additives, Mrs. E. Rosenberg, Mr. Bisiaux and Y. Huiban for their work in XPS and Castaing microprobe.

3.8.5   References

(1)   Toulhoat, H.: Potentiel et limite d'utilisation, aux températures élevées, des lubrifiants liquides dans les moteurs. Revue de l'IFP V 44, 3, 371—385 (Mai—Juin 1989).

(2)   Morizur, M.F.; Briant, J.: Electrical phenomena associated with boundary lubricated friction. Proceed. Vol. 5, 272—280 Eurotrib 89.

(3)   Clason, D.L.: Metalworking additives. A new approach to E.P. properties. Proceed. 11, 1.1 — 1.10 Esslingen (1988).

(4)   Morizur, M.F.; Briant, J.: Modifications of electron properties of friction surfaces in boundary lubrication. Proceed. IMechE 447—454 (1987-5).

(5)   Adams, J.H.; Godfrey, D.: Borate gear lubricant EP film analysis and performance. Lubr. Eng. 37, 1, 16—21 (1981).

(6)   Dacre, B.; Bovington, C.H.: The effect of metal composition on the adsorption of zinc di-isopropyldithiophosphate. Asle Trans. V26, 3, 33—43 (July 1983).

(7)   Adams, J.H.: Borate — A new generation EP gear lubricant. Lubr. Eng. 33, 5, 241—246 (1977).

(8)   Dong, J., Huang, Y.; Luo, X.M.: The research for the preparation and action mechanism of boron-type antiwear additives. Proceed. 11, 1.1—1.10, Esslingen (1988).

(9)   Christ, C.L.; Clark, J.R.: A crystal-chemical classification of borate structures with emphasis on hydrated borates. Phys. Chem. Minerals 2, 59—87 (1977).

(10)   Farmer, J.B.: Metal borates. Adv. Inorg. Chem. Radiochem. 25, 187—237 (1982).

3.9 An Investigation of Effects of Some Motor Oil Additives on the Friction and Wear Behaviour of Oil-soluble Organomolybdenum Compounds

D. Wei, H. Song and R. Wang,
Research Institute of Petroleum Processing, Beijing, P.R. China

Abstract

The compatibility of oil-soluble organomolybdenum compounds with some motor oil additives (detergent, dispersant, rust inhibitor and ZDDP) has been investigated on a four ball machine and a SRV tester.

The preliminary results indicated that a combination of two agents might have either synergistic or antagonistic effects on the friction and wear performance of the organomolybdenum compounds although many combinations were shown to be synergistic. This interactions between two agents were primarily determined by the additive types, additive concentrations and test temperatures.

It was also found that the presence of calcium sulfonate detergent was particularly benificial to molybdenum dithiophosphate and molybdenum dithiocarbamate in terms of friction reduction and wear reduction at temperatures in certain range. The induction period of the organomolybdenum compounds were also reduced. Surface analysis suggested that the synergistic effects were closely related to the formation of thick films, which were rich in molybdenum and sulfur, on rubbing surfaces.

3.9.1 Introduction

In recent years attention has been increasingly directed towards the application of oil-soluble organomolybdenum compounds in motor oils. Their value as friction reduction additives for motor oils was assessed using a series of laboratory friction testers, engines and vehicles (1, 2). The results revealed that both fuel economy and transmission efficiency were improved obviously due to the addition of organomolybdenum compounds.

More recently, some work have demonstrated that the possible harmful effects of oil-soluble organomolybdenum compounds can be depressed successfully and do not cause any wear and lubricity problems (3). These experimental results suggest that oil-soluble organomolybdenum compounds are promising friction reduction additives for formulating fuel saving motor oils.

Since the compatibility between additives is very important in connection with the actual performance of molybdenum-containing motor oils, it is necessary to investigate the phenomena and mechanisms concerning the interactions between motor oil additives and molybdenum compounds (4, 5). In this study emphasis has been placed upon the synergistic additive action between the organomolybdenum compounds and calcium sulfonate detergent, others have been examined to a lesser extent.

3.9.2 Experimental Methods

3.9.2.1 Friction and Wear Test

A MQ-12 four ball machine and an Optimol SVR tester have been used to investigate the friction and wear behaviour of a variety of additive combinations. The four fall test was conducted using GCr-15 steel ball 12.7 mm in diameter and was run at a load of 30 kg (294N) and a speed of 1200 rpm (0.8 m/sec). The SVR tester employed in this study comprised a stationary ball 12.7 mm in diameter and a mobil flate disc. Both specimens were made of GCr-15 steel (containing 1.4 % chromium).

The SRV friction test conditions employed in present work are shown below: load 120N, stroke length 1 mm, frequency 50 Hz. A step heating friction test procedure was used to examine the influence of temperature on the interaction between molybdenum compounds and motor oil additives. According to this procedure each test was started at 20°C with a running-in process (carried out at a load of 29N for 30 min), and then the friction test was performed at different progressively higher temperatures until a test temperature as high as 170°C was reached. The temperature increase for each test step was 10°C. In each temperature step the temperature of the disc was held at the desired level for a few minutes to allow the friction to stabilize and to be recorded.

The SRV wear tests were conducted at different temperatures (i.e. 40, 80, and 120°C) for 60 minutes. The test load used was 200 N, the stroke length 0.5 mm and the frequency 50 Hz. Wear was measured from the diameter of wear scar on the ball specimen at the end of the test.

In this study an oil change-running-in procedure was also employed to examine the influence of running-in oils on the interactions between molybdenum compounds and motor oil additives.

3.9.2.2 Base Oil and Additives

A highly refined paraffin oil with a viscosity of 14.73 mm/s at 40°C was used as a base oil. Two types of oil-soluble molybdenum compounds were used for blending with this paraffine oil to formulate base blends. The base blends A and B consisted of 0.5 % molybdenum dithiocarbamate (MoDTC) and 1 % molybdenum dithiophosphate (MoDTP) respectively. Different concentrations of a given motor oil additive in a particular base blend were formulated for four ball test and SRV test. Table 3.9.1 lists the analytical data of these compounds and additives.

Table 3.9.1: Some Analytical Data of Commercial Additives

| Additives | Code | Element Content (wt %) | | | | | | | TBN | AN |
		Mo	Zn	Ca	Ba	S	P	N	(KOH mg/mg)	
Molybdenum di-2-ethylhexyldithiocarbamate	MoDTC	4.1								
Molybdenum di-2-ethylhexyldithiocarbamate	MoDTP	8.2				12.3	5.5			
Polyisobutenyl succinimides of alkylene polyamines	PSAP							0.9		
Barium thiophosphonate	BaTP				14	0.8	0.6			
Calcium sulfonate	CaSF			5.2					145	
Zinc butyloctyldithiophosphate	ZDDP		9.7							134.4
Barium dinonylnaphthenesulfonate	BaDN				8.3				48	

3.9.2.3 Analysis of Surface Films

Electron probe microanalysis, Auger electron spectroscopy (AES) analysis and photoelectron spectroscopy analysis were employed to study the mechanisms of synergistic additive action between organomolybdenum compounds and calcium sulfonate. AES analysis was conducted with ball specimens. In AES analysis the primary electron energy was 3 kv, beam current $2\mu A$ and beam size ca. $20\mu m$. In XPS analysis the X-ray excitation was the $K\alpha$ radiation and the 1s line of carbon was taken as a reference to correct surface charging. In both AES and XPS analysis argon ion etching techniques were used to investigate the depth distribtuion of surface species produced in wear process.

3.9.3   Experimental Results and Discussion

3.9.3.1 The Effect of Additive Concentration

3.9.3.1.1   Detergent and Dispersant

Detergent and dispersant additives are widely used in motor oils. Their function is to keep oil-insoluble combustion products in suspension or dispersion in the oil so that they will not settle out or adhere to surfaces. To investigate the effect of these important additives on the friction and wear of molybdenum compounds three typical detergent-dipsersant additives were evaluated in two molybdenum-containing base blends using a four ball machine and a SRV tester.

The effect of adding a calcium sulfonate detergent (CaSF) is illustrated in Fig. 3.9.1. Fig. 3.9.1 shows the friction and wear of MoDTC—CaSF blend as a function of calcium sulfonate concentration. It can be seen that both friction and wear decreased with an increasing in calcium sulfonate concentrations up to approximately 1 % (i.e. trom f = 0.052 and d = 0.38 mm at c = 0 to f = 0.31 mm at c = 1 %). As calcium sulfonate concentration was changed from 1.0 % to 2.0 % friction and wear remained at lowest level and then increased slowly at higher concentrations.

The effect of adding succinimide dispersant on the performance of base blend A is demonstrated by Fig. 3.9.2. Although the addition of succinimide dispersant also caused wear to decrease slightly at low concentrations and then to pass through a minimum, the wear at higher concentrations was substantially higher than that for the base blend A. On the other hand, a significant deteriorative effect was indicated by the fact that friction rose monotonously with an increase in succinimide concentrations and finally reached a value as high as nearly 0.1.

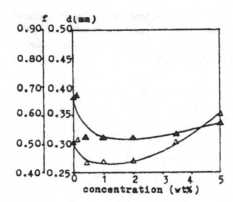

Figure 3.9.1: Effect of Adding Calcium Sulfonate Detergent on the Friction and Wear of Molybdenum Dithiocarbamate Blend. Upper: Wear Curve, Lower: Friction Curve. Four Ball Test Conditions: 30 kg, 1200 rpm, 20 ± 3°C

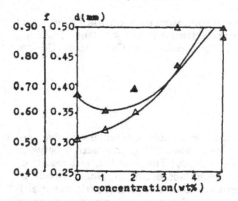

Figure 3.9.2: Effect of Adding Succinimide Dispersant on the Friction and Wear of Molybdenum Dithiocarbamate Blend. ▲ — Wear Diameter, △ — Friction Coefficient. Four Ball Test Conditions: 30 kg, 1200 rpm, 20 ± 3°C

Figure 3.9.3 displays the effect of adding barium thiophosphonate detergent (BaTP) on the friction and wear of blend A. If only the data of wear were considered, it would be noted that there was a substantial synergistic effect on wear between MoDTC and BaTP. The shape of wear curve was very similar to that obtained with the MoDTC/BaSF blend and the synergistic response on wear was nearly as strong as that of MoDTC/BaSF blend (see Fig. 3.9.1 and 3.9.3). However, a strong antagonistic effect on the friction was also observed. The

291

friction coefficient increased significantly as BaTP concentration was increased and the relation between friction coefficient and concentration was approximately linear. This factor suggested that there was a strong deteriorative effect on the friction reduction property of MoDTC by adding BaTP.

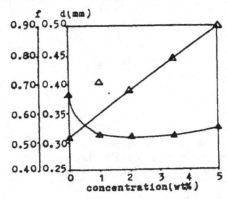

Figure 3.9.3: Effect of Adding Barium Thiophosphonate on the Friction and Wear of Molybdenum Dithiocarbamate Blend. ▲ — Wear Diameter, Δ —Friction Coefficient. Four Ball Test Conditions: 30 kg, 1200 rpm, 20 ± 3°C

3.9.3.1.2   Zinc Dialkyldithiophosphate

ZDDP has been widely used in engine oils as a multifunctional additive (i.e. antioxidant, corrosion inhibitor and antiwear agent). The interaction between MoDTC and ZDDP is also examined in this study. The four ball test results of MoDTC/ZDDP blend showed that the wear was essentially independent of ZDDP concentrations (Fig. 3.9.4). In contrast to wear results the friction coefficient rose significantly with an increase in ZDDP concentration and then passed through a maximum at a concentration of about 3.5 % and finally decreased slowly at higher concentrations. The harmful effect on friction due to the addition of ZDDP is substantial.

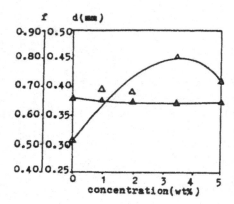

Figure 3.9.4: Effect of Adding Zinc Dialkyldithiophosphate on the Friction and Wear of Molybdenum Dithiocarbamate Blend. ▲ — Wear Diameter, △ — Friction Coefficient. Four Ball Test Conditions: 30 kg, 1200 rpm, 20 ± 5°C

3.9.3.1.3   Rust Inhibitor

Only one rust inhibitor, barium dinonylnaphthenesulfonate (BaDN), was examined. In contrast to the results of adding calcium sulfonate, a severe antagonistic interaction was observed with MoDTC/BaDN blend. Figure 3.9.5 provides corresponding data for MoDTC/BaDN blend. It is evident that the deteriorative effect of adding BaDN at concentrations above 1 % leads to the onset of scuffing. It is broadly understood that rust inhibitor BaDN works by forming a strongly absorbed, densely packed and hydrophobic film on the surface to be protected. The fact that BaDN markedly reduces the load-carrying capacity of MoDTC probably points out the drastic competition between both additives for the metal surfaces.

As illustrated above, for most motor oil additives the alternation of additive concentration generally has a remarkable influence on the friction and wear of MoDTC. However, the effect of adding these additives can be quite selective. A synergistic effect or a antagonistic effect depends on the nature of the additives and also additive concentrations. In the present study only one additive, calcium sulfonate detergent, was found to have a synergistic effect on both friction and wear of MoDTC in the transition concentration regions. Other additives evaluated had a significant harmful effect on the friction reduction of MoDTC, although some of them may have a detectable benificial effect on wear at certain concentrations as well. These phenomena probably can be explained by the difference in their relative reactivities towards MoDTC and the concentration dependence of surface reactions. Further study on adsorption/reaction thermodynamics and kinetics are needed for explaining these phenomena.

293

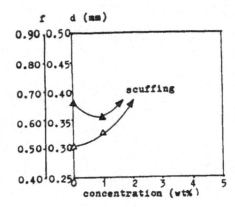

Figure 3.9.5: Effect of Adding Barium Dinonylnaphthenesulfonate on the Friction and Wear of Molybdenum Dithiocarbamate Blend. Upper: Wear Curve, Lower: Friction Curve. Four Ball Test Condition: 30 kg, 1200 rpm, 20 ± 3°C

3.9.3.2 Influence of Temperature

In addition to the effects of additive nature and additive concentrations, the experimental data also demonstrate that temperature also has an important effect on the interactions between molybdenum compounds and motor oil additives. The effect of temperature on friction in SRV tests are presented in Fig. 3.9.6 and 3.9.7.

Fig 3.9.6 illustrates the results of MoDTC base blend and MoDTC/CaSF blend. As shown in Fig. 3.9.6, three distinct stages were evident in both friction curves. With MoDTC base blend, a plateau with a very gentle slope appeared at temperatures lower than 50°C and the friction coefficient at this stage had a value of 0.08 − 0.09. The friction coefficient decreased slightly as the test temperature increased and reached a value of 0.07, and then levelled off to form the second plateau over the temperature range of 80 − 100°C. When the test temperature was further increased the friction coefficient increased gradually. The friction coefficient in the temperature range of 150 − 170°C was ca. 0.13. In case of MoDTC/CaSF, a similar friction curve is observed. However, the variation of friction was more sensitive to test temperature. At 170°C the friction coefficient reached a value as high as 0.17, which was equal to that for CaSF blend alone. The friction reduction action of MoDTC had been cancelled by adding CaSF at this temperature. It can be also seen from Fig. 3.9.6 that the synergistic effect between MoDTC and CaSF was restricted to a temperature range of 65 − 110°C and the corresponding minimum value of the friction coefficient was 0.06.

294

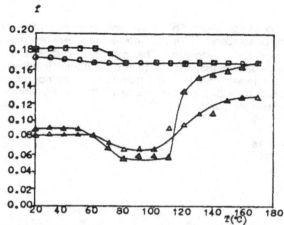

Figure 3.9.6: Friction—Temperature Plots for Paraffine Oil, ○; 1 % Calcium Sulfonate Blend, □; 0.5 % Molybdenum Dithiocarbamate Blend, △; and 0.5 % Molybdenum Dithiocarbamate plus 1 % Calcium Sulfonate, ▲. SRV Step Heating Test

Fig. 3.9.7 demonstrates the results of MoDTC base blend and MoDTP/CaSF blend. In contrast sharply to Fig. 3.9.6, the trends of both friction curves in Fig. 3.9.7 were quite different. A synergistic interaction between MoDTP and CaSF occurred over the whole temperature range evaluated. It can be seen that the synergistic effect was enhanced at elevated temperatures and the strongest effect appeared to take place at temperatures above 130°C.

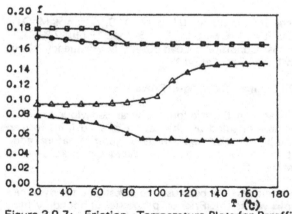

Figure 3.9.7: Friction—Temperature Plots for Paraffin Oil, ○; 2 % Calcium Sulfonate, □; 1 % Molybdenum Dithiophosphate, △; and 1 % Molybdenum Dithiophosphate plus 2 % Calcium Sulfonate, ▲. SRV Step Heating Friction Test

295

Besides the step heating friction tests, some SRV wear tests were also carried out to examine the influence of temperature on the interactions between molybdenum compounds and calcium sulfonate. The results are shown in Fig. 3.9.8. These limited wear data demonstrated that the synergistic effect in wear was significant and the interaction between molybdenum compounds and calcium sulfonate not only depended on the nature of the molybdenum compounds but also related to test temperature.

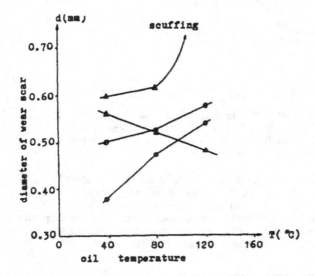

Figure 3.9.8:   Wear—Temperature Curves for 0.5 % MoDTC, Δ; 0.5 % MoDTC plus 1.0 % CaSF, ▲; 1 % MoDTP, ○; 1 % MoDTP plus 2 % CaSF, ● SRV Wear Test Conditions: Load 50/200 N, Frequency 50 Hz, Stroke length 1 mm, Test Period 1 h

3.9.3.3 The Influence of Oil Change—Running-in Process

It has been generally recognized that the friction and wear behaviour of an EP agent or a AW agent is strongly affected by the nature of solid films formed on rubbing surfaces. Essentially, the observed effects of additive nature, additive concentration and test temperature in the above paragraphs are closely related to the formation of antiwear films.

Running-in and oil change techniques have been used to investigate the dynamics of boundary films by previous works (6). Friction properties of boundary films and their lifetimes can be determined by changing base oil and test oil alternatively, and therefore it was quite useful in the study of film formation in boundary lubrication.

296

In this study a somewhat different oil change-running-in procedure was employed. The test was started with a running-in oil and performed for 15 — 60 minutes to establish an equilibrium wear scar, approximately 0.4 mm in diameter. Then the test was stopped and the running-in oil was poured out. Before changing to a sample oil both upper and lower test assemblies were rinsed carefully with the sample oil thrice. The test was started again and continued to run for 80 minutes.

Four different oil change procedures, used for the study of the interaction between MoDTC and CaSF, are shown in Table 3.9.2. The effects of oil change-running-in on friction are illustrated in Fig. 3.9.9 and 3.9.10.

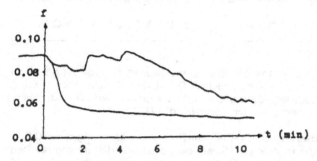

Figure 3.9.9: Effect of Running-in Oil on the Induction Period and Transition Period of Molybdenum Containing Oils. Upper Trace: 0.5 % MoDTC, Lower Trace: 0.5 % MoDTC plus 1 % CaSF. Running-in Oil, Liquid Paraffine. Four Ball Friction Test Conditions: Load 294 N, Speed 600 rpm, Initial Oil Temperature 20 ± 3°C

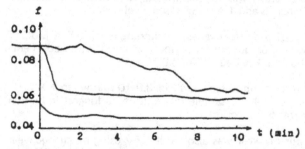

Figure 3.9.10: Effect of Running-in Oils on the Induction Period and Transition Period of Molybdenum Containing Oils. Upper Trace: Running-in Oil — 1 % CaSF, Sample Oil — 0.5 % MoDTC. Middle Trace: Running-in Oil — Paraffine Oil, Sample Oil — 0.5 % MoDTC plus 1 % CaSF. Lower Trace: Running-in Oil — 0.5 % MoDTC, Sample Oil — 0.5 MoDTC plus 1 % CaSF. Four Ball Friction Test Conditions: Load 294 N, Speed 600 rpm, Initial Oil Temperature 20 ± 3°C

Table 3.9.2: Oil Change Running-in for the Study of MoDTC-CaSF Interaction

Code of the Case of the Oil Change	Running-in Oil	Sample Oil
1	Paraffine Oil	0.5 % MoDTC
2	Paraffine Oil	0.5 % MoDTC + 1 % CaSF
3	0.5 % MoDTC	0.5 % MoDTC + 1 % CaSF
4	1 % CaSF	0.5 % MoDTC + 1 % CaSF

In the first case of oil change, the friction trace was erratic initially (induction period) then the paraffine oil was replaced by 0.5 % MoDTC and then friction coefficient decreased gradually during the test (transition period) and finally reached a constant level (steady low friction period) (7).

In the second case, however, when the paraffine oil was changed to 0.5 % MoDTC plus 1 % CaSF the friction coefficient dropped immediately and reached a constant level in only 2 minutes. These results suggest that the addition of CaSF in MoDTC containing sample oil has a significant beneficial effect on the diminution of both induction period and transition period.

The cooperative action of MoDTC and CaSF in the diminution of both induction period and transition period was further investigated with two different running-in oils, i.e. 0.5 % MoDTC and 1 % CaSF respectively.

In the third case, evidence of the effectiveness of running-in oil 0.5 % MoDTC was obtained from the very short induction period and low friction coeffienct of sample oil 0.5 % MoDTC plus 1 % CaSF (Fig. 3.9.10).

In the fourth case, it was interesting to see from Fig. 3.9.10 that when 1 % CaSF was employed as a running-in oil the sample oil exhibited a longest induction period as well as a longest transition period.

The effect of running-in oils on wear was also studied by using oil change wear tests. Fig. 3.9.11 compares the effects of different running-in oils (the different components of the sample oils) on wear behaviour of three sample oils. It can be seen that the running-in with CaSF blend resulted in increased wear but that the running-in with MoDTC (or MoDTP)/CaSF blend produced a marked wear reduction for sample oils.

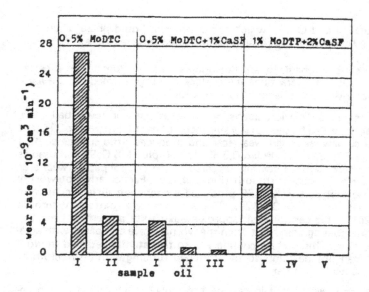

Figure 3.9.11: Comparison of the Effect of Different Running-in Oils on the Wear Reduction Efficiency of Molybdenum Containing Sample Oils. Running-in Oils: I – 1 % CaSF in Paraffine Oil, II – 0.5 % MoDTC in Paraffine Oil, III – 0.5 % MoDTC plus 1 % CaSF in Paraffine Oil, IV – 1 % MoDTP in Paraffine Oil, V – 1 % MoDTP plus 2 % CaSF. Four Ball Wear Test was Conducted under 294 N Load at a Speed of 600 rpm at Room Temperature

3.9.3.4 The Mechanisms of the Synergistic Effects Between the Molybdenum Compounds and the Calcium Sulfonate

In this paper so far the synergistic effects between the molybdenum compounds and the calcium sulfonate have been illustrated but the mechanism by which these effects arise has not been discussed. The mechanism of friction reduction and wear prevention of organomolybdenum compounds involves chemical reactions between the molybdenum compounds and the rubbing surfaces to form a molybdenum containing protective film. This study has shown that the formation of a stable protective film by molybdenum compounds is retarded significantly by the adsorption and/ or direct bonding of calcium sulfonate on rubbing surfaces in running-in stage. The antagonistic effect probably can be explained by competition between molybdenum compounds and calcium sulfonate for the rubbing surfaces. However, as demonstrated above, a significant synergistic effect between MoDTC (or MoDTP) and CaSF was observed when they both were used as sample oil additives. It thus appeared that some interactions between

MoDTC (or MoDTP) and CaSF, which occur either in liquid phase or on rubbing surfaces, are likely to be responsible for the synergistic response.

However, the precise mechanisms are not understood. In order to get a better understanding about the interactions the surface coatings formed by the additives were further investigated using surface analysis tools.

Fig 3.9.11 shows the SEM micrographs of the wear scars on lower ball specimens. At first sight, it can be seen that a single component additive, MoDTC and MoDTP, gave a somewhat rough wear scar and produced little surface coating (Fig. 3.9.12a and Fig. 3.9.12c). When 0.5 % MoDTC plus 1 % CaSF was used as a lubricant, patch-like surface coating appeared in the central part of wear scar and the worn surfaces become smooth (Fig. 3.9.12b). Particularly noteworthy was that 1 % MoDTP plus 2 % CaSF produced a crownlike surface coating in the centre of wear scar (Fig. 3.9.12d). The wear scar was extremely smooth (see Fig. 3.9.12d). A big central area (approximately sixty percent of wear scar area) with surface coating stand out from the worn surface compared with adjacent no-coating area. This surface coating was so persistently retained in worn surfaces that it can not be removed by stretched rubbing with a silk cloth or by washing with petroleum ether.

The EPMA analysis revealed that in case of 1 % MoDTP plus 2 % CaSF, molybdenum and sulfur elements were highly concentrated in the surface coating. Very little Mo and S elements were found in the no-coating area. The coverage of the surface coating, rich in Mo and S, in the most highly loaded area can build up a barrier to the direct contact of matching surfaces and therefore plays an important part in friction reduction and wear prevention.

The chemical composistions of these surface coatings were further investigated by means of AES analysis and XPS analysis.

AES analysis further revealed that the binary component additives behaved in a manner different from single component additives. When single component additives, MoDTC or MoDTP, were used the distributions of Mo and S elements (either in outest surfaces or in depth) in central area and boundary area were similar (Fig. 3.9.13). When CaSF was added, however, Mo and S elements were concentrated in the central area (Fig. 3.9.14). Depth profiling also demonstrated that reaction film (i.e. surface coating), given by 1 % MoDTP plus 2 % CaSF, was thick and had much higher Mo and S contents than that produced by 1 % MoDTP (Fig. 3.9.15). With 1 % MoDTP both the Mo and S concentrations increased gradually and reached a maximum of approximately 8 atomic percent with one minute of initial sputtering, and then kept at a constant level in depth. In case of 1 % MoDTP plus 2 % CaSF, both Mo and S concentrations were much higher (the maximums of Mo and S concentrations were 30 % and 23 % respectively) as compared with those observed for 1 % MoDTP alone. These results suggested that the synergistic effect between molybdenum compounds and calcium sulfonate detergents was closely related to the increase of Mo and S contents in reaction films.

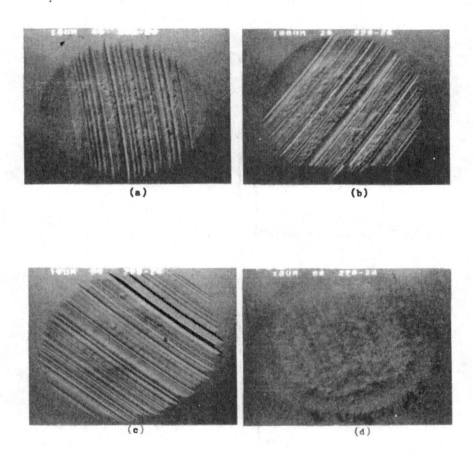

Figure 3.9.12: SEM Micrographs of the Wear Scar Obtained by Four Ball Wear Test. (a) 0.5 % MoDTC, (b) 0.5 % MoDTC plus 1 % CaSF, (c) 1 % MoDTP, (d) 1 % MoDTP plus 2 % CaSF. Four Ball Test Conditions: 30 kg, 600 rpm, 140 minutes, 20 ± 3°C

In XPS analysis the area of specimen examined in the spectrometer was ca. 1 mm by 2 mm and therefore it was necessary to produce a big enough wear scar to cover a substantial part of this area. In the present study an ALPHA model LFW-1 test was employed to prepare ring specimens for XPS analysis. The binding energy of the peaks observed are given in Table 3.9.3.

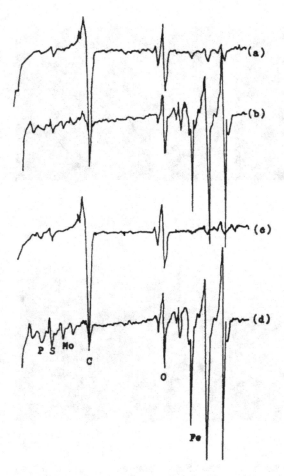

Figure 3.9.13: Auger Spectra of The Rubbing Surfaces. Wear Scars are Produced with 1 % MoDTP. Sampling Position: (a) and (b) — Near to the Boundary of Wear Scar, (c) and (d) — In the Centre of Wear Scar. Progress of Sputtering: (a) and (c) — Before the Sputtering (as received), (b) and (d) — 10 Minutes of Initial Sputtering

Comparison of these results showed that the binding energies for chief interest peaks, such as Mo_{3d}, S_{2p}, P_{2p}, N_{1s} and Fe_{2p}, were essentially not affected by the addition of CaSF in the test oils. The prominent peak occurred at binding energy between 161.2 kJ/mol and 161.8 is consistent with FeS (161.2 kJ/mol) but could not be also due to MoS_2 by comparison with reference compounds. Two rather broad Mo_{3d} peaks were observed in both cases (with and without

CaSF in the test oils), but the relative intensity of these two peaks was different. Molybdenum appeared in two forms. One probably can be corresponded to MoO_3 (232.5 kJ/mol) and the other was unsolved. It seems to be an organic molybdenum compound. The Fe_{2p} peak appeared at a position associated with Fe_2O_3 (711.0 kJ/mol) of Fe_3O_4 (711.2 kJ/mol) and P_{2p} peak was consistent with $FePO_4$ (133.5 kJ/mol).

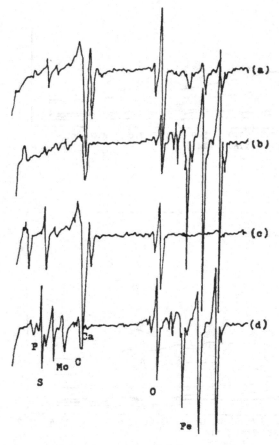

Figure 3.9.14: Auger Spectra of the Rubbing Surfaces. Wear Scars are Produced with 1 % MoDTP plus 2 % CaSF. Sampling Position: (a) and (b) — Near to the Boundary of Wear Scar, (c) and (d) — In the Centre of the Wear Scar. Progress of Sputtering: (a) and (c) — Before the Sputtering, (b) and (d) — 25 Minutes of Initial Sputtering

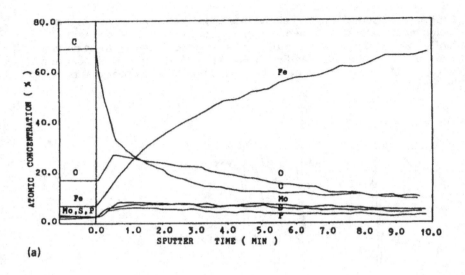

(a)

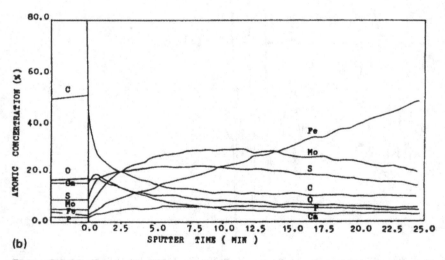

(b)

Figure 3.9.15: Depth Profiles at the Scarred Area of Test Ball. Wear Scars are
Produced with Additives. (a) 1 % MoDTP in Paraffine Oil, (b)
1 % MoDTP plus 2 % CaSF in Paraffine Oil

Table 3.9.3: XPS Analysis Results Obtained from the Wear Scar Tested with Molybdenum Containing Paraffine Oils.

Bounding Energy \ Additive / Element	MoDTC	MoDTC+CaSF	MoDTP	MoDTP+CaSF
C_{1s}	284.6	284.6	284.6 287.9	284.6 289.2
Fe_{2p}	710.9	710.9	711.3	710.1
Mo_{3d}	227.5 231.8	227.5 231.7	228.3 231.9	227.6 232.3
S_{2p}	161.4	161.3	161.8	161.2
O_{1s}	530.4 532.0	530.2 531.4	531.9	530.3 531.9
N_{1s}	398.0	398.1		
P_{2p}			133.1	132.7
Ca_{2p}		347.1		347.2

With the binary additive blends, MoDTC/CaSF or MoDTP/CaSF, two Ca_{2p} peaks were observed and one of them probably can be attributed to $CaCO_3$ (346.8 kJ/mol) of $CaSO_4$ (347.4). The N_{1s} peak was unsolved but it was more likely to be an organic nitrogen compound.

It is very interesting to see that the above analysis did not show the presence of MoS_2 in surface coating. This is consistent with the results obtained by Black but contradictory with those given by some other work (7 − 10). The contradictory reports on film compositions suggest that the reaction mechanisms of organomolybdenum compounds are very complicated. Organomolybdenum compounds can break down in different ways on rubbing surfaces according to test conditions.

In the present study the binding energies of the chief interest elements, such as Mo, S, P, N, and Fe, stayed constant inspite of the addition of CaSF in molybdenum containing oils, suggesting that the chemical environment of these elements in the reaction films do not change significantly due to the interaction between molybdenum compounds and calcium sulfonate. However, the interactions led to a profound increase of both Mo and S contents in worn surfaces. Calcium also appeared in surface coatings.

Based on the assumption that the large sprawling molecules of CaSF will block the adsorption of molybdenum compounds, at first sight, one may postulate that adding surfactant CaSF causes the deterioration of friction and wear properties of molybdenum compounds. Cameron and Spikes indicated that a small amount of "nondesorbed" calcium sulfonate that remained on metal surfaces was enough to prevent the EP agent from getting to the surfaces (11, 12). However, as described above, molybdenum compounds and calcium sulfonate detergents act synergistically in some circumstances, promoting the build up of both Mo and S elements in worn surfaces. These observations are in contradiction to the conventional view of surfactant interferring with the action of EP agents. This drove us to consider if other interaction mechanism were in operation.

It is possible that the interactions between molybdenum compounds and calcium detergents lead to the formation of a chelate complex. The increase of Mo and S in surface coating due to the addition of CaSF probably can be explained by the chelate complex formation, eliminating (or mitigating) the competition between molybdenum compounds and calcium sulfonate for the rubbing surfaces. The chelate complex probably can adsorb through its sulfonate parts onto rubbed surfaces and therefore, molybdenum compounds, as a part of the chelate complex, are also carried to surfaces.

In oil change running-in experiments running-in oils have quite different effects on friction reduction and wear prevention as well as diminution of induction period. This phenomenon is not fully understood, but the main factor governing the alternation of the induction period seems to correlated to the adsorption and reaction of the additives on the surface coating formed by running-in oils. As a matter of fact, molybdenum containing compounds adsorb and react slowly on the surfaces treated with calcium sulfonate detergents, leading to a long induction period and a long transition period.

The significant influence of test temperatures and additive concentrations on the friction and wear performance of binary additive blend, MoDTC/CaSF or MoDTP/CaSF, probably can be also attributed to the chelate complex formation, which occurs in certain ranges of temperatures and additive concentrations.

3.9.4 Conclusions

1. The effects of motor oil additives on the friction and wear performance of organomolybdenum containing oils can be quite selective, depending upon the nature of both the molybdenum compounds and the motor oil additive, as well as test temperatures and additive concentrations.

2. Running-in procedures appear to influence the reactions responsible for surface film formation in test stage, leading to the variation of reduction periods of organomolybdenum containing oils.

3. Combinations of organomolybdenum compounds and calcium sulfonate are often found to exhibit synergistic effects. The most significant synergistic effect on both friction reduction and wear prevention has been observed with molybdenum dithiophosphate plus calcium sulfonate. This phenomenon probably can be explained by chelate complex formation, eliminating or mitigating the competition between molybdenum dithiophosphate and calcium sulfonate for rubbing surfaces and also increasing the concentration of both molybdenum and sulfur in worn surfaces.

3.9.5　References

(1)　Bratthwatte, E.R.; Green, A.B.: "A critical analysis of the performance of molybdenum compounds in motor vehicles", Wear, 46 (1978) 2, p. 405—432.

(2)　Green, A.B.; Risdon, T.J.: "The effect of molybdenum-containing oil-soluble friction modifier on engine fuel economy and gear oil efficiency", SAE 811187.

(3)　Kuo, L.L.K.; Chang, S.T.; Chao, Y.T.: "Fuel economy engine oils via friction modifiers", Lubrication Engineering, 45 (1989) 2, p. 81—90.

(4)　Spikes, H.A.: "Additive-additive and additive-surface interactions in lubrication", Proc 6th Int. Colloq. Technische Akademie Esslingen, Jan 12—14 1988 "Industrial lubricants — Properties, Application, Disposal", ed W.J. Bartz, p. 4.2/1—4.2/8.

(5)　Inoue, K.; Watanabe, H.: "Interactions of Engine Oil Additives", ASLE Preprint No. 82-AM-1B-1.

(6)　Holinski, R.: "The influence of boundary layers on friction", Wear, 56 (1979) 1, p. 147—154.

(7)　Zheng, P.; Han, X.; Wang, R.: "The mechanism of friction reduction of oxymolybdenum Di-(2-ethylhexyl)-phosphorodithiotate under boundary lubrication", ASLE preprig No. 86-TC-4E-1.

(8)　Black, A.L.; Dunster, R.W.; Sanders, J.V.: "Comparative study of surface deposits and behaviour of MoS2 particles and molybdenum dialkyl-dithio-phosphate", Wear, 13 (1969) 2, p. 119—132.

(9)　Feng, I-M.; Perilslein, W.L.; Adams, M.R.: "Solid film deposition and non-sacrificial boundary lubrication", ASLE Trans., 6 (1963), p. 60—66.

(10)　Isoyama, H.; Sakurai, T.: "The lubricating mechanism of di-thio-dithio-bis (diethyl-dithiocarbamate) dimolybdenum during EP lubrication", Tribology International, 7 (1974) 4, p. 151—160.

(11)　Cameron, A.: "The role of surface chemistry in lubrication and scuffing", ASLE Trans. 23 (1980) 4, p. 388—392.

(12)　Spikes, H.A.; Cameron, A.: "A comparison of adsorption and boundary lubrication failure", Proc Roy Soc., A 336 (1974), p. 407.

3.10 The Study on the Antiwear Action Mechanism of Alkoxy Aluminium in Lubricating Oil

J. Dong, G. Chen and F. Luo
Institute of Logistics Engineering, Chongquing, P.R. China

Abstract

Aluminium is situated beneath boron in the same group of the periodic table. As the borate is a good antiwear agent of lubricating oils, it implies that aluminium compounds maybe possess similar antiwear property to borate. After testing seven alkoxy aluminium compounds, the results prove this assumption. The most efficient compound is triethoxy aluminium. When 2 mass % of triethoxy aluminium is added to ISO VG 68 oil, the load wear index of a four-ball machine increases from 219 N to 519 N and the weld point from 1568 N to 4900 N. This compound also exhibits good corrosion protection property.

After the steel balls of the four-ball machine have been run in aluminium-containing oil, its wear scars are analysed by scanning electron microscopy and X-ray energy dispersion analysis. The main compositions discovered on the scar surface are 5.7 − 48.3 atom % of iron, 4.4 − 11.2 atom % of aluminium, 37.1 − 59.1 atom % of oxygen, 8.7 − 23.5 atom % of carbon, and 0.3 − 1.1 atom % of chromium, under 3087 N − 39 200 N of axial loads. When the loads are lower than 2450 N, the contents of oxygen are almost equal to zero and the contents of iron increase to 68.6 − 93.8 atom %. Based on the results of surface analysis and thermodynamic principles, the antiwear action mechanism of triethoxy aluminium is suggested, under low loads the weaker bond C-C (350 KJ/mol) is broken, the broken fragments may form polymer film during friction processes. As the loads increase, the stronger bond Al-O (484.5 KJ/mol) is broken to create active aluminium and oxygen. The oxygen reacts with aluminium and iron to form a complicated solid solution containing AlxOY, FExOy on the surface. This complicated surface layer raises the weld point of the steel balls.

3.10.1 Introduction

The borates have been used as extreme pressure (EP) additive in lubricants since the 1960s. Especially, the derivatives of organo-borates were used as friction modifiers in the 1980s. Their excellent performances such as EP property, friction reducing ability, oxidation stability, and anticorrosion function motivates to study their action mechanism. Dong Junxiu and his workmates have proved that the organo-borates not only form adsorption film and friction polymer on rubbing surface, but also form a permeating layer of boron and carbon into the rubbing surface (1, 2).

As a result, the adsorption film, friction polymer and the permeating layer of boron and carbon reduce the friction and wear of rubbing pairs. The carburization and boronization during friction processes using organoborate additive have similar effects to chemical heat treatment of machine elements to modify their surface with carbonic and boric compounds (3). The permeating action of boron-type additives implies that the aluminium situated beneath boron in the periodic table may be used to synthesize antiwear additives, because aluminium has similar property to boron and has been used to improve the surface performance of machine elments in chemical heat treatment. Through synthesizing several organo-aluminiums and evaluating their properties, the triethoxy aluminium was found to be the best EP additive among them. The results of surface analysis show that aluminium can also permeate into rubbing surfaces under certain loads. According to the testing results, the assumption that aluminium can be used to synthesize antiwear additive is proved.

3.10.2 Performances of Alkoxy Aluminium

Seven kinds of alkoxy aluminium were synthesized and their antiwear properties evaluated on the four-ball machine. Experiments were carried out at room temperature, with running time being 10 seconds and the revolution of the upper ball being 1450 r/min. The results showed that triethoxy aluminium (TEAL) exhibited the most excellent EP characteristics (Table 3.10.1). Its maximum non-seizure loads are 804 N, and weld point loads 4900 N. Therefore our efforts focused mainly on the evaluation of other properties of TEAL.

Table 3.10.1: The EP property of alkoxy aluminium compounds in ISO VG 68 oil (concentration 2 %)

Compounds	Max. non-seizure load (N)	weld point (N)
glycol aluminium	< 696	1764
glyceric aluminium	< 696	1764
phenylmethanol aluminium	< 696	3087
para-diphenol aluminium	696	3087
ethanolamine aluminium	696	1764
α-naphtol aluminium	696	3920
triethoxy aluminium	804	4900

Firstly, the optimum concentration of triethoxy aluminium in the oil was determined. The figures listed in Table 3.10.2 show that the optimum concentration of triethoxy aluminium is 20 % in ISO VG 68 oil. If the concentration exceeds 2.0 %, the EP property of the oil cannot be improved anymore. Secondly, after determining the optimum concentration, the main properties of oil containing TEAL were measured. The friction coefficient of the formulated oil decreases by 11.7 % (Table 3.10.3). Its oxidation stability is obviously better than VG 68 oil (Table 3.10.4).

Table 3.10.2: The effect of TEAL's concentration (5) on EP property

C % (wt)	0.4	0.8	1.2	1.6	2.0	2.4	2.8
Max. non-seizure load (N)	< 549	< 549	< 549	647	804	804	804
weld point (N)	1960	2450	2450	3087	4900	4900	4900
Mean Hertzian load (N)	–	–	–	–	519	–	–
note			base oil is ISO VG 68				

Table 3.10.3: The effects of TEAL on friction coefficient of VG 68 oil under various loads (kg) on MM-200 machine

oils	60	80	100	120	140	160	180	200	average	decreases
ISO VG 68	0.058	0.060	0.061	0.062	0.062	0.066	0.068	0.070	0.0634	–
ISO VG 68 + 2 % TEAL	0.049	0.052	0.055	0.056	0.057	0.058	0.061	0.060	0.056	– 11.7 %

Table 3.10.4: The effects of TEAL on oxidation stability of VG 68 oil

Properties		ISO VG 68 oil	ISO VG 68 oil + 2 % TEAL
Oxidation stability (125° C, 8 h, oxygen rate = 200 ml/min) Total acid content mg KOH/g	before oxidation	0.1	26.3 (Total base numbers)
	after oxidation	2.9	1.1
	variation multiples	28	1.04
Copper corrosion rating, 100° C, 3 hs		1	1

3.10.3 Antiwear Action Mechanism

For studying the antiwear action mechanism, the oil containing 2 % TEAL was tested in the four-ball machine. Then the wear scars of the steel balls were observed by Scanning Electron Microscopy (SEM). The chemical elements on the wear scars were also measured through X-ray Energy Dispersion Spectroscopy (EDX). Figure 3.10.1 shows the topography of wear scars of steel balls. Under low loads (441 N), the wear traces of rubbing surface are rough, and some irregular pieces of film cover these traces (Figure 3.10.1a). Under higher loads (1960 N 3087), an even, plastic film distributes on the rubbing surface (Figure 3.10.1b, 3.10.1d), and some thin traces appear on the plastic film. The fact that wear traces become thinner under higher loads is certainly related to the action of the aluminium-type antiwear agent. Triethoxy aluminium contains C-C bonds, C-O bonds, and Al-O bonds. According to the value of bond energy, they are arranged in the following order:

$$Al\text{-}O \text{ (484.5 KJ/mole)} \quad C\text{-}O \text{ (381 KJ/mol)}$$
$$> C\text{-}C \text{ (350 KJ/mole)}$$

During friction process, TEAL adsorbs on the surface when the load is lower. As the chain of carbon is short in the molecule, it is difficult to form a continuous and tenacious protecting film. The film cannot separate the rubbing surfaces thoroughly, some asperities of the surfaces contact each other and abrades out rough traces during friction. When the load increases to the degree under which the mechanical energy is large enough to break down the Al-O bond, the Al will dissociate and permeate into the surface to substitute the atoms of iron. Table 3.10.5 shows the elements' content of wear scars analysed by EDX. The figures in the table demonstrate that the content of aluminium increases since the load is larger than 617 N. When the load increases to 3087 N, the contents of elements vary significantly, e. g. the amount of oxygen increases evidently and the plasticity of the film is weaker than that under 2450 N. This implies that when the loads are equal to or less than 2450 N, the formation of the protecting film is primarily due to the substitution of Al for Fe to create a substitutional solid solution. As the load rises to 3087 N, the protecting film contains a portion of oxide. The products on rubbing surfaces which are related to TEAL are listed as follows, and the changes of free enthalpy are also listed.

Products	ΔG (298° K) KJ/mole
$\alpha\text{-}Al_2 L_3$	− 1582
$Fe_2 O_3$	− 740.3
$Al_4 C_3$	− 196
$Fe_3 C$	18.8

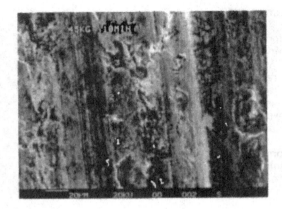

Figure 3.10.1a:
Load =441 N

Figure 3.10.1b:
Load = 1960 N

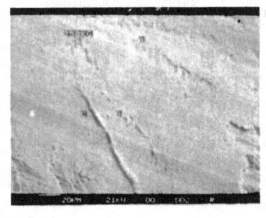

Figure 3.10.1c:
Load = 2450 N

312

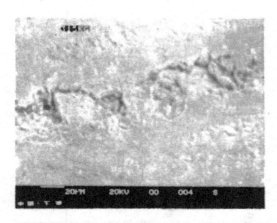

Figure 3.10.1d:
Load = 3087 N

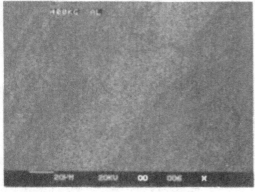

Figure 3.10.1e:
Two dimensional
distribution of alu-
minium under the
load of 3930 N

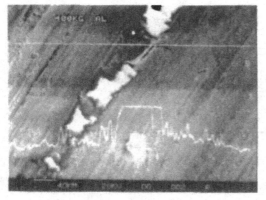

Figure 3.10.1f:
One dimensional distri-
bution of aluminium
under the load of
3920 N

Figure 3.10.1: Topography of wear scars of steel ball under different loads

Table 3.10.5: Atomic concentration (5) on the surface of steel ball's wear scars

atoms	surface of steel ball before test	loads (N)											
		617	784	980	1176	1569	1960	2450	3087	3920	4900		
Fe	95.3	93.8	79.8	80.4	82.1	81.7	76.2	68.6	48.3	5.7	96.4		
Al	0.89	3.7	3.0	11.9	8.5	4.7	5.9	12.1	4.4	11.4	1.19		
Si	0.91	0.94	1.4	0.74	0.64	0.59	0.64	0.23	0.35	0.06	0.7		
Cr	2.12	1.59	2.7	1.7	1.48	1.52	1.45	2.3	1.1	0.29	1.69		
O	t	t	t	t	t	t	t	t	37.1	59.1	t		
C	t	0.3	13.1	5.3	7.3	11.5	15.8	16.7	8.7	23.5	t		

note: t — trace

The changes of free enthalpy for α-Al$_2$O$_3$ are the smallest, which means that the α-Al$_2$O$_3$ may be formed more easily than the others. According to the assertion of literature (4), the film of aluminium oxide has excellent anti-weld ability. Literature (5) illustrates that if the concentration of aluminium in steel exceeds 8 %, a dense film of aluminium oxide can be formed on the elements' surface when the machine elements have been permeated with aluminium in chemical heat treatment. This aluminium oxide film possesses good oxidation stability and anti-corrosion property. The test results of oil containing TEAL are conformable with the description of above literatures. Therefore, it has been considered that the aluminium dissociated from TEAL may permeate into the rubbing surface, and some of them may form an oxide film which improves anti-weld ability, oxidation stability, and anti-corrosion property of the rubbing pairs during friction process. The figures in Table 3.10.5 also show the rising tendency of carbon content with the increase of loads. During friction process, the carbon dissociates from triethoxy aluminium and permeates into rubbing surface to form ferro carbide Fe$_x$C$_y$. The carburization of machine elements' surfaces may enhance their antiwear property.

3.10.4 Conclusion

3.10.4.1 Through tests, the assumption that aluminium, possessing similar properties to boron, can be used to synthesize EP agents has been proved.

3.10.4.2 The antiwear action mechanism of triethoxy aluminium is: Under low load conditions triethoxy aluminium forms an adsorption film on the rubbing surface. Under moderate load conditions, triethoxy aluminium dissociates to form Al-Fe solid solution and ferro carbide Fe$_x$C$_y$ with rubbing surface. Under high load conditions, the dissociated pieces of triethoxy aluminium form α-Al$_2$O$_3$ and Fe$_x$C$_y$ on the surfaces to raise anti-weld ability.

3.10.5 References

(1)   Dong Junxiu, Luo Xinming: "The Research for the Preparation and Action Mechanism of Boron-Type Antiwear Additives", 6th International Colloquium, Industrial Lubricants, Januar 1988, Esslingen, F.R. Germany.
(2)   Dong Junxiu, Chen Guoxu, Qian Jugen: "The Study on Glycol Borate Performance and its Antiwear Action Mechanism", The 5th International Congress on Tribology EUROTRIB 89, June 1989, Helsinki, Finland.
(3)   Dong Junxiu, Chen Ligong: "Study on Performance and Action Mechanism of B-N Antiwear Additive", International Symposium on Tribochemistry, August 1989, Lanzhou, P. R. China.
(4)   Solid Lubrication Handbook, 1978, p. 275.
(5)   Liu Jiajun: Introduction of Material Surface Technology, 1988, p. 336.

3.11 Functional Properties of EP-Additive Packages Containing Zn-Dialkyldithiophosphate, Sulphurized EP-Additive and a Metal Deactivator

G.S. Cholakov, K.G. Stanulov and I.A. Cheriisky, Higher Institute of Chemical Technology, Sofia, Bulgaria
T. Antonov, Petrochemical Combine Pleven, Bulgaria

3.11.1 Introduction

Zn-dialkyldithiophosphates (Zn-DADTP-s) are widely used in automotive lubrication being still among the most popular components of motor oil additive packages. They are also employed in plastic greases, hydraulic oils, metal working lubricants, etc. (1–3). Most desirable properties of the Zn-DADTP-s are their excellent antiwear and antioxidation effects, combined with certain anticorrosion, solubilization and friction modifying potential.

However, Zn-complex metal salts of organic dithophosphoric acids possess a relatively low thermal stability and may be decomposed hydrolytically though, of course, their stability depends on the type and length of organic radicals (4). Their performance in EP tribocontacts eventually working under high loads and temperatures is determined by the thiol sulphur atom (5), and because of that is relatively modest.

The deficiencies of the Zn-DADTP structure determine the additives with which they are combined for a particular application. For instance, Zn-DADTP-s have been introduced together with chlorine containing additives in the decade after the second World War for the manufacture of API GL-5 grade automotive gear oils, conforming to the then used requirements of the US MIL-L 2105B specification (6, 7). Later these packages, probably because of the inadequate training component, proved to be inadequate for the more severe requirements for moisture corrosion and thermal-oxidation stability accepted.

The performance of Zn-DADTP-s alone and in combination with other additives used in motor oil formulation — succinimides, sulphonates, salicylates, dithiocarbamates of different metals (including Mo), etc. have been extensively studied (8 – 13). Less work has been reported for other combinations, containing Zn-DADTP-s and especially such which may be useful in automotive gear oils of the higher grades.

The main object of the work described hereunder is to present experimental data on certain functional properties of compositions, containing Zn-DADTP, metal deactivator and sulphurized EP additive (sulphurized olefines and/or sulphurized esters) and discuss their eventual application.

Sulphurized olefines were chosen for their high EP properties and current use in modern S, P-ashless packages which satisfy the requirements of the highest grades of automotive (GL-4 and GL-5) and industrial EP gear oils (14). Sulphurized esters are well known friction modifiers, possess EP properties, and have also been used in gear oils (15, 16). The metal deactivator was required since EP additives containing "active sulphur" were chosen but also to help with antioxidation and antiwear properties of the package.

3.11.2 Experimental Details

Zn-dialkyldithophosphate (Zn-DADTP), dialkylpentasulphide (DAPS) and the S, P-ashless additive package (GL-pack) recommended at 6.5 % concentration for GL-5 gear oils employed in the investigation were commercial products of international additives companies. These additives are currently used in the regular production of lubricants in Bulgaria. The sulphurized methyl ester of fatty acides of lard (SME) was prepared in the authors' laboratory through methanolysis of lard and subsequent sulphurization of the methyl esters with 1.1 mols sulphur per mol ester in inert athmosphere at 140° C for 5 hours. The metal deactivator (MDA) is widely used in the rubber industry. All commercial additives were used as received. Table 3.11.1 presents the content of characteristic heteroatoms in the molecules of the additives used.

Two mineral base oils produced from phenol refined, solvent dewaxed, deasphalted and hydrorefined components were employed in the investigation. Both base oils (compounded to ISO 220 and SAE 90 grades, respectively) are currently used in the manufacture of AGMA 5 EP and GL-5 gear oils in Bulgaria. Table 3.11.2 presents the range of the main physicochemical properties of the oils as required by the corresponding Bulgarian mineral base oils standard specifications.

The main functional properties of the additive compositions investigated included lubricating properties as estimated in standard four ball machine tests; thermal stability, determined by thermogravimetric analysis, thermal oxidation stability, estimated in a laboratory gasometric apparatus. Selected samples were also tested for different properties in the FZG, thermal oxidation stability and Beaker static oxidation, demulsibility, foam ability and other standard tests, as required by the respective Bulgarian gear oil specifications.

Four ball machine tests were performed on a "Seta-Shell Four Ball EP Tester", using standard 12.7 mm stainless steel balls (manufactured by "SKF") and strictly following the procedures for the preparation of balls, load increments, calculation of results, etc. as recommended by the US ASTM D 2783 for EP tests and D 2266 — for wear tests.

Thermogravimetric analysis was performed in air with approximately 500 mg sample (heating rate 5° C per minute up to 500° C, inert reference substance — Al_2O_3).

Table 3.11.1: Content of Characteristic Heteroatoms in Lubricant Additives

Additive or package	Code name	Heteroatoms, %			
		Zn	S	P	N
Zn-dialkyldithiophosphate	Zn-DADTP	9.30	19.00	9.10	—
Dialkylpentasulphide	DAPS	—	38.65	—	—
Sulphurized methyl ester of fatty acids	SME	—	10.04	—	0.13
Metal deactivator	MDA	—	38.32		8.38
Combination of Zn-DADTP and MDA	Zn-MDA	6.13	13.81	6.03	0.25
Commercial S, P, N package for GL-5 oils	GL-pack.	—	30.04	1.94	0.30

Table 3.11.2: Characteristic Physicochemical Properties of Mineral Base Oils

Physicochemical Properties*	ISO 220 Viscosity Class	SAE 90 Viscosity Class
Kinematic Viscosity, $mm^2 s^{-1}$, at:		
40° C	220 ± 20	—
100° C	—	20 ± 2
Coke Residue (Conradson), % less than	0.55	0.60
Ash, %, less than	0.01	0.01
Total Acid Number, mg KOH/g, less than	0.04	0.04

* All properties determined as prescribed by the current Bulgarian State Standard Methods (19)

Table 3.11.3: Bulgarian State Standard Methods (BSSM-s) and Similar US or U.K. Methods

No.	Method	BSSM-s or Comecon Method (19)	US (ASTM, FTM) U.K. Method (20)
1.	Antiscuffing Properties by the Four Ball Machine	BSSM-9787 -82	ASTM D 2783
2.	Antiwear Properties by the Four Ball Machine (40 N, 120° C, 1 h)	BSSM-9786 -84	ASTM D 2266
3.	Copper Corrosion (3 h, 125° C)	BSSM-5747 -84	ASTM D 130
4.	Thermal Oxidation Stability Test (T.O.S.T., 163° C, 50 h)	BSSM-14212 -77	FTM 2504
5.	Storage Stability (60 days, centrifugation)	BSSM-14138 -79	FTM 3455.1
6.	Static Beaker Oxidation Test (150°C, 100 h, no catalyst)	BSSM-15613 -83	—
7.	Foaming Ability (Sequence: 25° C, 95° C, 25° C)	Comecon 3970-83	ASTM D 892
8.	Anticorrosive Properties in the Presence of Dist. H_2O (10 % H_2O, 60° C, 24 h)	BSSM-7806 -83A	ASTM D 665A
9.	Demulsibility	BSSM-15785 -83	ASTM D 2711
10.	Corrosion on Steel Bearing and Copper Plate ("Volkswagen" Method, 48 h, 100° C)	BSSM-14370 -77	—
11.	Antiscuffing Properties by the FZG Test	BSSM-9781 -72	IP 334

Thermal oxidation stability was determined in a gasometric laboratory glass apparatus (17, 18). The sample tested (10 ± 0.01 g) was oxidized in dishes of electrolytic copper in the presence of 1 % lauryl peroxide as initiator at 163 ± 2° C in oxygen atmosphere (0.1 MPa oxygen). Time for absorption of up to 20 cm³ oxygen was continuously recorded. Change of viscosity and catalyst lost after oxidation were also evaluated.

Other tests were performed as prescribed by the corresponding Bulgarian State Standard Methods (BSSM-s) (19). Table 3.11.3 lists these methods and similar US (ASTM, FTM) or U.K. (IP) standard methods (20).

3.11.3 Results and Discussion

3.11.3.1 Lubricating Properties

Results for the lubricating properties of the individual additives in SAE 90 mineral base oil are presented in Table 3.11.4.

The sulphurized hydrocarbons (DAPS) exhibit high EP properties even at the lowest concentration tested and at 6.5 % the Weld Load reaches the upper limits of standard four ball machine tests. Load Wear Indices of this additives are also the highest though seizure starts at relatively low loads. The properties of the reference GL pack at 6.5 % concentration, as expressed in the particular tests, are matched by DAPS even at 1 % concentration except for the Initial Seizure Load.

Zn-DADTP, on the other hand, ensures high Initial Seizure Loads and even below 2 % concentration achieves the results for the seizure loads of the reference package but Weld Loads with this additive are the lowest and the Load Wear Indices are rather modest. The metal deactivator does not influence very much the lubricating properties of the dithophosphate which is important since it is needed in further compositions to control the very high corrosion of copper by DAPS.

The sulphurized methyl esters have much lower EP potential than the sulphurized hydrocarbons (DAPS) as expressed in four ball machine tests. However, they are known to possess other useful properties, especially friction modification ability (15) which will not be accessed by the particular tests and parameters evaluated.

Performance of lubricant additives in EP tribocontacts is considered mainly to be dependent on the balance between their adsorption ability and their chemical activity towards the surfaces involved (5). This balance is set by the chemical structure of the additive (including the type of heteroatoms present), but is changing when the concentration of the additive and/or parameters in the contact change. The mechanism of action of Zn-DADTP-s and polysulphides alone have been studied and reported by different authors. Most of the approaches have been summarized and interpreted, for instance, by Kajdas (21).

320

Table 3.11.4: Copper Corrosion and EP Properties of Individual Additives in SAE 90 Base Oil*

Concentration of the Additive and Parameter	Additive or Package				
	Zn-DADTP	Zn-MDA	DAPS	GI-pack	SME
Base oil + 1 % Additive:					
— Initial Seizure Load, N	1000	1260	630	800	800
— Weld Load, N	2500	2500	5000	3150	2500
— Load Wear Index, N	529	526	676	449	385
Base Oil + 2 % Additive:					
— Initial Seizure Load, N	1600	1260	1000	800	1000
— Weld Load, N	2500	2500	6200	4000	3150
— Load Wear Index, N	618	530	853	566	502
Base Oil + 3 % Additive:					
— Initial Seizure Load, N	1260	1600	1260	1260	1000
— Weld Load, N	2500	2500	6200	4000	3150
— Load Wear Index, N	541	640	900	698	538
— Copper Corrosion, merits	–	1b	4c	1b	2a
Base Oil + 6.5 % Additive:					
— Initial Seizure Load, N	1600	1600	800	1000	800
— Weld Load, N	3150	3150	8000	5000	4000
— Load Wear Index, N	677	675	915	664	566
— Copper Corrosion, merits	–	1b	4c	2a	2a

* Properties of the SAE 90 Base Oil: Copper Corrosion, merits = 1a; Initial Seizure Load, N = 630; Weld Load, N = 1600; Load Wear Index, N = 266

In general, Zn-DADTP adsorbs and decomposes to inorganic phosphates and organic sulphur radicals, which are also deposited, forming a complex, presumably polymeric lubricating layer, even at relatively low flash temperatures. These layers are effective in preventing wear and passing higher loads without seizure (Table 3.11.4). After seizure, when the number of contacts of surface asperities increase, creating higher temperatures and more intensive surface interactions, the phosphate film is damaged while the organic sulphur radicals derived mainly from the thiol sulphur atom are not sufficiently reactive to decompose further and create an iron mercaptide film in due time. Thus, welding takes place at relatively low loads.

Polysulphides are able to decompose at relatively low temperatures to highly reactive towards iron lower molecular mass compounds and radicals (including H_2S and sulphur in an active form). So mercaptide films which are able to prevent welding but also lead to attritious corrosive wear of the metal surfaces, are formed. This may explain the low and decreasing with higher concentrations seizure loads and the high Weld Loads observed for DAPS (Table 3.11.4).

The behaviour of the sulphurized esters (SME) is similar to that of DAPS while the combination of metal deactivator and Zn-DADTP resembles the mechanism of action of Zn-DADTP itself.

Ther performance of the reference GL pack is possibly a combination of the effects of DAPS and Zn-DADTP, presumably because of the presence of sulphurized hydrocarbons and ashless P, S-compounds in the package.

Table 3.11.4 illustrates the general idea of the investigation — to combine the antiwear performance of the Zn-DADTP with the antiweld performance of DAPS while controlling the corrosive ability of DAPS, through Zn—DADTP and a metal deactivator, and making the package more versatile, especially for real contacts in which controlled slip through friction modification is needed by incorporation of sulphurized esters.

Table 3.11.5 presents the results for the lubricating properties of such additive combinations. The participation of DAPS in the experimentel packages is limited by the concentration and efficiency of the metal deactivator but still the properties of the reference package are achieved in all parameters evaluated. The Load Wear Indices of the experimental combinations are significantly higher which, with equivalent Initial Seizure Loads and Weld Loads, is due to smaller wear scar diameters.

The incorporation of SME leads to lower Weld Loads, probably because of competitive adsorption between DAPS and SME while the antiwear action of Zn-MDA remains intact.

It may be speculated that in the experimental combinations DAPS provides active low molecular mass sulphur compounds and radicals at higher temperatures than when tested alone.

322

Table 3.11.5: EP and Antiwear Properties of Combinations of DAPS and SME at Fixed Concentration of Zn-MDA in the Package

Additive Package*, %	Concentration in SAE 90 oil, %	Initial Seizure Load, N	Weld Load, N	Load Wear Index, N	Wear Scar Diameter, mm
80 Zn-MDA + 20 DAPS	3.0 6.5	1260 1260	4000 5000	723 805	0.54 0.49
80 Zn-MDA + 15 DAPS + 5 SME	3.0 6.5	1260 1260	4000 5000	668 765	0.49 0.50
80 Zn-MDA + 10 DAPS + 10 SME	3.0 6.5	1260 1260	3150 4000	618 740	0.50 0.51
80 Zn-MDA + 5 DAPS + 15 SME	3.0 6.5	1260 1260	3150 4000	588 626	0.47 0.50
80 Zn-MDA + 20 SME	3.0 6.5	1260 1260	3150 3150	549 583	0.50 0.51
100 Zn-MDA	3.0 6.5	1600 1600	2500 3150	639 675	0.49 0.48
100 GL-pack	3.0 6.5	1260 1000	4000 5000	699 664	– 0.56

* All experimental packages show better copper corrosion than the reference

323

3.11.3.2 Thermal Stability

Results for the thermal stability of the individual additives and the experimental combinations are presented in Table 3.11.6.

Zn-DADTP-s are considered to be additives of a relatively low thermal stability, which is a disadvantage for the mechanism of their antiwear action (21), and their overall performance in motor oils, since the latter have to fulfill their duties in zones of high temperatures as well.

However, results from Table 3.11.6 show that the reference GL-pack, designed for a specific application, is even less thermally stable. This is probably due to the participation of sulphurized hydrocarbons in the package which, as the thermal behaviour of DAPS indicates, are rather unstable. Surprisingly, the combinations of Zn-MDA with DAPS show higher thermal stability than Zn-DADTP and Zn-MDA for temperatures up to 250° C, which is even further increased when sulphurized esters are introduced in the package. The incorporation of the metal deactivator in Zn-DADTP leads to increased formation of volatile substances above 250° C.

Low thermal stability, while being a disadvantage for antiwear action, is in the same time favourable for antiweld action though it may lead to greater chemical corrosion at lower loads. So, the behaviour of DAPS and the GL-pack in the thermogravimetric analysis reflects also their ability to respond quickly in urgent siutations in the tribocontact by repairing damaged sulphide film, for instance, under shock loads. On the other hand, additives which decompose easily with temperature may be expected to have a greater rate of spending in practical lubrication. This may explain why relatively high amounts of additives with high sulphur content are needed to meet automotive gear oil specifications. Coming back to synergism in thermal stability of combinations of Zn-DADTP, MDA, DAPS and/or SME, it is known that some of the additives used in motor oil additive packages also improve the thermal stability of Zn-DADTP (10 — 12), which is altogether beneficial for motor oil application. The importance of the synergism, observed in Table 3.11.6 may be experimentally confirmed only in practical application.

It has to be pointed out that the results presented in Table 3.11.6 reflect only the formation of low molecular mass volatile decomposition products in the thermogravimetric analysis. Comparison of lubricating properties (Tables 3.11.4 and 3.11.5) and thermal stability (Table 3.11.6) shows a good correlation between Load Wear Indices and thermal stability.

3.11.3.3 Thermal Oxidation Stability

Results for thermal oxidation stability are shown in Table 3.11.7 and Table 3.11.8.

Table 3.11.6: Thermal Stability of Additives and Packages

Composition of the Package, % Weight Loss, % to Temperature					Temperature Corresponding to Weight Loss, °C	
	200° C	250° C	300° C	350° C	10 %	50 %
100 G L-pack	7.6	74.1	91.5	93.4	205	234
100 Zn DADTP	2.4	47.0	56.6	62.7	214	265
100 Zn-MDA	2.8	48.9	73.9	76.1	223	254
100 DAPS	36.3	71.6	83.2	92.6	184	208
100 SME	5.3	24.2	56.8	88.4	220	285
80 Zn-MDA + 20 DAPS	3.1	35.2	70.4	76.4	218	265
80 Zn-MDA + 15 DAPS + 5 SME	1.8	28.2	68.7	77.2	218	266
80 Zn-MDA + 10 DAPS + 10 SME	2.1	23.7	65.2	75.3	229	274
80 Zn-MDA + 5 DAPS + 15 SME	3.5	26.8	65.0	75.3	229	272
80 Zn-MDA + 20 SME	2.6	29.8	61.3	75.6	235	278

Table 3.11.7: Kinetics of Oxidation of SAE 90 Mineral Base Oil with Additive Packages

Composition of Doped Oil, %	Time for Absorption of cm^3 O_2/10 g Sample, min						
	3	6	9	12	15	18	20
100 SAE 90	31	39	48	57	64	71	77
97 SAE 90 + 3.0 GL-pack	68	85	99	114	128	137	142
93.5 SAE 90 + 6.5 GL-pack	91	103	120	140	161	172	180
97 SAE 90 + 3.0 Zn-MDA	80	99	115	125	136	151	155
97 SAE 90 + 2.4 Zn-MDA + 0.6 DAPS	93	107	127	147	164	181	188
97 SAE 90 + 2.4 Zn-MDA + 0.45 DPS + 0.15 SME	89	100	110	121	133	141	152
97 SAE 90 + 2.4 Zn-MDA + 0.3 DAPS + 0.3 SME	82	97	111	120	130	144	146
97 SAE 90 + 2.4 Zn-MDA + 0.15 DPS + 0.45 SME	73	91	106	118	132	143	150
97 SAE 90 + 2.4 Zn-MDA + 0.6 SME	53	62	70	77	89	95	99
97 SAE 90 + 3.0 Zn DADTP	85	102	113	122	131	141	145

Table 3.11.8: Change of Kinematic Viscosity at 100° C of the Oxidized Doped Oils and Loss of Copper Catalyst after Oxidation

Composition of Doped Oil, %	Change of Viscosity, %	Copper Catalyst Loss, %
100 SAE 90	5.0	0.1
97 SAE 90 + 3.0 GL-pack	4.9	0.5
94.5 SAE 90 + 6.5 GL-pack	2.8	0.5
97 SAE 90 + 3.0 Zn-MDA	6.1	—
97 SAE 90 + 2.4 Zn-MDA + 0.6 DAPS	4.0	0.2
97 SAE 90 + 2.4 Zn-MDA + 0.45 DAPS + 0.15 SME	2.5	0.1
97 SAE 90 + 2.4 Zn-MDA + 0.30 DAPS + 0.30 SME	2.2	0.2
97 SAE 90 + 2.4 Zn-MDA + 0.15 DAPS + 0.45 SME	4.9	0.2
97 SAE 90 + 2.4 Zn-MDA + 0.6 SME	3.8	0.3
97 SAE 90 + 3.0 Zn DADTP	5.9	0.1

The reference GL-pack at 6.5 % concentration ensures a rather high thermal oxidation stability of the SAE 90 mineral base oil, which is nevertheless matched or even surpassed by the best experimental combination containing Zn-MDA and DAPS at 3 % concentration. The reference package tends to a somewhat higher oxidation corrosion of copper, while oxidation in the presence of Zn-DADTP leads to greater changes of viscosity. The sulphurized methyl esters (SME), while imparing the oxidation kinetics of the best combination, are ensuring smaller changes of the viscosity of the oxidized oil (Table 3.11.8).

The introduction of the metal deactivator (MDA) together with the Zn-DADTP leads to a somewhat worse performance in the beginning of the oxidation but then, probably because MDA needs some time to react with the copper catalyst, the situation is rather improved.

On the whole, a synergistic combination of Zn-MDA, DAPS and a small amount of SME at concentration around 3 % would be able to ensure an adequate or better oxidation kinetics and change of viscosity, and somewhat better protection against oxidation corrosion of copper than the reference GL-pack at 6.5 % concentration under the specific conditions of the particular test for thermal oxidation stability.

It has to be pointed out that all additives participating in the experimental combinations are expected to influence oxidation in one way or another (22).

The Zn-DADTP-s are supposed to act as complex antioxidation additives throughout the whole oxidation process. They are known to react first with the peroxide radicals, then mercaptans, sulphides and disulphides, thus produced, decompose the hydroperoxides. Their high antioxidation effect is probably due to their sulphur content since dialkylphosphates are not antioxidants.

Alkylpolysulphides react with peroxides to thiosulphoxides and then — to thiosulphones.

Typical metal deactivators, while not directly reacting with the products of oxidation of hydrocarbons, are contributing to the antioxidation effect by controlling the influence of the copper catalyst on the oxidation process.

So, it is very interesting to interprete results from Tables 3.11.6, 3.11.7 and 3.11.8 in terms of the oxidation theory which, unfortunately, is not the purpose of this particular presentation.

3.11.3.4 Performance Related Tests and Ideas for Practical Application

Results from selected performance related tests are presented in Table 3.11.9. Table 3.11.10 lists the standard requirements for these tests in order to pass Bulgarian specifications for GL-5 automotive and AGMA 5 EP industrial gear oils.

The experimental results presented in the previous tables were intended to supply relatively easily obtainable information on properties which were a priori considered to be able to create problems. Work done at this stage included also extensive testing of storage (colloid) stability of additive combinations and oils doped up to 10 % level, since the present authors consider any separation of a second phase — solid or fluid — to be a very dangerous threat to tribosystems (23, 24). Some of these tests were run for a period up to 2 years. Fortunately, all combinations presented above proved to be stable. Another important part of this preliminary work was to screen all perspective combinations for copper corrosion till combinations were found which were equal to or better than the references.

Table 3.11.9: Results from Selected Performance Related Tests for Automotive (GL-5) and Industrial (AGMA 5 EP) Gear Oils

| Performance Related Test | Formulated Gear Oils | | | |
| | SAE 90 API GL-5 Automotive | | AGMA 5 EP Industrial | |
	Reference	Experimental	Reference	Experimental
Kinematic Viscosity, mm²s⁻¹ :				
– At 40° C	–	–	227.1	220.7
– At 100° C	19.8	19.9	–	–
Viscosity Index	95	95	93	95
Copper Corrosion, merits	1c	1b	1b	1a
Storage Stability	pass	pass	pass	pass
Anticorrosive Properties in the Presence of Dist. Water	pass	pass	pass	pass
"Volkswagen" Corrosion:				
– on steel bearing	pass	pass	pass	pass
– on copper plate, mg	– 1.2	– 0.7	+ 1.0	– 1.6
Foaming Ability, cm³ foam (5 min blow/10 min rest):				
– At 25° C	5/0	10/0	5/0	10/0
– At 95° C	20/0	5/0	50/0	5/0
– At 25° C	5/0	10/0	5/0	10/0
Demulsibility:				
– Total free water, cm³	80	80	80	80
– Water in oil, %	2.5	2.1	0.3	1.8
– Emulsion, cm³	1.0	1.0	0.1	1.0

Table 3.11.9 (continued)

Performance Related Test	Formulated Gear Oils			
	SAE 90 API GL-5 Automotive		AGMA 5 EP Industrial	
	Reference	Experimental	Reference	Experimental
Beaker Oxidation:				
— Viscosity Increase, %	—	—	15.2	8.1
— Loss of oil, %	—	—	1.8	2.8
— Insoluble in n-heptane, %	—	—	0.01	0.07
Thermal Oxidation Stability Test (T.O.S.T.):				
— Viscosity Increase, %	21.6	39.9	—	—
— Insoluble in n-pentane, %	0.74	0.20	—	—
— Insoluble in toluene, %	0.38	0.08	—	—
EP Properties (Four Ball Machine):				
— Initial Seizure Load, N	800	1000	800	800
— Weld Load, N	4000	4000	3150	4000
— Load Wear Index, N	586	618	502	570
EP Properties after Demulsibility Test:				
— Weld Load, N	—	—	2500	3150
Antiwear Properties (Four Ball Machine):				
— Antiwear Index, N	250	256	264	250
— Wear Scar Diameter, mm	0.56	0.50	0.50	0.58
Antiscuffing Properties (FZG Machine), stages	12[a]	11[a]	>12[b]	>12[b]

a 130° C; 16.6 m s^{-1}; b = 10 mm
b 90° C; 16.6 m s^{-1}; b = 20 mm

Table 3.11.10: Standard Requirements for the Selected Performance Related Tests for GL-5 Automotive (BSSM 14368-82) and AGMA 5 EP Industrial (BSSM 14367-82) Gear Oils

Performance Related Test	BSSM 14368-82	BSSM 14367-82
Kinematic Viscosity, mm^{2}s^{-1}:		
— At 40° C	—	200 – 240[b]
— At 100° C	18 – 20[a]	—
Viscosity Index, not less than	90[a]	90[a]
Copper Corrosion, merits, not higher than	3[b]	1[b]
Storage Stability	no separation	no separation
Anticorrosive Properties in the Presence of Dist. Water	no corrosion	no corrosion
"Volkswagen" Corrosion:		
— on steel bearing	no change	no change
— on copper plate, mg, not more than	5	5
Foaming Ability, cm^3 foam (5 min blow/10 min rest), not more than:		
— At 25° C	100/0	100/0
— At 95° C	50/0	50/0
— At 25° C	100/0	100/0
Demulsibility, not more than:	not required	
— Total free water, cm^3		80
— Water in oil, %	—	2.5
— Emulsion, cm^3	—	1.0
Beaker Oxidation, not more than:	not required	
— Viscosity Increase, %	—	20
— Loss of oil, %	—	5
— Insoluble in n-heptane, %	—	1
Thermal Oxidation Stability Test (T.O.S.T.), not more than:		not required
— Viscosity Increase, %	100	—
— Insoluble in n-pentane, %	3	—
— Insoluble in toluene, %	2	—

Table 3.11.10 (continued)

Performance Related Test	BSSM 14368-82	BSSM 14367-82
EP Properties (Four Ball Machine), not less than:		
— Initial Seizure Load, N	800[a]	700[b]
— Weld Load, N	4000[a]	3150[b]
— Load Wear Index, N	550[a]	500[b]
EP Properties after Demulsibility Test, decrease of Weld Load, not greater than	not required	one increment
Antiwear Properties (Four Ball Machine):		
— Antiwear Index, N, not less than	200	200
— Wear Scar Diameter, mm	—	—
Antiscuffing Properties (FZG Machine), stages, not less than	10[c]	12[d]

a   BSSM 9797-82
b   BSSM 13134-82
c   130° C; 25,0 m s^{-1}; b = 10 mm
d   90° C; 16.6 m s^{-1}; b = 20 mm

Since results presented above gave certain hopes for practical application, further investigation was dedicated to formulations which may be useful.

In principle, combinations as the ones discussed in this presentation may be experimented for a wide variety of lubricating products — metalworking lubricants, hydraulic fluids, stick-slip lubricants, fluids for different transmissions, etc. with or without the need of additional additives.

Table 3.11.9 presents the results for two experimental formulations. The first of the compositions tested for some of the requirements of the Bulgarian automotive GL-5/90 gear oil specifications did not contain SME, while the second — for application as an industrial 5 EP gear oil — included SME and a smaller amount of DAPS.

As seen from Table 3.11.9, results from previous tests were confirmed in general. Furthermore, properties not tested beforehand — demulsibility, foam ability, corrosion on steel under different conditions, etc. — were also satisfactory.

However, full implementation of the present results in practice needs much more extensive testing in conditions closer related to real service, such as, for instance, these required by the US MIL-21058 specification — CRC L-33, L-37, L-42, etc. and then — in fleet tests.

Hereunder will be summarized only some of the questions that await their experimental answers:

— Will Zn-DADTP create problems?
Certain hopes in this direction give results (reported previously (25)) of successful application of oils, containing a relatively high amount of Zn-DADTP in gear boxes of trucks and busses and the fact that even "new generation" borate packages (26) contain similar components (27).

— Will lubricating properties satisfy both the CRC L-37 and L-42 requirements?
Correlation between results from simple tribometers and these tests seems rather questionable. Still results have been reported that formulations with certain Zn-DADTP-s may pass L-37, while sulphurized hydrocarbons with radicals similar to those of DAPS may pass L-42 (16).

— Will the rather complex and multifacted moisture corrosion process in CRC L-33 be successfully dealt with?
The hope here is probably the better correlation between corrosion tests and the similarity of components of the experimental packages with the references. These and other similar problems remain to be solved by future investigations.

3.11.4 References

(1) Ishchuk, Iu.L.: Technology of Plastic Greases. Kiev: Naukova dumka 1986 (in Russian).
(2) Report No. K-3845: Presentation of Hydraulic and Gear Lubricants and Additives for Neftochim, Bulgaria. Lubrizol International Laboratories: Hazelwood 1978.
(3) Malinovskii, G.T.: Oil Based Lubricating Cooling Fluids for Metal Cutting. Moscow: Himia 1988 (in Russian).
(4) Vipper, A.B.; Vilenkin, A.V.; Gaisner, D.A.: Foreign Oils and Additives. Moscow: Himia 1981 (in Russian).
(5) Bratkov, A.A. (Editor): Theoretical Foundations of Himmotology. Moscow: Himia 1985 (in Russian).
(6) Papay, A.G.; Dinsmore, D.W.: Advances in Gear Additive Technology. Lubrication Engineering, 32 (1976) 5, 229—234.
(7) Papay, A.G.: Gear Lubricant Additive Technology. NLGI Annual Meeting: Chicago October 1974.
(8) Rounds, F.G.: Some Effects of Amines on Zn Dialkyldithiophosphate Antiwear Performance as Measured in 4 Ball Wear Tests. ASLE Transaction 24 (1981) 4, 431—440.

(9)   Hsu, S.M.; Pei, P.; Ku, C.S.; Lin, R.S.; Hsu, S.T.: Mechanisms of Additive Effectiveness. Ostfildern 1986. 3.14.1 — 3.14.10. Proc. Additives for Lubricants and Operational Fluids. Technische Akademie Esslingen. January 1986. Ostfildern.
(10)   Shirama, S.; Hirata, M.: Effects of Engine Oil Additives on Valve Train Wear, ibid. 4.4.1 — 4.4.13.
(11)   Rounds, F.G.: Changes in Friction and Wear Performance Caused by Interactions Among Additives. ibid. 4.8.1 — 4.8.21.
(12)   Barcroft, F.T.; Park, D.: Interactions of Heated Metal Surfaces Between Zn-Dialkyldithiophosphates and Other Lubricating Oil Additives. Wear 108 (1986) 3, 213—234.
(13)   Kuo, L.L.K.; Chao-Yuan Tung: Fuel Economy Engine Oils Via Friction Modifiers. Lubrication Engineering. 44 (1988) 2, 81—86.
(14)   Papay, A.G.: Industrial Gear Oils — State of the Art. ibid. 3, 218—229.
(15)   Papay, A.G.: Friction Reducers for Engine and Gear Oils — A Review of the State of the Art. Ostfildern 1986. 5.1.1 — 5.1.10. Proc. Additives for Lubricants and Operational Fluids. Technische Akademie Esslingen. January 1986. Ostfildern.
(16)   Damrath, Jr., J.G.; Papay, A.G.: Additives for Gear Oils — Function and Interactions, ibid. 4.2.1 — 4.2.8.
(17)   Emannuel, N.M.; Denisov, E.; Maizus, Z.K.: Chain Reactions of Oxidation of Hydrocarbons in Fluid Phase. Moscow: Nauka 1965 (in Russian).
(18)   Ivanov, Sl.; Karshalukov, K.: Manometric Installation for Measuring the Increase and Decrease of Gases in Chemical Reactions with Automatic Keeping of Constant Pressure. Chemistry and Industry (Sofia) (1974) 3, 127—131 (in Bulgarian).
(19)   Fuels and Lubricating Materials. Compilation of Bulgarian State Standards: Part I. Technical Requirements. Part II. Methods for Analysis. Sofia: Standartizatzia 1983 (in Bulgarian).
(20)   Vilenkin, A.V.: Oils for Gear Transmissions. Moscow: Himia 1982.
(21)   Kajdas, C.: Review of AW and EP Working Mechanisms of Sulphur Compounds and Selected Metallo-organic Additives. Ostfildern 1986. 4.6.1 — 4.6.19. Proc. Additives for Lubricants and Operational Fluids. Technische Akademie Esslingen. January 1986. Ostfildern.
(22)   Kuliev, A.M.: Chemistry and Technology of Additives for Oils and Fuels. Leningrad: Himia 1985 (in Russian).
(23)   Cholakov, G.S.; Kaishev, K.P.: Storage Stability of Sulphurized Sperm Oil Replacements: I. Origin and Nature of the Sludge. Wear 96 (1984) 2, 109—119.
(24)   Kaishev, K.P.; Cholakov, G.S.; Stanulov, K.G.; Shopova, M.D.: Solubility Problems with Sulphurized Sperm Oil Replacement EP Additives; Solubility of Additives in Mineral Base Oil Hydrocarbon Constituents. Wear 131 (1989) 3, 303—313.
(25)   Klucho, P.; Foltanova, S.; Szucs, L.; Matas, M.: Synthesis, Laboratory and Machine-Application Evaluation and Performance Tests of EP Additives on the Basis of Dialkyldithiophosphoric Acid Metal Salts in Gear Oils. Ostfildern 1986. 4.11.1 — 4.11.4. Proc. Additives for Lubricants and Operational Fluids. Technische Akademie Esslingen. January 1986. Ostfildern.
(26)   Adams, J.H.: Borate — A New Generation EP Gear Lubricant. Lubrication Engineering 33 (1977) 5, 241—246.
(27)   Adams, J.H.: US Patent 4 089 790/May 16, 1978.

ACKNOWLEDGEMENT

The authors are indebted to: The UNDP-UNESCO Laboratory for Small Scale Organic Products, Higher Institute for Chemical Technology, Sofia: The Committee for Science of Bulgaria and the Petrochemical Combine — Pleven, Bulgaria for their generous support and interest in this work.

3.12 Relationship between Chemical Structure and Effectiveness of Some Metallic Dialkyl and Diaryldithiophosphates in Different Lubricated Mechanisms

M. Born, J.C. Hipeaux, P. Marchand and G. Parc
Institut Francais du Pétrole, Rueil-Malmaison, France

Summary

The antiwear (AW) and extreme-pressure (EP) properties of some metallic dialkyl and diaryldithiophosphates (MDTP), including zinc, have been studied in different metal/metal contact conditions: four-ball and FZG rigs. We have noted:

● With pure dialkyldithiophosphates of different divalent metals (Zn, Cd, Cu, Pb, Ni, Co) prepared from the same alcohol:

  — on the four-ball rig, EP results, expressed in terms of load-wear index, show that the highest performances are obtained with ZnDTP but statistically the higher the ionic radius of the metal, the better the performance.
  EP results, expressed in terms of weld load, do not show any difference between these MDTP;

  — on the four-ball rig, AW results, expressed in terms of wear scar, show that the highest performances are obtained with ZnDTP; beyond a certain value these performances with Zn and Pb are almost independent of the concentration. For the other metallic DTP, there is a very close relationship between performance and concentration;

  — on the FZG gear rig, EP-AW results, expressed in terms of failure load stage and specific wear, show that statistically, the higher the ionic radius of the metal, the better the performance.

● With ZnDTP:

  — on the four-ball rig, EP results, expressed in terms of load wear index, do not depend on the chemical structure of the alcohol used for synthezised Zn-DTP but if results are expressed in terms of weld load, better performances are obtained with ZnDTP synthezised from secondary alcohols.
  AW results show that the effectiveness of ZnDTP from secondary alcohols is slightly higher when the alcohols used contain less than six carbon atoms. Beyond six carbon atoms in primary or secondary alcohol molecules performances decrease.

— on the FZG gear rig, EP-AW results show that performances do not depend on the primary or secondary nature of the alcohols used and that the smaller the hydrocarbon chain of these alcohols, the better the performances.

The results of this work indicate that ZnDTP is the most effective of the MDTP we have studied. For ZnDTP prepared from different alcohols and alkylphenols the performances depend not only on the chemical structure of the additive but also on contact conditions (kinematic, material types, etc.), and on the criteria considered (seizure load, weld load, wear, etc.).

3.12.1 Background

Metallic dialkyl and diaryldithiophosphates, particularly ZnDTP, have been extensively used in lubrication for over forty years. The advantage of these compounds is that several highly desirable lubrication-related properties such as antioxidant, detergent and more specifically antiwear properties, and to a lesser extend extreme pressure properties (load carrying capacity properties) can all be found in the same molecule. Other non-negligible advantages, from the practical point of view, include acceptable thermal stability, moderate toxicity, considerable case of storage and application, relatively low cost, and great effectiveness for a small dose since they are used in lubricants in concentrations not exceeding 1.5 weight %.

Although their use is decreasing for purely environmental reasons (they tend to deactivate the catalyst in automotive catalytic mufflers since they pass into the exhaust gases in very small quantities along with the crank-case oil) they are still one of the irreplaceable additives in modern lubricating oils and will probably continue to be so for several years.

3.12.2 Introduction

The number of studies performed for the purpose of linking the AW and EP properties of MDTP to their chemical structure is rather limited. Work by Rowe and Dickert (1) on a pin and disc machine has shown that AW effectiveness, measured in mild sliding conditions, is proportionately higher when the thermal stability of the MDTP is low. Effectiveness is rated as follows:

$$Ag > Pb > Co > Zn > Cu$$

Moreover, these authors have shown a correlation between the AW properties of the MDTP studied and the ionic radius of the corresponding metals.

According to Forbes (2), MDTP load carrying capacities vary considerably. If he is considering wear, he rates effectiveness as follows:

336

Zn > Ni > Fe > Ag > Pb > Sn > Bi

and if he is examining extreme pressure, his rating is

Ag > Bi > Sb > Ni > Zn

No metal can in fact provide all the optimum properties desired, but industrially speaking, the ZnDTPs are recognized as offering the best compromise between the different properties mentioned above.

A series of tests carried out by Larson (3) on a Chevrolet LS-5 engine with a cam/tappet system has shown that a ZnDTP, derived from two secondary alcohols, provides greater AW effectiveness when each phosphorus atom is linked to radicals of the same nature.

Forbes, Allum and Silver (4) noted, when studying extreme pressure and using a four-ball rig, that the nature of substitutes for P atoms in ZnDTP had very little influence on performance but that substitutes derived from secondary alcohols did have a slightly positive effect.

Jayne and Elliot (5) also studied EP-AW properties of ZnDTP on a four-ball rig and have noted no difference in AW effectiveness at moderate loads between ZnDTP derived from primary or secondary alcohols, which coincides with the results of Zamberlin Mikac-Cergolj and Bencetic (6). However, at severe loads, EP properties, expressed in terms of welding load, are better in the case of ZnDTP derived from secondary alcohols.

Tests carried out by Jayne and Elliot (5) on a Timken rig show that dialkyl ZnDTP have a higher load carrying capacity than diaryls, and that mixtures of alkyls and aryls have an intermediate capacity. Zamberlin's work leads to the same conclusions but in addition it points to the greater effectiveness of ZnDTP derived from secondary alcohols compared to those derived from primary alcohols.

Tests performed by Rowe and Dickert (1) on a pin and disc rig under mild sliding conditions show that differences in AW effectiveness of the ZnDTP studied are related to their thermal stability, and that the rate of wear of the copper pin is higher for secondary alcohol than for primary alcohol ZnDTP.

The diversity of the results obtained by the different authors shows that the EP-AW effectiveness of ZnDTP depends on a number of parameters involved when lubricated surfaces are in contact, i.e.:

— stress exerted on surfaces
— types of movement
— velocity of sliding surfaces
— temperature
— metallurgy and roughness of lubricated materials

337

and consequently on the measurement method used and perhaps even on the degree of purity of the MDTP of interest.

In this paper we propose first to examine the validity of industrial practices, in other words, to see whether ZnDTP are really the type of MDTP best suited to reducing wear in lubricated mechanisms. This led us to compare the EP-AW properties of a reference zinc dialkyldithophosphate to those of different metallic DTPs (prepared with the same alcohol as that used for the synthesis of the model ZnDTP) in which each metal atom is oxidized to the same degree, i. e. linked to the same number of sulfur and phosphorus atoms.

Second, we have attempted to point out the relationship between the AW-EP effectiveness of ZnDTP and the chemical structure of their organic chains, under different lubrication conditions.

In order to ensure the exact composition of the MDTPs studied and their degree of purity, the products were prepared in the laboratory by the two most common synthesis methods:

— by double decomposition, or metathesis, in the case of ZnDTP prepared with alcohols containing less than 8 carbon atoms (7); this method produces very pure metallic DTPs.

— by the zinc oxide method in the case of ZnDTP prepared from alcohols containing at least 8 carbon atoms and those prepared from alkylphenols (8).

3.12.3 Additives Studied

In order to study only the influence of the nature of the metal on the EP-AW properties of the metallic DTPs, the products were prepared from 4-methyl 2-pentanol:

$$\left[(CH_3-\underset{\underset{CH_3}{|}}{\overset{\overset{CH_3}{|}}{CH}}-CH_2-\underset{\underset{CH_3}{|}}{CH}-O-)_2-P\underset{S}{\overset{S}{\diagdown}} \right]_2 M$$

The metals studied were:

M = Zn — Cu — Ni — Co — Cd — Pb

The physical and chemical characteristics of these products are indicated in Table 3.12.1.

338

Table 3.12.1: Characteristics of the metallics DTP studied

$$\left[(CH_3-\underset{\underset{\displaystyle CH_3}{|}}{CH}-CH_2-\overset{\overset{\displaystyle CH_3}{|}}{CH}-O)_2 \quad P \overset{\displaystyle S}{\underset{\displaystyle S}{\diagdown}} \right]_2 M$$

Synthesis method = double decomposition
M = divalent Metal

| Metal | ELEMENTAL ANALYSIS | | | | | |
| | Metal % Mass | | Carbon % Mass | | Hydrogen % Mass | |
	Obs.	Theory	Obs.	Theory	Obs.	Theory
Zn	9.92	9.92	43.40	43.67	7.70	7.83
Ni	8.27	8.99	44.40	44.12	7.76	7.97
Cu	9.65	9.65	46.20	43.80	8.45	7.91
Co	8.75	9.02	44.32	44.12	8.12	7.96
Cd	15.90	15.60	42.60	40.77	7.70	7.36
Pb	25.80	25.85	36.02	35.96	6.52	6.49

The ZnDTP studied consisted of single versions based on a primary or secondary alcohol or an alkylphenol and a combined version based on an equimolecular mixture of two alcohols with different chemical functions.

SINGLE ZnDTP
Alcohols used   Alcohols
                                                                     or phenol

1-PROPANOL (n-propanol) — primary
2-PROPANOL (iso-propanol) — secondary
1-BUTANOL (n-butanol) — primary
2-METHYL 1-PROPANOL (iso-butanol) — primary
1-METHYL 1-PROPANOL (2-butanol) — secondary
iso-PENTANOL — primary
          alcohol mixture
2,2-DIMETHYL 1-PROPANOL — primary
   (neo-pentanol)
1-HEXANOL (n-hexanol) — primary
4-METHYL 2-PENTANOL — secondary
CYCLOHEXANOL — secondary
2-ETHYL HEXANOL — primary
OXO C_8 — primary
          alcohol mixture
1-METHYL 1-HEPTANOL (2-octanol) — secondary

339

2,6-DIMETHYL 4-HEPTANOL	secondary
2,6,8-TRIMETHYL 4-NONANOL	secondary
NONYLPHENOL	alkylphenol
DODECYLPHENOL	alkylphenol

COMBINED ZnDTP

The combined ZnDTP was prepared from an equimolecular mixture of iso-propanol (secondary alcohol) and iso-pentanol (primary alcohol).

The thermal decomposition temperature of the ZnDTP was determined by the following method: A 100 neutral solvent mineral oil containing 4.5×10^{-3} atom-gram of zinc (in the form of ZnDTP) per 100 g of mixture is gradually heated at a rate of $2-3°$ C/minute until the product clouds, indicating the thermal decomposition of the additive. The analytical characteristics of these compounds are indicated in Tables 3.12.2 and 3.12.3.

3.12.4 Experimental Procedure

The EP-AW performances are characterised by a pin-point hertzian contact occurring under rough elastohydrodynamic conditions (EHD) with pure sliding and under very high stress.

— FZG gear rig test

The moving metal surfaces are characterised by a linear hertzian contact occurring under EHD conditions with combined sliding and rolling and under very high stress. A mineral ARAMCO 400 neutral solvent base oil was chosen for these two experimental rig tests, their characteristics are as follows:

Viscosity at $100°$ C(mm^2/s)	=	9.51
Viscosity index	=	101
Sulphur content (mass %)	=	1.1

— Four-ball extreme-pressure test

The EP performance assessment tests were carried out in accordance with the ASTM D 2782-82 standard method.

The principal test characteristics are as follows:

Balls:	100 C6 steel
	12.7 mm diameter
Top ball revolution speed:	1425 — 1480 rpm
Sliding speed:	0.56 — 0.58 m/s
Oil temperature:	32.5° C
Duration of tests:	10 seconds

Table 3.12.2: Characteristics of the ZnDTP studied

$$\left[(RO)_2 - P \!\! \begin{array}{c} = S \\ - S \end{array} \right]_2 Zn$$

CHEMICAL STRUCTURE of R	CORRESPONDING ALCOHOL	Zn (% MASS) Obs	Zn (% MASS) Theory	Carbon (% MASS) Obs	Carbon (% MASS) Theory	Hydrogen (% MASS) Obs	Hydrogen (% MASS) Theory	Decomposition Temp. (°C)	SYNTHESIS METHOD ①
C-C-C-	n-PROPANOL	13.30	13.30	29.48	29.30	5.72	5.70	194	DD
C-C- C	iso-PROPANOL	13.30	13.30	29.38	29.30	5.77	5.70	183	DD
C-C-C-C-	1-BUTANOL	11.20	11.94	35.90	35.06	6.90	6.57	194	DD
C-C-C- C	2-BUTANOL	11.72	11.94	35.30	35.06	6.60	6.57	194	DD
C-C- C	iso-BUTANOL	13.00	11.94	35.00	35.06	6.40	6.57	225	DD
C-C-C-C- (75%) C ; C-C-C- (25%) C	iso-PENTANOL	10.87	10.83	39.67	39.78	7.32	7.29	225	DD
C-C- C	neo-PENTANOL	10.82	10.83	39.87	39.78	7.43	7.29	247	DD
(cyclohexane ring)	CYCLOHEXANOL	10.01	10.03	44.60	44.20	5.75	6.75	195	DD
C-C-C-C-C-	1-HEXANOL	9.37	9.91	43.89	43.70	7.81	7.88	225	DD
C-C-C-C- C	4-METHYL 2-PENTANOL	9.92 / 9.99	9.91 / 9.91	43.57 / 44.10	43.70 / 43.70	7.87 / 7.94	7.88 / 7.88	194	DD / —
C-C-C-C-C-C-C-C	2-OCTANOL	7.97	8.47	50.03	49.77	8.97	8.81	238	—
C8 OXO	OXO-OCTANOL	9.60	8.47	50.39	49.77	9.06	8.81	234	—

① DD = double decomposition
   I = industrial method

341

Table 3.12.3: Characteristics of the ZnDTP studied

$$\left[(RO)_2 - P {\overset{S}{\underset{S}{\diagdown}}} \right]_2 Zn$$

CHEMICAL STRUCTURE of R	CORRESPONDING ALCOHOL OR ALKYPHENOL	Zn (% MASS) Obs	Zn (% MASS) Theory	ELEMENTAL ANALYSIS Carbon (% MASS) Obs	Carbon (% MASS) Theory	Hydrogen (% MASS) Obs	Hydrogen (% MASS) Theory	Decomposition Temp. (°C)	SYNTHESIS METHOD ①
C-C-C-C-C-C- / C-C	2-ETHYL 1-HEXANOL	8.30	8.47	50.10	49.77	8.95	8.81	238	DD
	2,6-DIMETHYL 4-HEPANOL②	6.50	7.90	-	52.20	9.18	-	185	I
	2,6,8-TRIMETHYL 4-NONANOL	5.25	6.56	-	-	-	-	.	I
(75% / 25%) 50% / 50%	ISOAMILIC ALCOHOL / ISOPROPYLIC ALCOHOL COMBINED	12.09	12.06	34.51	34.46	6.54	6.48	195	DD
C₉-	4-NONYLPHENOL③	5.41	5.78	63.50	63.84	7.90	8.13	> 260	I
C₁₂-	4-DODECYLPHENOL④	3.38	3.16					> 260	I

	Phosphor Found	Phosphor Theory	Sulphur Found	Sulphur Theory
	3.06	3.00	6.00	6.20

① DD = double decomposition
   I = industrial method
② contains 5% mass excess alcohol
③ contains 25% mass fluxing mineral oil for easier handling

The criteria used to assess the results were the load-wear index figure (LWI) and the welding load of the ball.

The amount of additives used for these tests was designed to provide an oil with following metal contents:

$[Metal] = 1.52 - 3.04 - 7.6 - 15.2 \, (10^{-3})$ atom gram/kg

— Four-ball wear test

The NF E 48-617 method, whereby a 40 kgf load is normally applied for one hour, was supplemented by three successive loads of 60 — 80 — 100 kgf.

The criterion used to assess the antiwear (AW) performance of each additive is the mean wear diameter (d*), i. e. the arithmetic average of the wear diameters for the successive loads of 40 — 60 — 80 — and 100 kgf.

$$d^*(mm) = \frac{d_{40} + d_{60} + d_{80} + d_{100}}{4}$$

In addition, a mean value is calculated for each metallic DTP, based on the mean wear diameter (d*) corresponding to the arithmetical average of the d* for the 4 concentrations studied:

$$\overline{d}^*(mm) = \frac{d^*1.52 + d^*3.04 + d^*7.6 + d^*15.2}{4}$$

Further, these values d* are expressed in terms of relative antiwear effectiveness ϵ, determined in relation to the performance of a base oil ($\epsilon = 0$ for d*B = 1.90 mm) and to the performance of an ideal oil whose wear diameter d* would be minimal and would correspond to the average of the compensation diameters d*$_c$ for loads of 40 — 60 — 80 — 100 kgf. The compensation diameters used in the ASTM methods are as follows:

40 kgf	60 kgf	80 kgf	100 kgf
0.33 mm	0.38 mm	0.42 mm	0.46 mm

Consequently

$$d^*C = \frac{d_c40 + d_c60 + d_c80 + d_c100}{4} = 0.40 \text{ mm}$$

The relative antiwear effectiveness determined by the relation:

343

$$\epsilon = \frac{d^*_B - d^*_X}{d^*_B - d^*_C} \times 100$$

Thus becomes:

$$\epsilon = \frac{1.90 - d^*_X}{1.50} \times 100$$

The amount of additives used was identical to that used for EP four-ball rig tests:

Metal = $1.52 - 3.40 - 7.6 - 15.2 \ 10^3$ atom-gram/kg

— FZG gear rig tests:

The tests were based on the CEC-L-07-A-85 method, namely:

Gears: "A" type
Sliding speed: 5.56 m/s
Initial oil temperature: 90° C

The amount of additives used was designed to produce an oil with the following additive content:

Metal = $0.76 - 1.52 - 3.04 - 7.6 \ 10^{-3}$ atom-gram/kg

The criteria used to assess performance were the failure load stage and the gear specific wear.

3.12.5 Test Results and Discussion

3.12.5.1 Metallic DTP

3.12.5.1.1 Four-ball EP test

The EP performance, expressed in terms of Load Wear Index (LWI), show that among the different metallic DTPs studied, those prepared from heavy metals (characterised by a large ionic radius), are statistically more effective than those prepared from metals with a lower ionic radius.

But ZnDTP is an exception with a very high LWI particularly at low concentration (Tables 3.12.4 and 3.12.5). Decreasing order of EP effectiveness is as follows:

Zn ➤ Cd > Cu > Pb > Co > Ni

344

EP performances, expressed in terms of Welding Load, differ little from one additive to another. Decreasing order of effectiveness is as follows:

$$Zn = Cd = Cu > Co = Ni > Pb$$

Table 3.12.4: Four ball extreme pressure test load wear index versus metal concentration

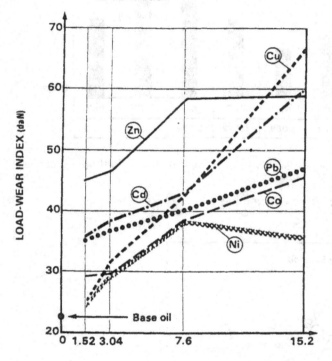

Table 3.12.5: Four ball extreme pressure test load wear index versus metal ionic radius

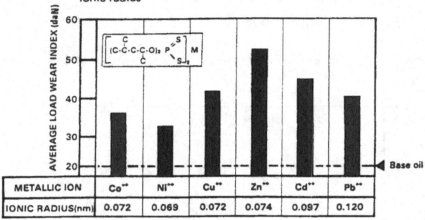

METALLIC ION	Co⁺⁺	Ni⁺⁺	Cu⁺⁺	Zn⁺⁺	Cd⁺⁺	Pb⁺⁺
IONIC RADIUS(nm)	0.072	0.069	0.072	0.074	0.097	0.120

Table 3.12.6: Four ball wear test, average wear diameter (d*) versus concentration of metal

$$\left[(C-\overset{\overset{C}{|}}{\underset{\underset{C}{|}}{C}}-C-C-O)_2 \; P\overset{S}{\underset{S}{\diagdown}} \right]_2 M$$

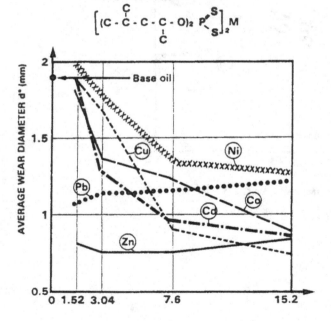

METAL CONCENTRATION (10⁻³at/kg)

346

Table 3.12.7: Four ball wear test, relative antiwear effectiveness (ϵ) versus metal ionic radius

METALLIC ION	Co⁺⁺	Ni⁺⁺	Cu⁺⁺	Zn⁺⁺	Cd⁺⁺	Pb⁺⁺	Base oil ($\mathcal{E}$ = 0)
IONIC RADIUS (nm)	0.072	0.069	0.072	0.074	0.097	0.120	

3.12.5.1.2 Four-ball AU tests

Zn DTP stands out as having the highest AW action, particularly at lower concentrations. Together with PbDTP, it is exceptional in that AW performance is practically independent of concentration within the range of concentrations studied. This is not the case for the other metallic DTPs studied since their AW properties vary considerably with the quantities of additive used (Tables 3.12.6, 3.12.7).

3.12.5.1.3 FZG EP tests

The results (Table 3.12.8) show that the heavy metal DTPs have the highest EP effectiveness in terms of failure load stage and that of the light metal DTPs studied, ZnDTP is the most effective. Decreasing order of effectiveness is as follows:

Cd > Pb > Zn > Cu > Co > Ni

The AW results (specific wear, Table 3.12.9) show that there is little change in the ranking order of the effectiveness previously indicated, except in the case of CuDTP which has particularly low antiwear properties:

Cd > Pb > Zn > Co > Ni ≫ Cu

347

Table 3.12.8: FZG extreme pressure test, extreme pressure properties and versus metal ionic radius

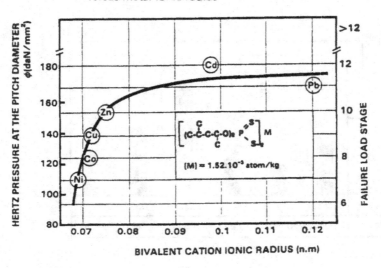

BIVALENT CATION IONIC RADIUS (n.m)

Table 3.12.9: FZG extreme pressure test, specific wear versus metal ionic radius

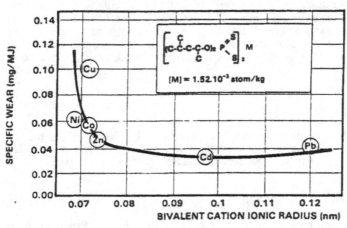

BIVALENT CATION IONIC RADIUS (nm)

In view of the toxic nature of heavy metal salts, we may conclude from the tests on FZG and 4-ball rigs that, of the metal DTP studied, ZnDTP is the most effective safe EP-AW additive, which fully corroborates industrial practice.

3.12.5.2 Single ZnDTP

3.12.5.2.1 Four-ball EP Tests

It can be seen from Table 3.12.10 that the EP effectiveness (LWI) of the ZnDTP studied is not related to the primary or secondary nature of the alcohols used in their preparation, and that the organic chain length R does not have a significant influence on their effectiveness.

The ZnDTP with the most effective EP properties are those prepared from 4-methyl 2-pentanol, isobutanol and 2-butanol. The aromatic ZnDTP differs little from the other ZnDTP studied.

Table 3.12.10: Four ball extreme pressure test, four ball index versus organic chain length

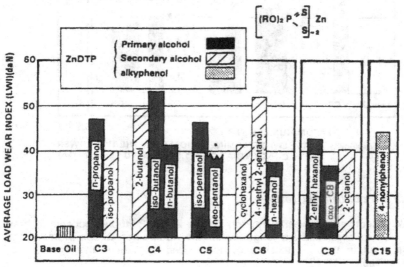

C ATOM NUMBER IN R
* Low solubility ZnDTP : LWI DETERMINED FROM ONLY THE TWO LOWER CONCENTRATIONS (1.52–3.04.10^{-3} atom Zn/kg)

On the other hand, EP effectiveness, expressed in terms of welding load (Table 3.12.11), is unrelated to the organic chain length R and depends only on the primary or secondary nature of the alcohols used in its synthesis. The ZnDTP prepared from secondary alcohols are more effective (results conform to those obtained by Jayne and Elliot).

Table 3.12.11:   Four ball extreme pressure test, welding load versus organic structure

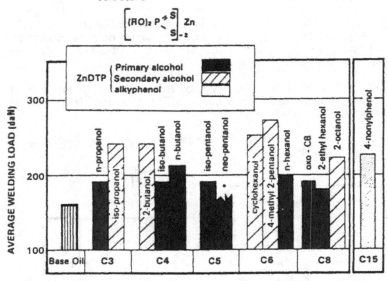

C ATOM NUMBER IN R

* Low solubility ZnDTP : LWI DETERMINED FROM ONLY THE TWO LOWER
CONCENTRATIONS (1.52—3.04.10^{-3} atom Zn/kg)

3.12.5.2.2 Four-ball AW Tests

The AW effectiveness (wear diameter) of the Zn DTP studied appears to have no relation to organic chain length when the number of carbon atoms in the alcohols used in their synthesis does not exceed six. Above this figure, wear increases (Table 3.12.12).

The effectiveness of low concentrations of ZnDTP derived from primary alcohols varies greatly: results are excellent for those derived from n-butanol, isobutanol and n-propanol, medium for that derived from iso-amylic alcohol, and very poor with even a pro-wear effect for those derived from octanol and neo-pentanol.

350

Table 3.12.12: Four ball wear test, four ball wear versus organic chain length

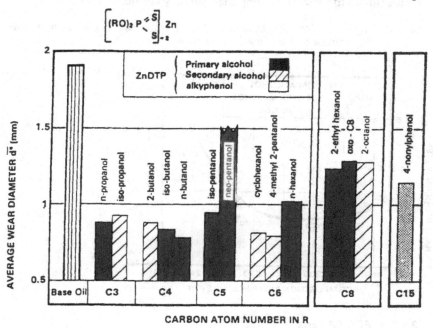

On the other hand, ZnDTP derived from secondary alcohols, with the exception of that prepared from 2-octanol, have similar and very effective AW properties.

If these results are expressed in terms of the thermal decomposition temperature of the ZnDTP studied (Table 3.12.13), the following conclusion may be drawn:

— ZnDTP derived from secondary alcohols have low thermal stability and generally speaking result in similar and very low rates of wear, whatever the organic chain lengths (with the exception of the ZnDTP derived from 2-octanol).

— The AW effectiveness of ZnDTP derived from primary alcohols varies considerably:

  — it is excellent when thermal stability is low and when the alcohols used have short organic chains;
  — it is very poor when thermal stability is high and when the organic chain is long;
  — it is mediocre when thermal stability is medium and when the organic chain is of medium length.

— The alkylphenol ZnDTP studied is thermally very stable but has AW properties similar to those of the long chain primary alcohol ZnDTP.

Table 3.12.13:   Relationship between ZnDTP thermal stability and AW properties on four ball wear test

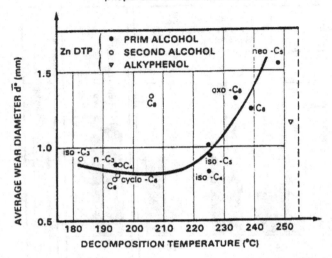

3.12.5.2.3 FZG EP Tests

Table 3.12.14 shows that, at low concentration, the shorter the organic chain of the alcohol used in preparing the ZnDTP, the greater the EP effectiveness (failure load stage).

In Table 3.12.15 it can be seen that the higher the concentration, the smaller the difference in EP effectiveness between the different ZnDTP studied.

With regard to EP effectiveness, AW results expressed in terms of specific wear depend on the organic chain length of the alcohol used. The shorter the chain, the better the performance (Table 3.12.16).

The primary or secondary nature of the alcohol does not affect results.

Table 3.12.14: FZG extreme pressure test, load capacity versus organic chain length (low concentration)

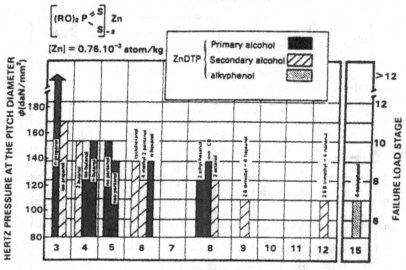

Table 3.12.15: FZG extreme pressure test, load capacity versus organic chain length (high concentration)

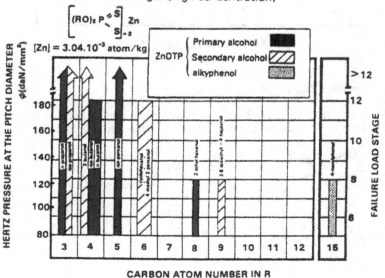

Table 3.12.16: FZG extreme pressure test, specific wear versus organic chain length (low concentration)

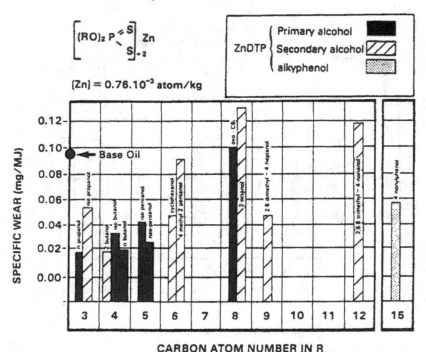

3.12.5.3 Combined ZnDTP

The rig test results obtained with ZnDTP (AB) are compared to those obtained with the ZnDTP derived from the corresponding pure alcohols (A and B):

$$\begin{pmatrix} \text{Iso-C}_3 - \text{O} \diagdown \\ \qquad\qquad P \diagup\!\!\!\!\diagup S \\ \text{Iso-C}_5 - \text{O} \diagup \quad \diagdown S \end{pmatrix}_2 \quad - \text{ Zn} \quad \text{A B}$$

$$\left[(\text{Iso-C}_3 - \text{O} -)_2 P \diagup\!\!\!\!\diagup S \atop \diagdown S \right]_2 \quad - \text{ Zn} \quad \text{A}$$

354

$$\left[(Iso\text{-}C_5 - O-)_2 P \underset{S}{\overset{S}{<}} \right]_2 - Zn \qquad B$$

3.12.5.3.1 Four-ball EP Tests

As can be seen in Table 3.12.17, the combined ZnDTP produces a marked EP synergy effect (LWI), particularly in small concentrations, but the effect is much less pronounced when higher concentrations of the additive are used. The EP properties (LWI and welding load) of the combined AB additive are generally at least equal to that of the most effective A or B additive.

Results of wear tests (wear diameter) show little difference in effectiveness between the three ZnDTP studied. Nevertheless a slight AW synergy effect is noted in the case of the combined ZnDTP.

3.12.5.3.2 FZG EP Tests

There is little difference between the EP effectiveness of the combined ZnDTP additive (failure load stage) and that of the pure ZnDTP A or B additives. However, a definite antiwear synergy effect is noted in that case of the combined Zn DTP (specific wear), particularly at low concentrations (Tables 3.12.17 and 3.12.18). This antiwear synergy effect coincides with that observed by Larson (3) in cam and finger follower wear test.

3.12.6 Conclusion

The first part of this study is devoted to an examination of the load carrying capacity properties of different metallic dialkyl and diaryldithiophosphates. Tests were performed on a four-ball rig operating under predetermined standard conditions and on an FZG gear rig simulating the contacts that occur in automotive gear boxes and differentials. The following comments may be made:

— Of the different metallic DTP studied, the lead and cadmium DTP — i. e. the large ionic radius heavy metal DTP — have the best EP-AW properties.

— Although zinc is not considered a heavy metal, it nevertheless produces a DTP that is practically as effective as the abovementioned DTP, whatever the concentration.

— In view of the toxic nature of the heavy metal salts, the conclusions drawn fully corroborate industrial practice, namely that the use of zinc leads to AW DTP additives that are not only very effective but also extremely safe.

Table 3.12.17: EP-AW synergy effect of combined ZnDTP

Zn: A
Zn: B
Zn: AB

FOUR BALL EXTREME PRESSURE TEST

EVALUATION CRITERIA	Zn 10⁻³ atom/kg	Base Oil	Zn DTP STUDIED		
			A	B	AB
LWI (daN)	0	22.6			
	1.52		36.8	34.5	43.8
	3.04		37.5 } 40.2	43.8 } 46.1	44.0 } 47.4
	7.60		39.6	53.0	55.0
	15.20		47.0	53.2	46.7
WELDING LOAD (daN)	0	160			
	1.52		200	160	200
	3.04		200/250	200	200
	7.6		250	200	250
	15.2		250	200	250

FZG EXTREME PRESSURE TEST

EVALUATION CRITERIA	Zn 10⁻³ atom/kg	Base Oil	Zn DTP STUDIED		
			A	B	AB
FAILURE LOAD STAGE	0.76	6	11	10	11
	1.52		>12	12	12
	3.04		>12	>12	>12
SPECIFIC WEAR (mg/MJ)	0.76	0.317	0.08	0.06	0.04
	1.52		0.04	0.03	0.02
	3.04		0.03	0.03	0.03

Table 3.12.18: FZG extreme pressure test, synergy effect of combined ZnDTP

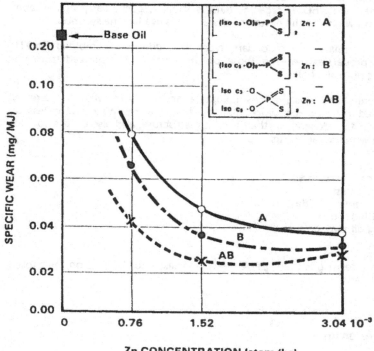

Zn CONCENTRATION (atom/kg)

The second part of the study attempts to establish the relationship between the chemical structure of the DTP and their EP-AW effectiveness, based on the results of rig tests using a considerable number of compounds. The conclusions reached vary greatly depending on the method of investigation used.

The results of four-ball EP and AU tests (100 C6 steel balls, pinpoint contact during pure sliding) do not point to any relationship between chemical structure and EP properties if performance is based solely on LWI. On the other hand, if the welding load of the balls is taken into account, the ZnDTP derived from secondary alcohols prove the most effective, whatever the length of the organic chains linked to the P atoms.

ZnDTP derived from primary and secondary alcohols (with low thermal stability) have the highest AW properties. When the number of carbon atoms in the organic chain of the alcohols exceeds six, thermal stability increases but AW properties deteriorate.

357

The results of FZG EP test (hardened, quenched 20 MC 5 steel, combined sliding and rolling) show that EP-AW protection of transmission mechanisms is greater when the ZnDTP organic chain is shorter, particularly at low concentrations, and whatever the nature of the alcohols used in the synthesis.

A mixture of primary and secondary alcohols produces a combined ZnDTP with EP-AW properties close to those of the best ZnDTP derived from the corresponding pure alcohol.

All these results and observations complete and confirm what has already been suggested by research work, namely that the EP-AW effectiveness of ZnDTP does in fact depend on their chemical structure, and also on the contact conditions in which wear occurs, i. e.:

- type of contact
- stress on moving surfaces
- surface roughness
- surface movement (sliding, rolling)
- sliding/rolling ratio
- lubricating oil temperature
- etc.

Lastly, when a test rig is used, EP-AW effectiveness also depends on the chosen assessment criteria.

3.12.7 References

(1)   Rowe, C.N.; Dickert, J.J.: The Relation of Antiwear Function to Thermal Stability and Structure for Metal 0,0-dialkylphosphorodithioates. American Chemical Society preprint 10 (1965) D 71-D 83.
(2)   Forbes, E.S.: Antiwear and Extreme-pressure Additives for Lubricants. Tribology (1970) 145—152.
(3)   Larson, R.: The Performance of Zinc Dithiophosphates as Lubricating Oil Additives. Scientific lubrication (1958) 12—20.
(4)   Forbes, E.S.; Allum, K.G.; Silver, K.B.: The Load Carrying Properties of Metal Dithiophosphates: Application of Electron Probe Microanalysis. Institution of Mechanical Engineers. Gothenburg (1969) 188.
(5)   Jayne, G.J.J.; Elliott, J.S.: The Load Carrying Properties of some Metal Phosphorodithioates. Am. Chem. Soc. preprints (1960) 139—149.
(6)   Zamberlin, I.; Mikac-Cergolj, I.; Benecetic, M.: The Effect of the Chemical Structure of Zinc Dithiophosphates on Antiwear and Extreme-pressure Properties of Lubricating Oils. Nafta (Zagreb) 20 (1069) 351—358.
(7)   Brazier, A.D.; Elliott, J.S.: The Thermal Stability of Zinc Dithiophosphates. J. Inst. Petr 53 (1967) 518, 63—76.
(8)   Elliott, J.S.; Jayne, G.J.J.; Barber, R.I.: Evaluation of Antioxidants for Automotive Lubricants Using the Rotary bomb. Inst. Petrol. London 55, (1969) 544, 219—226.

3.13 Evaluation of the Antiwear Performance of Aged Oils through Tribological and Physicochemical Tests

G. Monteil, A.M. Merillon and J. Lonchampt, Peugeot S.A., Voujeaucour, France
C. Roques-Carmes, ENSMM, Besancon, France

Summary

Extending automotive engine drain intervals requires a good knowledge of the antiwear level of lubricant efficiency. These performances are often closely linked to the oxidative degradation of the oils.

Physicochemical analysis and tribological tests were carried out on several oils oxidized by laboratory oxidation tests at different stages.

An interesting correlation was found between the antiwear efficiency and some characteristics of the electrochemical impedance spectra of such oxidized lubricants.

A very promising way of understanding the mechanisms of wear mitigation by aged engine oils is proposed.

3.13.1 Introduction

The trends observed in the recent automotive engine developments have led to a significant increase in the constraints imposed on the lubricants.

The new constraints which contribute to a perceptible speeding up of the effects of the oil aging are, among others:

— an increase in the oil drain intervals,
— the lowering of the engine oil consumption
  and thereby its regeneration,
— an increase in the crankcase oil temperatures,
— the lowering of the crankcase capacity...

For these reasons and some others, it becomes very important to achieve a better knowledge of the residual ability of the lubricants to carry out their functions, particularly the antiwear function, during the aging process in field service.

Oxidation aging phenomena are mainly responsible for the loss of antiwear efficiency which is dramatically critical for valve train systems.

Consequently, the aim of the present study is to make its contribution to a better knowledge of the influence of the lubricant oxidation upon valve train wear.

For this purpose, some tests were carried out on a serial of lubricants of different levels of performance.

A controlled accelerated aging of these oils was achieved by a laboratory air oxidation test. That test led to a simulation of the behaviour of the lubricants in service.

Their remaining antiwear efficiency was assessed by means of different tribological tests. Last of all, an electrochemical impedance spectrum measurement technique was used.

3.13.2 Experimental

3.13.2.1 Lubricants

The main characteristics of the engine oils used in this work are listed in Table 3.13.1. These oils are mineral oils (B and D), or semi-synthetic (A and C). Each of the tests presented here has been done on a similar oil batch.

Table 3.13.1: Characteristics of the lubricants under study.

Reference of the lubricant	SAE Grade	Kinematic viscosity at 100°C (cSt)	Performance Level
A	10 W 30	11.9	API SF/CD
B	20 W 50	17.9	API SF/CC
C	15 W 50	19.5	API SF/CC
D	10 W 40	14.5	API SF/CC

3.13.2.2 Oxidation Test

The laboratory method used to oxidize the lubricants is the one currently in service in the laboratories of Peugeot S.A. and other French car manufacturers. It is referred to as the method GFC TO21 x 88.

Briefly described, it consists in oxidizing in an erlenmeyer a quantity of 300 ml of the lubricants to be tested at a temperature of 160°C and under an air flow of 10 liters per hour.
The oxidation test was performed without any catalytic element. Its duration was variable in our experiments.

3.13.2.3 AC Impedance Technique

The AC impedance technique for studying the electrochemical properties of the lubricants consists in subjecting the lubricant contained in an electrochemical measurement cell to a low level excitation alternative voltage superimposed or not at a DC polarisation potential.

In the work presented here, the measurement cell is of the two electrode type. The AC peak to peak voltage is 140 mv and the bias polarisation voltage is equivalent to zero DC, in other words, to the rest potential of the system.

The impedance Z of this electrochemical cell is equal to the ratio voltage/current. It is measured over a large range of frequencies (1 mHz to 30 kHz). The detail of the electrochemical circuit and the different apparatuses is shown on Fig. 3.13.1 and Fig. 3.13.2.

Figure 3.13.1: Electrochemical impedance measurement test stand

It consists of an electrochemical measurement stand; a potentiostat solartron 1286 (A) controlling the polarisation DC voltage, a frequency response analyser solartron 1250 (B) allowing the generation of the sinusoidal alternative excitation signals applied to the cell and the analysis of its response.

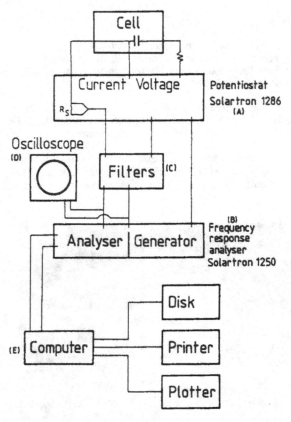

Figure 3.13.2: Block diagram of the electrochemical measurement circuit

The response signals are filtered by means of a computer software driven filter box (C) and visualized on an oscilloscope (D). The whole electrochemical stand is wholly computerized (E) via a IEE 488 connection. The electrochemical cell used has been specifically built up for this study. It is shown on Figures 3.13.3 and 3.13.4. The cell is made of an insulating material and it can be used with oil temperatures up to 130°C. Its design allows the variation and the control of the distance between the two electrodes down to very low values (25 μm) by means of a micrometer. The thermostatic regulation is achieved by means of a heating coil immersed in a special compartment of the cell.

A mechanical stirrer allows the temperature to be uniform. A platinum temperature probe controls its value.

362

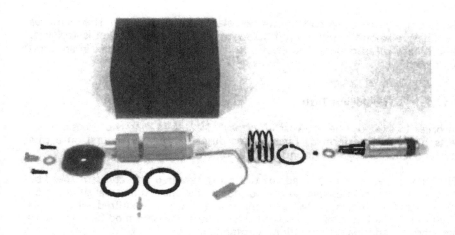

Figure 3.13.3: General view of the components of the cell

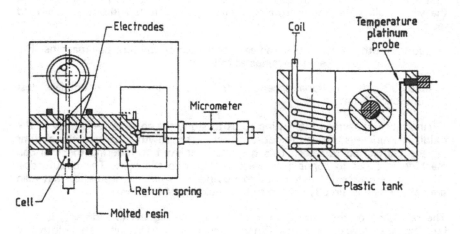

Figure 3.13.4: Description of the electrochemical cell

The electrodes are made from calibrated metal bars molded by gravity in a special resin.

The truncating of these bars allows two electrodes to be made. Their surfaces are finely diamond polished until a roughness lower than 0.5 μm is achieved. Before each measurement, the electrodes are polished and washed in an ultrasonic bath.

3.13.2.4 Tribological Tests

In order to evaluate the antiwear efficiency of oxidized oils, two types of wear tests have been performed. Firstly, long duration tests (50 hours) on a cam tappet friction simulation rig and secondly on a four-ball machine.

The cam tappet wear rig is made of two cams, powered at a constant speed by an electric motor, in contact with two tappets designed as in a real situation on the Peugeot XU engines. Owing to the low quantity of oxidized oil available (300 ml), the lubricating principle of this contact has been modified. The cams are lubricated during their rotation by rotating in a small oil bath.

The contact load and rotation speed have been respectively set at 83 daN and 2000 t/mn corresponding to an oil bath temperature of 75°C.

This kind of test has been used taking into account its good correlation with the valve train wear results from fired engine bench tests and automotive field tests.

In addition, other wear tests have been conducted on the four-ball machine. The test conditions have been determined as follows:

load: 30 daN, rotational speed: 1400 t/mn, oil temperature: ambient, test duration: 30 mn.

Furthermore, it seemed interesting to us to study the antiwear efficiency of the oxidized lubricants through the observation of the growth of the reactional films on the contacting surfaces. This has been achieved by altering the four-ball machine in order to connect an electric circuit to the balls. This circuit, described in reference (1), allows us to observe the electric insulation between the balls, which is directly related to the thickness of the films.

The recording of the voltage among the balls during the experiments is done by a digital storage oscilloscope. Filtered signals are thus available on a plotter.

Let us bear in mind that in such an electric scheme, a short circuit (0 volt) means a metallic contact between balls without wear protection films on it, and a full voltage, as is applied, indicate a continuous antiwear film.

The wear evaluation on the four ball machine is classically achieved by the measurement of the wear scar diameter.

3.13.3 Results

3.13.3.1   AC Impedance Spectra

There are only a few papers available in the literature concerning the application of the electrochemical impedance spectrum techniques to the study of lubricating oils (2 — 5). Consequently, in order to understand the phenomena observed in this kind of experiments, some preliminary tests have to be done. More particularly, some of them have been related to the study of the influence of specific parameters.

Influence of the Inter-electrodes Distance

For this serial of experiments, the electrode material was 35CD4 steel: the oil A oxidized on three levels was chosen as the electrolytes (0.96 and 264 hours of oxidation).

Figures 3.13.5a and 3.13.5b show the aspect of the impedance diagrams plotted in the Nyquist plane of the oil A after 96 h in the oxidation test for different electrodes spacings.

In this work, it could be assumed that the electrolyte resistance Re, in our case the lubricant, is represented by the value of the real part of the impedance at the low frequency limit of the first loop (6).

Then it is possible to plot the curve illustrating the evolution of this resistance Re with respect to the electrode spacing (cf. Figure 3.13.6).

It can be seen that the value of Re increases quite linearly with the electrode spacing, whatever the aging level of the oil. However, if this curve is extrapolated to the null value of the spacing which is the "short circuit" one, a value of Re equal to zero may be found.

This is not the case and a residual value of the resistance is noticed and is different for the degree of oxidation of the oil as shown in table 3.13.2.

This table also contains the values of the specific resistance of these oils calculated from the slopes of the curves in Figure 3.13.6.

Likewise, it is possible to assume that the capacitance created by the bulk electrolyte (lubricant) between the electrodes is given by the relation (1).

$$C = \frac{\epsilon\epsilon_0 \; S}{1} \tag{1}$$

where
C : capacitance of the cell
ε : dielectric constant of the lubricant
ε₀ : dielectric constant of the vacuum
S : surface of the electrodes
1 = electrode spacing

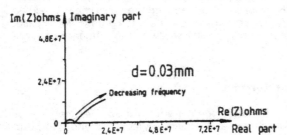

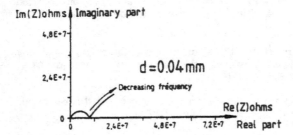

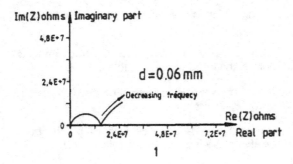

1

Figure 3.13.5a: Nyquist plots of the impedance spectra for different electrodes spacings

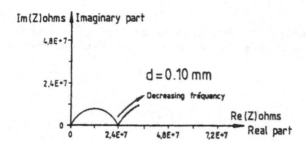

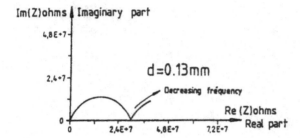

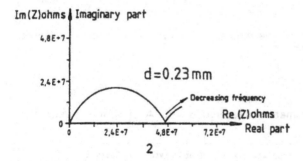

Figure 3.13.5b: Nyquist plots of the impedance spectra for different electrodes spacings

In the hypothesis where we consider that the first loop of the impedance diagram is representative of the capacitive behaviour of the electrolyte, relation (2) applies:

$$C = \frac{1}{2\pi f R_e} \tag{2}$$

R_e: electrolyte resistance
f : frequency corresponding to the minimum value of the imaginary part of the impedance. This is equivalent to the frequency position of the summit of the first loop.

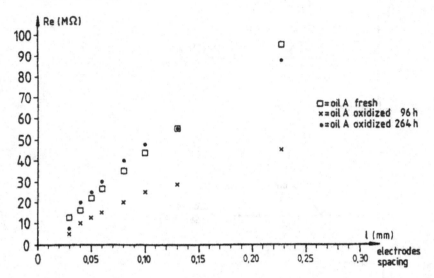

Figure 3.13.6: Evolution of the resistance R_e with respect to the interelectrodes distance

Table 3.13.2: Electric parameters measured on oil A at different oxidation levels

Reference of the lubricant	Extrapolated value of Re for zero spacing (M Ω)	Electrolyte Resistivity (G Ω.cm)	Dielectric constant
Oil A fresh	7	12	5.4
Oil A after 96 h of oxidation	5	5.1	5.6
Oil A after 264 h of oxidation	15	9.6	6

The curves in Figure 3.13.7 illustrate the dependence of the capacitance versus the inverse of the electrodes spacing. One can conclude that relation (1) applies well to the lubricant even if the latter has been aged by oxidation. The calculated values of the dielectric constants of the oils are also listed in table 3.13.2.

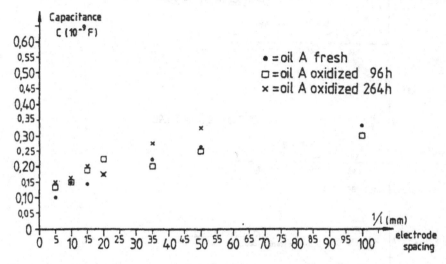

Figure 3.13.7:   Evolution of the capacitance C with respect to the interelectrodes distance

Influence of the electrode metallurgy

The nature of the materials in contact with the lubricants in automotive engines is extremely different. Consequently we have tried to evaluate their influence upon the trends of the impedance diagrams. Different metallurgies have been used in addition to the 35CD4 steel: a GS cast iron and two aluminium alloys AS7GO.3 and AS5U3. Only one lubricant has been used as a reference oil in the cell (oil B after 144 hours of oxidation).
The electrodes spacing was set at 0.025 millimeter.

The Nyquist diagrams illustrating the results of these impedance measuring conditions are shown on figure 3.13.8. The first loop of the diagrams is almost not affected by the nature of the electrode material. The capacitance and resistance values deduced from the characteristics of the first loop are indicated in table 3.13.3. Bearing in mind the very slight differences noticed, the global behaviour of the electrolyte can be assumed to be similar.

369

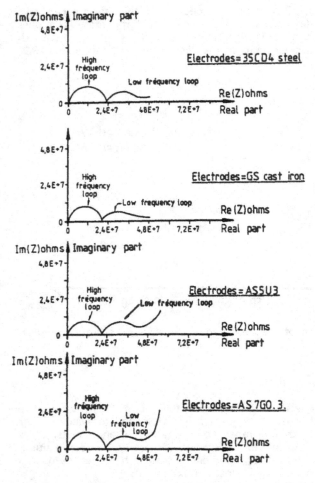

Figure 3.13.8: Nyquist plots of the impedance spectra for different electrodes materials

370

Table 3.13.3: Values of the capacitance and the resistance of the cell for different electrodes materials

Electrode material	Resistance Re (M Ω)	Capacitance C (pF)
35CD4 Steel	22.4	537
GS cast iron	19.2	509
AS7G0.3	21.6	518
AS5U3	20	530

In return, the trends of the diagrams for the low frequencies of excitation, which correspond to the electronic exchange mechanisms at the interfaces, is very different for all the tested materials. Two types of diagrams can be drawn, one corresponding to the ferrous materials (cast iron and steel) whose low frequency part is roughly represented by a capacitive loop and the other, corresponding to aluminium-based alloys whose diagram aspects are represented by a low frequency capacitive loop followed by a divergent straight line.

Further investigations are needed to explain these phenomena. However, it must be noticed that the reactional mechanisms can be brought out by this technique. In addition, they seem to be different for the ferrous and aluminium based materials.

This report is supported by the physicochemical examinations on the surfaces of friction materials which show that the chemical composition of the reactional films analyzed by XPS were very different for these two kinds of materials (7).

Influence of the level of Oxidation

Oils A and B have been used for this study. They have been submitted to different durations of the oxidation test (0.50, 96, 144, 200 and 264 hours). The electromechanical cell was set up at two spacings (0.1 and 0.03 mm) at room temperature (23°C).

The corresponding diagrams are represented in figure 3.13.9a and 3.13.9b for the large inter-electrodes distance.

These diagrams show that the dielectric properties of the oils are dramatically affected by the degree of oxidation depending upon the chemical formulation of the lubricant.

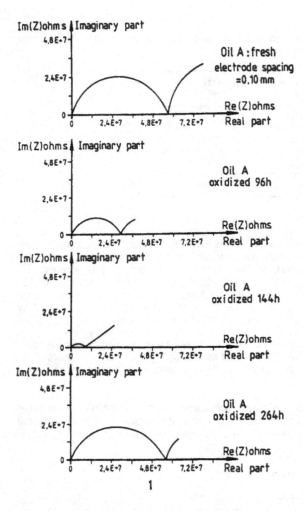

Figure 3.13.9a: Nyquist plots of the impedance spectra for different oxidation levels

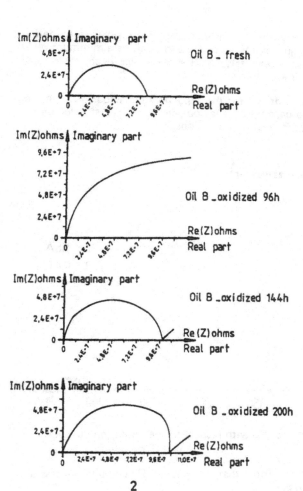

Figure 3.13.9b: Nyquist plots of the impedance spectra for different oxidation levels

On figure 3.13.10 are represented the variations of Re during the oxidation process. Oil A shows very slight variations of Re and exhibits a minimum value with respect to the oxidation. Conversely, oil B shows a significant increase of this value and exhibits a maximum in the curve.

No equivalent plots have been drawn for the capacitive behaviour. As a matter of fact, the variations of the capacitive behaviour are very little affected by the oxidation, at least in the high frequency excitations.

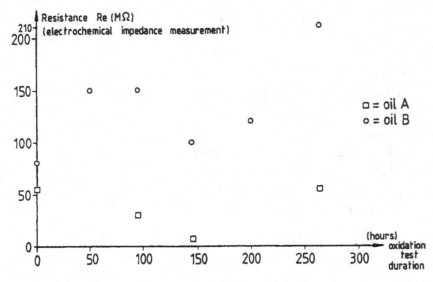

Figure 3.13.10:   Dependance of Re with respect to the oxidation duration

All the reports given previously show that the value of Re could be successfully used to evaluate the oil degradation.
Thus it is possible to monitor the variations of the degradation of the oil by means of the simplest electric way: a DC voltage cell.
For this purpose we have made such measurements in a specially designed cell with stainless steel electrodes.
The DC voltage was set to 1v, the electrodes spacing to 0.3 mm in order to measure the global resistance of the cell.
Figure 3.13.11 illustrates the results obtained with this device. The relative variation (in percent) of the resistance of the cell measured on an oxidized oil with respect to the same measurement on the corresponding fresh oil is drawn versus the oxidation time. This total resistance cell is the sum of the different resistances associated to the bulk electrolyte and to the two electrochemical electrode/lubricant reactions.

374

One can immediately establish that the general trends of the curves of oils A and B are very similar to the corresponding ones previously drawn with the results of the impedance spectrum technique.

In addition, it can been seen that the curve trends related to the oxidation behaviour of oils A,C,D are very close to one another but very different from those of oil B.

In order to translate all these observations into an antiwear performance criterion, some wear tests have been done. The results of these tests are examined below.

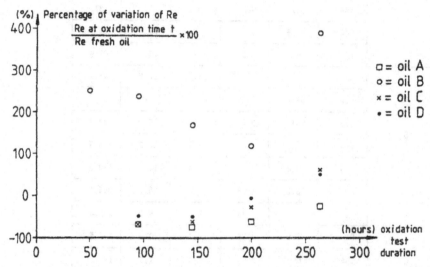

Figure 3.13.11: Evolution of the electrolyte resistance during the oxidation process as measured in DC voltage

3.13.3.2   Wear Tests

Table 3.13.4 collects all the tribological results obtained with the different oils which are the subject of the present work. The wear results on four-ball machine are expressed by the mean value of the wear scar diameters of three different tests.

Wear values resulting from the cam-tappet endurance tests are expressed by the mean value of the weight loss of the cams of two tests (two cams for one test). Systematic microscopic observations were made of the worn faces of cams and tappets with different oils to make sure that the wear process involved the same mechanism and differed only in their respective intensity.

Table 3.13.4: Physicochemical results on oxidized oils

Oil reference	Oxidation test duration (h)	Weight loss of cams (mg)	Wear scar diameter (mm)	TAN (mg KOH /g)	Viscosity at 100° C (cSt)	Viscosity at 40° C (cSt)
A	0	-	0.34	2.5	11.9	81.8
	96	4	0.32	4.5	11.6	82.4
	144	1.7	0.36	5.2	12	87
	200	2.35	0.36	6	12.9	99.5
	264	5	0.87	7	16.6	147.9
B	0	-	0.34	2.3	17.9	161.4
	50	709.6	0.37	-	-	
	96	604	0.42	4	16.9	158.7
	144	1	-	5.1	17.3	166.5
	200	9.8	0.38	6.4	18.7	189.2
	264	16.3	0.38	8.6	25.5	366.6
C	0	-	0.33	2.8	19.5	148.8
	96	2.9	0.39	4.7	20.2	158.5
	144	6	0.40	5.8	21.1	172.8
	200	1.9	0.36	6.9	25.1	231.4
	264	4.62	0.38	8	37	356.8
D	0	-	0.34	2.6	14.5	95.7
	96	1.4	0.37	4.6	14.7	99.4
	144	4.6	0.37	5.3	16.1	111.4
	200	5.5	0.38	6.5	19.2	153.2
	264	3.1	0.39	8.2	29.4	249.9

376

In this table are also listed the values of the physicochemical characteristics measured on the same oils; the total acid number (TAN: method ASTM D664), the kinematic viscosities at 40 and 100°C, traditionally used to qualify the oil degradation following oxidation tests.

Cam tappet Test Result

Figure 3.13.12 shows the evolution of the wear of the cams with respect to the degree of oxidation of the oil. It shows that the general profile of these curves are very close to those illustrating the evolution of electrochemical measurements on the oils (cf. figures 3.13.10 and 3.13.11).

Two different kinds of behaviour can be noticed, one related to a progressive increase in wear with respect to the oxidative level (oils A, C, D) and the other related to the presence of the maximum in the wear curve (oil B).

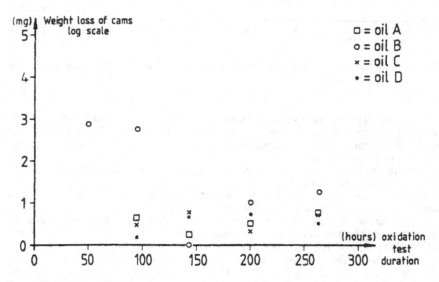

Figure 3.13.12:   Relation between the wear of cams and oxidation test duration.

Four-ball Test Results

Figure 3.13.13 illustrates the wear results at the end of the 30 min run. Roughly, the trends visible with respect to the oxidation level are the same as those that can be seen at the end of the endurance test.

377

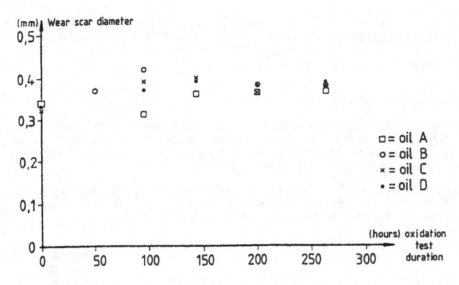

Figure 3.13.13:   Wear scar diameter on four-balls machine versus oxidation
duration time.

The relative lack of correlation between the two tests in not surprising and not very significant. More attention is to be paid to the wear results of the motored valve train wear which are more closely linked to the real valve train wear problems of automotive engines.

More interesting is the study of the formation of the reactional films during these four-ball wear tests.

Curves in figure 3.13.14 show, for all the oils, the evolution of the contact potential between the balls related to this formation.

All the lubricants show a similar trend such as, a lowering of the insulation state of the balls corresponding to a slowing down of the growth kinetic of the protective layers with a progressive increase in the oxidation process.

Some of these films are very unstable and, moreover, worn out, as is shown by the curves of oils A oxidized for 144 h and oil C for 96 h and 144 hours, which are decreasing beyond a certain duration.

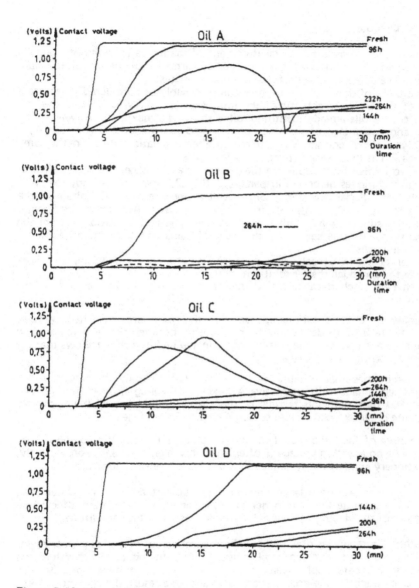

Figure 3.13.14: Evolution of the contact potential between the balls during the wear test.

3.13.4 Discussion

The last results relative to the measurement of the insulating state of the balls show a good correlation with the antiwear performance of the oils after oxidation tests evaluated by means of a cam tappet wear test.

The evolution of this insulation, which can reasonably be considered as representative of the antiwear activity of the lubricants (8), allows us to find the two groups of the oils already identified; oil B showing a maximum of wear degradation and percentage of metallic contacts and, on the other hand, oils A, C and D showing a continuous decrease in their insulation state, in the same classification as their wear mitigation performance.

These two kinds of behaviour are the same as the one observed by the electrochemical techniques described in paragraph 3.13.2.3. We have already seen that the value of "global resistance" of the cell as measured in DC, voltage or the specific resistance R_e of the electrolyte deduced from AC Impedance Spectra exhibited a similar trend, with respect to the type and oxidation level of the lubricants, to the one shown by the wear of cams and the contact resistance between the balls.

No other physicochemical value measured on the lubricant during this work was able to reproduce these evolutions.

Thus, this electrochemical technique seems to be promising in the study of the oil activity.

A number of other questions appear from the result presented here. Among others, figure 3.13.15 describing the relationship between the relative variation of the resistance Re and the wear of cams in the endurance test shows that the correlation is not satisfactory enough.

This relation becomes notably better if we consider only the results of the measurements made with alternative AC excitation voltages. It can be explained by the fact that, with this technique, there is no polarisation of the electrodes perturbing the interfacial system response as with DC voltage.

As a matter of fact, the global resistance value of the cell is a complex parameter. The polarisation voltage is often of a very high level and, consequently, the stationary state of the system cannot be reached before a long time.

Conversely, the AC Impedance measurements do not perturb the system and allow us to separate the electrochemical reaction effects which are affected by the nature of the electrodes material, the polarisation voltage amplitude.

The other discrepancies observed for the highest values of the cam wear might be easily explained by mechanical considerations. In this test, such values can be qualified as "catastrophic wear." and the cam profile is quite completely worn away. Consequently, the acceleration curve and, as a result, the contacting force law, becomes erratic, giving rise to an autocatalytic increase in the wear kinetics. There ist a phenomenon of "geometrical activation" of the wear which shifts the curve towards the wear value axis. Mild and medium wear do not suffer from this kind of problem.

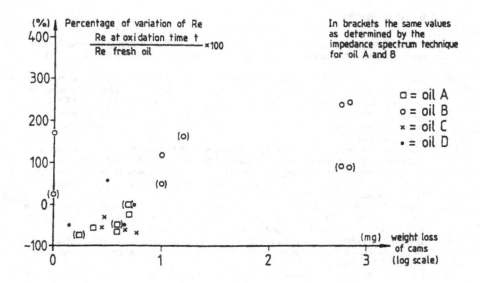

Figure 3.13.15: Relation between electrochemical impedance determination of Re and wear performance

3.13.5 Conclusion

It has been shown that the electrochemical techniques applied to the study of the lubricants allowed a forward-looking evaluation of their antiwear capability during a laboratory simulated oxidation process.

This evaluation is shown to be well assessed by means of an electrochemical impedance spectrum technique. This method allows us to display the differences in the chemical reactions with respect to the oil oxidation for various lubricants. For example, the real part component of the impedance measured at high frequencies can already be a very powerful tool for such studies. In addition, interesting potentialities of studying the reactional mechanisms of the lubricants have been offered by extending these measurements to the low frequency domain.

Nevertheless, an important work has to be done for a better knowledge and understanding of the strictly interfacial reactions between various metal surfaces and the active species of lubricants.

This research is being carried out.

381

3.13.6 References

(1)   Monteil, G.; Lonchampt, J.; Roques-carmes, C.: Etude tribologique du système came-poussoir. Amsterdam 1985. 2.2 p 1—8. Proceedings Eurotrib 85.9.12.09.85. Ecully.

(2)   Wang, S.; Tung, S.: Electrochemical phenomena in lubricants. I, potential measurements and analysis of metal/additive interactions — 1985 p 642—643. Extended Abstract No. 431, Fall Meeting of the Electrochemical Society. Las Vegas 13-18.10.85.

(3)   Wang, S.; Maheswari, S.; Wang, Y.; Tung, S.: An electrochemical technique for characterizing metal-lubricant interfacial reactions. ASLE transactions 30 (1987) 3, p. 394—402.

(4)   Wang, S.; Maheswari, S.; Tung, S.: AC Impedance measurements of the resistance and capacitance of lubricants. ASLE transactions 30 (1987) 4, p. 436—443.

(5)   Wang, S.; Maheswari, S.; Tung, S.: The nature of electrochemical reactions between several zinc organodithiophosphate antiwear additives and cast iron surfaces. Tribology transactions 31 (1988) 3, p. 381— 389.

(6)   Barral, G.; Diard, J.P. le Gorrec, B.; Dac Tri, L.; Montella, C.: Impedance de cellules de conductivité I. Determination de plages de fréquence de mesure de la conductivité. Journal of applied electrochemistry 15 (1985), p. 913—924.

(7)   Denizot, D.; Monteil, G.; Roques-Carmes, C.; Lonchampt, J.: A tribological study of synchronizing devices used in car gearboxes. Journal of automobile engineering Part D2.203 (1989), p. 111—115.

(8)   Cameron, A.: Thick boundary lubrication. Helsinki. 1989 not published. EUROtrib 89 12-15.6.89. Helsinki.

3.14 Mathematic Model for the Thickening Power of Viscosity Index Improvers. Application in Engine Oil Formulations

H. Bourgognon and C. Rodes, Centre de Recherche Elf Solaize, Lyon, France
C. Neveu and F. Huby, Rohm & Haas European Operations, Paris, France

Summary

It is essential for a lubricating oil blender to be able to quantify the contribution of each component entering in the formulation to the viscosity of the finished product.

In this context several studies have been devoted to the modelization of the thickening power of different types of polymers at both high ($100°$ C and $40°$ C) and low ($- 15°$ C) temperature in base oils of various origin and viscosity.

To complement these earlier investigations, it is necessary to quantify the effect of the additive package on the viscosity of the formulation. For this purpose, the effect of several DI packages of different performance levels has been investigated in order to derive appropriate mathematical models representing their contribution of the oil viscosity.

3.14.1 Introduction

It is necessary for both economical and technical reasons (compatibility with seals, thermal stability, . . .) to optimize the composition of an engine oil in order to minimize the quantity of VI improver and to use the most viscous base oil blend. This is possible in a laboratory with no time constraint but it is impossible in a blending plant. The latter is not in a position to adjust the formulation to take into account the variability of the components (VI improver, base stocks, DI package).

In order to be able in a blending plant to produce oil exhibiting consistent viscosity, it is necessary to develop precise models describing the thickening power of a polymer and of a package in various base stocks at different temperatures.

3.14.2 Background

In a previous study (reference 1) it has been shown that the Kraemer and Huggins equations which are normally applicable only in the field of dilute solutions could be used:

a) to describe the thickening power of a PMA VI improver in typical engine oil blends at temperatures ranging between 38° C and 175° C, and

b) to provide an estimate of the intrinsic viscosity of a PMA polymer in a given solvent and at a given temperature.

They are of the form:

LN(VF/VB)=ETAK*c−beta*(ETAK*c)**2   Kraemer

VF/VB=1+ETAH*c+k*(ETAH*c)**2   Huggins

VF is the viscosity of the blend containing c % by weight of polymer in a base oil of viscosity VB.

ETAH and ETAK are the intrinsic viscosities calculated using respectively the Huggins and Kraemer equations. Beta and k are the second order coefficients of these equations.

In addition, the effect of the solvent (paraffinic mineral oil, solvent refined) on the intrinsic viscosity ETA(T) could be approximated by the equation:

ETA(T)=ETA0(T)*(1−delta(T)*VB)

where VB is the solvent viscosity at 100° C and delta(T) is a constant for PMA's at temperature T. ETA0(T) corresponds to the intrinsic viscosity of the polymer at temperature T extrapolated to a zero centistoke base oil.

In a second study (reference 2), it was shown that the Kraemer and Huggins equations could be used at both 100° C and 40° C outside of the field of dilute solution of OCP VI improvers, providing one takes into account the dilution oil.

At − 15° C the equations could be simplified for the dPMA and OCP VI improver by neglecting the second order term.

The solvency of the base stock can be represented by a model of the form:

ETA(T)=ETA0(T)*(1−delta(T)*VB+gamma(T)*(d−.88))

where VB is the solvent viscosity at 100° C, delta(T) and gamma(T) are constants for PMA's and OCP's at temperature T and d is the specific gravity of the base oil at 15° C. Finally, ETA0(T) corresponds to the intrinsic viscosity of the polymer at temperature T extrapolated to a zero centistoke base oil of specific gravity d = 0.88.

384

3.14.3 Objective of the Study

In order to optimize the composition of engine oils in a production plant, it is necessary to complement the previous results by equations which can represent:

a) the contribution of a package to the kinematic viscosity at 100° C and 40° C and in the CCS (ASTM D 2602) at — 15° C.

b) the effect of the presence of a package on the contribution of the polymer to viscosity and on the solvent power of the base stocks.

3.14.4 Contribution of Package Components to Viscosity

A DI package is made of several components which all have a contribution to the blend viscosity. Before embarking on the modelization of the effect on the viscosity of the complete package, we have examined the variability which may result from the fluctuations of the package composition. For this purpose, we have evaluated:

a) the contribution of each of the main DI components to the blend viscosity.

b) the variability of blends based on samples of packages delivered to a blending plant.

3.14.4.1 Contribution of Package Components to Viscosity

To define a model describing the contribution of each of the main components of a package to the viscosity of a lubricant, we have used a greco-latin design. Such a model will be simple but sufficient for our purpose.

We selected 6 components which were combined in such a way that the sum of the concentrations was kept equal to 8 %. The formulations are detailed in Table 3.14.1. V1 et V2 are ashless dispersants, V3 is a ZDTP, V4, V5, V6 are detergents.

The analysis of the viscometric data shown in Table 3.14.2 showed that V1, V2 and V3 had the largest influence on the blend viscosity. In order to obtain a more precise model, three additional blends shown in Table 3.14.1 were made. The viscometric data are shown in table 3.14.2.

Table 3.14.1: Contribution of package components to viscosity Blend compositions

	V 4	V 5	V 1	V 2	V 6	V 3	Ref.
Blend 1	5.6	0.48	0.48	0.48	0.48	0.48	92
Blend 2	0.48	5.6	0.48	0.48	0.48	0.48	92
Blend 3	0.48	0.48	5.6	0.48	0.48	0.48	92
Blend 4	0.48	0.48	0.48	5.6	0.48	0.48	92
Blend 5	0.48	0.48	0.48	0.48	5.6	0.48	92
Blend 6	0.48	0.48	0.48	0.48	0.48	5.6	92
Blend 7	1.33	1.33	1.33	1.33	1.33	1.33	92

a) Greco-Latin Design

	V 4	V 5	V 1	V 2	V 6	V 3	Ref.
Blend 8	0.48	0.48	3.04	3.04	0.48	0.48	92
Blend 9	0.48	0.48	3.04	0.48	0.48	3.04	92
Blend 10	0.48	0.48	0.48	3.04	0.48	3.04	92

b) Additional blends

Table 3.14.2: Contribution of package components to viscosity viscosimetric data

Name	Viscosity 100° C (cSt)	Viscosity 40° C (cSt)	Viscosity − 15° C (cPo)
Blend 1	12.78	81.41	2800
Blend 2	13.07	85.59	2950
Blend 3	13.74	92.15	3200
Blend 4	13.77	91.05	3200
Blend 5	12.36	82.51	2950
Blend 6	11.86	76.11	2700
Blend 7	12.86	84.49	2900

a) Greco-Latin Design

Name	Viscosity 100° C (cSt)	Viscosity 40° C (cSt)	Viscosity − 15° C (cPo)
Blend 8	13.72	91.72	3100
Blend 9	12.70	82.29	2900
Blend 10	12.67	82.89	2900

b) Additional blends

The complete analysis of the data enabled us to obtain the following models:

KV   100° C = 12.63 + 0.191*V1 + 0.194*V2 − 0.187*V3

KV   40° C = 82.35 + 1.66*V1 + 1.54*V2 − 1.5*V3

DV   −15° C = 2863 + 51.5*V1 + 51.5*V2   − 48.5*V3

in which Vi is the concentration in percent weight of component Vi. This model appears to fit correctly the ten data points of our design. However, it should not be valid outside the range of composition studied.

It is interesting to note that:

a) the two ashless dispersants V1 and V2 give essentially the same level of viscosity increase.
b) each percent of ZDTP (V3) decreases the viscosity by almost the same amount as each percent of ashless dispersant increases it.

Using these models, we can estimate the effect which would result from a 5 % increase and a 5 % decrease of the concentration of the ashless dispersants and ZDTP respectively.

We have taken, for example, a SAE 15 W 40 containing 5 % of ashless dispersants and 1 % of ZDTP. Increasing the ashless concentration to 5.25 %, reducing the ZDTP content to 0.95 % and rebalancing the package to maintain a treat rate of 8 % would result in the following increase of viscosity:

KV $100°$ C increase = 0.06 cSt
KV $40°$ C increase = 0.5 cSt
DV $-15°$ C increase = 15 cPo

3.14.4.2 Analysis of Packages Used by a Blending Plant

We have obtained 16 and 12 samples of respectively an SF/CC and an SHPDO package which were retained at the time of delivery.

Using each of the gasoline package samples, we have prepared a formulation using a given base oil mixt and a given VI improver sample. A similar exercise was repeated with the Diesel package samples.

The complete results are gathered in Table 3.14.3. The mean viscosity and the standard deviation at $100°$ C, $40°$ C and $-15°$ C have been calculated for each package. It can be seen in Table 3.14.3 that the standard deviation obtained for the Diesel and gasoline packages are essentially identical irrespective of the temperature. Consequently, an average value can be used to estimate at each temperature the variability associated with the use of different DI package deliveries.

Standard deviation $100°$ C = 0.04 cSt
Standard deviation $40°$ C = 0.2 cSt
Standard deviation $-15°$ C = 110 cPo

To each of these three standard deviations corresponds a variance which can be broken down into:

VP: variance due to fluctuation in package composition
VB: variance due to the preparation of the blend
VM: variance due to the precision of the viscosity measurement.

388

Table 3.14.3: Package thickening effect reproducibility

OBS	SP.GRAV	KV.100	KV.40	CCS.15
1	0.8838	13.380	88.15	3000
2	0.8838	13.360	87.84	31.50
3	0.8838	13.380	88.04	2925
4	0.8838	13.360	87.83	3150
5	0.8839	13.425	88.19	3150
6	0.8838	13.375	88.00	3000
7	0.8838	13.370	88.04	3000
8	0.8836	13.340	87.98	3000
9	0.8838	13.380	88.00	3150
10	0.8838	13.420	88.10	3150
11	0.8838	13.420	87.89	3200
12	0.8838	13.390	87.89	3200
13	0.8838	13.410	88.55	3150
14	0.8838	13.420	88.33	3300
15	0.8839	13.440	88.10	3000
16	0.8839	13.370	87.88	3150

a) Gasoline package

OBS	KV.100	KV.40	CCS.15
1	14.15	98.87	3850
2	14.06	98.32	3850
3	14.10	98.26	3850
4	14.07	98.15	3500
5	14.11	98.68	3850
6	14.10	98.66	3850
7	14.10	98.47	3850
8	14.13	98.51	3700
9	14.19	98.29	3600
10	14.08	98.65	3700
11	14.10	98.29	3700
12	14.18	98.69	3700

b) Diesel package

Temperature	Package	Moyenne	Ecart type
100	Essence	13.39	0.03
	Diesel	14.11	0.04
40	Essence	88.05	0.19
	Diesel	98.49	0.22
−15	Essence	3105	103
	Diesel	3750	118

c) Mean Viscosity, Standard deviation

Knowing that the standard deviation associated with the precision of the viscosity measurement corresponds to approximately:

Standard deviation 100° C = 0.02 cSt
Standard deviation   40° C = 0.12 cSt
Standard deviation −15° C = 75 cPo

it can be concluded that the sum of the variabilities due to the fluctuations of the package composition and to the preparation of the blends still accounts for more than half of the variance irrespective of the temperature.

However, we must consider that the residual standard deviations of the models developed in prior studies are significantly higher than those calculated in this section. Consequently, we will neglect in our calculations the variations which could be associated to the variability of the package.

3.14.5 Model Describing the Contribution of Package to Viscosity

In order to define a model which could be used to represent the contribution of a package to the viscosity as a function of temperature, we have used data generated according to the following factorial design:

Factors	Levels
Base stock origin	3
Package type	SF/CC (8.0 %) SF/CD (11.8 %) SHPD (14.5 %)
% PMA	None 6 % in SF/CC and SF/CD formulations 5 % in SHPD formulation
Temperature	− 15° C, 40° C, 100° C

Considering that the intrinsic viscosity of a polymer depends on the viscosity of the base oil, we have completed our blending study in three different base oil mixtures having a kinematic viscosity of 8 cSt at 100° C. These base stocks were selected because of the difference in aromatics content and, consequently, specific gravity as shown below:

Base code	Aromatic carbons		Specific gravity
	HPLC	IR	
B1	21.2	6.25	.8767
B2	22.9	5.8	.8720
B5	39.7	8.4	.8835

3.14.5.1 Model Describing the Contribution of Packages to Viscosity

Our analysis is based on the blends containing no VI improver. The viscometric results are gathered in Table 3.14.4. We have decided to use a model on the Kraemer equation with no second term order:

$$LN(VP/VB) = K1 * Xp'$$

VP and VB are the viscosity of the base oil with and without package respectively. Xp' is the concentration of the package corrected for the addition of VI improver according to the formula:

$$Xp' = Xp/(100 - XVI)$$

with:
XVI = 6 for the SF/CC and SF/CD packages
XVI = 5 for the SHPD package
Xp = recommended package concentration in fully formulated oil.

This correction was made in order to take into account the presence of a VI improver in the fully formulated oil. When using Xp' the relative concentration of the package in the base oil plus package only is the same as in the fully formulated oils.

The values of K1 for each of the packages and each of the base oils are detailed in Table 3.14.5. It can be seen that, irrespective of the temperature, K1 depends significantly on the package type but that it is essentially independent on the base oil origin.

3.14.5.2 Effect of the Package on the Thickening Power of the Polymer

Using the Kraemer and the Huggins equations we have calculated the intrinsic viscosity of the dPMA polymer for each of the blends at each of the three temperatures. When the blend contained a package, we use VP (base oil + package) in the equations instead of VB (base oil no package).

The results of this exercise are gathered in Table 3.14.6. We completed a variance analysis which indicated that:

a) the package type has a significant effect on the intrinsic viscosity at both 100° C and 40° C

b) the base stock origin has a significant effect on the intrinsic viscosity at 40° C

c) at − 15° C neither the package type nor the base stock origin have a significant effect on the intrinsic viscosity.

Table 3.14.4: Thickening effect of polymers, effect of DI package performance level

Package	Pack-rate	Ba-blend	VB 100	VB 40	VBCCS	VP 100	VP 40	VPCCS	VF 100	VF 40	VFCCS
SF/CC	8.00	B1	8.01	61.12	48	9.42	75.31	63.5	16.52	127.89	79.0
SF/CC	8.00	B2	8.01	57.56	34	9.41	70.79	45.0	16.41	117.93	55.0
SF/CC	8.0	B5	7.99	59.00	35	9.44	73.08	47.0	16.52	125.72	59.3
SF/CD	11.85	B1	8.01	61.12	48	9.68	79.28	72.0	17.20	135.56	90.0
SF/CD	11.85	B2	8.01	57.56	34	9.63	74.30	52.0	16.92	124.58	65.0
SF/CD	11.85	B5	7.99	59.00	35	9.70	76.93	54.5	17.19	132.80	69.0
SHPD	14.50	B1	8.01	61.12	48	9.77	80.96	74.0	16.12	130.18	88.5
SHPD	14.50	B2	8.01	57.56	34	9.80	78.00	52.5	16.10	122.0	62.0
SHPD	14.50	B5	7.99	59.00	35	9.86	78.17	55.0	16.16	127.87	67.0

Table 3.14.5: Thickening effect of packages

		K1 100 Mean	K1 40 Mean	K1 CCS Mean
Package	Bablend			
SF/CC	B1	0.0191	0.0245	0.0329
	B2	0.0189	0.0243	0.0329
	B5	0.0196	0.0251	0.0346
	All	0.0192	0.0247	0.0335
SF/CD	Bablend			
	B1	0.0150	0.0206	0.0322
	B2	0.0146	0.0203	0.0337
	B5	0.0154	0.0210	0.0351
	All	0.0150	0.0206	0.0337
SHPD	Bablend			
	B1	0.0130	0.0183	0.0284
	B2	0.0132	0.0199	0.0285
	B5	0.0138	0.0184	0.0296
	All	0.0133	0.0189	0.0288

Table 3.14.6:   Thickening effect of polymers, effect of DI package performance
level

VIITYPE PMAD

Package	Bablend	ETAH 100	ETAH 40	ETAH CCS	ETAK 100	ETAK 40	ETAK CCS
SF/CC	B1	0.2021	0.1834	0.1101	0.2066	0.1908	0.0952
	B2	0.2000	0.1762	0.0960	0.2045	0.1834	0.0843
	B5	0.2013	0.1874	0.1106	0.2058	0.1949	0.0955
SF/CD	B1	0.2073	0.1864	0.1147	0.2120	0.1939	0.0986
	B2	0.2031	0.1790	0.1084	0.2076	0.1862	0.0939
	B5	0.2064	0.1902	0.1150	0.2110	0.1968	0.0988
SHPD	B1	0.2145	0.1974	0.1086	0.2193	0.2055	0.0961
	B2	0.2126	0.1855	0.0960	0.2174	0.1930	0.0861
	B5	0.2117	0.2034	0.1125	0.2164	0.2117	0.0991

We have plotted in Figures 3.14.1 to 3.14.3 the dependence of the intrinsic
viscosity on the package concentration and base stock origin. It can be seen
that:

a) the addition of a package reduces the effect of the base oil origin on the
   intrinsic viscosity of the polymer especially at 100° C. We will, therefore,
   neglect it in our models.

b) at 100° C and 40° C the intrinsic viscosity increases with the concentration
   of DI package sometimes in a nonlinear manner.

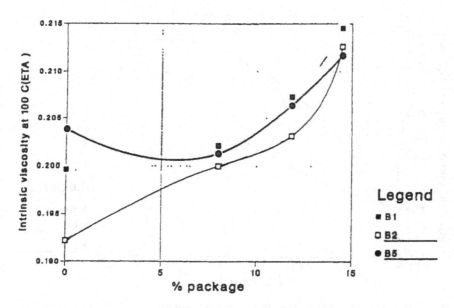

Figure 3.14.1: Effect of packages on intrinsic viscosity

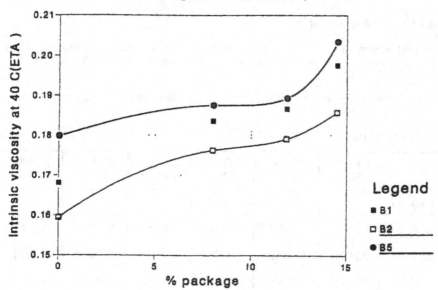

Figure 3.14.2: Effect of packages on intrinsic viscosity

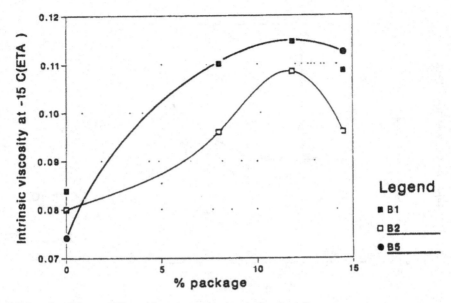

Figure 3.14.3: Effect of packages on intrinsic viscosity

3.14.5.3 First model

A first model can be obtained by using the average intrinsic viscosity of the polymer (all the package concentrations and all base stocks), see Table 3.14.7.

Table 3.14.7: Thickening effect of polymers, effect of DI package performance level

VIITYPE PMAD

	ETAH-100	ETAH-40	ETAH-CCS	ETAK-100	ETAK-40	ETAK-CCS
Package						
SF/CC	0.2011	0.1824	0.1056	0.2057	0.1897	0.0917
SF/CD	0.2056	0.1849	0.1127	0.2102	0.1923	0.0971
SHPD	0.2129	0.1954	0.1057	0.2177	0.2034	0.0938
ALL	0.2065	0.1875	0.1080	0.2112	0.1951	0.0942

The equations are of the form:

$$LN(VF/VP) = Etamean*c - beta*(Etamean*c)**2 \qquad Kraemer$$
$$VF/VP = 1 + Etamean*c + k*(Etamean*c)**2 \qquad Huggins$$

Using these equations, we have recalculated the viscosity of each of the blend at each of the temperatures. We have plotted the predicted versus the actual viscosities in figures 3.14.4 to 3.14.6.

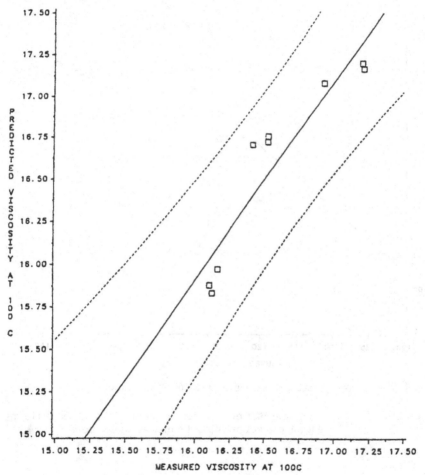

MODEL BASED ON GENERAL AVERAGE OF ETA HUGGINS

Figure 3.14.4: Thickening effect of polymers, effect of DI package performance level, prediction using the mathematical model

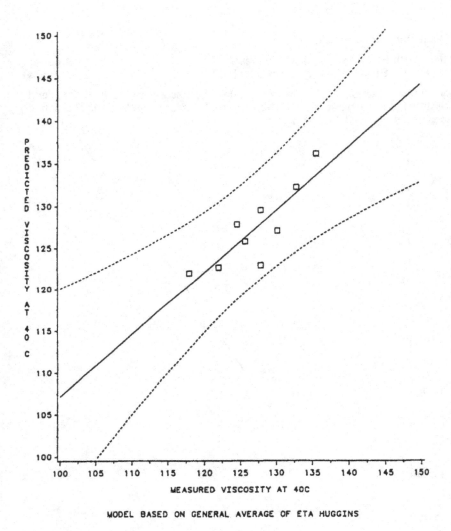

MODEL BASED ON GENERAL AVERAGE OF ETA HUGGINS

Figure 3.14.5:   Thickening effect of polymers, effect of DI package performance level, prediction using the mathematical model

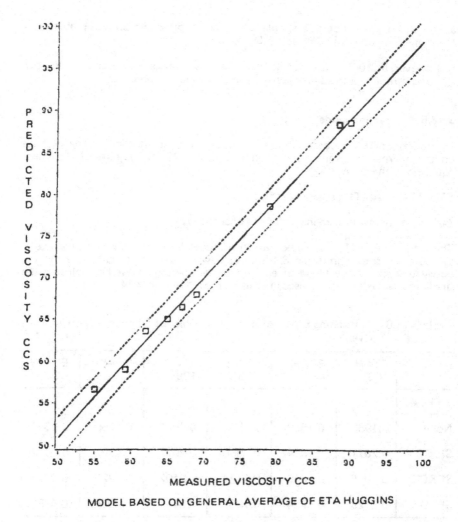

MODEL BASED ON GENERAL AVERAGE OF ETA HUGGINS

Figure 3.14.6: Thickening effect of polymers, effect of DI package perfomance leve, prediction using the mathematical model

a) it can be seen that this simple approach is providing an acceptable level of precision at − 15° C in the CCS.

b) however, at 100° C and 40° C it is necessary to take into account the type/concentration of the package to obtain a satisfactory precision.

3.14.5.4   Second model

In order to take into account the effect of the concentration of the package on the polymer thickening power at 100° C and 40° C, we have used a linear equation of the form:

ETA(T) = ETA0(T)*(1−alpha(T)*Xp)

Alpha is a constant independent of the package type.

The value of ETA(T) calculated with this model for each package at each temperature are shown in Table 3.14.8. We have used them to calculate the viscosity of each of the blend at each of the temperatures. We have plotted the predicted versus the actual viscosities in Figures 3.14.7 to 3.14.9.

Table 3.14.8:   Thickening effect of polymers, effect of DI package performance level

	ETAH-100	ETAH-40	ETAH-CCS	ETAK-100	ETAK-40	ETAK-CCS
Package						
None	0.1983	0.1694	0.0623	0.2021	0.1756	0.0641
SF/CC	0.2011	0.1824	0.1056	0.2057	0.1897	0.0917
SF&CD	0.2056	0.1849	0.1127	0.2102	0.1923	0.0971
SHPD	0.2129	0.1954	0.1057	0.2177	0.2034	0.0938

It can be seen that this model is more precise than the previous one at 100° C and 40° C but less at − 15° C in the CCS.

We have summarized in Table 3.14.9 the residual standard deviations corresponding to the 2 models for each of the equations and each temperature.

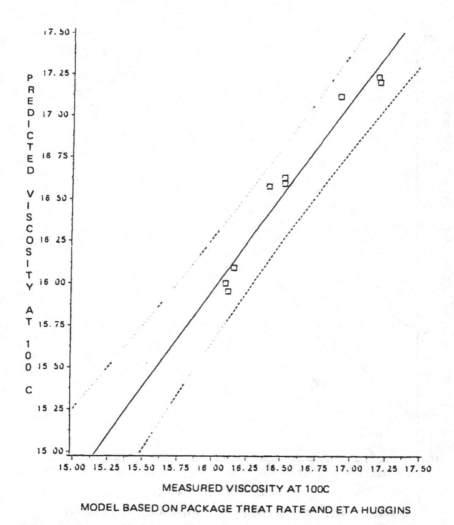

Figure 3.14.7:   Thickening effect of polymers, effect of DI package performance level, prediction using the mathematical model

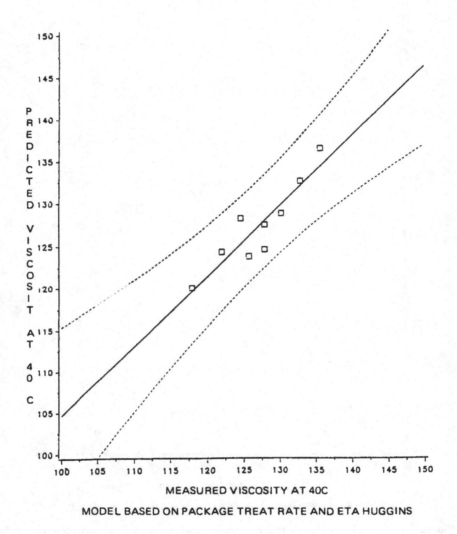

MODEL BASED ON PACKAGE TREAT RATE AND ETA HUGGINS

Figure 3.14.8:   Thickening effect of polymers, effect of DI package performance level, prediction using the mathematical model

402

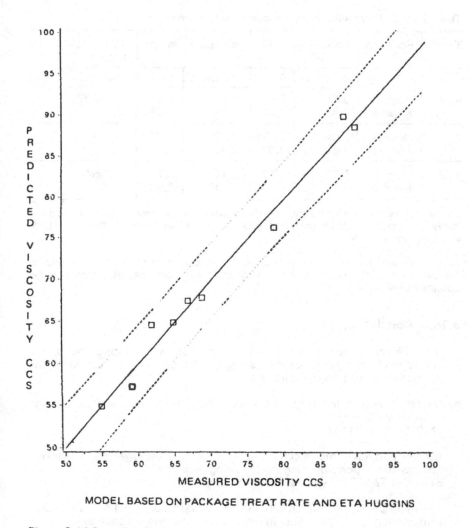

MODEL BASED ON PACKAGE TREAT RATE AND ETA HUGGINS

Figure 3.14.9:   Thickening effect of polymers, effect of DI package performance level, prediction using the mathematical model

Table 3.14.9: Compared model residual standard deviations

T	Equation	General average of η		$\eta = \eta_{ox}\,(1-\alpha\,xp)$		Without package
		Absolute	%	Absolute	%	Study CEC
− 15	Huggins	0.87	1.23	1.78	2.54	1.7 %
	Kraemer	0.80	1.13	1.45	2.07	
40	Huggins	2.64	2.07	2.19	1.72	1 %
	Kreamer	2.66	2.09	2.21	1.73	
100	Huggins	0.214	1.29	0.118	0.71	0.6 %
	Kraemer	0.213	1.29	0.113	0.68	

These results can be compared to those which were generated in a previous study (reference 2) in which no package was used but in which the base stock viscosity was variable.

It can be seen that at all temperatures, the precision of the models describing the presence of the package is not very different from that of the models we obtained previously.

3.14.6 Conclusion

a) It can be concluded that in a first approximation we can neglect the variability in thickening power of the package which could be associated to the fluctuation of the package composition.

b) We can represent the contribution of a package to the viscosity using an equation of the form:
LN (VP/VB) = K1*Xp

c) We can also take into account the effect of the package on the intrinsic viscosity of the polymer using the following equation:
ETA(T)=ETA0(T)*(1−alpha(T)*Xp)

d) At − 15° C one can alternatively use an average value of ETA(T) calculated with different packages for use in the Kraemer and Huggins equations.

e) We have obtained models which can be used to estimate the effect of a package on viscosity. Their precision is similar to that of models defined in a previous study in which no package was used (reference 2).

f) This study should be complemented by the evaluation of different polymer chemistries, using base stocks of different viscosities and at different CCS temperatures.

3.14.7 References

(1)   Neveu, C.; Huby, F.: Solution Properties of Polymethacrylate VI Improvers, 5th International Colloquium, 1986, pp 8-3-1/3-3-14.

(2)   Neveu, C. Huby, F.; Rodes, C.; Bourgognon, H.: Modelisation mathématique du pouvoir épaississant des améliorants d'indice de viscosité, IIIrd international CEC Symposium (Paris 1989) 11 LM.

3.15 Surface Morphology and Chemistry of Reaction Layers Formed Under Wear Test Conditions as Determined by Electron Spectroscopy and Scanning Electron Microscopy

Y. de Vita, I.C. Grigorescu and G.J. Lizardo
Intevep S.A., Caracas, Venezuela

Abstract

Friction and wear tests were performed in the Optimol SRV unit in the presence of engine lubricant oils A and B with different formulation at a constant load and at different temperatures.

Microanalytical studies on wear were accomplished using photomicrographs. In those samples where the wear scar was not detected by SEM, a digital inter-ferential optical profiler was employed. The behaviour of friction coefficient vs. time was analysed and metal-metal contact was detected in some test samples. Under severe wear conditions and lubrication failure, the morphologies of the wear scars suggested that several wear mechanisms were possible: adhesive, delamination, plowing and oil corrosion. It was observed smooth worn scars formed under continuous lubrication conditions.

On the basis of semiquantitative analysis performed by AES and on the results of other researchers, it could be suggested that on those worn scars, with very smooth surface, a glassy compound could be formed with amorphous structure containing phosphorus and sulphur as major components.

Similitude was observed between the worn area in valve lifter sides, from a taxi fleet test, run with oils A and B, and those corresponding to the SRV test samples morphology. This fact could suggest that the same wear mechanism occurred, as it was described above.

3.15.1 Introduction

Theoretical (1) and experimental research (2) have shown that valve lifters often operate in either the boundary or the partical elastohydrodynamic (EHD) lubri-cation regime.

For this reason and in order to assess possible correlation between the results of wear obtained on valve lifters, from a taxi cab fleet test and those from SRV test device, morphological and chemical surface studies have been carried out under boundary lubrication conditions.

Much effort has been directed towards the understanding of wear mechanism in the presence of two different lubricant formulations; A and B. Therefore, the present study was conducted to investigate the relationship between the severity of the damaged surface to the variation pattern of friction coefficient versus time, and to state the chemical composition of the worn surface at a given temperature and a constant load.

Finally, the worn areas at the top and at the sides of two valve lifters from taxi cab test were examined by SEM and EDX.

3.15.2 Experimental

The wear experiments were carried out on a modified commercial cylinder plane test device (Optimol SRV), consisting of a small cylinder (15 x 22 mm) scillating on a static plane chip (234 x 7.85 mm).

The experimental set-up has been described elsewhere (3). A schematic diagram of the test equipment appears in Fig. 3.15.1.

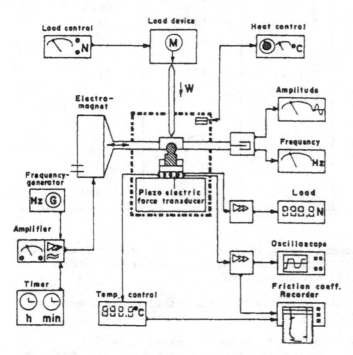

Figure 3.15.1: Diagram of the test equipment (3)

Both cylinder and test sample were 100CR 6 steel with 62–63HRC hardness. The contact surface of the test flat specimens were polished to 0.1 μm grit before tested, while the cylindrical contact surface was grinded as received. Both, cylinders and plane chips, were carefully cleaned in pure acetone.

The tests were performed at a constant load of 500 N, oscillation amplitude 1000 mm, temperature range 70–110°C, frequency 50 Hz and time 60 minutes. All of them were controlled piezoelectrically while the friction coefficients were monitored continuously. The friction and wear behaviour of oils A, B with different formulations were evaluated. Some inspection properties are given in Table 3.15.1.

Table 3.15.1: Properties of the oils A and B

Properties	Method	A	B
Viscosity Grade	–	SAE 10W30	SAE 10W30
Kinematic Viscosity 40°C (cSt) 100°C (cSt)	ASTM D–445	66.21 10.17	86.68 10.39
Sulfated ash (% p)	ASTM D–874	1.02	0.82
Element Content (% p) Mg Ca p Zn S		0.057 0.131 0.121 0.135 0.463	– 0.188 0.086 0.097 0.450

Previously, a repeatability study of ten run test were performed, and ± 4 % experimental error of friction coefficients was calculated from standard deviation of the mean at 95 % confidence level.

During the present investigation, all friction coefficient and wear measurements were made in triplicate.

The recorded diagrams of friction coefficient-time were studied. The mean value of the steady state friction coefficient was calculated and the "running-in" phase and transitory fluctuations of this paper were observed.

In order to estimate the wear performance, an approximate assessment of the wear scar was determined by a photographic method.

The worn surface morphology was examined by Scanning Electron Microscopy (SEM) (ISI-SS40) and Digital Interferential Optical Profiler. The result was compared with the microstructure of the worn valve lifters from a taxi cab fleet test where oil A and B were used.

The chemical composition of the surface layers along the width of the wear scar centre was determined by Energy-Dispersive-Xray Analysis (EDAX-9100) and by Auger Electron Spectroscopy (AES). The AES analysis were performed in the Leybold Heraeus surfaces analyses system at INTEVEP (Energy analyser 11) at 5×10^{-8} mbar.

The electron gun operated at 3KeV. Several areas were selected for depth profile analysis and a Argon ion gun was used at 3KeV, a current of 10 mA and 10^{-6} mbar.

The Auger spectra were obtained by digital differentiation of the intensity spectra. The data were collected at an energy pass of 200 eV with a step energy of 1eV. Fig. 3.15.10 shows a typical Auger spectra for specimens subjected to a SRV wear test.

A semiquantitative approach based on elemental sensitivity factors was employed (4). The atomic concentration C_x may be given by:

$$C_x = \frac{I_x/S_x}{\sum_i I_i/S_i}$$

where I_x is the Auger electron peak-to-peak height for a specific Auger transition, and S_x is the relative elemental sensitivity factor.

Additionally, qualitative corrosion tests were performed by immersion of the chip samples in oils A and B, heated at 150°C during six hours. Surface corrosion damage was observed by SEM in order to compare it with scar morphology.

3.15.3   Results and Discussion

3.15.3.1   Friction and Wear Behaviour

In order to study friction and wear, the experiments were carried out in a range of temperatures 70-110°C for oils A and B. It was observed that at temperatures below 70°C both oils presented a similar friction behaviour, while above 100°C the friction coefficient showed a quite different tendency. In oil A it increased while in B it decreased.

Fig. 3.15.2 shows the friction coefficient vs. temperature diagram for the A and B oils. The friction coefficient of A is mostly constant into the test temperature range (70–110°C). The observed variations fall in the error range. The oil B showed a more evident increment up to 100°C with a decreasing tendency at 110°C.

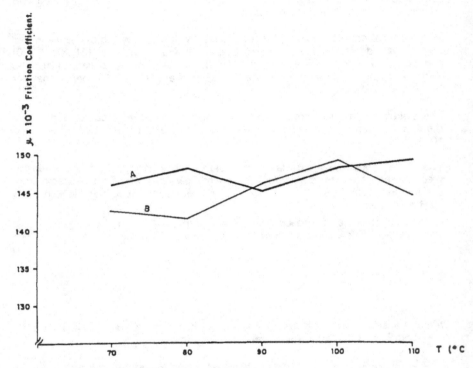

Figure 3.15.2: Friction coefficient variation vs. temp., oil A and B

Different patterns of the friction-time trace were recorded during test. Fig. 3.15.3a and 3.15.3b show quasiconstant and slow variation pattern respectively. The friction behaviour is related with continuous fluid film lubrication conditions. The Fig. 3.15.3c presents high and dense peaks at the initial stage related by R. Schumacher with the "running-in" phase (3). Less frequent transitory peaks appear also during the steady state (Fig. 3.15.3d). As A. Jahanmir states, under unidirectional sliding conditions, this type of friction coefficient variation corresponds to the metal-metal contact wear induced by the transient fluid film rupture and restoration (5). Further morphological studies on worn scars samples with similar behaviour seems to confirm this assumption, as well as in the case of the reciprocating sliding.

410

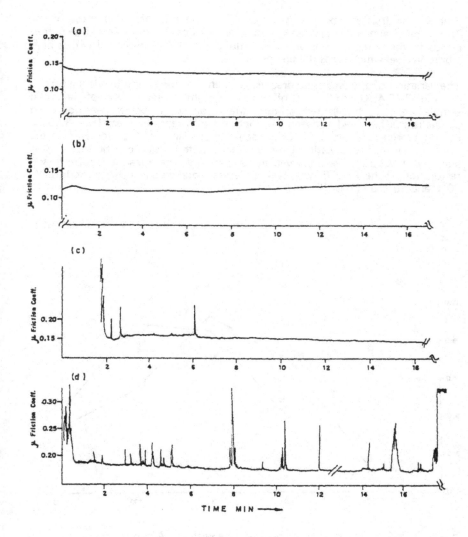

Figure 3.15.3: Friction coefficient vs. time
a) A-oil, 100°C; b) B-oil, 100°C
c) A-oil, 70°C; d) A-oil, 90°C

411

Finally, the friction coefficient increases to values higher than those of the "running-in" and transitory peaks, and then remains constant. This stage corresponds to the complete rupture of the fluid film and is associated with a high vibration level which stops the test device.

The variation of the wear scar area vs. temperature for A and B oils appears in the Fig. 3.15.4. Up to 90°C B oil induces a slightly inferior wear of the metal sample while, at higher temperature, the A oil shows better behaviour. No parallelism is observed between the friction coefficient and wear (Fig. 3.15.5), as it has been reported by R. Schumacher in similar tests (3). However, in the present case, higher deviation between steady state friction coefficient vs. time and wear-time trace was observed at temperature where transient rupture and restoration of the fluid film, as well as severe metal-metal wear occurred (oil A-80°C and 90°C and B-90°C).

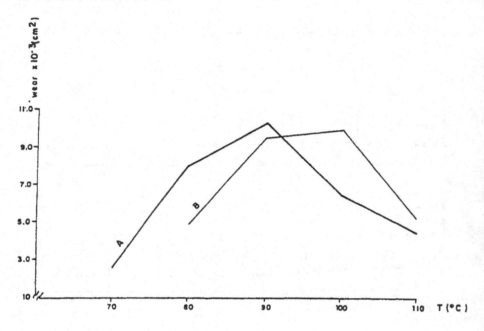

Figure 3.15.4: Wear variation vs. temperature with oils A and B

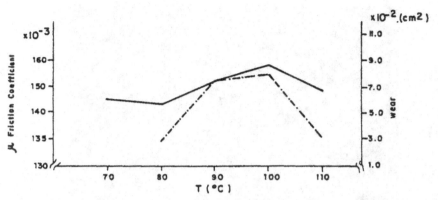

Figure 3.15.5a: Correlation between friction coefficient and wear for oil B

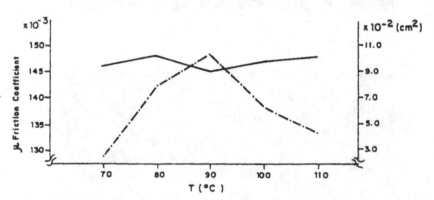

Figure 3.15.5b: Correlation between friction coefficient and wear for oil A

3.15.3.2   Wear Morphology vs. Friction Coefficient Pattern

Severe surface modifications were observed by Scanning Electron Microscopy
only in test samples where fluid film rupture and restoration were appreciated
by means of transitory friction coefficient peaker after a "running-in" phase,
i.e. A-80°C, B-90°C. These samples presented quite similar fretting wear morphologies, characterized by polished abrasive tracks alternating with rough areas
(Fig. 3.15.6). The latter, more extensive in the centre of the scar, are probably
surface fractures produced by adhesive and/or delamination wear mechanism
(5, 6, 7). At higher magnification, some morphological differences were appreciated, mostly in the rough zones. They were rather related to the oil quality
than to the test temperature and attributed to the oil corrosivity (Fig. 3.15.7).

413

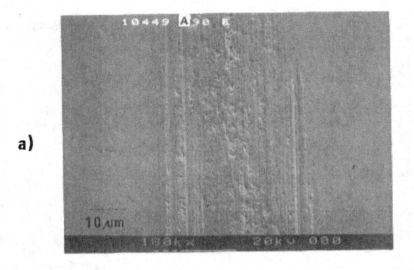

a)

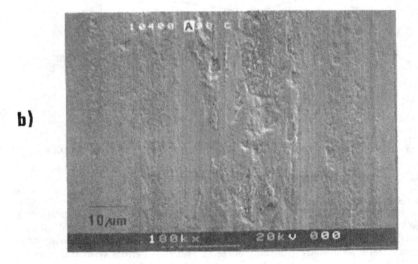

b)

Figure 3.15.6: Morphology of the wear scar sample A-90°C, a) scar end; b) scar center

414

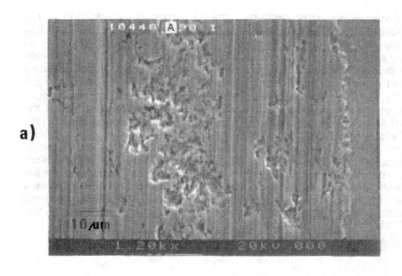

a)

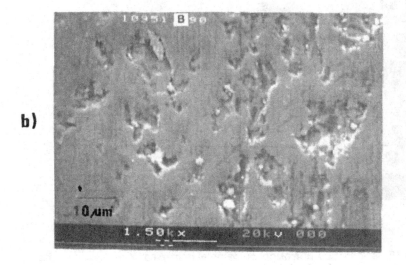

b)

Figure 3.15.7: Morphological details on worn scar rough area. a) sample A-90°C; b) sample B-90°C

Some experimental evidences supported the oil corrosion participation in the wear mechanism. In all the samples statically immersed in oils at 150°C, pitting corrosion rapidly developed and, based on the pits density, oil corrosivity was qualitatively evaluated higher for A oil and lower for B oil (Fig. 3.15.8). Additionally, corrosion fatigue cracking was observed in the surface on an oscillating cylinder, out of the contact area but relatively close to that, at approx. 0.1 μm distance (Fig. 3.15.9). The observed pits were probably produced by the corrsive action of the A oil spread in this area. Under the combined effect of the normal load and oscillating movement inducing cyclically variating stress in the material, fatigue crack was originated in the pits. It seemed that oil corrosivity favoured the corrosion fatigue, inducing delamination, and as well as the preferential attack of the fracture area in the wear scar, which is chemically more active (8). The effect is more evident in the A-oil samples, where pores have similar morphology to statical corrosion pits (Fig. 3.15.7b). In these samples, a low concentration of sulphur and phosphorus was found, suggesting that the additive reactive film has been mechanically removed by rubbing (Fig. 3.15.10b and 3.15.10c).

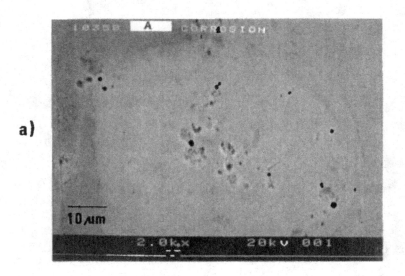

a)

Figure 3.15.8a: Pitting corrosion on sample immersed in oil during 6 hours, at 150°C; a) A-oil; b) B-oil

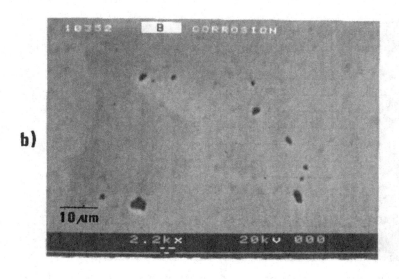

b)

Figure 3.15.8b: Pitting corrosion on sample immersed in oil during 6 hours, at 150°C; a) A-oil; b) B-oil

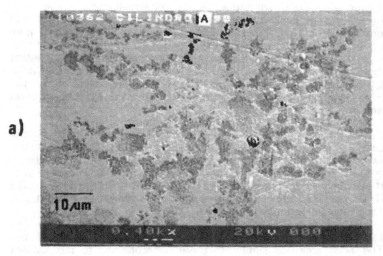

a)

Figure 3.15.9a: Fatigue corrosion on cylindrical sample B-90°C out of the contact area
a) Pits distribution

417

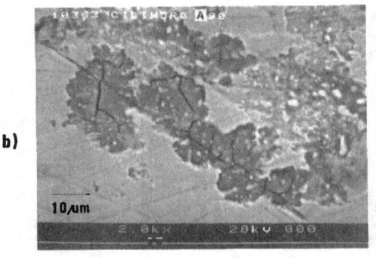

b)

10 μm

Figure 3.15.9b: Fatigue corrosion on cylindrical sample B-90°C out of the contact area.
b) Cracks initiated inside the pits.

One of the samples, A-70°C, showed a friction behaviour characterized by a high friction coefficient peak during the "running-in" phase only, followed by quasi-constant value (Fig. 3.15.3c). The results of this running test were not included in the diagrams of Figs. 3.15.2 and 3.15.4. The wear scar area has a close value to the former samples which presented metal-metal contact wear morphology. However, the worn surface was relatively smoother. Digital interferential optical profiler showed the presence of asperities of 0.1 μm height, distributed along the sliding direction (Fig. 3.15.11a). No concentration of sulphur and phosphorus was detected by AES.

A higher degree of smoothness was observed in those samples which did not show any peak in the diagram of friction coefficient vs. time. They also presented a higher concentration of sulphur, phosphorus, calcium and oxygen in all the worn scar. Most of the samples presented a quasi-flat worn surface with asperities of less than 0.5 μm. In Fig. 3.15.11b it is shown the transversal profile of the A-70°C wear scar with isolated angular peaks, denser on the left side and with a longitudinal continuous wave form.

In the B-110°C sample, a relatively deep and very polished wear scar was observed. The worn area was similar to those corresponding to test samples where metal-metal contact occured in Fig. 3.15.11c.

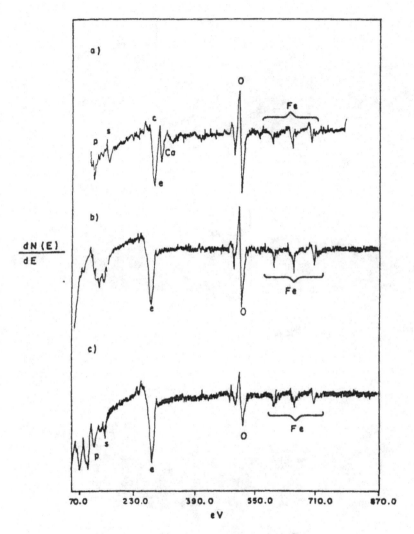

Figure 3.15.10:   AES spectra of wear scar center
a) A-oil, 70°C; b) A-oil, 90°C; c) B-oil, 90°C

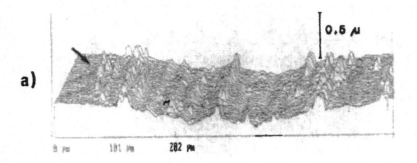

a)

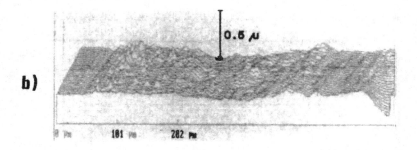

b)

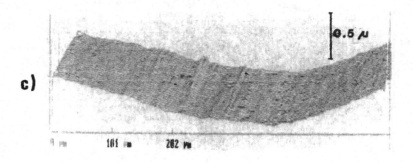

c)

Figure 3.15.11:   Interferential profiles
          a) Oil-A, 70°C with rupture of lubricant film during "running-in" phase and probable deposits, indicated by an arrow
          b) Oil-A, 70°C without rupture of lubricant film
          c) Oil-B, 100°C

The chemical composition of the borders and center of the worn scars was determined by AES depth profile analyses. For comparison purposes, sulphur and phosphorus content is reported in Table 3.15.2, corresponding to A-70°C and B-100°C wear scars.

In spite of it being very difficult to discern between inorganic deposits and metallic asperities, it could be suggested that isolated angular peaks could correspond to concentrated deposits, since such morphology has been observed even out of the wear scar area in originally flat surface. In Fig. 3.15.11a it is indicated by an arrow.

Table 3.15.2: AES and AES depth profiles analyses (5 min of sputtering time) along the width of different wear scars in the presence of oil A and B

Samples	Elements	Auger wt %			Auger Depth Profiles wt %		
		Scar Border1	Scar Border2	Scar Center	Scar Border 1	Scar Border 2	Scar Center
A−70°C	S	20	19	16	16	41	17
	P	19	14	19	31	21	20
	0	13	10	15	21	38	19
	Ca	11	0	9	31	3 %	24
B−100°C	S	19	15	16	33	39	18
	P	8	16	11	6	17	11
	0	19	29	18	48	34	35
	Ca	4	8	13	0	0	26

Since it has been found a non-uniform distribution of sulphur, phosphorus and other elements, it could be supposed that an amorphous smooth film has covered the contact area (9, 10). Barcroft et al. (11) proposed a mechanism of iron sulfide phosphide nucleation in the region where high temperatures were generated by local asperities contact, mostly in the "running-in" phase, followed by the extension of an amorphous variable composition thiophosphate layer in regions at lower temperatures. In our case, the presence of a continuous amorphous layer seemed more probable to occur in the smooth area (sample B-100°C and right side of the A-70°C sample), where a phosphorus/sulphur relation in the external surface was higher than in the sub-surface layer. This fact could suggest that primarily richer sulphurous compounds were formed.

However, a more complex oxidation reaction is supposed to occur in the early stage, as the oxygen content was higher in the corresponding sub-surface layers. In the isolated peaks area the S/P ratio was higher in the surface, which could be related with the sulphide-phosphide local nucleation mentioned above. Further studies will be followed in our group in order to elucidate the possible process of desposits formation.

3.15.3.3 Comparison of Worn Surface Morphology between SRV Test Specimens and Valve Lifters

The morphology of the top and side worn surface of valve lifters of taxi cab fleet test was also studied. Both surfaces were characterized by relatively uniform distributed fatigue corrosion pits (Fig. 3.15.12). The density of pits was higher in the case of the more corrosive A oil. On the surface of the valve lifters, areas between pits presented plowing tracks similar to that observed in the SRV test flat samples (Fig. 3.15.12b). This could be related to the higher sliding amplitude on this surface, which induces a major participation of abrasion in the wear mechanism (6). On the top surface, a uniform polish was observed in areas between pits (Fig. 3.15.12a). This morphology is similar to that reported by Jahnamir (12).

Based on the morphological features of the worn areas, a better correlation can be established between the wear mechanism of the cylinder flat configuration SRV test and the side surface behaviour. Further tests with lower sliding amplitude and/or ball flat SRV test, will be carried out in order to approach the valve lift top morphology.

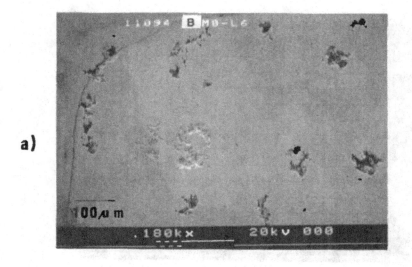

a)

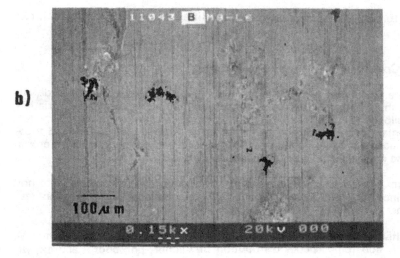

b)

Figure 3.15.12: Wear morphology of valve lifters from taxi cab fleet test, run with B-oil.
a) top surface;
b) side surface;

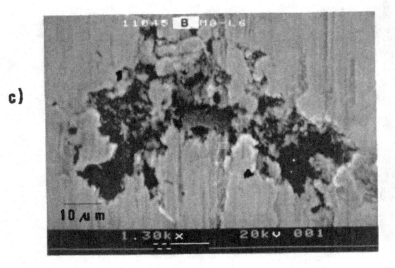

c)

Figure 3.15.12: Wear morphology of valve lifters from taxi cab fleet test, run
with B-oil
c) detail of fatigue corrosion pit from Fig. b)

3.15.4 Conclusions

During the performed tests, different wear mechanism were developed depending on the oil formulation and temperature. Friction coefficient variation
pattern allowed a quite reliable identification of cases, where transitory metal-
metal contact occurred during the "running-in", steady state phases and those
of continuous fluid film conditions. This relation was corroborated by morphological and chemical analysis of the worn surface.

Parallelism between wear scar areas and friction coefficient values were not
found, since different wear mechanisms were developed depending on test
conditions.

The additive reactive film was more evident in continuous fluid film test
samples. A non-homogeneous distribution of sulphur, phosphorus and oxygen
concentration in the sub-surface layer, suggesting a more oxidative reaction
at the beginning of the film formation. It appears that a higher relation between
phosphorus and sulphur on the surface could be related with the preferential
formation of an morphous smooth film. However, more studies are required on
the reactive film nucleation, growth mechanisms and the corresponding surface
micromorphology.

3.15.5 Acknowledgement

The authors are indebted to INTEVEP for the permission to publish these results. They thank Dr. Calatroni, (Simón Bolívar University), Dr. Gilbert Tribillon (Laboratoire d'Optique, Conté Université de Franche, Besancon, France) who performed the profiles measurements reported here and, Dr. Vladimir León for his collaboration to this work.

3.15.6 References

(1)   Dysan, A.: Elastohydrodynamic Lubrication and Wear of Cams and Finger Follower Automotive Valve Gear. Trib. Int., 13 (1980), 121–132.

(2)   Hamilton, G.M.: The Hydrodynamics of Cam Follower. Trib. Int., 13 (1980), 113–120.

(3)   Shumacher, R.: Surface Reaction Behaviour of Isogeometrical Phosphorus Compounds. 37th ASLE Annual Meeting, Cincinnati, Ohio, May 10-13, 1982. Preprint No. 82-am-5A-3, 7p.

(4)   Davis, L.E.; MacDonald, N.C.; Palmberg, P.E.; Riach, G.E.; Weber R.E.: Handbook of Auger Electron Spectroscopy, Physical Electronic Division (1978).

(5)   Johnanmir, S.: The Relationship of Tangential Stress to Wear Particle Formation Mechanisms Wear. 10 (1985), 233–252.

(6)   Bill, R.C.: Review of Factors that influence Fretting, Wear, Materials Evaluation Under Fretting Conditions. ASTM Special Technical Publication 780 Ann Arbor, 1982, Michig. 192, 165–181.

(7)   Kusner, D.: A New Machine for Studying Surface Damage due to Wear and Fretting. ASTM Special Technical Publication 780, Ann Arbor, Mich. 1982, 17–29.

(8)   Poon, C.; Hoeppner, D.W.: The Effect of Environment on the Mechanism of Freeting Fatigue, Wear, 52 (1979), 175–191.

(9)   Martín, J.M.; Mansot, J.L.; Berbezier, I.; Belin, M.; Balossier, G.: Microstructural Aspects of Lubricated Mild Wear with Zinc Dialkyldithiophosphate. Wear, 107 (1986), 355–366.

(10)   Martín, J.M.; Mansot, J.L.; Berbezier, I.; Dexpert, H.: The Nature and Origin of Wear Particles from Boundary Lubrication with Zinc Dialkyl-Dithiophosphate. Wear, 93 (1984), 117–126.

(11)   Barcroft, F.T.; Bird, R.J.; Hutton, J.F.; Park, D.: The Mechanism of Action of Zinc Tiophosphates as Extrem Pressure Agents, Wear. 77 (1982), 335–384.

(12)   Johnanmir, S.: Examination of Wear Mechanism in Automotive Camshafts, Wear. 108 (1986), 235–254.

4. Engine Oils and Their Evaluation/ Engine Lubrication Aspects

Engine Oils and Their Evaluation

Sludge Deposits in Gasoline Cars

Special Aspects of Engine Lubrication

4.1 The Changing Requirements of the 1980s–Automotive Oil Evaluation by Bench and Field Testing

A. Quilley
Adibis BP Chemicals (Additives) Ltd., Redhill, Great Britain

4.1.1 Introduction

In this paper it is my intention to evaluate the main changes that have occurred in gasoline engine oils during the 1980s. To do this I propose first to look at those changes and the reasons for them and, second, to compare in terms of engine performance the technology of today with that of the early 80s.

4.1.2 Engine Oil Development

If we were to go back to around 1980, then we would find the performance level of a good quality oil would be, typically, SF/DB 226.1. This oil would contain a number of components to make it function:

— Dispersant
— Detergent
— Antiwear Additive
— Base Oil

The dispersant would probably be of low molecular weight mono-succinimide type; the detergent would be based on phenate/sulphonate or salicylate, the antiwear additive would be a zinc dithiophosphate based largely on long chain primary alcohols; and the base oil would probably be conventional solvent refined.

As regards in-service performance there were, even in the early 80s, a number of problems reported. The most severe related to sludge performance and anti-wear (both of which were considered by some to be less severely tested by the SF specification than by SE). There were also changes made to the types of rubbers used for making seals, which caused difficulties because the new rubber types were more readily degraded by commonly used dispersant types. In order to overcome these problems there arose, as might be expected, certain individual customer requirements which were more severe than the official SF specification (including a double length VD and a VD with higher pass limits).

All in all, therefore, in the early 80s the specified level of oil quality was SF/DB 226.1 but there was a market requirement for oils whose performance was in excess of these specifications. It follows that to meet this need there was a lot of development work going on and some significant changes in technology. This was concerned mainly with dispersants and anti-wear additives.

Then, around the mid-80s coincident with, though not necessarliy due to, the movement towards leadfree fuel and the introduction of alternative fuels additives there arose another major problem, namely, black sludge. There may have been a number of ways of dealing with this problem, one of which may have been modified engine design but there was a need for an early solution. A CEC Committee was set up to investigate the problem. From this came the now well-known M.102E sludge test. This test introduced new demands on the oil additives and especially on the dispersant. Further developments were therefore needed to cope with it and still meet all the other requirements (especially seals).

With the growing number of problems in-service, it was evident that new specifications were required sooner rather than later. Thus, during the last couple of years the SG specification and DB 226.5 have been introduced to define an oil which is capable of dealing with all requirements of current engines. These specifications naturally include severe sludge and wear tests. To complicate matters further new seals tests have been introduced by, first VW and, more recently, CCMC. The combination of the new sludge, wear, and seals requirements has resulted in a need for some radical changes in technology and, in some cases, the specifications have proven very difficult to meet.

Other changes which have occurred during the 80s include a trend towards low viscosity oils together with a requirement for low volatility. This has led to the use of synthetic and extra high VI base stocks to an ever-increasing extent. This is a trend which shows no signs of abating and which has led to the widespread use of antioxidants and stabilising additives.

At one time there was also a trend towards low phosphorus oils which required the use of alternative anti-wear additives but this seems to have disappeared for the time being.

An SG/DB 226.5 oil, then, will contain nominally the same component types as SF/DB 226.1, but the actual chemistry is really quite different. The dispersant will typically be high molecular weight bis-succinimide or a cross-linked product in order to meet seals and sludge requirements. The zinc dithiophosphate will be more active in order to meet the severe wear requirements and will therefore contain a significant proportion of short chain secondary alcohol based product. Additionally, the base oils may well include some non-conventional stocks and the additive package, therefore, some antioxidant. In terms of the detergent there may be no significant changes in the type of products used but it is likely the proportion of phenate will be higher in order to enhance high temperature stability.

4.1.2.1 SG/DB 226.5 vs SF/DB 226.1

As indicated in the previous section, the oils going into gasoline engines have changed quite significantly over the last 10 years. The question is how do today's oils compare with those of the early eighties. Let us consider first the Sequence VD:

	SF	SF+	SG	Limit
Average Sludge	9.4	9.6	9.6	9.4
Average Varnish	7.4	7.3	7.2	6.7
Piston Varnish	8.2	7.2	6.7	6.6
Cam Wear, Average, 10^{-3} in	0.4	0.4	0.3	1.0
Maximum, 10^{-3} in	0.8	0.5	0.4	2.5

(In this comparison SF+ represents a level of performance in excess of SF, typical of the market requirement of the early to mid-eighties.)

It can be seen from these data that the Sequence VD has no ability to distinguish between the oils. The reason, of course, is that an SF quality performs well in the test and there is no facility to measure any significant improvement in performance. This demonstrates that there was indeed a requirement for a new test.

The IIID illustrates the same point but in a slightly different way:

	SF	SF+	SG	Limit
Viscosity increase %	299	27	33	375
Average Sludge	9.6	9.6	9.6	9.2
Piston Varnish	9.2	9.4	9.2	9.2
ORLF Varnish	5.7	7.5	7.2	4.8
Cam & lifter wear Average, 10^{-3} in	1.7	1.9	1.5	4.0
Maximum, 10^{-3} in	3.2	3.3	4.5	8.0

Here we see no discrimination between SF+ and SG but we do see a higher viscosity increase in the SF oil. The overall picture, however, still suggests the need for a new test.

The L-38 shows some interesting results:

	SF	SF+	SG	Limit
Bearing weight loss, mg	25	19	7	40

The reduction in bearing weight loss from SF to SF+ to SG reflects the changes that have occurred in dispersant chemistry. The higher molecular weight succinimides with much lower levels of nitrogen found in SG oils are much less aggressive towards the bearings than the low molecular monosuccinimides.

That gives a comparison in some of the tests existing in the early eighties. What about the tests brought in during the mid to late eighties? Let us consider the Daimler Benz M.102E sludge test:

	SF	SG	Reference Oil
Average Sludge 150 hrs	8.4	9.4	9.0
Average Sludge 225 hrs	7.7	9.0	7.9
Fuel	B3	B3	B3

(For simplicity and consistency I am referring to SF and SG quality oils rather than DB 226.1/226.5.)

In this test we can see a clear difference in performance between SF and SG, indicating that there is discrimination.

With respect to the VE, the following comparison also indicates a definite ability to distinguish sludge performance.

	SF	SG	Limit
Rocker Cover Sludge	4.6	9.4	7.0
Average Sludge	6.4	9.4	9.0
Piston Varnish	6.9	6.6	6.5
Average Varnish	5.9	6.2	5.0
Cam wear			
Average, mm	0.17	0.10	0.13
Maximum, mm	0.35	0.24	0.38

By comparison to the Sequence VD it is evident that this test has introduced a far more severe sludge assessment criterion. It is interesting to note, however, that the SF oil is only a borderline fail on wear.

Another interesting test to look at is the VW seals test:

	SF	SF+	SG	Limit
Elastomer 1				
Tensile strength	− 17	+ 3.8	+ 14	+ 15/− 10
Elongation at break	− 42	− 17	− 6.2	+ 10/− 15
Cracks at 120 %	Yes	No	No	No
Elastomer 2				
Tensile strength	− 33	+ 2.8	+ 5.3	± 20
Elongation at break	− 55	− 24	− 16	± 25
Cracks at 120 %	Yes	No	No	No
Elastomer 3				
Tensile strength	− 61	− 36	+ 4.8	± 20
Elongation at break	− 50	− 33	± 1.6	± 25
Cracks at 120 %	Yes	No	No	No

This illustrates the changes that have occurred in dispersant chemistry, with a trend towards high molecular weight succinimides with relatively low levels of nitrogen and, consequently, lower aggression towards the seals rubbers.

Thus the bench tests indicate a definite change from SF to SG, particularly in terms of sludge prevention. How is this reflected in field testing? The following is a test carried out on 1.6 litre Ford Escorts. The test duration was 58,000 km.

	SF	SF+	SG
Rocker Cover Sludge	9.4	9.5	9.9
Average Sludge	9.75	9.8	10
Piston Merit	21	21	32
Cam Wear, mm	0.038	0.033	0.016
Cam Pitting	Light	Light	None
Follower Wear, mm	0.144	0.098	0.010
Follower Pitting	Light	V.Light	None
Top Ring Weight Loss, g	0.23	0.34	0.057
Top Ring Gap inc., mm	0.29	0.17	0.051

433

Here we see, as we would hope, an all-round improvement in performance with the SG oil. There is a notable difference in sludge performance and piston merit, but perhaps most interesting is the very significant difference in wear performance which was not evident in the Sequence VE test.

Similar conclusions are drawn from the following field trials on Vauxhall Astras:

	SF	SF+	SG
Rocker Cover Sludge	9.9	10	10
Average Sludge	9.9	10	10
Piston Merit	19	23	29
Cam Wear, mm	0.012	0.005	0.004
Cam Pitting	None	None	None
Follower Weight Loss, g	0.0035	0.0037	0.0017
Follower Pitting	None	None	None
Top Ring Weight Loss, g	0.44	0.42	0.08
Top Ring Gap inc., mm	0.78	0.65	0.20

Here, however, the differences in performance between SF and SG are much less marked.

Further field trials with Peugeot 205 and Nissan Sunny cars demonstrate the all-round performance of SG oils:

	Peugeot 205 SG	Nissan Sunny SG
Rocker Cover Sludge	10	10
Average Sludge	10	10
Piston Merit	33	42
Cam Wear, mm	0.003	0.002
Cam Pitting	Trace	None
Follower Wear	0.000	0.000
Follower Pitting	None	None
Top Ring Weight Loss, g	0.02	0.03
Top Ring Gap inc., mm	0.00	0.17

4.1.3 Conclusions

There have been very many changes in the performance demands on gasoline engine oils during the 1980s. These have been reflected primarily in some very significant changes in dispersant types required to meet the new requirements. There have also been notable changes in the anti-wear additives used. It would appear from the data presented that the current engine test requirements do result in finished engine oils that meet the demands of the marketplace in terms of performance.

4.2 An Investigation into the Lubricating Engine Oil's Mechanical and Chemical Behaviour

Part I: Experimental Findings

S.L. Aly, M.O.A. Mokhtar, Z.S. Safar, A.M. Abdel-Magid, M.A. Radwan and M.S. Khader, Cairo University, Egypt

The present work presents experimental study of the mechanical wear in engine components and analysis of the basic chemical and physical properties of lubricating engine oils as new and used for prolonged intervals of time in running vehicles under field and laboratory conditions. Properties are measured and analysed as being influenced by engine type, life and running distance. Simulated wear tests are conducted to identify the effect of oil condition, whether new or used, on the wear rate of engine sliding pairs. With viscosity being the basic property which controls the hydrodynamic performance of engine bearings, special attention has been given to its variation with engine running time. Based on experimental data, some correlations could be deduced between oil viscosity and running time.

Results concluded that although the oil viscosity varies, but slightly, with running distance, the change in oil properties and the existance of contaminants may be more disastrous as they are affected by and in the same time affect engine performance, wear and life.

4.2.1 Introduction

Due to the rapid demands for developments in high speed efficient engines, much of the current researches have been devoted to the study of the fundamentals of friction and wear with a special consideration to lubrication technology in internal combustion engines. In this context engine lubricants and oils have gained a good deal of attention in an endeavour to promote their working life, to enhance their tribological properties and to ensure minimal frictional losses in engines by limiting surfaces direct interaction.

Oils viscosity has proven theoretically and experimentally to be the predominant factor in assessing the lubricant behaviour (1, 2). In practice, however, viscosity is only one of many other characteristics of the lubricating oils, and it is essential that the oil should maintain an unbroken film between moving surfaces and exhibit physical as well as chemical properties which ensure safe operation and storage with cleaning and anti-corroding action.

Under normal engine operating conditions, the behaviour of lubricants may follow almost the basic hydrodynamic theories (1 — 4) as in crankshaft bearings and piston/cylinder lubrication, whereas in cam shaft lubrication, elastohydrodynamic theories prevail (5).

During service the properties of oils change and may come to a state that the oil fails to keep engines performing efficiently and safely. A portion of the heat generated during combustion is eventually transferred to the oil which with prolonged time of operation, deteriorates the oil properties by firstly reducing the viscosity at high temperatures and secondly by increasing oxidation rate with a subsequent formation of organic compounds which may attack bearings' materials. Moreover, other reactions may occur between oil molecules that result in asphaltenic compounds which contributed in the formation of sludge (2). Water, fuel, soots, residues, metallic wear particles and environmental dirt particles may contaminate the oil as a result of combustion, engine wear and filtration. These contaminants are actually functions of the engine life, condition, use and atmospheric conditions. Under these situations the oil deteriorates and should be replaced by new oil.

In the present work, it has been decided to investigate experimentally some samples of new and used oils by analysing the possible changes on oil properties during operation in filed and laboratory atmosphere and to assess the corresponding engine wear behaviour. The test scheme comprises two sets of tests, namely, analysis of the basic physical and chemical properties of oils and wear tests.

4.2.2   Experimental Work and Results

To determine the properties of new and used oils, different tests are standardised (3). Physical tests are frequently used to characterize petroleum oils since the lubricant performance often depends upon or may be related to such properties, e. g. viscosity, density, pour point, gravity, flash and fire points, demulsibility, odor and colour.

Chemical tests include tests for carbon and other residues, oxidation, corrosion, acidity, oiliness, extreme pressure, ash and precipitation number. Add to these tests mechanical tests to study lubrication efficiency, wear or moving surfaces, frictional losses, engines efficiency and performance.

In the present scheme of work samples of used oils have been extracted from running vehicles and tested. Also planned tests were carried out to show possible changes in lubricant properties compared to its original ones. A simulated wear test is also included in the scheme.

4.2.2.1   Field Tests

These tests are mainly concerned with the investigation of the effect of running distance, size and model of passenger cars on the lubricant properties. Tests are confined to small passenger cars with petrol engines.

Table 4.2.1: Analysis results for new and used oils in an 1100 cc passenger car working in normal atmospheric conditions

Oil Condition	New	Used (in same car model and manufacturer)						
Engine Service Class.	SF/CC							
Car km Reading	20000+	12000*	17500*	25500*	36000*	77500*	122500**
Oil Change Period (km)		11000	5500	5000	3000	1500	2000
Analysis: Flash Point °C	198.9	162.8	148.9	140.6	185.0	148.9	151.7
Kinematic Viscosity cSt: at 40°C	160.5	133.3	138.9	140.6	144.5	134.5	124.6
at 100°C	20.2	15.73	17.20	18.87	18.03	17.72	14.75
Total Base No.	9.942	3.824	2.551	4.587	4.612	4.776	4.635
Strong Acid No.	Nil	Nil	Nil	Nil	Nil	Nil	Nil
Insoluble 0.3 micron (wt %)	Nil	0.919	0.963	0.961	1.29	0.601	0.630

* same car
** reconditioned
+ recommended by manufacturer (15 w/50 multigrade oil)

Table 4.2.2: Analysis results for new and used oils in different passenger car models and manufacturers working in normal atmospheric conditions

Oil Condition	New	Used					
Engine Service Class	SF/CC						
		Manufacturer (1)			Manufacturer (2)		
Engine Size (cc)		900	1100	1600	900	1100	1600
Car km Reading		10900	68000	53000	52000	84000	88000
Oil Change Period (km)	7500*	2000	1500	2000	1500	1500	1500
Analysis: Flash Point °C	215.6	173.9	160	168.3	154.4	165.6	160.0
Kinematic Viscosity cSt: at 40°C	152.1	145.2	132.4	146.2	108.6	129.9	143.9
at 100°C	14.2	17.82	15.89	17.11	13.67	13.98	16.96
Total Base No.	2.494	2.664	5.49	4.057	4.529	1.757	3.770
Strong acidity No.	Nil	Nil	Nil	Nil	Nil	Nil	Nil
Insoluble 0.3 micron (wt %)	Nil	0.781	0.999	0.909	0.732	0.862	0.962

* recommended by manufacturer (20 w/50 multigrade oil)

439

The results, as given in Table 4.2.1, have shown that the lubricant, within the recommended safe working period, is still retaining its properties with an acceptable reduction in the flash point and viscosity. Other properties have shown, however, insignificant variations. These results are also emphasized, as shown in Table 4.2.2, to show that the size of petrol engines used is of negligible effect on the deterioration of oil properties irrespective of car models or manufacturers.

Under controlled conditions in a laboratory test carried out on an 1100 cc petrol engine, the viscosity variation with equivalent running distance, Table 4.2.3, has been found to display a behaviour in which the viscosity drops firstly and then starts to increase as shown in Figure 4.2.1.

Table 4.2.3: Kinematic viscosity variation with running distance under laboratory conditions

Equivalent Travel Distance (km)	0	1000	2500	4200	5900	7500	8200
Kinematic viscosity (cSt)							
at 40°C	178.4	168.2	176.3	176.6	180.4	180.7	178.8
at 100°C	20.40	19.12	19.63	19.88	20.10	20.30	20.00

The variation of the kinematic viscosity (ν) with running distance (D) can fit into a polynomial in the form:

$$\nu = \nu_0 + \nu_1 D + \nu_2 D^2$$

where ν_0, ν_1, and ν_2 are constants depending on oil type.

It is expected that the decrease and subsequent increase in the oil viscosity is the result of interplay between the oil and its contaminants by foreign particles or oxidation products on one hand and the degradation of oil constituents on the other hand.

4.2.2.2 Wear Tests

Simulated wear tests are conducted on a modified crossed cylinders test apparatus (6). By rubbing a stationary cast iron piston ring against a hard steel rotating shaft under dry and lubricated situations, comparative results have been recorded and herein displayed. Tests under lubricated conditions are carried out using drop lubrication with either new or used oils. Different applied loads at various running speeds are used for test durations up to 15 seconds to avoid excessive wear under dry contacts.

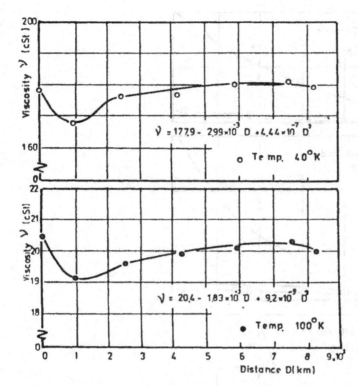

Figure 4.2.1: Variation of oil kinematic viscosity with running distance

Referring to attained results, Figure 4.2.2, it can be seen that the general expected feature of wear variation with load, speed, duration and contact condition could be fairly recognized from graphs. Any increase in the applied load and/or test duration results in a corresponding increase in wear volume. The use of lubricated contacts resulted in lower wear, approximately one tenth of the wear attainable under dry conditions. It is worth mentioning that in preliminary tests using floaded lubrication instead of drop feeding, within the limited test duration no appreciable wear was detected. Careful examination of attained results reveals that the use of new oils as lubricants renders less wear than using used oils.

With the increase in either the running speed or test duration, Figure 4.2.2, it is clear, as would be expected, that the increase in the travel distance affects the wear by increasing its worn volume.

441

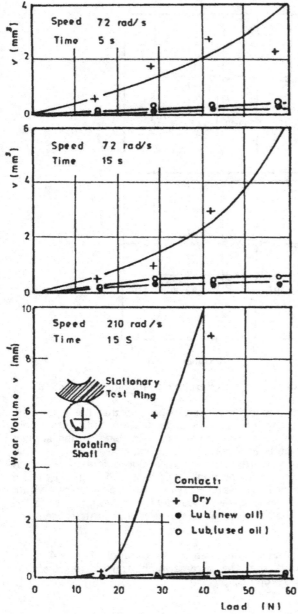

Figure 4.2.2: Effect of load, speed and contact condition on wear volume — comparative test

442

The role of a lubricant in any sliding contact is to limit metallic interaction by forming an oil film capable of sustaining the applied loads. Hence the frictional losses and wear should be minimized, less friction and lower wear rates are thus a character of a sliding pair lubricated with good quality oil. In the present test, although the contact situation is not representative of piston/cylinder lubrication, the results can be considered of some importance to cam lubrication and wear. However, accepting that these results are of comparative nature, the attained results would throw some light on the effect of quality of oils on engine components' wear; new oils render less wear than used oils.

The use of used oils as a lubricant indicated that under all test conditions the wear increases compared to new oils' test results, Figure 4.2.2. This can be attributed to both the deterioration of oil properties by continuous use and the presence of abrasive wear debris. Higher speeds have shown under lubricated contacts to yield slightly lower wear volumes. This can be explained in the light of hydrodynamic theories where higher speeds help in generating higher pressures, with a consequential lower wear.

The presence of contaminants, e. g. wear debris and dust particles, carried with the oil are surely of bad influence on the surfaces' wear. With the oil being flowing under pressure in the crank shaft bearings and splashed between piston and cylinder, the kinetic energy of the contaminants may erode sliding surfaces. Moreover, the hard wear and dust particles would work as abrasive media.

From the wear point of view, it can be seen that engine wear is of a complicated nature and depends not only on the physical and chemical properties which control the lubrication behaviour but also on the nature of contaminants.

4.2.3 Conclusion

Although the physical and chemical properties of oils are affected by prolonged use, it is expected that the mechanical wear tests would be of some importance to determine oil wear prevention quality.

Results emphasized that the presence of hard contaminating particles in the lubricating oils may be more dangerous and limit the use of oils than the slight variations in oil viscosity.

443

4.2.4 References

Part I

(1)   Pinkus, O.; Sternlicht, B.: Theory of Hydrodynamic Lubrication, New York, McGraw Hill, 1961.

(2)   Schilling, A.: Automobile Engine Lubrication — Two Parts, G.B.; Scientific Pub. (GB) Ltd., 1972.

(3)   Hahn, M.W.: Das zylinderische Gleitlager endlich breiter unter zeitlich Veränderlicher, Dr.-Ing. Thesis, T.H. Karlsruhe, 1957.

(4)   Fantino, B.; Godet, M.; Frene, J.: Dynamic Behaviours of an Elastic Connecting Rod, SAE Paper No. 830307, 1983.

(5)   Dowson, D.; Higginson, G.R.: Elastohydrodynamic Lubrication, Oxford, Pergamon, 1966.

(6)   Mokhtar, M.O.A.; Radwan, M.A.E.; Safar, Z.S.; Khadar, M.S.: A Study of Engine Wear as Influences by Lubricant Condition, Proc. 2nd Applied Mech. Engineering (AME) Conf., Military Tech. College, May 6-8, 1986, Cairo, Egypt.

4.2 An Investigation into the Lubricating Engine Oil's Mechanical and Chemical Behaviour

Part II: Theoretical Interpretation

S.L. Aly, M.O.A. Mokhtar, Z.S. Safar, A.M. Abdel-Magid, M.A. Radwan and M.S. Khader, Cairo University, Egypt

Based on previous experimental data, it could be seen that the changes in lubricating oil properties and accompanied wear of sliding surfaces during engine operation affect the performance characteristics of the engine. In the present work, the basic hydrodynamic lubrication theory has been applied to identify the possible variations in bearings lubrication behaviour as influenced by bearing wear and oil deterioration. Load capacity, side leakage, oil flow rate and coefficient of friction are given for different values of radial clearances (due to wear), new and used oils, full power and idling speeds and at different running temperatures. By way of comparison, the load carrying capacity has been computed for non-newtonian oils, a situation which is expected to describe polymer thickened oils more rationally.

Results revealed that the changes in clearance dimensions due to wear would be of more influence on the engine performance than the changes in viscosity; increased clearances lead to lower load capacity, greater side leakage and oil flow rate and higher frictional losses.

4.2.1 Introduction

Lubricating oils may behave in different mechanisms according to operating conditions and type of components to be lubricated. In general, the different lubrication regimes which exist in engines are: boundary, hydrodynamic and elastohydrodynamic lubrication (1).

Under boundary lubrication regime, tribological behaviour is determined by surface properties as well as lubricant properties other than viscosity, namely polarity and chemical reactivity (1, 2). This regime can occur in engine components when pressures get high and velocities are relatively low or when an oil film cannot be formed. This can possibly be seen during starting and stopping, during normal running at the ring/cylinder interface at top and bottom dead centers and between highly loaded parts such as valve stems and rocker arms. However, when the temperature of the surfaces increases, the thermal agitation energy becomes greater than the adhesion energy due to the polarity and the molecules become very mobile and disoriented and are thus unable to prevent metal to metal contact.

Hydrodynamic lubrication is a regime in which the shape and relative motion of sliding surfaces form an oil film having a hydrodynamic pressure sufficient enough to sustain applied loads without rupture (3, 4). In this case the oil viscosity is the predominant property which affects the generated hydrodynamic pressures. This regime is present whenever a wedge shaped film configuration can exist with relatively high relative speeds between surfaces forming the wedge, e. g. cylinder liner and piston rings, crankshaft bearings, con-rod bearings and all sliding bearings present. Theoretically, if the engine is kept running without stop and start and assuming a continuous sufficient oil supply, the moving surfaces will be fully separated by the oil film and frictional resistance becomes very low with almost no appreciable wear (5). Recent trends are declined towards using multigrade oils which exhibit non-newtonian viscosity behaviour due to polymer thickening additives (viscosity index improvers) (6). The influence of viscosity index improvers on the lubrication behaviour is actually twofold; firstly to nearly stabilize and enhance the viscosity of lubricant oil within working range of temperature; this, however, may come on the expense of the nature of oil behaviour to convert it from newtonian to non-newtonian fluid, and secondly to reduce wear depending on additive polarity of problems of volatility and stability are discarded; at equal viscosity polar products such as dispersant polymethacrylates create four times less wear than pure mineral oils whereas non-polar additives such as polyisobutan fail to provide base oils with any particular antiscuffing properties (6).

Elastohydrodynamic lubrication, the third lubrication regime present in engines, is an application of hydrodynamic theories in line or point contact (7). In which case the contact pressures and consequently the induced temperatures influence the lubricant viscosity and surfaces' elastic deformations which in turn governs the generated pressures and temperatures (7, 8). This regime is a characteristic feature of cams, gears and rolling elements lubrication.

In this paper, analysis will be confined to handle the effect of wear on the crankshaft bearings' hydrodynamic performance. This can be understood as a change in bearings radial clearance. Load capacity, flow rates and friction losses have been determined for the safe radial clearances limits using new or used oils (9). The effect of adopting oils of possibly non-newtonian behaviour is also investigated to show the changes in the load capacities.

4.2.2 Analysis

The effect of dimensional changes on lubrication behaviour of sliding contacts can be accounted for by applying the basic Reynolds equation for hydrodynamic lubrication (4). However, for non-newtonian fluids assuming a power low model of fluid behaviour (10), the relation between the shear stress τ and the rate of deformation rate du/dy is given by:

$$\tau = \mu_o \, (du/dy)^n$$

$$\tau = \mu \, (du/dy)$$

446

where:

$$\mu = \mu_o \, (du/dy)^{n-1}$$

For newtonian fluids, the flow behaviour index n will be equal to unity (n = 1) whereas for non-newtonian fluids n will be greater than one (n > 1) for dilatent fluids and less than one (n < 1) for pseudoplastic fluids.

The modified Reynolds equation to be applied in the general case of newtonian or non-newtonian fluids will be (11, 12):

$$\frac{\partial}{\partial x} \left(\frac{h^{n+2}}{\mu_o \, n} \, \frac{\partial p}{\partial x} \right) + \frac{\partial}{\partial z} \left(\frac{h^{n+2}}{\mu_o} \, \frac{\partial p}{\partial x} \right) = 6 \, U^n \left(\frac{dh}{dx} \right)$$

where:

h: film thickness
n: flow behaviour index
p: hydrodynamic pressure
u: sliding velocity
U: journal speed
x, y, z: coordinates in directions along and across the oil film and axially respectively
μ: viscosity
μ_o: consistancy constant ($\mu = \mu_o$ at n = 1)
τ: shear stress

This equation has been solved numerically using finite difference technique (12) for the pressure distribution and hence other performance characteristics could be computed numerically.

4.2.3 Application

The present analysis is applied to the bearings of a 4 cylinder European automobile petrol engine with the following specifications (2, 9):

No. of Cylinders:	4 (1160 cc)
Power:	55.5 BHP
Speed at full power:	6000 rpm
Speed at idling:	1500 rpm
Nominal Crank bearing diameter:	48 mm
Radial clearance:	0.035 to 0.070 mm
	(max. 0.100 mm)
Length to diameter ratio:	0.61

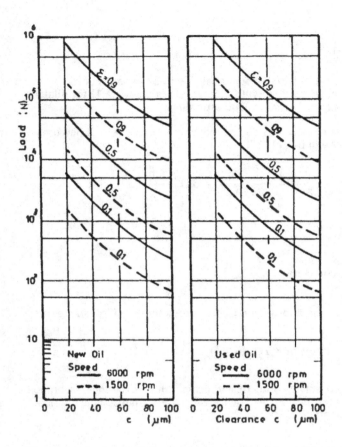

Figure 4.2.1: Variation of bearing load with radial clearance c for different eccentricity ratios ε

For other bearing diameters, the present results will hold qualitively true, while the values of the parameters variation should be corrected and recomputed for exact quantitative results.

The oils' specifications adopted in the present analysis are confined to tested oils (9), 15 w/50 multigrade oil for use 20000 km.

448

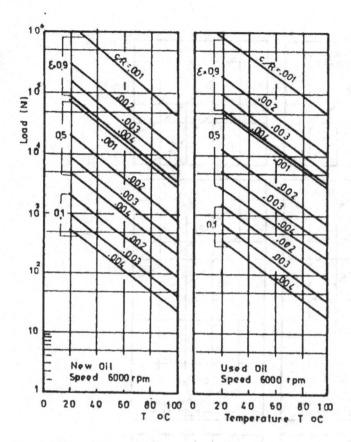

Figure 4.2.2: Effect of oil temperature on the bearing load for different clearance ratios c/R and different eccentricity ratios ε

4.2.4 Results and Discussion

The performance characteristics, in dimensional form, of the engine bearings are plotted in figures 4.2.1 – 4.2.5 as computed from attained data (9). The graphs give the variation of the expected load capacity values as a function of the eccentricity, ε, clearance ratio c/R, and journal speed for actual oils in use and practical ranges of radial clearances c. Flow rate, side leakage and friction coefficients are also plotted under same situations.

449

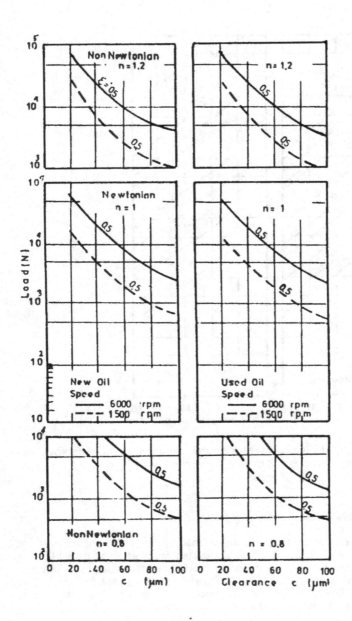

Figure 4.2.3: Variation of the bearing load with radial clearance c assuming newtonian (n = 1) and non-newtonian (n > / < 1) oil

450

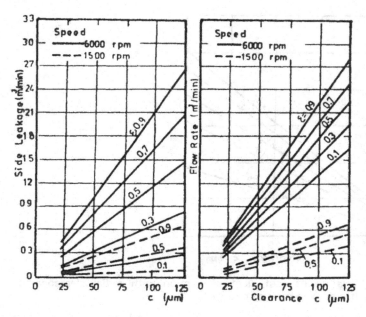

Figure 4.2.4: Variation of oil side leakage Q_s and circumferential flow rate Q with radial clearance c

Figures 4.2.1 and 4.2.2 have shown that the lubricant oil is slightly affected by use through its recommended running life and within the range of operating temperature. Figure 4.2.1 gives the values of the load capacity as a direct function of the radial clearance. As general recommendation (2), the oil film thickness should not go below 0.0025 mm or beyond 0.0042 mm for main bearings and 0.002 mm to 0.004 mm for big end bearings. Hence, for the present engine bearings computations, the range of radial clearances is taken practically to be 0.025 to 0.070 mm and it is further recommended that the clearance should not exceed 0.100 mm even after engine bearings' wear for safe running without the probability of oil film failure. Add to this that the choice of the values of the clearances is based on the concept of maximum load carrying capacity and minimum power loss with minimum side leakage assuming a steady state hydrodynamic solution (4, 12). However, in actual bearings subjected to dynamic loadings, the effect of squeeze action on bearings' performance is expected to be predominant (13). In this context strict limitations should be put on bearings' wear as excessive increase in radial clearance, as graphs indicate, would deteriorate the hydrodynamic performance; this is evident by the drop in load capacity under full power or idling speeds using either new or used oils.

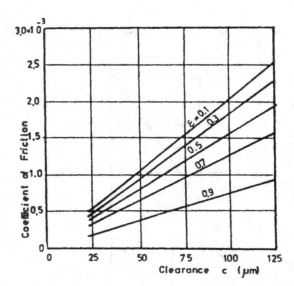

Figure 4.2.5: Variation of coefficient of friction with radial clearance for different eccentricity ratios ϵ

Due to the change in oil viscosity with temperatures during engine running, Figure 4.2.2 shows that, even for the data of multigrade oils, the load capacity may vary markedly.

Lubricants, due to polymeric additives, are expected to behave non-newtonian as pseudoplastic fluids (n = 1) when new. However, by continuous use the flow behaviour index n may vary. As there is little information and experimental data on the actual behaviour of lubricants, it has been herein decided to plot comparative data to show the effect of flow behaviour index variation on the load capacity values. Figure 4.2.3 gives the possible effects of mode of fluid flow on the load capacity. It is clear that, although the use of polymeric additives enhances the viscosity variation with temperatures, the use of non-newtonian fluids with n < 1 influences the bearing capacity by reducing it. As there is no experimental evidence on the way the viscosity may vary with frequent use, depending on type of additives, base oil, condition of application and possible contaminants, the graphs in Figure 4.2.3 show that dilatent fluids may be more superior than pseudoplastic ones (12).

Figures 4.2.4 and 4.2.5 reveal that the radial clearance between bearing surfaces has a direct impact on oil flow rate, side leakage and frictional resistance. The increase in radial clearance due to possible engine bearings' wear would require higher rates of oil delivered to bearings to ensure sufficient oil film formation without starvation and to take account of the side leakage. The coefficient of friction displays an increasing trend with the increase in radial

452

clearance. This situation affects the bearing performance by increasing the power losses which render higher temperature rise with a consequential variation in oil behaviour.

4.2.5 Conclusion

It can be concluded that restrict limitations on oils and hence on engine components wear rate should be put forward to assure efficient and longer engine life. The proper assignment of lubricant properties and flow behaviour (whether newtonian or non-newtonian) and its corresponding working life would not only safeguard the engine aginst excessive wear but would also guarantee high performance with minimal power losses.

The interaction between lubricant properties and engine wear can be summed up as follows: excessive engine wear due to improper oil properties or prolonged use of oils would lead to greater bearings' clearances, which in turn lead to reduced load capacity, higher frictional losses and greater oil flow rates.

4.2.6 References

Part II

(1)   Braithwaite, E.R.: Lubrication and Lubricants, Amsterdam, Elsevier Pub., 1967.
(2)   Schilling, A.: Automobile Engine Lubrication. G.B., Scientific Pub. (GB) Ltd., 1972.
(3)   Pinkus, O.; Sternlicht, B.: Theory of Hydrodynamic Lubrication, New York, McGraw-Hill, 1961.
(4)   Cameron, A.: The Principles of Lubrication, London, Longmanns Green Ltd., 1966.
(5)   Mokhtar, M.O.A.; Howarth, R.B.; Davies, P.B.: Wear Characteristics of Plain Hydrodynamic Bearings During Repeated Starting and Stopping, ASLE Trans., 20, 1977, 191–194.
(6)   Lubricating Oil and Additives-Lubrizon Tech Presentation, Report C-7528, The Lubrizon Corp., April 1983.
(7)   Dowson, D.; Higgenson, G.R.: Elastohydrodynamic Lubrication, Oxford, Pergamon, 1966.
(8)   Cheng, H.S.; Sternlich, S.: A Numerical Solution for the Pressure, Temperature and Film Thickness Between Two Infinitely Long Lubricated Rolling and Sliding Cylinders under Heavy Loads, Trans. ASME, Journal of Basic Engineering, 87, 1965, 695–707.
(9)   Mokhtar, M.O.A. et.al.: Investigation into the Lubricating Engine Oils' Mechanical and Chemical Properties — I. Experimental Findings, a paper to be presented in the 7th Inter. Colloquium Tribology on Automotive Lubrication, Tech. Akademie Esslingen, January 16-18, 1990, Esslingen (FRG).
(10) Tanner, R.I.: Study of Anti-isothermal Short Journal Bearing with Non-Newtonian Lubricants, Trans ASMEJ. Applied Mechanics, 32, 1965, 781–787.

(11)   Dien, I.K.; Elrod, H.: A Generalized Steady State Reynolds Equation for Non-Newtonian Fluids — With Application to Journal Bearings, Trans. ASME, Journal of Lubrication Tech., 105, 1983, 385–390.

(12)   Abdel-Latif, L.A.; Safar, Z.S.; Mokhtar, M.O.A.: Behaviour of Non-Newtonian Lubricants in Rough Bearings Applications, Proc. 14th Leeds/Lyon Symposium on Tribology, INSA, Sept. 8-11, 1987, Lyon — France.

(13)   Shawki, G.S.A.; Mokhtar, M.O.A.: Computer Aided Study of Journal Bearing Performance under Cyclic Loads, Part I: Theory, ASME paper no. 71-Vibr-86; Part II: Application, ASME paper no. 71-Vibr-87.

4.3 Development of Superior Engine Oils for Diesel Locomotives in India

J.R. Nanda, G.K. Sharma, R.B. Koganti and P.K. Mukhopadhyay, Indian Oil Corporation Ltd., Faridabad, India
R.M. Sundaram, Ministry of Railways, Lucknow, India

Abstract

Performance of IC engines largely depends on their design, operation, maintenance and also on the quality of fuels and lubricants. In this paper the development of superior crankcase lubricants with a view to enhance the performance of rail road diesel engines operated by the Indian Railways has been presented. Starting from the '70s with LMOA Generation-II performance level, oil quality has moved up to Generation-IV level oil and subsequently to high TBN Generation-IV oil. In the literature, there are indications of superior performance multigrade oils particularly with respect to fuel economy. Such benefits have also been reported for rail road diesel engines.

A programme to draw specifications and develop multigrade lubricants was taken up jointly with the Indian Railways. Laboratory evaluation of different V.I. improvers in combination with appropriate DI package was carried out to develop formulations with desired and viscometric characteristics including shear stability, temporary viscosity loss (TVL) and high temperature high shear viscosity (HTHSV). Subsequently, candidate oils were evaluated for fuel economy in Petter AV-1 followed by test bed evaluation on a stationary locomotive engine with load box facility. Field trials are currently under progress for establishing the fuel economy, engine durability, lube oil life and other performance criteria. Fuel econommy data for multigrade oils as observed in laboratory and stationary loco engines have been presented.

4.3.1 Introduction

Large scale dieselisation of the Indian Railways commenced in the early sixties. At present Indian Railways operate a fleet of about 3000 diesel locomotives, the majority of which (approx. 87 %) are of ALCO design. With the advancing technological development in the country, the communication and transport system has also to move at a faster pace. Growth of traffic has been increasing over the years and Indian Railways has to keep pace with this.

Locomotive engines with higher loadings developing higher bmeps impose high mechanical and thermal stress on various engine components including the lubricant which therefore has to withstand severe operating requirements viz: (a) higher temperatures, (b) higher unit loading, (c) corrosive acids and increased level of insolubles on account of higher rate of fuel burning. This has necessitated upgradation of oil quality over the years like elsewhere in the world. Indian Railways which consume about 1.3 million tons of diesel per annum and about 30,000 kl of crankcase oil have been taking adequate measures to conserve petroleum products as the expenditure on fuel and lubricants forms a major part of their working/operating expenses. The objectives of conserving petroleum products are being tackled from all possible directions and some of the projects which Indian Railways are persuing are as follows (1).

— Optimising Engine design to improve combustion and thermal efficiency including adoption of well matched piston top profile and ring configuration, better engine-turbo match.

— Locomotive design incorporating aerodynamic profile to reduce drag

— Track modernization

— Driving/operating techniques using microprocessor controls

— Fuel and lubricant quality upgradations

In this paper, it is intended to cover the efforts undertaken jointly by Indian Railways and Indian Oil Corporation to upgrade the lubricant quality over the years from Generation-II level to Generation-IV plus High TBN level and to move towards a fuel efficient multigrade 20 W — 40 lubricant.

4.3.2 Lubricant Performance Levels

Major steps in railroad improvements have been categorised by Locomotive Maintenance Officers Association, USA (LMOA) in four groupings called "Generations". Table 4.3.1 gives this classification system.

Indian Railways, according to their fleet requirements have been conventionally using two types of lubricants — one for GM-EMD Locos and another for ALCO locos. GM locos require zinc-free oils due to use of silver bearings in the gudgeon pin area, whereas for ALCO locomotives this restriction on additive chemistry does not exist. ALCO locomotives have two popular models denoted as WDM2 and YDM4, which employ 16 cylinder 251B diesel engine and 6 cylinder 251D diesel engine respectively. The engine characteristics for these two models are given in Table 4.3.2.

Table 4.3.1: Railroad oil classification

LMDA* Designation	Group (TBN levels by D-2896)	Service Requirements	Formulation Characteristics	Introduction Year
Generation II	A (7 TBN)	Extended oil filter life.	Dispersancy first added.	1964
Generation II	B (7 TBN)	Extended ring sticking protection in 4 cycle engines.	Improved detergency for upper ring belt control deposits	1968
Generation III	C-1 (10 TBN)	Improved base reserved to Group B for 2 cylce engines.	Higher alkalinity included with high detergency package.	1968
Generation III	C-2 (10 TBN)	Improved alkalinity of Generation III and improved insoluble control of Generation II.	Higher alkalinity and dispersancy.	1975
Generation IV	D (13 TBN)	Added protection under adverse operating condition.	Much higher alkalinity with improved dispersancy and detergency.	1976

Generation I: Pre — 1964
* LMDA: Locomotive Maintenance Officer's Association, U.S.A.

Table 4.3.2:   Details of popular types of ALCO engines in Indian Railways

Performance Characteristics	WDM2 (16-251B)	YDM4 (6-251D)
Cycle	4-Stroke	4-Stroke
Aspiration	Turbo-super charged with charge aircooling	Turbo-super charged with charge aircooling
Bore, mm.	228	228
Stroke, mm.	267	267
Compression Ratio	12.5 : 1	12.5 : 1
Horse power (gross) BHP	2600	1350
Power per cylinder BHP	162.5	225
Speed: Full load, rpm. Idle (normal), rpm.	1000 400	1100 400
Piston speed, m/s	8.89	9.78
BMEP, kg/cm^2	13.57	17.02
Fuel consumption at full Horse Power, litre/min.	8.2	4.5

Following is the chronological order of different types of lubricants used in the past by Indian Railways.

Prior to 1964, Indian Railways were importing proprietary brands of lubricating oils as recommended by ALCO/GM. Under the Marketing cum Distributorship Agreement of Indian Oil with Mobil, Indian Railways switched over to the use of Delvac S-140 and Delvac-1140. After the expiry of this agreement in mid 1974, Indian Oil R & D Centre developed Servi RR 402 which was accepted by Indian Railways for use in 16-251B and 6-251D diesel engines after extensive field evaluation extending over a period of one year in 24 locos in 4 loco sheds. GM-EMD locos in Indian Railways were all these years using Esso's Diol RD-78 which was blended using special Naphtenic MVI base stock (imported). Servo RR 402 was considered to be an interim substitute and the need was felt to have a further superior oil to improve oil life especially of 251D diesel engines. To reduce the wear rate of engine components on account of high sulphur operation, a high TBN oil to combat the deleterious effects was considered to be the best solution to bring down the maintenance cost. Apart from this advantages in other areas, namely increased filter life, long oil life, reduced carbonization on power assembly, etc. could also be derived with the use of high TBN oil.

Table 4.3.3: Physico-chemical characteristics of railroad oils Servo RR 402, 405 and 407 (Typical Data)

Properties	Servo RR402	Servo RR405	Servo RR407
LMOA Generation	Gen. II	Gen. III	Gen. IV
K. Viscosity, cSt			
at 40° C	168.0	171.0	165.6
at 100° C	15.90	15.90	15.80
Viscosity Index	97	96	97
Pour Point, °C	− 15	− 15	− 15
Flash point, COC, °C.	260	260	260
TBN (D-2896) mg KOH/g	8.69	10.91	13.34
Sulphated Ash, % wt	1.10	1.25	1.55
Phosphorus, % wt	0.075	−	−
Zinc, % wt	0.078	−	−
Calcium, % wt	0.28	0.34	0.45
Nitrogen, % wt	0.028	0.060	0.075
Formulation:			
Base oil, % wt	94.05	90.50	86.50
Additive dosage, % vol.	5.95	9.50	13.50
Total:	100.00	100.00	100.00

As a result of joint endeavours, two products viz. Servo RR 405 and Servo RR 407 were developed by Indian Oil R & D. These products were developed using indigenous paraffinic base stock and zinc free modern railroad additive systems (imported) and could be categorised as Generation-III and Generation-

459

IV levels. The oils were approved by GM-EMD for field trials after successful evaluation in their inhouse tests and also met the specification/requirements stipulated by ALCO. Based on GM-EMD's approval and other technical back up, Indian Railways accepted Servo RR 405 and 407 for full scale field trials with a view to have common formulation for both ALCO and GM-EMD locos. Trials were undertaken on 56 locos fitted with GM-EMD 16-567D3 diesel engines, ALCO16-251B and ALCO 6-251D diesel engines for a period of one year at 7 locosheds spread over different zones of the country. Superior performance of these grades for various parameters could be clearly established over Servo RR 402 in this trials. Table 4.3.3 gives the typical physico-chemical characteristics of these oils — RR 402, 405 and 407 and Table 4.3.4 lists the percentage of improvements which could be established during the course of trials with the use of Generation-IV lubricant over Generation-II lubricant. The benefits include longer oil drain life, filter life, reduced wear, reduced oil/fuel ratio, improved piston deposit control, better alkalinity retention and reduced level of insolubles. The benefits observed by Indian Railways with the use of Generation-IV level oil were directionally similar to those reported elsewhere in the world with the use of superior quality lubricants.

Table 4.3.4: Benefits with the use of Servo RR 407 (Gen. IV oil) over Servo RR 402 (Gen. II oil) Field Data

Performance Characteristics	% Improvement	
	16-251B	6-251D
Liner wear per 100,000 km	21.9	− 7.5*
Top ring wear per 100,000 km	61.5	16.0
% Lube/Fuel ratio	19.2	− 12.9*
Average merit rating	4.2	4.9
Filter life	27.0	12.8
Oil life	more than double	

* Some locomotives have used different combinations of liners and also from different sources and this has affected the ultimate wear data. Increased liner wear also seems to be the probable cause of high oil consumption.

4.3.3 Development of an Indigenous Formulation

On the successful completion of these trials, Indian Railways started using Servo RR 405 and Servo RR 407 in various regions defined on the basis of sulphur levels in fuel and finally adopted Servo RR 407 as a common oil for the integrated operation of diesel locomotives. Although Indian Railways were fully satisfied with the performance of these oils, in an effort to indigenise additive system, hitherto imported, in the years 1981—82 Indian Oil Company started working on a 'component based formulation' namely Servo RR 409. The following two factors were recognised jointly by Indian Oil and Indian Railways before taking up the development of Servo RR 409.

a) For the integrated operation of railways in various zones with different sulphur level in fuel, higher TBN formulation to achieve high TBN stabilization would be developed. It was well recognised that TBN provides additional protection against corrosive wear. 16 TBN formulation was thus aimed at.

b) The majority of locomotives were of ALCO design, hence a zinc based formulation for which additive components are locally available will be worked out.

A zinc based formulation Servo RR 409 was thus developed and then tried out in 20 locos at 3 locosheds. Typical test data of Servo RR 409 are given in Table 4.3.5. Percentage improvement as obtained in field trials with the use of Servo RR 409 over Servo RR 407 is listed in Table 4.3.6.

Table 4.3.5: Data on monograde rail road oil Servo RR 409 based on indigenous component-approach developed by Indial Oil for railways

Characteristics	Typical Data
K. Viscosity, cSt @ 40° C	171.13
@ 100° C	15.87
Viscosity Index	95
TBN mg KOH/g (D-2896)	16.0
Pour Point, °C	− 18
Colour, ASTM	6.5
Flash Point, COC, °C	254
Sulphated Ash, % wt	2.05
Foam Test (Tendency/Stability)	
Sequence I	20/nil
Sequence II	nil/nil
Sequence III	20/nil
Panel Coker Test (300° C)	
Rating	2—0
Weight gain, mg.	1.04

Table 4.3.5: continued

Characteristics	Typical Data
CLR L-38 Bearing Weight Loss, mg.	23.9
Petter AV-B test overall Merit Rating	60.4
CAT 1G2 Test % TGF WTD	28 232.3
Oxidation Test (Modified IP-280), (150° C for 150 hours, Air flow rate 0.5 l/min.)	
% Change in K. Viscosity @ 40° C Weight change in copper foil, mg. Change in TAN value	+ 28.71 − 0.60 + 1.26
Dispersancy Test (in-house) 40° C viscosity increase, cSt.	
with 1 % carbon black with 2 % carbon black with 3 % carbon black	1.0 1.1 0.5

Table 4.3.6: Benefits with the use of Servo RR 409 (Generation IV plus oil) over Servo RR 407 (Generation IV oil) Field Trial Data

Performance Characteristics	% Improvement	
	16-251B	6-251D
Liner wear per 100,00 km	61.2*	16.75
Top ring wear per 100,000 km	− 2.7	29.9
% Lube/Fuel ratio	52.7*	− 24.3
Average Merit Rating	comparable	
Filter life	comparable	
Oil life	Comparable	Needs further monitoring.

* Control locomotives have shown abnormal liner wear and hence cannot be taken as representative for comparison. Probably this factor has influenced lube consumption also.

Table 4.3.7: Published fuel economy benefits of multigrade engine lubricants in medium speed locomotive engines

Company	Reference	Engine	Test Apparatus	Reference	Test	Fuel savings %	Conditions
				Lubricant Viscosity Grade (SAE)			
Chevron	4	EMD 16-645 E3B	Locomotive on Load Box	40	20W-40	1.1	MRDC
Chevron	9	GE 12-7FDL	Test Stand	40	20W-40	2–13	Low idle
Chevron	9	EMD-12-645 E3	Test Stand	40	20W-40	2–7	Low idle
Exxon	10	CRSD 157	Generator Stand	40	20W-40	0–2	Idle
General Motors	11	EMD8-645E3	Test Stand	40	20W-40	4–9	Idle
Gulf Canada Ltd.	12	EMD 16-645 E3	Locomotive on Load Box	40	20W-40	1.3	MRDC
Amoco	3	EMD2-567C	Engine Test Stand	40	20W-40	0.8	MRDC
Amoco	3	EMD 12-645 E3B	Engine Test Stand	40	20W-40	0.7–0.9	MRDC
British Rail	6	DMU's and Class 56	Operating Locomotive Fleet	30	15-W-40	0–5	6 months Normal Service
Indian Railways	1	ALCO 251B	Locomotive on Load Box	40	20W-40	5.9 3.9	Mainline Switcher

MRDC – Medium Road Duty Cycle (GM-EMD's)

Table 4.3.8: Requirements of a fuel efficient multigrade railroad oil for Indian Railways

Characteristics	Targets
SAE Grade	20W-40
Appearance	Clear
Colour ASTM	8 max.

K. Viscosity, cSt
   at 100° C
   at 40° C
	15.5–16.3
	Report

Viscosity Index	Report
Apparent viscosity, cP, max. at 10° C by CCS	4500
Apparent viscosity cP (Borderline pumping) at − 15° C by MRV, max.	30,000
TBN (D-2896), mg KOH/g, min.	13
Flash point, °C, COC, min.	225
Pour Point, °C, max.	− 21

Foaming Tendency/Stability
   Sequence I
   Sequence II
   Sequence III
	25/nil
	150/nil
	25/nil

Diesel Performance, min.	API 'CD' level
Shear Stability (Bosch rig)	Stay- in SAE 40 grade
100° C viscosity loss after 30 passes, cSt, min.	i.e. 12.5
High Temperature High Shear Viscosity, cP at 150° C, 10^6 sec^{-1}, min.	3.5
% Temporary viscosity loss at 150° C, 10^6 sec^{-1}	Report
Noack test, % Evaporation loss after 1 hour at 250° C, max.	15

Table 4.3.8: continued

Characteristics	Targets
Oxidation stability, 150 hours at 150° C, 0.5 lt air/min.	
a) % viscosity increase at 40° C. max.	10
b) weight change of copper foil, mg max.	10
c) increase in TAN, mg KOH/g, max.	2.5
Filter plugging test (in-house)	
time to filter oil at 17–23°C	10 % increase
at 0–2°C	max, over standard oil RR 409
Fuel economy test	
a) % basic reduction in Petter AV-1 test over RR 409, min.	2.5
b) Demonstration of fuel efficiency in stationary loco engine ALCO 251 B fitted with load box, min.	1 %

4.3.4   Development of Multigrade Fuel Efficient Railroad Oil

Following the successful development of wholly indigenous Servo RR 409, multigrading of this formulation was undertaken. Multigrade oils have been found to give significant fuel economy benefits in medium speed railroad engines all over the world. Although friction modifier (FM) additives have shown promising results on medium speed railroad engines with oils containing FM did not, however, show any benefits (3, 4). Some of the published data on fuel economy benefits with the use of multigrade engine lubricants in medium speed locomotive engines are listed in Table 4.3.7.

Keeping in view the large fuel consumption in the diesel locomotive of the Indian Railways costing about $ 3.5 billion per annum, development of multigrade fuel efficient railroad oil was undertaken. In joint consultation between Indian Oil Corporation and Indian Railways, the targets of proposed multigrade oil were evolved. Table 4.3.8 summarises these requirements. British Rail had pioneered the work on development of multigrade rail road oils (6, 7) and it was agreed to follow a similar approach. It was therefore decided to keep the viscosity at high operating temperature of sump at the same level as that of the monograde oil. Although lowering of viscosity is expected to give better fuel economy, yet it was feared at the outset that it may result in increased wear. British Rail had also adopted the same viscosity for multigrade as that of monograde oil at the operating temperature. Since in Indian Railways sump operating temperatures are about 90 — 95° C, a 20W-40 grade was

465

considered to be appropriate with $100°$ C kinematic viscosity being similar to monograde, i.e. 15.5 to 16.3 cSt. The same D.I. package as in Servo RR 409 was used and a number of V.I. improvers were screened. Table 4.3.9 presents the viscometric data of a number of formulations with different V.I. improvers. The shear stability data on various formulations and the fuel efficiency data in comparison with monograde oil determined in Petter AV-1 engines are covered in Table 4.3.10. In our studies all V.I. improvers gave TVL in the range of 10.0 — 13.5. PVL variations are somewhat wider (9.0 — 19.0). The data do not indicate any general correlation between fuel economy and TVL or PVL as claimed by some researchers (2), nevertheless, candidate oil A0-26 gave a fuel efficiency of 4.5 % while satisfying all other requirements. Subsequently, test bed trials to determine the fuel economy were undertaken by the Indian Railways in a stationary engine (ALCO 16-251B engine) in Golden Rock Workshop of Southern Railways. The observed fuel economy with the use of A0-26 was 5.9 % and 3.9 % for the main line and switcher cycle respectively over the monograde (1). In a subsequent test at the R & D Centre of Indian Railways (13), fuel economy of around 1 % was seen with some of the multigrades. Field trials are currently under progress with three multigrade oils in 29 locomotives fitted with 16-251B and 6-251D diesel engines in 2 locosheds. Except for one formulation which has shown increased viscosity, the other two are behaving satisfactorily. Detailed assessment of performance for different parameters including fuel savings and lube consumption shall be made on completion of trial. Regular adoption of multigrade railroad oil in Indian Railways is shortly expected once the logistics are worked out including techno-economical viabilities.

4.3.5 Future Activities

In future, the following will be investigated:

(i)   To evaluate and adopt indigenous V.I. improvers.

(ii)   To study the extent of improvement of fuel economy with the use of FM oils in Indian locomotives.

4.3.6 Acknowledgement

The authors wish to thank the management of Indian Oil Corporation Limited and Research, Design & Standards Organization (RDSO) of Indian Railways for publishing this paper.

Table 4.3.9: Viscometrics data on candidate multigrade oils using various V.I. improvers

| Formulations | Apparent Viscosity cP | | Kinematic Viscosity | |
	(CCS) @ −10° C	(MRV) @ −15°C	cSt @ 40°C	@ 100° C
AO-17	3100	14,905	128.91	15.85
AO-18	3400	17,892	136.51	15.80
AO-19	3600	19.272	137.78	15.81
AO-20	3200	13,769	137.40	15.80
AO-25	4200	18,084	138.56	15.85
AO-26	3400	15,455	131.87	15.80
AO-27*	4200	1,61,112	139.04	15.85
AO-28	3900	17,422	137.66	15.85
Limits for SAE 20W-40 Railroad Oil	4500 max.	30,000 max.	−	15.5—16.3
AO-17, 19, 20, 26	Styrene Isoprene Copolymer			
AO-19, 27, 28	Olefin Copolymer			
AO-25	Styrene Isoprene Copolymer + Styrene ester Copolymer			

* does not meet MRV test requirement.

Table 4.3.10: Shear stability and fuel economy data on candidate multigrade oils

Formulations	% Permanent VISCOSITY LOSS (Bosch-injector rig 30 cycle test)	% Temporary VISCOSITY LOSS (Tapered bearing simulator 150° C, 10^6 Sec^{-1} Test)	% Fuel Economy over RR 409 BASE LINE DATA (Petter AV-1 fuel economy test)
AO-17	9.72	11.69	2.09
AO-18	10.63	9.96	2.29
AO-19	11.32	10.80	3.75
AO-20	9.0	12.96	3.76
AO-25	13.0	12.42	2.13
AO-26	19.0	11.74	4.51
AO-28	11.6	13.50	2.86

4.3.7 Abbreviations

TBN	=	Total Base Number
CCS	=	Cold Cranking Simulator
MRV	=	Mini Rotary Viscometer
cP	=	Centipoise
cSt	=	Centistokes
V.I	=	Viscosity Index
V.I.I	=	Viscosity Index Improver
F.M	=	Friction Modifiers
HTHSV	=	High Temperature High Shear Viscosity
PVL	=	Permanent Viscosity Loss
TVL	=	Temporary Viscosity Loss
MRDC	=	Medium Road Duty Cycle

4.3.8 References

(1) Research Designs and Standards Organisation: Development of multigrade lubricant oil for diesel locomotive — Results of satisfactory engine tests conducted at Golden Rock Workshop of Southern Railways. RDSO investigation Report No. MP-295/86, December 1986.

(2) Baltersley, J.; Hillier, J.E.: The prediction of lubricant related fuel economy characteristic of gasoline engines by laboratory bench test. Proc. Fuel efficient engine oil for improving the economy of vehicles symposium, Ed. W.J. Bartz, June 1984.

(3) Stauffer, R.D.; Zahalka, T.L.: Fuel savings with multigrade engine oils in medium speed diesel engines. Lub. Engg. 40 (1984), 12, 744—751.

(4) Logan, M.R.; Parker, C.K.; Pallesen, L.C.: Improved fuel economy through lubricant technology in medium speed rail road diesel engines. Lub. Engg. 37th Annual meeting. May 10-13, 1982, Cincinnati, Ohio.

(5) Sharma, G.K.; Mukhopadhyay, P.K.: Development of fuel efficient engine oils. Proc. Indo-OAPEC Seminar on Hydrocarbon Industry. Feb. 1987, New Delhi.

(6) Morley, G.R.; Eland, J.E.; Dunn, K.: British Rail switch of multigrade. Ind. Lubrication and Tribology, July/August 1984, 124—130.

(7) Morley, G.R.; Eland, J.E.: Development of fuel efficient multigrade oils for B R traction uses. Report Ref. TR Lub. 2, May 1983.

(8) Nanda, J.R.; Kashyap, A.K.: Railroad Diesel Engine Oils, 3rd LAWPSP Symposium, 1982, India.

(9) Thomas, F.J.; Ahluwalia, J.S.; Shamah, E.: Medium speed diesel engine lubricants, their characgeristics and evaluation. ASME paper no. 84-DGP — 17, 1984.

(10) Younghouse, E.C.: Lubricants with improved frictional properties for medium speed diesel engine applications. ASME paper No. 82-DGEP-6, 1982.

(11) Pratt, T.N.: Discussion of Reference 10.

(12) Hamilton, G.D.: Reduced locomotive fuel consumption using a multigrade friction modified engine oil. Annual meeting of ASLE, 1984.

(13) Research Designs and Standards Organisation: Evaluation of Fuel economy with the use of multigrade lubricating oil. Engine Development Directorate, RDSO Report No. TR/ED/88/4 August '88.

4.4 Very High Shear Rate, High Temperature Viscosity Using the Automated Tapered Bearing Simulator-Viscometer

T.W. Selby, Savant Inc., Midland, USA
T.J. Tolton, Dow Corporation, Freeland, USA

Abstract

While the automation of the Tapered Bearing Simulator Viscometer (TBS) has been dependent on several state-of-the-art developments, its ability to be used as an absolute viscometer with relatively high precision was a first requirement. In view of the ease of changing and measuring shear rates while in operation, the TBS was chosen to produce the data for engine bearing oil-film-thickness correlation through use with the empirical Cross Equation. Very good correlation is reported in the literature and these results confirm to the use of the TBS in both automated and non-automated modes. A new test method shows considerable reduction in analysis time and an equally marked improvement in precision. The paper presents the background of the instrument, the steps of its automation, and its application to trenchant problems and new opportunities in the area of very high shear viscometry.

4.4.1 Introduction

The Tapered Bearing Simulator Viscometer (TBS), shown in Figure 4.4.1, has been used commercially since the early 1980s (Ref. 1, 2). During the intervening years, several changes have been made reflecting developments in the art of temperature control, and its benefit in simplifying and automating the instrument. Much of this work has been done by the close cooperation of investigators in the Tannas Co. and in Savant, Inc. with whom both authors have been associated.

This paper presents further information on the development of automation for the TBS Viscometer as well as recent information on the original reason for its development — correlation between the TBS and the engine bearing. However, for full understanding, the paper first presents some of the background factors leading to the development of TBS automation.

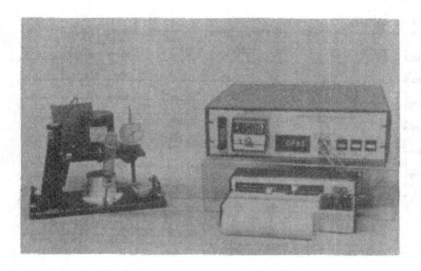

Figure 4.4.1

4.4.1.1 Background

A number of technical papers have documented the development of the TBS Viscometer (Ref. 1 — 8). Essentially, the instrument was designed with a geometry simulating that of the automotive journal bearing since this was one of the important potential applications for information from the instrument. The normally concentric cylinder arrangement for rotational viscometers was modified to have a slight taper along the lines of the Kingsbury Tapered Plug Viscometer (Ref. 9) and particularly, the work of Pike, et. al. (Ref. 10).

In the development of the TBS Viscometer, a number of design requirements were set, the most important of which were to reduce extraneous friction to a minimum in order to increase the sensitivity of the viscous torque signal. The instrument has always performed well as a true viscometer by showing very linear torque/shear-rate calibration curves and coincident intercepts with Newtonian reference oils, as shown in Figure 4.4.2.

471

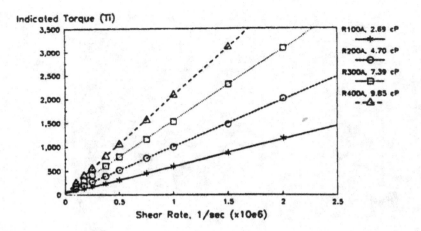

Figure 4.4.2: Newtonian Fluid Performance

4.4.1.2 Importance of the Tapered Coaxial Configuration

The tapered design was primarily chosen for the development of the TBS Viscometer to permit vertical displacement of the rotor and stator and the ability
to thereby vary rotor/stator clearances and, thus, shear rate. Adjustment of
height was so easy because of the relatively light weight of the platform holding the motor that it was quickly found possible to do this while the instrument was running. (As far as is known, at least among commercial viscometers,
the TBS is unique in this regard.)

As a consequence, relatively early in the use of the instrument, independent
studies by one of the authors and an associate showed (Ref. 3) that the reciprocal of torque, 1/t, varied linearly with the rotor height, H, as shown in Figure
4.4.3, as Newtonian theory would require. (Much earlier, unknown to Selby
and Piasechi (Ref. 3) at the time, Kingsbury (Ref. 9) had demonstrated the
same relationship with his Tapered Plug Viscometer which provided confirmation of the authors work.) This linear relationship of 1/t vs. H was found
to exist over a fairly broad range of rotor/stator displacement. Thus, this relationship indicated that not only was the TBS Viscometer effectively an absolute Viscometer (an instrument with which viscosities can be calculated from
the rotor/stator dimensions and rotor speed) but also that temperature effects
were demonstrably negligible over a rotor/stator gap ranging up to about 8
microns.

From these findings, it was possible to determine the operating shear rate
quickly and experimentally on an absolute basis. That is, the theoretical contact height (TCH) of the rotor and stator could be determined where 1/t became

472

zero. From 1) the TCH, 2) the actual position of the rotor in the stator, and 3) knowledge of the rotor taper, the operating shear rate could be calculated. The unique capacity of the TBS to determine operating shear rates "on the run" was an important factor in simplifying the calibration of the instrument and automating it, as will be shown.

Indicated Rotor Depth, mm

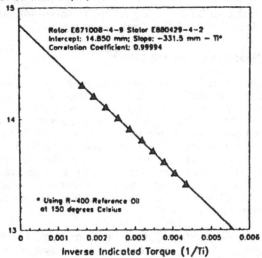

Figure 4.4.3: Application of Absolute Method to Determine Rotor/Stator Gap Relationship

Table 4.4.1

(Shear Rate) $\dot{\gamma}$	(Error) $2\dfrac{T_f}{T_c}$	(Shear Rate) $\dot{\gamma}$	(Error) $2\dfrac{T_f}{T_c}$
1 000 s^{-1}	23.9 %	100 000 s^{-1}	2.99 %
2 000	21.6	200 000	1.75
5 000	17.0	500 000	0.83
10 000	12.8	1 000 000	0.47
20 000	8.9	2 000 000	0.26
50 000	4.9		

The presence of two flats on the rotor raised a question about the assumption of absolute viscometry from the mechanical data (Ref. 11). However, a theoretical analysis by DuParquet (Ref. 12) indicated that the flats would have negligible effect at shear rates above 500,000 sec^{-1} where the error would be less than the repeatability of the instrument (see Table 4.4.1). As previously noted, the linearity of 1/t vs. H in Figure 4.4.3 extends at least to a gap of 8 microns with a Correlation Coefficient of 0.9999+. In the TBS Viscometer, an operating gap of eight microns at 3600 rpm is equivalent to about 440,000 sec^{-1} which experimentally tends to confirm DuParquet's theoretical work.

4.4.1.3 Thermoregulator and Heater Development Effects

It is perhaps stating the obvious to note that a high level of temperature control is a necessity for the practice of viscometry, particularly in high shear viscometry. The higher the shear rate, the more care which must be taken by design of the viscometer to control the effects of heat generated by viscous friction. In the case of the tapered geometry, higher shear rates are obtained by narrower gaps rather than by higher speeds. For example, the TBS Viscometer works at a gap of only 3.5 microns to generate a shear rate of 1,000,000 sec^{-1} at 3600 rpm. The thinness of the sheared film thus offers little opportunity for heat retention by the fluid and distortion of the reasonably linear shear gradient across the gap is avoided. (Similarly, heat transfer to the oil film from the stator heater is essentially immediate.)

Much thought and effort has been expended to design the optimum thermoregulation for the TBS Viscometer. This effort has been encouraged by the rapid evolution of thermoregulators from simple on/off switches, to proportional bandwidths, to automatic reset, to derivative controls during the last few years and progress is continuing. This evolution has had a major impact on TBS development. Each new advance in thermoregulation has been incorporated as available and it must be emphasized that the present level of simplification and automation of the TBS is significantly dependent on the aforementioned advances in thermoregulator development, as will be evident.

The heating source has similarly gone through three stages of modification as technology progressed. At present, using modern high capacity heating membranes, heat is applied to the stator rapidly and uniformly. However, the thin membrane also permits excess heat to be quickly "dumped" through the membrane as well.

4.4.1.4 Continuous or Long-Duration Operation of the TBS

One of the field observations made earlier in the development of the TBS was that over a period of days or weeks, the indicated contact height determined by the 1/t vs. H relationship increased slowly (the gap became smaller). While the slow change in indicated contact height did not affect the gathering of data (since the TBS is calibrated daily), the effect was puzzling.

474

At the outset, the effect was variously attributed to slow expansion of the wire-wound, flexible shaft coupling the rotor to the motor, deposits forming in the gap, changes in the housing holding the stator, or some combination. To eliminate the firstmentioned possibility and to decrease the rate of heat transfer up the relatively thick, wire-wound, flexible shaft, a thin, single-wire, flexible shaft was developed, which, at times, seemed to correct some of the phenomenon. However, it became obvious that there was a more important factor to be considered.

Ultimately, it was found that deposits on the stator wall facing the rotor were the culprit. These deposits apparently formed slowly from decomposition of the base oil and/or additives at the high temperature of viscometric analysis. Certain strong solvents were found to be capable of removing the deposits, after which the TCH would drop back to original value.

As a consequence of this experience and in anticipation of the long-duration operation of the TBS Viscometer when automated, special fluids were chosen for reference oils and a so-called "idling fluid" was developed by the Mobil Oil Company for specific use in the TBS. This idling oil is recommended for use at any time when the instrument is waiting for further work. Tests have shown that the fluid will withstand weeks of exposure at 150° C and full rotor speed with insignificant wall effects or operating problems with the instrument when again used for viscosity determination. Most importantly, the TCH remained reasonably constant. Simultaneously with the development of the idling fluid, protective circuitry was developed for the TBS viscometer to shut down the unit in the case of overheating or power outages (the latter is important since the TBS Viscometer should not be started up with cold fluid in the gap set for operation at $1,000,000 \text{ sec}^{-1}$).

4.4.2 Standardization of the TBS Viscometer

4.4.2.1 ASTM D4863-87 — Relative Rotor Position Method

The initial laboratory utilization of the TBS Viscometer led to the formation of a Rotational Viscometer Task Force under the leadership of Robert B. Rhodes (Ref. 8) within the appropriate group in the ASTM, namely Committee D2, Subcommittee 7, Section B. Reports on the activities of the Section and the Task Force have been recently published (Ref. 2,8). Essentially, this first method (Ref. 6) employed a relative technique of comparing the viscosities of a Newtonian and a non-Newtonian fluid which at $1,000,000 \text{ sec}^{-1}$ had identical viscosities. (However, the absolute technique possible with the TBS was used to establish the viscosity of the non-Newtonian oil at 10^6 sec^{-1}.) The roundrobin study gave a repeatability of 3.1 % and a reproducibility of 3.9 % at the 95 % confidence level.

The method, unfortunately, required plotting torque, t, versus rotor height, H, curves for the Newtonian and non-Newtonian fluids and interpreting their point of interception. Such interpretation could be difficult as shown in Figure 4.4.4. Coupled with the limitations of technology at the time (reflected by a manual-reset thermoregulator and a resistance-wired, silicon-rubber pad heater), the method was relatively slow and laborious. After the time required for calibration, relatively few samples (8 to 12) could be run in a day since sample-to-sample turn-around was a minimum of about 1/2 hour.

Indicated Rotor Depth, mm

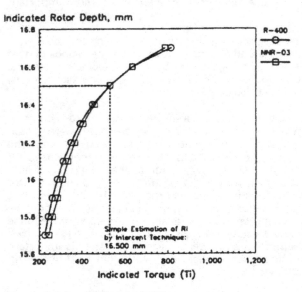

Figure 4.4.4: Ri Determination

Even this minimum turn-around time was possible, in fact, because the TBS Viscometer was designed so that no sample cleanup is necessary. That is, each sample "chases" the previous sample from the test cell while the rotor is spinning which helps to speed return to analysis. (However, the "chase" procedure was primarily chosen to avoid using solvents with their attendant problems of solvent contamination, odors, and the serious potential for either flash fires or toxic exposure conditions produced by some solvents at high temperatures.)

Fifty-ml injections of each fluid were standard but 30-ml, injected 10-ml at a time — with a few seconds pause between injections, was found sufficient to give complete interchange of oils in the shearing zone.

4.4.2.2 ASTM D4863-90 — Absolute Rotor Position Method

It was experimentally found that the reciprocal torque technique also produced an essentially straight line with the non-Newtonian calibration oil. The

476

significance of this finding was that the intercept of the Newtonian and non-Newtonian oil could be easily calculated as a unique point, rather than hand plotted and interpreted. This technical development, coupled with the previously discussed availability of

1. Advanced thermoregulators which, with automatic reset, could handle widely different viscosities,
2. High capacity heaters which don't block the heat flow from the stator,
3. Idling oil which could be left in the cell for weeks at a time without adverse effects, and
4. The ability of the TBS Viscometer to determine the TCH "on the run",

combined to make possible a faster, simpler method as well as to open the opportunity of automating the instrument.

The new version of the method requires bringing the instrument to operating temperature for about an hour to permit the thorough warming of the equipment. However, with the use of idling fluid and the safeguards built into the latest models of the instrument for unattended idling and automatic operation, a preferred alternative is to leave the instrument on all the time at operating temperature and with the rotor spinning at the desired gap so that the instrument is immediately ready for use at any time.

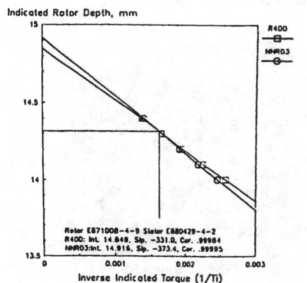

Indicated Rotor Depth, mm

Inverse Indicated Torque (1/Ti)

Figure 4.4.5: Reciprocal Torque Intercept Technique for Setting Gap

477

When the instrument is at temperature, the reciprocal torque vs. height technique is used to determine both the TCH of the Newtonian and non-Newtonian oils and their straight-line intercept as shown in Figure 4.4.5. The height of the rotor is adjusted to this intercept value and the instrument calibrated with four Newtonian reference oils. The method requires a modern thermoregulator. Using it, analysis time is now less than 10 minutes for sample turn-around. Significantly improved precision was shown when the ASTM round-robin conducted on the method in 1989 gave repeatability of 0.96 % and a reproducibility of 2.59 % at the 95 % confidence level. The method and pertinent information on the round-robin is presented in ASTM Research Report D02-1253 (Ref. 13).

4.4.3 Automation of the TBS Viscometer

4.4.3.1 First Stage — Automatic Sampling

With studies showing success in
1. thermoregulation,
2. close heat control without operator attendance,
3. simple determination of the rotor position,
4. obtaining availability of a stable idling fluid, and
5. the incorporation of safe, continuous operation,

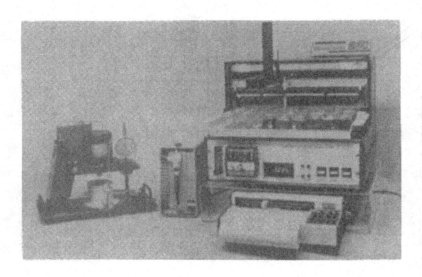

Figure 4.4.6

478

automation of the TBS Viscometer was now quite feasible. The first step was to set up a programmable sampler, as shown in Figure 4.4.6. This work has been covered in a past paper (Ref. 5). Essentially, the TBS Viscometer was calibrated as usual after which the sampler was activated to progressively analyse the unknown samples and reference fluids comprising the loaded sampler. In all, the sampler holds 70 tubes of oil. Use of a strip-chart recorder helped in the assimilation of data.

4.4.3.2 Second Stage — Semi-Automatic Calibration

One of the more technically difficult parts of the use of any viscometer is calibration and, since the TBS Viscometer is usually calibrated (despite its absolute nature), the instrument is no different. While the new (absolute) technique considerably simplified the calibration, there was still a need for even simpler approaches. Fortunately, the microcomputer is just right for such applications and one of the systems operating in Japan is pictured in Figure 4.4.7.

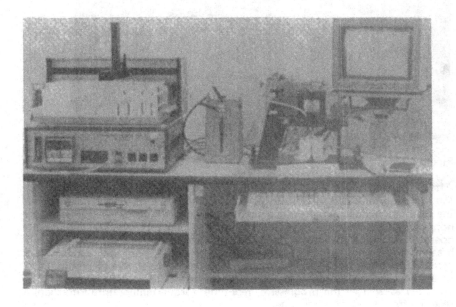

Figure 4.4.7

The second stage of automating the TBS Viscometer was to semi-automate the calibration. Using the computer keyboard, the operator is asked to answer certain questions on the computer keyboard regarding sample identification and location of calibration fluids on the sampling rack. The computer program then directs the automatic sampler to pick up certain reference oils for intercept analysis. When this is accomplished, the program then directs the operator to set the rotor at certain positions to generate necessary data to calculate the intercept of the Newtonian and non-Newtonian reference oils which plot is shown on the cathode ray tube (CRT) as pictured in Figure 4.4.8.

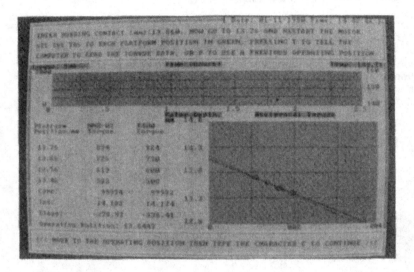

Figure 4.4.8

After obtaining the intercept value, the computer selects calibration oils from the sampling rack and automatically runs, computes, displays, and prints out the calibration data. At the same time, the computer checks the value of the non-Newtonian reference fluid and, if out of a preselected range, requires the operator to adjust the gap slightly and then reruns the calibration. After calibration, the computer displays the calibration curve, and the related values for the intercept, slope, and correlation coefficient, as shown in Figure 4.4.8. It then moves on to make the run, adding to the display the viscosities determined as well as the progressive, real-time, torque/temperature information received from the TBS Viscometer.

480

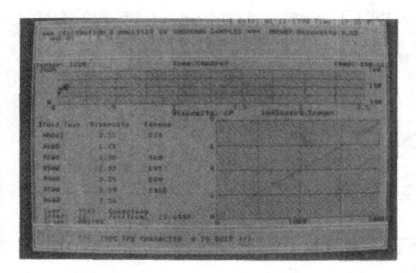

Figure 4.4.9

4.4.4 Applications of the TBS/Automated-TBS Viscometer

4.4.4.1 General

One of the obvious benefits of automation is that it permits the TBS Viscometer to be used with relatively untrained personnel who need to spend considerably less time in attendance. However, the un-automated TBS equipped with the proper thermoregulator can be used just as effectively as the automated version, albeit with considerably more attention. At the extraordinarily high shear rates possible with the instrument (above 2×10^6 sec^{-1}), automation may improve precision but this has not yet been studied.

4.4.4.2 Singular Temperature — Multiple Shear Rate Data

One area of work using the TBS Viscometer for a portion of the necessary data, has been the study of the relationship between temporary and permanent viscosity loss (TVL and PVL). The work requires determination of the dynamic (absolute) viscosities of both the fresh and shear-degraded oils and the results, when plotted, form so-called Viscosity Loss Trapezoids (VLT) as shown in Figure 4.4.10 (from Ref. 14). These VLTs tend to be unique for each combination of VI improver type, concentration, and MW distribution, as well as varying to a lesser degree with the viscosity and solvency of the base stock. The patterns shown in Figure 4.4.10 are for four experimental SAE 10W-40

481

grade oils at 150° C. It has been found that further data obtained on the same oils at 100° C adds considerable information to the understanding of the behavior of and distiction between VI improvers as applied to lubricating oils and hydraulic fluids.

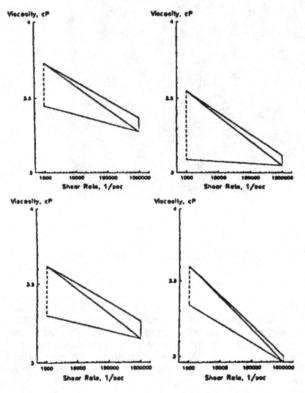

Figure 4.4.10: Engine Oil Viscosity Loss Trapezoids

The VLT approach was extended (Ref. 7) using multiple shear rates on a different set of oils with the results shown in Figures 4.4.11, 4.4.12, and 4.4.13 for a single-grade SAE 40, a somewhat shear-labile SAE 10W-40, and a shear-stable SAE 10W-40, respectively — the latter reflecting European blending and additive practices. (It should be remarked that this work was done at 100° C using a Model 600 TBS Viscometer coupled with a Tannas liquid bath capable of handling the torque required and heat produced with the considerably higher viscosities encountered.)

As expected, the straight-grade SAE 40 shows a horizontally collapsed trapezoid (reflecting essentially no PVL or TVL). In unsurprising contrast to the latter oil, there are marked differences between the two SAE 10W-40s in regard to the *low-shear* PVL values. However, it is interesting that there seems to be relatively little difference between the two oils in regard to the *initial low shear and final high shear* viscosity values (taken at 2×10^6 sec^{-1}). This suggests that low shear measures of permanent viscosity loss may be of little value in comparison to high shear viscosity data as related to bearing operating conditions. However, before this statement can be seriously considered the effect of higher operating temperatures on viscosity must be determined. The next section treats this question.

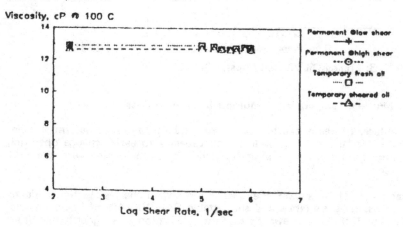

Figure 4.4.11: Engine Oil Viscosity Loss, 40

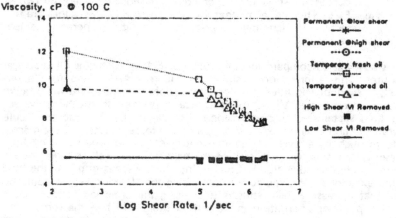

Figure 4.4.12: Engine Oil Viscosity Loss, 10W-40

483

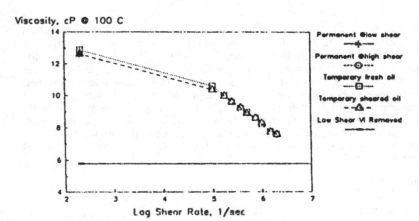

Viscosity, cP @ 100 C

Log Shear Rate, 1/sec

Permanent @low shear

Permanent @high shear

Temporary fresh oil

Temporary sheared oil

Low Shear VI Removed

Figure 4.4.13: Engine Oil Viscosity Loss, 10W-40

4.4.4.3 Multiple Temperature — Multiple Shear Rate Data

The advantages of being able to readily change shear rates (and determine them accurately) "on the run" as well as the capacity to easily change operating temperature, permitted some interesting data to be generated with the TBS Viscometer.

For example (and to answer the question raised in the last section), Figure 4.4.14 shows the shear-labile and shear-stable SAE 10W-40s of Figures 4.4.12 and 4.4.13 at 150° C. It is evident that the application of a higher temperature does not alter the similarity between the two in regard to their high shear viscometric properties, in spite of the obvious difference between the two in their low-shear PVL. Thus, it seems that low shear viscosity loss values are not relevant or predictive of high shear response of bearings at operating temperatures.

Another study was to compare the TVL of an SAE 10W-40 engine oil at several temperatures. This information is shown in Figure 4.4.15. TVL (or "shear-thinning", "pseudoplastic flow", or "orientation phenomena") is a well-known phenomenon occurring in VI improved lubricants (as well as in other solvated-polymer fluid systems). The phenomenon is explained by the macromolecule being stretched and oriented as flow occurs by the forces exerted on the macromolecule through its embedment in the matrix of the moving solvating fluid — in other words, the "viscous grip" of the base fluid (as well as other macro-molecules in simultaneous flow). Considering this "viscous grip" of the base oil on the expanded macromolecules comprising the VI improver, it would be expected that increasing shear stresses (increasing the orienting forces) would create more polymer orientation in flow (i. e. more TVL) at the same shear

484

rate. This behaviour is evident in Figure 4.4.15. The study suggests that it would be of interest and perhaps of value to study macromolecular orientation at very high shear rates with different VI improvers and VI improved systems using widely different base oil viscosities and polymer solubilizing ability. For many years the theory and practice of polymer solution dynamics was wanted for very high shear viscometric data. That time would seem to have come.

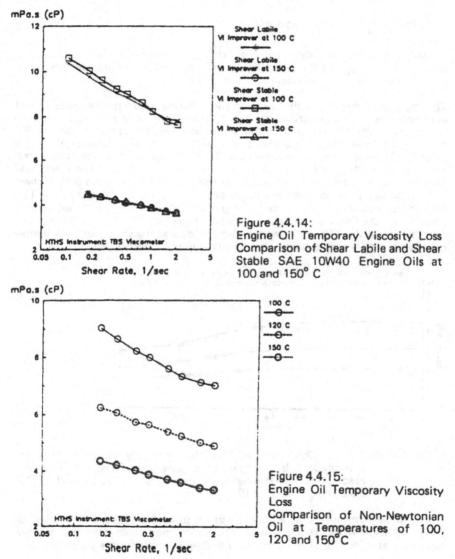

Figure 4.4.14:
Engine Oil Temporary Viscosity Loss Comparison of Shear Labile and Shear Stable SAE 10W40 Engine Oils at 100 and 150° C

Figure 4.4.15:
Engine Oil Temporary Viscosity Loss
Comparison of Non-Newtonian Oil at Temperatures of 100, 120 and 150°C

485

4.4.4.4 Correlation with the "Cross Equation"

Since engine bearings frequently operate at shear rates well in excess of 10^6 sec^{-1}, an empirical equation has been applied called the Cross Equation (Ref. 15):

$$\log \frac{\eta_0 - \eta}{\eta - \eta_u} = a + b \log\gamma + c/T$$

In which η is the high-shear viscosity at a temperature, T ($^\circ$K), and a shear rate, γ; η_0 and η_u are the viscosities at zero shear rate and infinite shear rate, respectively. a, b, and c are empirical constants.

This equation, chosen on the basis of curve fitting, permits interpolation and extrapolation of viscometric data — the latter being very significant because of the extraordinarily high shear rates produced by the operating engine. Figure 4.4.16 (from Ref. 16) shows the agreement with the Cross Equation by several HTHS viscometers including the TBS. On the basis of the close agreement between the TBS Viscometer and the Cross Equation shown here and with other oils in the set, and because of the ease of spanning the desired shear rate range using the absolute technique available with the instrument, the TBS Viscometer was chosen to represent the group of HSHT viscometers. Subsequent solution of the Cross Equation by statistical regression of the viscositity/shear-rate data from the TBS provided the viscosity data at the shear rates and temperatures needed for engine-bearing/oil-film-thickness correlation.

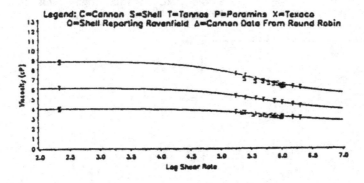

Figure 4.4.16: Plot of Observed and Predicted Viscosity vs. Shear Rate for Multigrade Oils, Comparison of Available Phase III Data with Predictions from Tannas Data, only 100° C, 120° C and 150° C Data Used, OIL-BFT-19

It should be mentioned that the close agreement of the TBS with the empirical Cross Equation tends to substantiate the information from both and, further, suggests that it may be of value to seek a theoretical basis for the equation.

4.4.4.5 Correlation with Engine Oil-Film Thickness Studies

The original motivation to develop the TBS Viscometer was to relate high-shear, high-temperature viscometric results to engine operation. During the time that the TBS Viscometer was being developed and improved by automation, highly significant work was being done by investigators into bearing oil film thickness. Their work and that of many others in the field was summarized relatively recently in a status report by the Engine Correlation Task Force of ASTM D2, Subcommittee 7, Section B (Ref. 17). It is recommended to the reader, that among other applications of the information contained, to use this report for a rich bibliography spanning a number of years.

Very recently an ASTM symposium on the subject brought to the force the most recent developments and concerns of those responsible for cooperatively specifying engine oils (Ref. 18). The need for high-temperature, high-shear viscometry was frequently mentioned as one of the major issue to be faced — one that could bring about changes in the way engine oils are presently specified.

With so much effort expended on developing the TBS Viscometer as a useful instrument, including work on the methods and the round-robins, by so many highly dedicated individuals, it is heartening to see that data from the instrument has been frequently used and proven helpful in generating dependable and consistent viscometric data in several of these studies. Surprisingly high levels of correlation (R^2 = 0.98+) between the oil-film thickness and TBS-derived viscometric data were reported (Ref. 19) which tend to confirm and underscore the long accepted assumption that high shear viscosity is the primary factor in developing bearing oil-film thickness.

From another point of view, the TBS Viscometer has met its original goal of simulating the bearing. At the same time, it has proven to be an instrument capable of broader viscometric applications to the determination of polymer solution dynamics.

4.4.5 Summary

The Tapered Bearing Simulator Viscometer has, for a number of years, been used in the field for producing high-shear viscosities but with a somewhat labor-intensive "relative" technique. The instrument has shown classic viscometric response and, because of the ability to set a variable rotor/stator gap while in operation, yields classical reciprocal torque versus gap relationship with Newtonian fluids. Thus, over a fairly broad range of rotor/stator gap, the instrument has been shown to be free of temperature effects and to operate as an absolute viscometer.

As a consequence of the discovery that a non-Newtonian reference oil will also give linear reciprocal torque versus rotor height curves, the method has been considerably simplified by more fully utilizing the TBS Viscometer's ability to be applied as an absolute instrument. As a consequence, calibration is faster and the precision of the instrument has been improved to 0.96 % repeatability and 2.59 % reproducibility at the 95 % confidence level. Sample-to-sample turn-around time is now less than 10 minutes.

These developments have encouraged the incorporation of automation to further simplify its use by relatively untrained personnel and reduce the attention required to produce information.

Automation of the TBS Viscometer required the incorporation of advances in thermoregulation, heating membranes, safety controls for continuous running, and an idling oil permitting long, unattended runs with the instrument. Automation has been successfully accomplished and such units are now being used in the field.

Applications of the TBS to a range of studies and problems showed its use in generating multiple shear rate/multiple temperature data. One of the most important applications of this latter ability was in generating multi-shear, multi-temperature viscosities on oils used in engine-bearing oil-film-thickness correlation to determine the level of correlation. The results gave a Coefficient of Determination (R^2) of 0.98+ for a combination of single-grade and multigrade reference oils, confirming the value and pertinence of high-shear viscometry in general and the TBS Viscometer in particular. A high level of correlation between the TBS Viscometer data and the empirical Cross Equation suggested that the Cross Equation may have a theoretical basis.

Another area of study was opened in regard to the relation of Permanent Viscosity Loss (PVL) and Temporary Viscosity Loss (TVL). It was shown that two oils may have considerably different PVLs at low shear rates but this may not make any significant difference in their high-shear viscometric properties. This, in turn, suggested that, as far as bearing lubrication is concerned, perhaps PVL should be measured only at high shear rates. Further information related to this was in the production of Viscosity Loss Trapezoids which are an interesting and revealing way to examine the contributions and shear susceptibilities of VI improvers.

4.4.6   Acknowledgments

The authors wish to thank Prof. Dr.-Ing. Bartz and his colleagues for making such a colloquium possible.

Anyone involved in the development and standardization of an instrument and its attendant method knows the manifold effort required from many individuals. The authors do not have space to record these colleagues from industry individually who worked together in yeoman manner to bring about lasting technical accomplishments. In particular the authors wish to note the strong and effective leadership given by Mr. Robert B. Rhodes who Chaired the ASTM Rotational Viscometer Task Force through the trenches and minefields.

In the working environment of Tannas and Savant, there were again a number of individuals who helped either in the development of this paper or of past papers some of which data are presented herein. Our sincere thanks to them. In particular the authors would like to thank Dave Piasecki who has been deeply involved in TBS work virtually since the instrument was born. He has unselfishly given of his free time to smooth the way. Chiara Selby's supportive and skilled efforts in pulling the pieces of paper together to make this finished publication are gratefully acknowledged.

Taken together, the prayers and good wishes of friends and family were the deciding factor, and are most humbly appreciated.

4.4.7 References

(1) Piasecki, D.A.; Selby, T.W.; Smith, M.F.: Development and Performance of the Tapered Bearing Simulator for High Shear, High Temperature Viscometry. 1981. PP. 11. SAE Fuels and Lubricants Meeting. 81.10.19.—22. Tulsa, Oklahoma, USA.
(2) Smith, M.F.: History of ASTM Involvement in High-Temperature, High-Shear Oil Viscosity. High-Temperature, High-Shear Oil Viscosity — Measurement and Relationship to Engine Operation (ASTM STP 1068). 1. Edition. Philadelphia, USA: ASTM 1989, P. 3—13.
(3) Selby, T.W.; Pasecki, D.A.: The Tapered Bearing Simulator — An Absolute Viscometer. SAE Technical Paper Series, No. 830031. 1983. PP. 9. SAE International Congress, Cobo Hall. 83.02.29.—03.04. Detroit, Michigan, USA.
(4) Hoshino, M.: High Shear-Rate Viscosity and Its Measuring Technique. Lubrication (Japan). 30 (1985), 7—12.
(5) Tolton, T.J.: The Tapered Bearing Simulator — Now an Automated Viscometer. SAE Technical Paper Series, No. 872045. 1987. PP. 5. SAE Fuels and Lubricants Meeting. 87.11.02.—0.5., Toronto, Canada.
(6) ASTM Standard Test Method for Measuring Viscosity at High Temperature and High Shear Rate by Tapered Bearing Simulator, D4683-87. In: 1989 Annual Book of ASTM Standards. Part 5.03. Philadelphia, Pennsylvania: 1989, P. 518—523.
(7) Tolton, T.J.: The Viscosity Losses of New and Sheared Multiweight Oil Formulations Determined at Multiple Shear Rates. SAE Technical Paper Series, No. 890727. 1989. PP. 10. SAE International Congress. 89.02.27.—03.03. Detroit, Michigan, USA.
(8) Rhodes, R.B.: Development of ASTM Standard Test Methods for Measuring Engine Oil Viscosity Using Rotational Viscometers at High-Temperature and High-Shear Rates. High-Temperature, High-Shear Oil Viscosity — Measurement and Relationship to Engine Operation (ASTM STP 1068). 1. Edition. Philadelphia, USA: ASTM 1989, P. 14—22.
(9) Kingsbury, A.: Heat Effects in Lubricating Films. Mechanical Engineering. 55 (1933), 685—688.

(10) Pike, W.C.; Banks, F.R.; Kulik, S.: A Simple High Shear Rate Viscometer — Aspects of Correlation with Engine Performance. The Relationship between Engine Oil Viscosity and Engine Performance — Part VI (SAE SP-434). 1 Edition. Warren, PA.: P. 47—55.

(11) Taylor, J.C.: Letter to CEC CL-23 members. 1984.

(12) DuParquet, J.P.R.: A Few Steps in the Development of Rotary Viscometry. PP. 17. Ravenfield Conference on Viscosity. 85.09.24.—26. Bolton, England.

(13) ASTM Research Report D02-1253. Available from ASTM Headquarters, Philadelphia, Pennsylvania, USA.

(14) Selby, T.W.; Peoples, M.C.: Applications of Dynamic Viscosity at Low Shear and Higher Temperatures. STLE. Society of Tribology and Lubrication Engineers 1988 Annual Meeting. 88.05.9.—12.

(15) Cross, M.M.: Rheology of Non-Newtonian Fluids; A New Flow Equation for Pseudoplastic Systems. Journal of Colloid Science 20 (1965), P. 417.

(16) Report to ASTM by Melanie Tobias of Texaco, Inc. Used with permission of Serge Cryvoff, Chairman of Bearing Oil Film Thickness Task Force, Section B, Subcommittee 7, ASTM D2. 89.12.04. PP. 41.

(17) Report from Engine Correlation Task Force; ASTM D2, Subcommittee 7, Section B: The Relationship Between High-Temperature Oil Rheology and Engine Operation — A Status Report. ASTM DS-62. ASTM, Philadelphia 1985.

(18) Spearot, J.A.; Editor: High-Temperature, High-Shear Oil Viscosity — Measurement and Relationship to Engine Operation, STP 1068. 1. Edition ASTM, Philadelphia 1989.

(19) Spearot, J.A.; Murphy, C.K.; Deysarkar, A.K.: Interpreting Experimental Bearing Oil Film Thickness. SAE Technical Paper Series, No. 892151. 1989. PP. 14. SAE Fuels and Lubricants Meeting. 89.09.25.—28. Baltimore, Maryland.

490

4.5 Literature Survey on Sludge Deposits Formation in Gasoline Engines

C.D. Neveu, Rohm & Haas European Operations, Paris, France
W. Böttcher, Röhm GmbH Chemische Fabrik, Darmstadt, West Germany

Abstract

In 1985, IGL 20 was established by CEC to study "black sludge" formation in modern gasoline engines. This investigation group was later changed to a project group identified as PL-37. A subgroup of PL-37 was then formed to support the activities of the main group by 1) surveying the literature and 2) completing used oil analysis on selected samples.

The objective of the literature survey was to establish the level of current knowledge related to deposit formation in gasoline engines. A total of 69 documents up to 1986, were thoroughly reviewed by four members to extract information concerning:

The mechanism of deposit formation with particular emphasis on sludge.

The most useful methods for determining the composition of deposits and used oils.

4.5.1 Introduction

In 1985, IGL 20 was established by CEC to investigate sludge formation in modern gasoline engines. These deposits were called "black sludge" because their appearance was significantly different from those previously observed.

This investigation group was later changed to a project group identified as PL-37. A subgroup of IGL 20/PL-37 was formed to support the activities of the main group by:

1. surveying the literature
2. completing used oil analysis on selected samples

The present report summarizes the outcome of the literature survey. Used oil analysis is still in progress.

4.5.2 Objective of the Literature Survey

The objective of our literature survey was to establish the current level of knowledge related to deposit formation in gasoline engines. A total of 69 documents up to 1986 were thoroughly reviewed by four members to extract information concerning:

— the mechanism of deposit formation with particular emphasis on sludge
— the most useful methods for determining the composition of deposits and used oils

Findings from this survey have been organized according to the following scheme:

— mechanism of sludge formation
— review of the usefulness of analytical methods.
— list of the main factors in deposit formation.
— appendices:
    — list of the 10 most useful papers with abstracts
    — list of all papers reviewed with brief abstracts.

4.5.3 Summarized Findings on Sludge Formation Mechanism

1) Definition of "Black Sludge"

A general definition embracing most types of sludge was given by a BTC sub-group in 1979:

"A deposit primarily composed of oil and combustion products which does not drain from surfaces but can be easily wiped off with a cloth."

Sludge deposits vary widely in terms of consistency, water content, color, solids content, location where found, and operating conditions under which they are formed.

"Black sludge" would according to the above definition correspond to a thick to solid material with low water content, of dark color, light oil insolubles content, typically found in the rocker cover, cylinder head, timing chain cover, oil sump, oil pump screen and oil ring in variable quantities.

Furthermore, the formation of "black sludge" deposits was reported to be promoted by mixed driving conditions alternating between stop and go, and highway operations. None of the papers examined dealt specifically with "black sludge". However, the literature survey provides us with useful insights on sludge and oil insolubles composition, on mechanism of formation, and on methods of characterization which we recommend should be considered when investigating the "black sludge problem".

492

2) Mechanism of Sludge Formation

The literature survey suggests that the mechanism of sludge formation is an extremely complex progress involving three main steps.

Step 1: The first stage is the formation of sludge "precursors" in the blow-by gases as a result of a radical reaction between nitrogen oxides and fuel derived olefins.

Step 2: After condensing in the engine, the precursors undergo many transformations in the crankcase oil to form a "binder". This is a complex mixture of components carrying chemical species which are relatively rich on oxygen, nitrogen and sulphur coming either from air or the fuel.

Step 3: The sludge binder causes agglomeration and subsequent deposition of water, oil soluble, and insolube matters. Oil soluble components are derived from both fuel and lubricant. Oil insoluble matter can be divided into organic and inorganic materials. Again, the organic fraction results from fuel and lubricant degradation products. The oil insoluble fraction consists of oil and fuel additive derivatives as well as wear metals and contaminants.

2a) Sludge Precursors

At low crankcase temperatures, lubricating oils do not appreciably contribute to deposit formation. Oxidation proceeds at a negligible rate. Deposits from low temperature operating conditions chiefly consist of products which pass the piston as blow-by gases, condense in the oil film on the cylinder wall and flow with the oil into the crankcase.

Blow-by gases are made up of combustion products, partially oxidized fuel components, inorganic salts, and of nitrogen containing compounds. The latter have a low molecular weight and are mainly difunctional. Alpha-beta nitro-nitrates were the predominant chemical class detected. These products are formed by a radical reaction of fuel or fuel derived unsaturated hydrocarbons such as olefins and di-olefins with nitrogen oxides. This is shown as stage (a) below.

$$R\text{---}\underset{\underset{R}{|}}{C}H\text{---}CH_2\text{---}R \xrightarrow{\text{heat}} R\text{--}H + R\text{---}CH=CH\text{---}R$$

olefin

$$(a)\ R\text{---}CH=CH\text{---}R \xrightarrow{NO,\ NO_2} R\text{---}\underset{\underset{NO_2}{|}}{C}H\text{---}\underset{\underset{X}{|}}{C}H\text{---}R \quad X = OH\ or\ ONO_2$$

fuel olefin

Nitro-nitrate if X = ONO$_2$

Nitrogen oxides are formed from atmospheric nitrogen during combustion. This reaction is promoted by high temperature. In gasoline engines, it usually reaches a maximum with fuel/air mixture on the lean side of stoichiometric. A part of the nitric oxide is subsequently converted into nitrogen dioxide: $2NO + O_2 = 2NO_2$. The latter is more reactive than nitric oxide. However, because they are interconvertible and because separate analysis is difficult, the term NO_x is often used to designate a mixture of the two.

Note also the composition of blow-by gases is markedly different from those of exhaust gases. This difference occurs because the amount of blow-by passing the rings is greater at time of maximum cylinder pressure when the piston is near to TDC but before all the charge adjacent to the piston has been burnt.

Several experiments confirmed that the formation of the nitro-nitrate species is an essential step in the mechanism of sludge formation. Running an engine with CO_2 in place of N_2 in the intake air resulted in a clean engine even with a "dirty" gasoline. Alternatively, using a fuel such as methane which did not contain any olefin resulted in a clean engine. Also, diverting blow-by gases to atmosphere eliminated sludge formation.

Further studies demonstrated that fuel was controlling the degradation pattern of the lubricant. Under identical conditions, infrared analysis showed that lubricants of different quality produced similar degradation patterns (as measured by the concentration of carbonyl, nitro-nitrate and nitro compounds).

2b) Sludge "Binders"

The condensable blow-by products can be divided into a hydrocarbon phase containing a large amount of nitro-nitrate species and an aqueous phase containing formaldehyde and small amounts of products which appear to be the result of the hydrolysis of nitro-nitrate species. After condensing on the cold part of the engine, the hydrocarbon phase undergoes complex reactions in the crankcase medium. These reactions are promoted by higher operating temperatures when oil is circulated over the pistons.

It has been shown that on heating the hydrocarbon phase abstracted from the condensable blow-byproducts, resins are gradually deposited. At intermediate stages of heating, chromatographic fractionation yields a family of nitro-hydroxy carbonyl compounds of increasing molecular weight and decreasing solubility in hydrocarbons. This is shown as step (b) below and apparently involves a so called "aldol type" condensation by which carbonyl and nitro compounds react to form polymeric species of varying molecular weight.

(b)
$$R—CH———CH——R \xrightarrow{\text{heat}} \text{Nitro-Hydroxy-carbonyl compounds}$$

with NO_2 and ONO_2 groups

Nitro-nitrate

494

The low molecular weight products of step (b) are hydrocarbon soluble, but they have some limited solubility in water and have been shown to be fairly potent surfactants. Submitting them to prolonged heating causes the production of insoluble resins which can be seen as varnish deposits in the engine.

2c) Sludge deposition

The sludge "binder" is an essential ingredient which the oil must contain before organic and inorganic compounds are emulsifield/agglomerated to form sludge. The organic portion of sludge consists of carbonyl, sulfur, nitrogen derivatives, polymerized hydrocarbons, soot and oil. The inorganic portion comprises salts, wear particles, and water. Sludge is then deposited in the engine where it undergoes further transformations.

4.5.4   Analytical Test Methods

A large number of analytical techniques have been used in an effort to characterize the many compounds involved in the process of sludge formation. These can be divided into six categories.

a) routine drain analysis
b) separation techniques
c) spectroscopic methods
d) thermal methods
e) analysis of insoluble matters
f) others

1) Most Useful Test Methods

Test methods disclosed in the paper reviewed are listed in Table 4.5.1 of Appendix 2. They have been ranked by order of overall usefulness. The number of citations for each method and the reference in which they were used are shown in Table 4.5.2 of Appendix 2.

The four methods which have been found to be the most useful are:

1. Blotter spot test for insolubles/dispersancy of oils during bench or field tests.
2. Differential IR for oxidation, nitration, fuel content, etc. of used oils. It can also be used to examine sludge samples directly.
3. Dialysis, filtration, centrifuging and solvent extraction procedures for the separation of oils and sludges prior to further chemical analysis.
4. Carbon-14 tracers were useful in mechanistic studies of deposit formation, though there was little recent work.

The following four methods were commonly used and considered to be useful:

1. Elemental analysis of used oils and sludges (additives, wear metals and contaminants).
2. Routine drain-oil analytical methods, including kinematic viscosity, TBN, TAN/SAN and insolubles content (various methods, including centrifuging and membrane filtration).
3. Microscopy (optical and electron) for the examination of used oils and sludges. Other means of measuring particle size or molecular weight were also helpful.
4. TGA of used oils and sludges to estimate volatility.

It should be noted that chromatographic techniques, such as GC, GPC, HPLC and column chromatography were not widely used but had proved useful.

2. Analysis Scheme of Used Oils Selected by PL-37 Subgroup

The following scheme was adopted by the PL-37 subgroup for the analysis of used and fresh oil samples made available to PL-37 in mid 1987.

1. Kinematic viscosity @ 40 and 100° C before removal of fuel diluent, using a mini-viscometer (or ASTM D445/IP71 if sufficient sample is available).
2. TBN and TAN by ASTM D664/IP177 using buffer end-point.
3. Total solids (both heptane and toluene diluents) by IP 316 @ 10,000 g.
4. Blotter Spot dispersancy before and after heating (240° C, 5 min). Method is detailed in Appendix 3.
5. Difference IR, reporting A/cm for oxidation (near 1710), nitration ($RONO_2$: near 1630, but only if 1270 peak is also present), nitro compounds (RNO_2: near 1560) and aromatics content (near 1600).
6. Fuel dilution by simulated distillation GC method (most laboratories use proprietary methods, but ASTM D2887 is applicable).

Of these methods, IR, Blotter Spot and TBN proved the most useful in comparing field oils and lubricants evaluated in the PL-37 sludge test.

4.5.5 Causal Factors

Causal factors have been extensively investigated over the last thirty years. They can be divided into three categories. The first includes factors which have an effect on the formation of the sludge precursors. The second contains the parameters which have an effect on the amount of sludge precursors which can condense in the engine. Finally, the parameters which have an influence on the concentration in the crankcase oil of the sludge binder and of the various materials entering into the composition of sludge play a significant role in the process of sludge formation and deposition.

1. Formation of Sludge Precursors

- High fuel olefin content (thermally or catalytically cracked gasoline) promotes the formation of sludge precursors.

- The presence of lead in the gasoline reduces the amount of sludge precursors formed in the combustion chamber. Lead compounds act as radical traps which limit the radical reaction between nitrogen oxides and fuel olefins.

- Because they produce increased nitrogen oxides, lean burn engines are more prone to sludging than power plants operated under rich or stoichiometric conditions.

- Spark advance, temperature of combustion, as well as all other parameters which control the temperature and the pressure in the combustion chamber have an effect on amount of nitrogen oxides formed during the combustion.

2. Condensation of Sludge Precursors

- The amount of sludge precursors which can condense in the engine is dependent on the operating conditions and on the ventilation system. Foul air ventilated systems and low temperature operating conditions are known to be severe in terms of sludge formation.

- Mixed driving conditions which promote the condensation of sludge precursors and their subsequent transformation into sludge "binder" are also severe.

3. Concentration of Sludge Binder and Insolubles in the Oil

- Low crankcase capacity and/or long drain intervals increase the amount of sludge binder and of other components entering into the composition of sludge. Additionally, the presence of components which reduce the solvent capability of the lubricant, like unburnt methanol for example, might promote the deposition of sludge.

- Fuels with low volatility increase the concentration of unburnt hydrocarbons in the crankcase oil which can enter into the composition of sludge. In this aspect, gasolines with aromatic content are known to be severe.

4.5.6 Conclusions of the Literature Survey

"Black Sludge" is different from sludges previously observed and, therefore, none of the papers reviewed relates specifically to this phenomenom.

497

The mechanism of sludge formation derived from the literature survey is thus likely not to be directly applicable to "Black Sludge". However, some of the causal factors identified (e.g. fuel quality and type of service) are also known to be important in the formation of "Black Sludge".

The methods used in the past for the analysis of sludges and oils are believed to be equally applicable to the study of "Black Sludge".

It is recommended that used oil analysis is employed to study the effect of operating conditions in the M-102E sludge test (e.g. fuel batch, test duration) and to compare oils and sludges from bench engine tests and field tests.

4.5.7 References

(1) West, C.T.: "Black sludge & noxidation inhibitors", Mail Station J-8, 16/09/86. Unable to locate reference.
(2) Coates, J.P.; Setti, L.C.: "IR spectroscopic methods for the study of lubricant oxidation products", (Spectra-tech. Inc. & Perkin-Elmer Corp.) ASLE Trans., 29 (3), 394–401, March 86.
(3) Kadam, A.N.; Zindge, M.D.: "IR spectroscopic analysis of used crankcase oil", (Nat. Inst. Oceanog. Res. Cent., Bombay), Res. Ind., 30 (4), 382–385, 1985.
(4) Ebert, L.B.; Davis, W.H.; Dennerlein, D.R.: "The chemistry of i/c engine deposits. I – microanalysis, TGA & IR spectroscopy", (Exxon, NJ), ACS Div. Pet. Chem., Preprints, 26 (2), 593, 1981.
(5) Davis, W.H.; Ebert, L.B.; Dennerlein, J.D.; Mills, D.R.: "The chemistry of i/c engine deposits. II – extraction, mass spectroscopy & NMR", (Exxon, NJ), ibid. p. 603.
(6) Carey, L.R.; Stover, W.H.; Murray, D.W.: "Extended drain passanger car engine oils", (Imperial oil Res. Dept.) SAE 780952.
(7) Asseff, P.A.: "Used engine oil analysis – review", (Lubrizol, Ohio) SAE 770642.
(8) Zeelenberg, A.P.; Wortel, J.M.: "More information on oil and engine from sludge analysis", (Shell, Amsterdam) SAE 770643.
(9) Stambaugh, R.L.; Kopko, R.J.; Franklin, T.M.: "Effect of unleaded fuel & exhaust gas recirculation on sludge & varnish formation", (Rohm and Haas) SAE 720944.
(10) Williams, A.L.: "Lacquer precursors from a paraffinic lubricant traced by carbon-14", (Mobil R & D) ACS Div. Pet. Chem., Preprints, 14 (4), A7, 1969.
(11) Dmitroff, E.; Moffitt, J.V.; Quillian, R.D.: "Why, what and how: engine varnish", (US Army Fuels & Lubes, SWRI) Trans. ASME, J. Lub. Technol., 91. 406–416, July 1969.
(12) Dimitroff, E.; Moffitt, J.V.; Quillian, R.D.: "Aromatic compounds in fuels identified as main precursors of engine varnish", (US Army Fuels & Lubes, SWRI) SAE Journal. 77 (7), 52–59, 1969. As ref. 11 above.
(13) Newhall, H.K.: "Control of NOx by exhaust recirculation – a preliminary theoretical study", (Dep. of Mech. Eng., Univ. of Visconsin) SAE 670495.
(14) Bowden, J.N.; Dimitroff, E.: "Mechanism studies of polymeric dispersants", (US Army Fuel & Lubes, SWRI) ACS Div. Pet. Chem., Preprints, 7 (4), B45, 1962. Ordered from British Library.
(15) Bennett, P.A.; Jackson, M.W.; Murphy, C.K.; Randall, R.A.: "Crankcase gas causes 40 % of auto air pollution", (Gen. Motors, Res. Labs.) SAE Journal, 68 (3), 30–36, March 1960.

(16) Biswell, C.B.; Catlin, W.E.; Robbins, G.B.: "New polymeric dispersants for hydro-carbon systems", (EI du Pont de Nemours & Co.) Ind. Eng. Chem., 47 (8), 1598—1601, 1955. Early investigations of dispersant action. 1 ref.

(17) Spindt, R.S.; Wolfe, C.L.: "How gasoline helps to form engine deposits", (Gulf R & D, Mellon Inst.) SAE Journal, 60 39—44, May 1952.

(18) Porst, A.: "Erfahrungen mit Gasmotorenölen", Beratungsgesellschaft für Mineral-ölanwendungstechnik (Experience with gas engine oils).

(19) Janssen, O.: "Aussagewerte der Ergebnisse von Analysen gebrauchter Motorenöle", Erdöl und Kohle — Erdgas — Petrochemie 23, 4 (1970), p 216 (Statements based upon results from analysis of used engine oils).

(20) Bartl, P.: "Moderne analytische Methoden zur Beurteilung von neuen und gebrauch-ten Motorenölen (I)", Tribologie + Schmierungstechnik 33, 3 (1986), p 146.

(21) Bartl, P.: "Moderne analytische Methoden zur Beurteilung von neuen und gebrauch-ten Motorenölen (II)", Tribologie + Schmierungstechnik 33, 4 (1986) p 228 (Modern analytical methods to judge new and used engine oils).

(22) Wedepohl, E.; Hildebrandt, U.: "Schlammbildung in Dieselmotoren", Erdöl und Kohle, Erdgas, Petrochemie Bd. 37, 6 (1984), 254.

(23) Spearot, J.A.; Gallopoulos, N.E.: "Concentrations of Nitrogen Oxides in crank-case Gases", SAE Paper 760563.

(24) Rodgers, D.T.; Rice, V.W.; Jonach, F.L.: "Mechanism of Engine Sludge Formation and Additive Actions", SAE Transactions Vol. 64 (1956) 782ff.

(25) Krenz, K.L.: "Gasoline Engine Chemistry", Lubrification Vol. 55, 6 (1969).

(26) Geyer, J.: "The Mechanism of Deposit Formation and Control in Gasoline Engines", ACS Symposium Division Petroleum Chemistry, Sept. 1969.

(27) Vineyard, B.D.; Coran, A.Y.: "Gasoline Engine Deposition: I. Blowby collection and the idendification of Deposit Precursors", ACS Symposium Division Petroleum Chemistry, Sept. 1969.

(28) Coran, A.Y.; Vineyard, B.D.: "Gasoline Engine Deposition: II. Sludge Binder", ACS Symposium Division Petroleum Chemistry, Sept. 1969.

(29) Mastin, R.G.; Gorry, L.J.: "The effects on Nitrogen Fixation on Gas Engine Oper-ation", Jour. Am. Soc. Eng. Dec. 1692, p 517.

(30) Winner, D.B.; McReynolds, L.A.: "Nitrogen Oxides and Engine Combustion", SAE Transaction 1962, p 733.

(31) Berry, E.; Webster, A.B.: "Emulsion Formation in Gasoline Engines", Jour. Inst. Petr. Vol. 55, No. 544, p 245.

(32) Covitch, M.J.; Humphrey, B.K.; Ripple, D.E.: "Oil Thickening in the Mack T-7 Engine Test — Fuel Effects and the Influence of Lubricant Additives on Soot Aggre-gation", SAE paper 852126.

(33) Dotterer, G.O.; Hellmuth, W.W.: "Differential Infrared Analysis of Engine Oil Che-mistry in Sequence V Tests, Road Tests, and other Laboratory Engine Tests", Lubri-fication Engineering 42, 2 (1985), p 89—97.

(34) Ku, C.S.; Hsu, S.M.: "A Thin-Film Oxygen Uptake Test for the Evaluation of Auto-motive Crankcase Lubricants", ASLE Preprint No. 82-LC-ID-1.

(35) Spindt, R.S.; Wolfe, C.L.; Stevens, D.R.: "Nitrogen Oxides, Combustion and Engine Deposits", SAE Transaction 64 (1956), 797—811.

(36) Rüdinger, V.: "Ein Modell der Ablagerungsbildung am Dieselmotorkolben", Erd-öl und Kohle-Erdgas-Petrochemie 27, 353 (1974).

(37) Staffehl, R., et al: "Kennzeichnung der Alterungsneigung von Motorenölkomponen-ten, gewonnen durch Hydroraffination von Erdölvakuumdestillaten mit Hilfe der UV-Spektrometrie", Schmierstoffe Schmierungstechnik 36, 62—67.

(38) Kozakowski, G.; Sobanska, K.: "Anwendung spektroskopischer Methoden zur Un-tersuchung der Alterungsprozesse von Motorenölen", Schmiertechnik 16, 278—282 (1985).

(39) Spilners, I.J.; Hedenburg, J.F.: "Effect of Fuel and Lubricant Composition on Engine Deposit Formation", American Chemical Society, Division of Petroleum Chemistry, Preprint, Atlanta 1981, pp 632–638.

(40) Harris, S.W.; Eggerding, D.W.; Udelhofen, J.H.: "Analysis of the PV-1 Test via Used Oils", Lubrication Engineering 38 (8), 487–496 (1982).

(41) Farley, F.F.; Greenshields, R.J.: "Deposition of Lacquer and Sludge in Passenger Car Service", Industrial and Engineering Chemistry 41, 902 (1949).

(42) Verdura, T.M.: "Infrared spectra of Lubricating Grease Base Oils and Thickeners Part II", NLGI Spokesman 1971, 168–296.

(43) Shechter, H.: "The Chemistry and Mechanism of Reaction of Oxides of Nitrogen and Olefin", Record of Chemical Process 25, 55 (1964).

(44) Fenimore, C.P.: "The Ratio NO_2/NO in Fuel-Lean Flames", Combustion and Flame 25, 85–90 (1975).

(45) Dets, M.M.; Chermenin, A.P. Zhurba, A.S.; Erokhina, G.T.: "Motor Oil Test for Low-Temperature Sludge Formation", Chemistry and Technology of Fuels and Oils 12, 626 (1976).

(46) England, C.; Corcoran, H.: "The rate and Mechanism of the Air Oxidation of Parts-per-Million Concentrations of Nitric Oxide in the Presence of Water Vapor", Ind. Eng. Chem. Fundam 14, 55 (1975).

(47) Davies, H.M.: "Low Temperature Sludge", IP Review p 425–433, Dec. 1965.

(48) Brand, J.C.D.; Stevens, I.D.R.: "Mechanism and Stereochemistry of the Addition of Nitrogen Dioxide to Olefins", Journal Chemical Society, p 629 (1958).

(49) Mahoney, L.R.; Korcek, S.; Hoffmann, S.; Willermet, P.A.: "Determination of the Antioxidant Capacity of New and Used Lubricants; Method and Applications", Ind. Eng. Prod. Res. Dev 17, 250 (1978).

(50) Badiali, F.L.; Berti, F.; Cassiani Jugoni, A.A.; Pusateri, G.: "Evaluation of Dispersancy by Analytical Methods", SAE 780932.

(51) Allman, L.J.; Brehm, A.E.; Colyer, C.C.: "The ABC of Motor Oil Oxidation", SAE 700510.

(52) Agins, P.J.; Mulvey, D.: "The Mechanism of Sludge Suspension in Engine Oil", J. Inst. Petr. 44, 229 (1958).

(53) Papok, K.K.; Susjew, B.J.: "Chemical Composition of Lacquer Deposits", Khim. i. Techn. Topliv. i. Massel, Vol. 8 (9), 1963 (p 64).

(54) Kohl, K.B.; Frame, E.A.: "Development of Methodology for Engine Deposit Characterisation", National Bureau of Standards Special Publication 674 (1984).

(55) Stavinoha, L.L.; Wright, B.R.: "Spectrometric Analysis of Used Oils", SAE 690776 (1969).

(56) Otte, O.M.: "A Practical Application of Engine Inspection Observations and Used Gas Engine Oil Analysis", SAE 730744 (1973).

(57) Sibenaler, E.: "Exploitation Photometriques des Epreuves a la Tache", Collect. Colloq. Semin. Inst. Fr. Petrole. No. 21, 35–59 (1971).

(58) Amprimoz, L.: "Surveillance des Moteurs par Spectrographie", Collect. Colloq. Semin. Inst. Fr. Petrole. No. 21, 143–56 (1971).

(59) Broman, V.E.: "Factors Affecting the Formulation of Engine Oils for LP-Gas Service", LP-Gas Engine Fuels, ASTM STP 525, pp. 3–17 (1973).

(60) Cartwrigh, S.J.; Carey, L.R.: "Control of Engine Oil Acidity", SAE 801366 (1980).

(61) Forbes, E.S.; Wood, J.M.: "Development of a Bench Detergency Test for Automotive Oils and its Correlation with the MS sequence V engine Test", Ind. Eng. Chem. Prod. Res. Develop. 8 (1), 48–54 (1969).

(62) Korcek, S.; Mahoney, L.R.; Johnson, M.D.; Hoffman, S.: "Antioxidant Decay in Engine Oils During Laboratory Tests and Long Drain Interval Service", SAE 780955 (1978).

(63) McGeehan, J.A.; Rynbrandt, J.D.; Hansel, T.J.: "Effect of Oil Formulations in Minimising Viscosity Increase and Sludge Due to Diesel Engine Soot", SAE 841370 (1984).
(64) Dimitroff, E.; Quillian, Jr., R.D.: "Low Temperature Engine Sludge What? Where? How?", SAE 650255 (1965).
(65) Daniel, W.A.; Wentworth, J.T.: "Exhaust Gas Hydrocarbons-Genesis and Exodus", SAE 486B (1962).
(66) Asseff, P.A.: "Engine Performance as Influenced by Lubricant Deterioration", SAE 680760 (1968).
(67) Nicholls, J.E.; El-Missiri, I.A.; Hewhall, H.K.: "Inlet Manifold Water Injection for Control of Nitrogen Oxides — Theory and Experiment", SAE 690018 (1969).
(68) Oberdorfer, P.E.: "The determination of Aldehydes in Automobile Exhaust Gas P.E.", SAE 670123 (1967).
(69) Daniel, W.A.: "Engine Variable Effects on Exhaust Hydrocarbon Composition" (A single-cylinder Engine Study with Propane as the Fuel), SAE 670124 (1967).

APPENDIX 1

Summary of the 10 Most Important Papers
Reviewed by PL-36 Subgroup

1. "MORE INFORMATION ON OIL AND ENGINE FROM SLUDGE ANALYSIS"

A.D. Zeelenberg, J.J. Wortel (Shell Amsterdam). SAE 770643

Analysis of suspended and deposited material from engines (unspecified) with emphasis on the organic fraction from a simple separation procedure. TGA/DTG with FID detection or GLC identification of evolved gases. (TGA with MS, GC — MS and IR was not useful.) Suspended solids were very similar to sludges. Differentiation between sludges from different sources was claimed using these methods which could, therefore, be useful for evaluation of sludge simulation tests (glassware or bench engine). Addition of NO_x increases the amount of sludge but does not change its nature. Water is important in sludge formation reactions. Sludge particle size (by Disc Centrifuge) depends on dispersant: typically 100 nm. 37 refs (inc. nos. 9, 11, 12, 17) — some of possible relevance.

2. "LACQUER PRECURSORS FROM A PARAFFINIC LUBRICANT TRACED BY 14C"

A.L. Williams (Mobil R & D)

Complex experiments: labelling of n-octadecane + 1st stage oxidation products (—OH, —CHO,—COOH). Oxidation in Mobil Thin Film Oxidation test 274C. Functions which are NOT lacquer precursors include: primary alcohols, ketones, carboxylic acids. Only 0.1 % of oxygen consumed is incorporated into lacquer. Esters contribute to lacquer, but bifunctional molecules (ketoacids, hydroxyacids) which can polymerise are most important. Very little on analysis. In a purely paraffinic lube oil, antioxidants which control acids or peroxides also control lacquer.

501

3. "WHY, WHAT AND HOW: ENGINE VARNISH"

E. Dimitroff, J.V. Moffitt, R.D. Quillian (US Army F & L, SWRI). Trans, ASME, J. Lub. Technol., 91, 406 – 416, July 1969

Experiments using a single-cylinder, COT engine with different fuels, lube oils, engine conditions. Aromatic compounds (e.g. alkylbenzenes) in fuel are major varnish precursors. Lead octane improvers and antioxidants had no effect, although inorganic lead compounds were prominent in the varnish. Lubes are NOT varnish precursors, but do have a major effect on varnish control. Engine conditions considered (speed, temperature, compression ratio, timing, air-fuel ratio) had only a minor effect. Varnish composition by IR, NMR, elemental analysis show it is composed mainly of alkane with condensed hydrocarbon + hydroxy compounds, ketones, organic nitrates and lead salts. Elemental methods were rudimentary by modern standards. Detailed mechanism proposed: effect of NO_x is minor; varnish is more dependent on reactions with blow-by than on lube-oil degradation. A good approach, needs repeating under modern conditions.

4. "THE MECHANISM OF DEPOSIT FORMATION AND CONTROL IN GAS-OLINE ENGINES"

J. Geyer., Symposium Division-Petroleum Chemistry Sep. 1969

For this investigation a cyclic temperature engine sludge test was employed to produce the basic material (sludge and used oil). The used oil was centri-fuged and the supernatant oil treated in the laboratory in various ways either simply with heat or with various gases to keep the deposit production going. A major effort was to investigate the contribution of gasoline to sludge formation.

5. "GASOLINE ENGINE DEPOSITION: I. BLOW-BY COLLECTION AND THE IDENTIFICATION OF DEPOSIT PRECURSOR"

B.D. Vineyard, A.Y. Coran. ACS Symposium Division Petroleum Chemistry Sep. 1969

The paper describes a sophisticated system to collect blow-by (tube fitted at the inner side of the piston). The gases were scrubbed with dodecane. The washing liquid was removed and the remaining substance analysed via GPC/IR. The nitro-nitrate derivates were identified to be the major deposit precursor. The precursor seems to be of low molecular weight and difunctional to a large extent. With this knowledge a synthetic precursor was prepared and IR cross checked with material obtained from the engine.

6. "GASOLINE ENGINE DEPOSITION: II. SLUDGE BINDER"

B.D. Vineyard, A.Y. Coran, ACS Symposium Division Petroleum Chemistry, Sep. 1969

The second part of the investigation looked at the sludge binder, a material defined as the non-volatile acetone soluble part of engine sludge and varnish. The solubility of this binder is investigated using various solvents including water at different pH to gather information about functional groups. The interpretation of the IRs combined with the results from part I were used to draft a theory of sludge binder formation.

7. "EFFECT OF FUEL AND LUBRICANT COMPOSITION ON ENGINE DEPOSIT FORMATION"

I.J. Spilners and J.F. Hedenburg, American Chemical Society, Division of Petroleum Chemistry Preprints, Atlanta 1981, pp. 632–638

Deposits formation was studied in a modified single cylinder CLR engine. Fuel effects are more pronounced than base oil stability effects in terms of deposit formation. Olefins cause heavier carbonyl compounds than by nitrated intermediates. But both precursor types are leading towards deposits via oxidation and condensation. Primary reactions of fuel/blow-by component promote reactions with oxidatively unstable base oil components.

8. "ANALYSIS OF PV-I TEST VIA USED OILS"

S.W. Harris, D.W. Eggerding and J.H. Udelhofen, Lubrication Engineering 38 (8), 487 – 496 (1982)

MSVC and VD are compared. Bench spot dispersancy tests show that PV-1 drain oils from tests with very good sludge ratings but poor varnish ratings contain incipient sludge which can be coagulated. Leaded fuel causes lower sludge but higher varnish rating than unleaded fuel. Presumably due to the lower temperatures ($\neq 10°$ C). PV-1 drain oils consist of higher levels of $R_2 C = O, RONO_2$, and RNO_2. Varnish samples contain hydroxyl, carbonyl, carboxylate, enolize beta-diketone, nitrite ester and nitro goups, average molecular weight is 770. Average varnish rating appears to correlate with the amount of insolubles below 1.2 μm.

9. "LOW TEMPERATURE ENGINE SLUDGE: WHAT? WHERE? HOW?"

E. Dimitroff and R. D. Quillian Jr., SAE-650255 (1965)

Under low temperature engine operations sludge manifests itself in the engine oil or forms deposits on various engine components. Sludge formation is initiated by liquid oxidation products, inorganic salts, and polymerized organic compounds that pass by piston rings. The liquid oxidation products undergo further chemical reactions in the crankcase oil forming solid "sludge binders". These "binders" are essential ingredients which the oil must contain before the organic solids, inorganic salts, wear particles and soot can be deposited as sludge.

10. "CONDITION MONITORING OF CRANKCASE OILS USING COMPU- TER AIDED INFRARED-SPECTROSCOPY"

J.P. Coates and L.C. Setti. SAE 831681 (1983)

Modern computer assisted Infrared instrumentation enable detailed analyses on the condition of engine oils, e.g. depletion of ZDTP, oxidative degradation, water and glycol contamination, unburnt gasoline and oil nitration.

Analytical Method: IR by direct and differential methods used to study lubri- cant contaminants (water, gasoline) and lubricant degradation.

Engine tests: III D oils including a sludge fail examined.

APPENDIX 2

Table 4.5.1

ANALYTICAL TECHNIQUE	No. of Refs	OVERALL USEFULNESS	USEFULNESS BY REVIEWER			
			#1	#2	#3	#4
ROUTINE DRAIN ANALYSIS						
Blotter spot	8	H	H	H/M	–	H
Elemental analysis	10	M	L	H/L	M	H
Kinematic viscosity	11	M	M	H/M/L	M	M
TAN & SAN	10	M	L	H/M	M/L	M/L
Insolubles (various)	10	M	–	M	M/L	H
Fuel dil & flash point	7	M	L	M	M	M/L
TBN	6	M	–	H	M	M
Water/glycol	4	M	–	M	M	M/L
SEPARATION TECHNIQUES						
Dial./filter/centrifuge	11	H	H	M	M	H
Solvent extraction	7	H	–	M	M	H
GC	4	M	L	H/M	–	–
GPC/HPLC	2	M	–	–	H/M	–
SPECTROSCOPIC METHODS						
IR	26	H	H	H/M	H/M	H
UV/vis spectroscopy	3	M	M	–	–	–
TLC-IR	2	M	–	–	H	L
13-C NMR and 1-H NMR	4	L	L	M	M	L
Mass spectrometry	2	L	–	–	M	L
Chemiluminescence	1	L	L	–	–	–
THERMAL METHODS						
TGA volatility	3	M	–	M	M	H
TGA-FID & TGA-GLC	1	M	–	–	–	M
TGA-MS/GCMS & TGA-IR	1	L	–	–	–	–
Ramsbottom carbon residue	1	M	–	H	–	–
ANALYSIS OF INSOLUBLES						
Optical microscopy	4	M	H	M	–	M
Particle size	3	M	–	–	H/M	M
Gravimetric analysis	3	M	M	–	–	–
Solubility tests	1	M	M	–	–	–
Electrophoresis	2	L	L	–	–	M
Electron microscopy	2	L	L	–	–	M
OTHER METHODS						
Carbon-14 tracers	2	H	–	–	H	H
Titration* (ketones/radicals . . .)	4	M	L	H/M	–	–
Vapour phase osmometry (molecular weight)	3	M	–	M	–	–
Density	1	L	L	–	–	–
Peroxide number	1	L	L	–	–	–
Iodine value (unsat'n)	1	L	L	–	–	–

KEY TO DEGREE OF USEFULNESS:
 H = HIGH; M = MEDIUM; L = LOW

Table 4.5.2

ANALYTICAL TECHNIQUE	No. of Refs	REFERENCES 69 papers reviewed
ROUTINE DRAIN ANALYSIS		
Blotter spot	8	7, 41, 48, 51, 55, 58, 60, 69
Elemental analysis	10	4, 7, 24, 41, 46, 49, 55, 57, 59, 67
Kinematic viscosity	11	7, 20-21, 24, 28, 53, 62-63, 65-66, 69
TAN & SAN	10	6-7, 24, 28, 30, 40, 59, 63, 65, 69
Insolubles (various)	10	7-8, 21, 24, 59, 62, 65-67, 69
Fuel/flash	7	7, 28, 30, 42, 46, 52, 67
TBN	6	7, 28, 30, 36, 63, 69
Water/glycol	4	7, 26, 28, 67
SEPARATION TECHNIQUES		
Dial./filter/centrifuge	11	8, 28-29, 32, 40, 42, 48, 51-52, 55, 58
Solvent extraction	7	5, 8, 18, 29, 33, 35, 37
GC	4	45, 67, 68, 69
GPC/HPLC	2	26, 34
SPECTROSCOPIC METHODS		
IR	26	2-4, 7, 14, 22, 24, 26, 30, 32-36, 39, 41, 49, 52, 55-56, 58-59, 61, 63, 66, 69
UV/vis spectroscopy	3	42, 46, 51
TLC-IR	2	3, 34
13-C NMR and 1-H NMR	4	5, 26, 41, 55
Mass spectrometry	2	5, 26
Chemiluminescence	1	45
THERMAL METHODS		
TGA volatility	3	4, 20, 66
TGA-FID & TGA-GLC	1	8
TGA-MS/GCMS & TGA-IR	1	8
R'bottom carbon residue	1	59
ANALYSIS OF INSOLUBLES		
Optical microscopy	4	14, 48, 51, 66
Particle size	3	8, 20, 28
Gravimetric analysis	3	40, 42, 69
Solubility tests	1	41
Electrophoresis	2	14, 53
Electron microscopy	2	14, 51
OTHER METHODS		
Carbon-14 tracers	2	10, 31
Titration	4	41, 50, 55, 65
Vap. phase osmometry	3	55, 57, 67
Density	1	46
Peroxide number	1	42
Iodine value (unsat'n)	1	46

APPENDIX 3/1

BLOTTER SPOT TEST

OBJECTIVE

To evaluate the "residual dispersive power" of crankcase oil, after subjecting it to heating so that the insoluble products derived from combustion begin to flocculate.

MATERIALS

Test tubes: 10 x 75 mm — Pyrex

Thermostatic bath capable of holding test tubes and offering sufficient thermal inertia (i. e. a block heater)

Thermometer

Special paper (French: Durieux n° 122)
            (British: Whatman Chromo 1)

Mount for special paper

Syringe or disposable pipettes

Drying closet or oven

METHOD

Preparation of samples

Place 2 to 3 cm^3 of oil to be examined in a test tube.

Heat for 5 min at 240° C in a thermostatic block heater.

Allow to cool to room temperature (30 min)

Production of the spot:

Place the special paper in its holder (make sure that it is flat)

Approximately 20 microlitres (one drop) of the test oil is placed on the paper. It may be necessary to touch the paper lightly with the drop of oil if it does not fall on its own accord.

Dry the spots for an hour at 80° C, being sure that the paper is flat. (The evolution of the oil spot stain is accelerated by passage through the drying closet.) The exterior diameter of the spot should measure 32 to 35 mm.

For permanent records, photograph within one hour.

APPENDIX 3/2

EVALUATION POSSIBILITIES

Valuable information can be obtained by comparing blotter spots on oil samples before and after heating at 240° C.

Used oil is in good "dispersive" condition if the distribution of the carbonaceous material is uniform before heat treatment.

"Residual dispersive power" can be estimated by the oil's reluctance to coagulate after heat treatment.

4.6 Development and Application of an "On the Road" Test Method for the Evaluation of Black Sludge Performance in Gasoline Passenger Cars

P.G. Carress
Adibis — BP Chemicals (Additives) Ltd., Redhill, Great Britain

4.6.1 Introduction

The increasing number of incidence of black sludge formation in the field prompted the author's company to investigate the formation of a gasoline car test fleet to reproduce this problem.

This paper describes the methods used and the results obtained.

4.6.2 Test Vehicles

Make:	Ford
Model/Type:	Escort 1.3 C.V.H.
Fuel:	Gasoline
Engine:	Normally Aspirated Carburettered C.V.H.
Special Features:	Transverse Engine — Front Wheel Drive — Overhead Camshaft Hydraulic Followers

4.6.3 Test Fuel

Density at 15°C	mg/m^3	795
RON		100.7
MON		88.8
Lead Content	g/l	0.15
Existent GHM	mg/100ml	1
RVP	bar	0.68
E 70	% V	25.5
E 100	% V	39
E 180	% V	91.5
FBP	°C	221
Benzene	% V	5.5
Toluene	% V	7.1
FIA Aromatics	% V	56.5
Olefins	% V	15.0
Saturates	% V	28.5

509

Methanol	% V	2.9
TBA	% V	1.7
MTBE	% V	0.0

4.6.4 Manufacturers Oil and Service Recommendations

Engine Oil: 10W-30 — 15W-30 dependent on temperature, or meeting Ford Specification SSM-2C 9001-A-A

Engine Oil Capacity: 4.0 litres (7.0 pts)

Engine Oil and Filter change period: 10,000 km

4.6.5 Test Schedule

Engine Preparation: (at each test stage)
Engine flushed, filter replaced, top deck washed clean, rocker cover renewed, test oil added, ignition checks, timing check, timing belt replaced carburetter diaphragm replaced, CO corrected, breather system checked and value replaced.

Drivers:
Ten drivers selected who commute between 5 and 20 miles twice daily, i.e. morning and evening. One driver selected who commutes 70 miles twice daily, i.e. morning, evening and weekends.

Driving Cycle:
The ten drivers to change cars weekly every Monday.
Cars to be driven Monday to Friday on commuter journeys and for approx. 300 miles each weekend with different drivers.

Idle Time:
All vehicles to be idled for 15 minutes mid morning and mid afternoon

Test Duration:
10,000 miles (9 cars)
19,000 miles (1 car)

Oil Changes:
None other than at test start

Oil Filter Changes:
None

Test Oils:
A B C D

Test Inspection:
   At mid point and at test end

Photographs:
   At mid point and at test end

4.6.6   Test Oils

Oil A Poor CEC Sludge Reference Oil

Oil B Good CEC Sludge Reference Oil

Oil C SG/CD, DB 226.5 VW 500/505 10W-40 Oil

Oil D SG/CD, DB 226.5 VW 501/505 15W-50 Oil

4.6.7   Results

Figures 1 to 12 inclusive

4.6.8   Conclusions

All vehicles showed the ability to produce black sludge in varying amounts on the low reference oil. Since no two vehicles give exactly the same results the significance of the ratings and photographs should be assessed individually for each vehicle.

Other than for vehicle C464 TLC which showed a high oil consumption all other vehicles showed similar low oil consumption and therefore this can be discounted as a result influence.

Across all six vehicles all test oils except for Oil D on vehicle B469 XRH showed better rating results and sludge control than Oil A reference oil. Minimum reference Oil B gave better sludge control than the two test oils.

The varying climatic conditions during the tests are effective on the control of black sludge.

The used oil analysis and the aging of the cars are also considered a contributory factor.

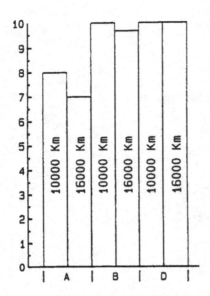

Figure 4.6.1: FORD ESCORT 1.3 C.V.H. (Ref: Fig. 1 and Fig. 2) B811 WRH

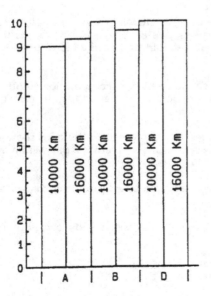

Figure 4.6.2: FORD ESCORT 1.3 C.V.H. (Ref: Fig. 3 and Fig. 4) C464 TLC

512

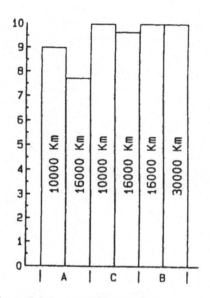

Figure 4.6.3:   FORD ESCORT 1.3. C.V.H (Ref: Fig. 5 and Fig 6) B277 WCK

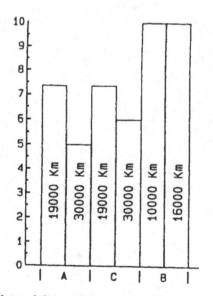

Figure 4.6.4:   FORD ESCORT 1.3 C.V.H (Ref: Fig 7 and Fig 8) B318 XKH

513

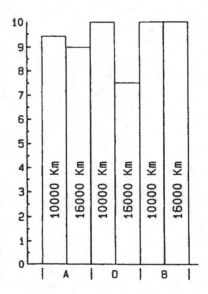

Figure 4.6.5: FORD ESCORT 1.3 C.V.H (Ref: Fig 9 and Fig 10) B469 XRH

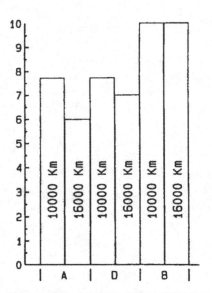

Figure 4.6.6: FORD ESCORT 1.3 C.V.H (Ref: Fig 11 and Fig 12) A505 MKH

514

4.7 Review of Oil Consumption Aspects of Engines

D.C. Roberts, Esso Petroleum Co. Ltd., Abingdon, Great Britain

Abstract

The key performance feature of an engine oil readily visible to the motorist or truck driver is its control of its own consumption. If oil consumption is high it is soon noticed by dip stick checks, oil deposits on parking places, and high levels of exhaust emissions such as blue smoke. If as a result the oil level in the engine sump gets too low, engine damage will result. To most people low oil consumption is synonymous with the oil maintaining good engine performance with low levels of engine wear and low running costs.

Control of oil consumption has always been important, but today it is even more so because of its possible adverse effect on exhaust emissions and the life of exhaust catalysts and particulate traps.

An oil helps control oil consumption by both its additive chemistry and its physical properties. The additive chemistry helps by minimising piston ring and cylinder liner wear, the prevention of bore polishing and piston ring sticking, and ensuring oil seals on the valve stems and crankshafts continue to work effectively. Tests to ensure good oil performance in these areas were developed over many years and they are briefly reviewed in the paper. Of the physical properties of an oil, viscosity and volatility are the two key characteristics known to be important in the control of oil consumption. Over many years viscosity per se was held to be the all important factor, but with the continued trend to lower viscosity oils for improved fuel economy and lubrication at engine start, control of volatility is recognised as the key factor in achieved low oil consumption in today's and tomorrow's hotter running engines.

The paper reviews the key factors affecting oil consumption and the various tests developed to ensure acceptable levels in service in the field under arduous operating conditions. It also reports on a programme of work investigating the relative influence of viscosity and volatility of multigrade oils in different passenger car engine types. The principal findings were that volatility was a more important factor influencing oil consumption than viscosity and that this applied to most passenger car engine types, not just to air-cooled engines. Linear equations were developed for each engine type relating oil consumption to fresh oil volatility at $375°C$ (ASTM-D 2887) and fresh oil kinematic viscosity at $100°C$. These gave an idea of the relevant importance of volatility and viscosity in the different engine types tested.

4.7.1   Introduction and Overview

Since the Otto and Diesel engines were invented over 100 years ago there has always been concern over their oil consumption. Because of this, considerable developments in engine design and engine oil technology have taken place in order to minimize oil consumption as, over the years, operating conditions have become increasingly more severe, oil sump capacities have been reduced, and oil change intervals extended up to 20,000 km for gasoline engine passenger cars and 50,000 km for heavy duty DI diesel engines.

Engine oil is consumed in an automotive engine in a number of ways:

— Past the piston rings into the combustion chamber where it is volatilized and/or burnt and passes to atmosphere via the exhaust system.

— Down the inlet and exhaust valve stems into the combustion chamber and/or the exhaust system.

— Past the crankshaft and camshaft oil seals into the atmosphere.

— Into the atmosphere or engine air/fuel induction system via the crankcase venting system.

Oil flow past the pistons is largely controlled by the piston oil control ring and its design, but the other piston rings, cylinder wall honing, and the fit of the pistons in the engine block all play their part. The oil flows down the valve stems and out of the crankcase via the crankshaft are controlled by valve stem and crankshaft oil seals. In older engine designs most noticeable oil loss was via leaks from the crankcase, particularly via worn crankshaft seals and the air breather, and blue smoke in the exhaust as the result of stuck piston rings. In modern engines with improved crankshaft and valve stem seals, and closed crankcase ventilation system, the reduced amount of oil that is still consumed is lost mainly via the combustion chamber to the exhaust.

The flow of oil through the narrow passages presented by the piston oil control ring and the oil seals is largely controlled by the oil's viscosity. However, piston and ring design, cylinder wall honing, and oil seal designs are all important. The amount of oil that volatilizes in the combustion chamber and other hot parts of the engine is governed by the temperature and pressure the oil is exposed to and its own volatility.

As piston rings and cylinder walls wear, cylinder bores become polished, piston rings tend to stick and oil seals harden and become ineffective, their restriction on oil flow into the combustion chamber or the atmosphere diminishes and oil consumption increases. This can be a gradual long term effect but with the sudden sticking of a piston ring or breakage of an oil seal, oil consumption will suddenly rise.

It is well known that engine oil technology, particularly additive chemistry, can significantly reduce long-term piston ring wear, cylinder liner wear, and bore polishing. Also it can prevent piston ring sticking and maintain oil seals in good condition. Use of such technology can help maintain over the life of the engine excellent control of oil consumption, which is largely governed by the engine oil's viscosity and volatility for a given set of engine operating conditions. Because engine oil technology can influence oil consumption directly or indirectly in various ways, a number of different engine and oil seal tests have been developed over the years aimed at specifying engine oils maintaining long engine life and continued good control of oil consumption.

Well known engine tests that look directly at oil consumption control are:

50hr VW 1302 DKA 6/79 Air Cooled Gasoline Test.

200hr Cummins NTC 400 Turbo Diesel Test.

600hr Mack T6 Turbo Diesel Test.

Engine tests which look at secondary factors affecting oil consumption control are:

Ring Stick	CEC L-03-A-78 100hr Cortina Test
	CEC L-35-T-84 50hr VW 1431 Turbo Diesel Test
Liner Wear	CEC-L-17-A-78 216hr OM 616 Diesel Test
Bore Polish	CEC L-27-T-79 200hr Ford Tornado Turbo Diesel Test
	CEC PL29 300hr OM 364A Turbo Diesel Test

Of the above all but the Cortina and VW 1302 engines are diesel engines. This is because the diesel is more prone to piston ring and liner wear, and bore polishing due to corrosive wear arising from the diesel fuel sulphur content. To guard against oil seal problems CCMC and VW have special seal compatibility requirements covering a number of different standardized elastomer materials. These seal requirements are listed in "Appendix 1".

Except for the VW 1302 test, where basestock volatility is the key factor, engine oil additive chemistry strongly influences the results of the above engine tests and CCMC/CEC & VW seal tests. To formulate oils to pass these other engine tests and seal tests requires careful selection and balancing of the detergent, dispersant, anti-wear, anti-oxidant and viscosity index improver chemistries used. Some generalised comments on the engine oil formulation factors affecting the above engine tests and CCMC/CEC & VW seal tests are as follows:

517

VW 1302	Engine oil volatility the key factor affecting oil consumption.
Cumins NTC 400	Likes lower ash (1.0 %), dispersants, and lower viscosity multigrade oils in preference to monograde oils.
Mack T6	Satisfied with API CE quality 1.0 % ash oils.
Cortina	Satisfied with CCMC GI quality 1.0 % ash oils.
VW 1431 TD	Requires CCMC PD2 1.5 % ash oils.
OM 616	Satisfied with CCMC PD2 quality 1.5 % ash oils.
Ford Tornado	Likes high ash SHPD oils and 15W40 multigrade oils in preference to monogrades.
OM 364A	Likes high ash SHPD oils.
CEC & VW Seals	Sensitive to dispersant type, but also affected be detergent type, sulphur containing additives and certain synthetic basestocks.

Appendix

Elastomer Seal Compatibility Requirements—Engine Oils

(All tests on fresh engine oils)
Limits on Changes in Elastomer Properties

	Hardness Units	Volume %	Elongation at Break %	Tensile Strength %
1. CCMC (CEC L-39-T-87: 168 hours at 150°C)				
RE1 (Fluorelastomer)	0/+5	0/+5	−60/0	−50/0
RE2 (Acrylate)	−5/+5	−5/+5	−35/+10	−15/+10
RE3 (Silicone)	−25/0	0/+30	−20/+10	−30/+10
RE4 (Nitrile)	−5/+5	−5/+5	−50/0	−20/0
2. VW (96 hours at 150°C)				
Fluorelastomer FKM E-281	N/A	N/A	−25/+25	−20/+20
Fluorelastomer FP 7501	N/A	N/A	−25/+25	−20/+15
Fluorelastomer WO 781	N/A	N/A	−15/+10	−10/+15

4.7.2 Factors Affecting Oil Consumption

The mechanism of engine oil consumption is very complex, and there are many factors which influence it. Some factors have a primary influence while others secondary, particularly those affecting the long term changes in oil consumption as the engine ages. These many factors fall into three main categories:

Engine Design;
Engine Operating Conditions;
Engine Oil Technology.

4.7.2.1 Engine Design

Engine Block	— Rigidity and bore diameter
	— Bore roundness/tolerance
	— Cylinder wall honing quality
Piston	— Height and stroke
	— Roundness and tightness in bore
	— Land and skirt clearances
	— Secondary motion and maximum tilt angle
	— Ring positions
	— Grove width/depth
Piston Rings	— Profiles and tensions
	— Gaps and side clearances
	— Oil control ring type and effectiveness
Connecting Rod	— Misalignment and twist
	— Big end/small and offset
Crankshaft	— Oil lip seal effectiveness
Cylinder Head	— Valve stem clearance
	— Effectiveness of valve stem and camshaft oil seals
Crankcase	— Closed ventilation

By good engine design significant reductions can be made in oil consumption. The typical oil consumption figures of modern heavy duty diesels are 0.7 but improved designs can achieve as low as 0.2 g/KW (1)*. Today's engine designers are striving to achieve this and lower levels of oil consumption in moves towards meeting 1991 and 1994 diesel particulate emissions targets. Lower oil consumption will also help reduce bore polishing by reducing the build-up of crown land carbon deposits, and this in turn will reduce the long term increase in oil consumption as the engine ages.

* Numbers in parenthesis designate references at the end of the paper.

4.7.2.2 Engine Operating Conditions

Severe operating conditions, particularly high speed high temperature operation of gasoline engines at full load can cause high oil consumption. In diesel
the fuel sulphur content can have a long term effect due to liner and ring wear
and the build-up of top ring groove deposits. Today's diesel tend to be more
prone to wear under high load medium speed operations.

The influence of operating conditions on oil consumption became abundantly
clear in Europe in the early 1960s with the rapid development of the autobahn,
autoroute, autostrada, and motorway networks. These high speed roads opened
up the opportunity for a large number of cars to be driven for long periods at
high speeds and maximum power. These conditions produced high oil temperatures and often catastrophic damage to the engine primarily as the result of high
oil consumption. This occurred because the cars at the time were mostly lubricated with SAE 10W-30 oils formulated with solvent 100 and 150 neutral base-
stocks of, by today's standards, high volatility. Such oils had given very satisfactory lubrication and oil consumption when vehicle speeds were generally 50
to 80 kmh or lower, and maximum power was rarely used. However, once the
motorists had the chance to put their feet down on accelerator pedals for long
periods, trouble was often in store. The situation was saved by the oil industry
introducing the high viscosity 20W-50 multigrade oils. Their use spread from the
UK, where it was first introduced to satisfy the thirst of the Mini, to all over
Europe and beyond. Now 25 years later many motorists still prefer to use them
as, in their eyes, SAE 20W-50 is synonymous with quality.

4.7.2.3 Engine Oil Technology

The low or improved oil consumption obtained with SAE 20W-50 oils,
compared to the 1960s SAE 10W-30 oils, was attributed by most people at the
time purely to their higher viscosity. However, the penny eventually dropped
amongst oil formulators that such oils had significantly lower volatility, and
that this was really the important factor controlling oil consumption, especially
in hot running engines. As a result oil companies in the 1970s again began to
favour marketing 10W multigrade oils but now formulated with improved
narrow cut mineral or special non-conventional basestocks that gave finished oil
volatilities close to matching that of the SAE 20W-50 oils. These improved low
volatility 10W multigrades give low oil consumption in today's lean or fast burn
engines even under arduous conditions. This is due to the combination of tight
control on oil volatility together with improvements in engine design and engine
oil formulation technology. The latter has led to better control of engine wear,
particularly piston ring and liner wear, less tendency to ring sticking, and
prolonged oil seal life through the use of dispersant addtitives and multifunctional viscosity index improvers compatible with the oil seal elastomers.

Volkswagen, with their long experience of hot running air-cooled engines and preference for monograde SAE 30 or 40 oils, were the first car maker to appreciate the significance of oil volatility as the key factor controlling oil consumption in hot running engines. They have largely led the way in the promotion of European low viscosity, low volatility fuel efficient oils by the introduction of a very tight oil consumption target of 0.4 g/kWh in the DKA 79/6 VW 1302 air-cooled engine test. This limit is one of the key requirements of their VW 500.00 service fill specification for fuel efficient 5W & 10W multigrade oils, and the specification now also includes a severe Noack volatility requirement of 13 % maximum and severe oil seal elastomer requirements.

CCMC (Committee of European Common Market Constructors) have followed Volkswagen's lead and have adopted the 13 % maximum volatility requirement and seal test requirements with other elastomer seal materials for their new CCMC G5 service fill specification for low viscosity oils. Ford and GM have also recently introduced GC distillation volatility requirements into their latest factory fill requirements, but these are less demanding than the VW/CCMC requirements and equate to approximately 20 % maximum Noack.

The Volkswagen & CCMC seal requirements greatly limit the choice of oil additive chemistries that can be used in engine oil formulations, particularly that of the dispersant. The Volkswagen & CCMC volatility requirement and the Volkswagen VW 1302 oil consumption requirement rule out the possibility of formulating 5W and 10W multigrade oils using only conventional mineral basestocks. The lowest volatility that can be achieved with a 10W40 mineral based oil is 14 to 15 %, which gives a typical VW 1302 oil consumption result of 0.65 g/kWh. Typical Noack volatility and VW 1302 oil consumption results for 10W30 and 10W40 oils with different basestock components are shown in Figs. 4.7.1 and 4.7.2 (2). The 10W30 oil was formulated with a conventional solvent 150 neutral basestock; the 10W40 (1965) oil with a mixture of solvent 100 and 150 basestocks; the 10W40 (1975) with a narrow cut solvent 130, and the 10W40 (1985) with a narrow cut solvent 140. The last oil met the CCMC G3 15 % maximum volatility requirement but not the VW 1302 oil consumption requirement or the current VW 500.00 and new CCMC G5 13 % maximum volatility requirement. To meet these it is necessary to use a part synthetic basestock. Lower volatility and oil consumption can be achieved by going to a fully synthetic product. However, it should be noted that some synthetic basestock such as 4 cS PAO are still quite volatile and cannot be used alone to meet the severe volatility and oil consumption requirements.

In European gasoline engines it is generally true that higher oil viscosity gives lower oil consumption. In European turbo charged DI diesels higher ash SHPD oils give lower bore polishing and hence less tendency for oil consumption to increase as the engine ages. There is also evidence that multigrade mineral based oils give lower bore polishing than monograde oils of lower volatility. This is supported by results from Cummins NTC 400 tests (3) depicted in Fig. 4.7.3 that show that multigrade mineral oils give lower oil consumption than monesgrades and that oil consumption falls as viscosity is reduced. Further Cummins

NTC 400 results (3) depicted in Fig. 4.7.4 also show that oil consumption tends to increase as ash content is increased. This is contrary to experience in European engines (Mercedes Benz, Volvo, Scania, DAF, RV1, Perkins Eagle) for which high ash oils (1.8–2.0 %) are recommended for extended drain service.

As far as engine oil technology factors are concerned they can be equivocal and there are no hard and fast rules. What holds for one engine type/design does not necessarily hold for another, particularly with regard to long term effects in DI Diesels.

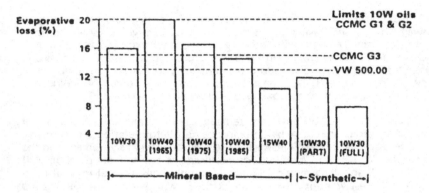

Figure 4.7.1: Volatility Comparison – Noack, DIN 51581, 250°C for 1 Hour

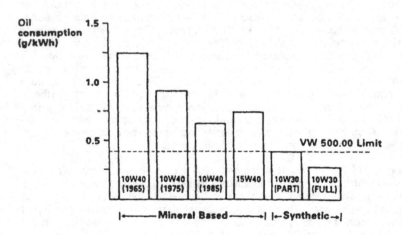

Figure 4.7.2: Oil Consumption Comparison VW 1302, Test, Air Cooled Engine: 50 Hour Duration

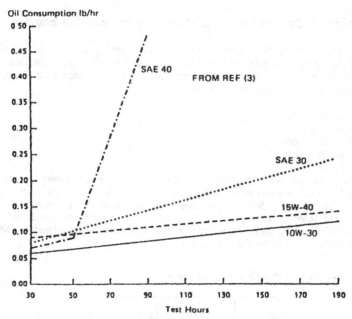

Figure 4.7.3: Cummins NTC 400, Viscosity Grade Effect, API SF/CE

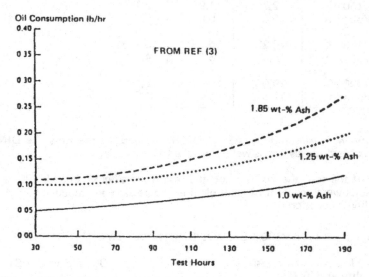

Figure 4.7.4: Cummins NTC 400, Effect of Ash, Wt-%, API SF/CD: 15W-40

523

4.7.3   Influence of Oil Viscosity and Volatility in Gasoline Engines

In 1975 Esso carried out a field test programme at 3 test sites to study the influence of oil viscosity and volatility on oil consumption. 52 gasoline engined cars were used covering volume produced models from Europe, Japan and North America. Four basic engine types were covered:

— 4 cylinder water cooled
— 4 cylinder air cooled
— 6 cylinder water cooled
— 8 cylinder water cooled

and the oil consumption was determined under typical mixed speed driving conditions. The 9 specially prepared multigrade mineral based test oils used, all formulated with the same additive package and VI improver system, were 10W30, 10W40 and 10W50 products. For each viscosity grade there were High, Medium, and Low volatility variants, with KV 100 kinematic viscosities/ASTM D2887 375°C volatilities as shown in Table 4.7.1.

Table 4.7.1:   Test Oil Viscosity&Volatility Data

	High	Medium	Low
10W30: KV100 cS	12.78	11.95	12.15
%Volatility	19.0	6.0	4.0
10W40: KV100 cS	16.23	15.19	16.01
%Volatility	18.0	6.2	4.1
10W50: KV100 cS	20.05	19.40	19.43
%Volatility	17.5	6.4	4.0

The volatility was measured by ASTM D2887 because at the time the DIN 51581 Noack Method was not as well established as it is today.

Full details of the test, the test oils and the results are all given in SAE 890726 (4). The results were analysed by multiple linear regression and modelled on the generalised linear equation

$$\text{Oil Cons.} = B_0 + B_1 . KV100 + B_2 . VOL\ 375. (1/1000\ km)$$

where KV100 is kinematic viscosity cS at 100°C and VOL 375 is % ASTM D2887 volatility at 375°C.

524

The same linear form of equation was found suitable for correlating the data for each engine type. Viscosities and volatilities at other temperatures were investigated but the best correlations were obtained using the values shown. No justification was found for using anything other than linear viscosity and volatility terms, or for the incorporation of a viscosity/volatility interaction term.

The analysis of all the data produced the following oil consumption equations:

Four cylinder air cooled engines

Oil consumption = 0.117 − 0.00124 KV100 + 0.00369 VOL 375

Four cylinder water cooled engines

Oil consumption = 0.191 − 0.00324 KV100 + 0.0049 VOL 375

Six cylinder water cooled engines

Oil consumption = 0.333 − 0.016 KV100 + 0.00412 VOL 375

Eight cylinder water cooled engines

Oil consumption = 0.196 − 0.00498 KV100 + 0.00712 VOL 375

The contributions of viscosity and volatility to the oil consumption of the four engine types are shown in the upper and lower parts of Figs. 4.7.5, 4.7.6, 4.7.7 and 4.7.8. The oil consumption as a function of volatility for oils of 12, 15 and 18 cS, corresponding to XW30, 40 and 50 multigrade, are shown in Figs. 4.7.9, 4.7.10, 4.7.11 and 4.7.12. Also shown are confidence bands for each set of lines. The constant terms (B_0) in the above equations account for variables other than viscosity and volatility held constant during the test programme.

The following conclusions can be drawn from Figures 4.7.5 through 4.7.12 over the range of viscosities and volatilities examined.

- For all four engine types tested oil consumption decreased with increasing viscosity and increased with increasing volatility.

- The influence of viscosity and volatility on oil consumption in the four engines types varied as follows:

	4 cyl. Air Cooled	4 cyl. Water Cooled	6 cyl. Water Cooled	8 cyl. Water Cooled
Viscosity	Very low	Low	Very High	Low
Volatility	High	High	High	Very high

525

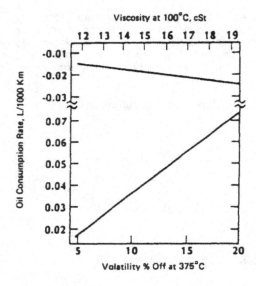

Figure 4.7.5: The Separate Contributions of Viscosity and Volatility to Oil Consumption in Four Cylinder Air Cooled Engines

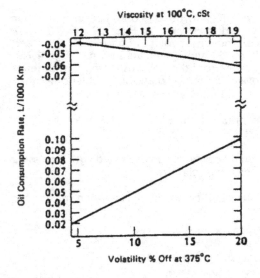

Figure 4.7.6: The Separate Contributions of Viscosity and Volatility to Oil Consumption in Four Cylinder Water Cooled Engines

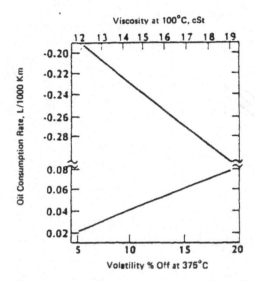

Figure 4.7.7: The Separate Contributions of Viscosity and Volatility to Oil Consumption in Six Cylinder Water Cooled Engines

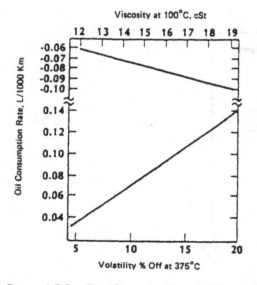

Figure 4.7.8: The Separate Contributions of Viscosity and Volatility to Oil Consumption in Eight Cylinder Water Cooled Engines

527

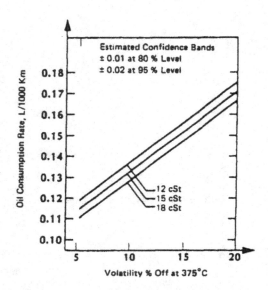

Figure 4.7.9: Oil Consumption as a Function of Volatility in Air Cooled Engines

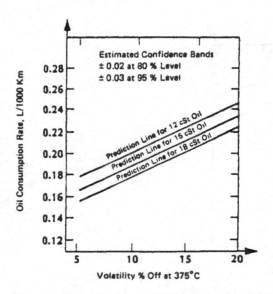

Figure 4.7.10: Oil Consumption as a Function of Volatility in Four Cylinder Water Cooled Engines

528

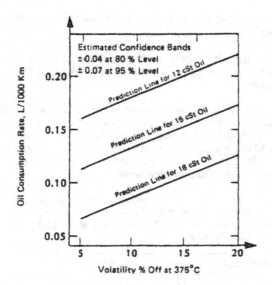

Figure 4.7.11: Oil Consumption as a Function of Volatility in Six Cylinder Water Cooled Engines

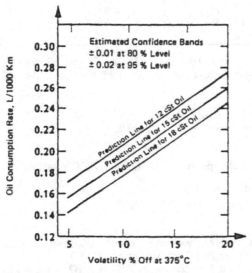

Figure 4.7.12: Oil Consumption as a Function of Volatility in Eight Cylinder Water Cooled Engines

- Viscosity per se at constant volatility was only a significant factor in 6 cylinder water cooled engines

- Volatility per se at constant viscosity was a significant factor in all four engine types, not only in air cooled engines, and had a large effect in the 8 cylinder water cooled engines.

The field test findings for the 4 cylinder air cooled engines were confirmed by carefully controlled VW 1302 oil consumption bench tests carried out on the 10W30 and 10W50 high and low volatility oils used in the field test. These results are shown in Fig. 4.7.13 and show viscosity to have little or no influence, and volatility to be the significant factor. The two low volatility multigrade oils were close to matching a low volatility SAE 30 monograde oil.

Analysis of the field test data showed that it was not possible to generate one equation to correlate the data for the four engine types tested. However, it is possible to calculate overall weighted fleet averages for viscosity and volatility effects in multigrade oils in the following manner.

Engine Type	Percent Change in Oil Consumption	
	1cS Change in Viscosity	1 % Change in Volatility
A.4 cyl.Air Cooled	0.99	2.67
B.4 cyl.Water Cooled	1.75	2.51
C.6 cyl.Water Cooled	11.94	2.99
D.8 cyl.Water Cooled	2.59	3.63

The test fleet consisted of 7 type A cars, 18 type B, 4 type C, and 23 type D. Using these numbers gives a weighted fleet average for 1 cS change in KV100 viscosity of:

$$\frac{7\,(0.99) + 18(1.75) + 4(11.94) + 23(2.59)}{52\ \text{Cars}} = 2.80\,\%$$

and a weighted fleet average for 1 % change in ASTM D2887 375°C volatility of:

$$\frac{7\,(2.67) + 18(2.51) + 4(2.99) + 23(3.63)}{52\ \text{Cars}} = 3.06\,\%$$

The results of the above analysis compare well with other published data (5) which reported a fleet average change in oil consumption of 3 % for a 1 cS KV100 viscosity change, and 2 % for a 1 % change in ASTM 2887 volatility at 375°C.

530

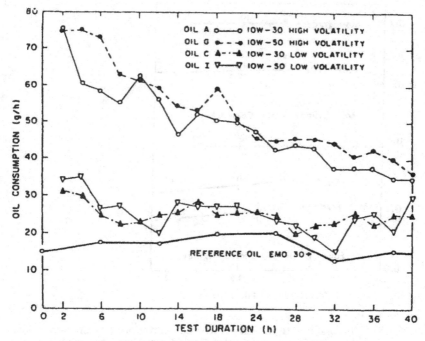

Figure 4.7.13: VW 1302 Weighed Sump Oil Consumption Test

In the 4 cylinder air and water cooled engines very low volatility monograde SAE 30 (K), 40(L) and 50 (N) oils were also evaluated. The results for these three oils are compared against the oil consumption results for the medium volatility 10W30 (B), 10W40 (E) and 10W50 (H) oils in Figure 4.7.14. In the 4 cylinder water cooled engines the multigrade oils gave significantly lower oil consumption than the monogrades; the reverse was true for the 4 cylinder air cooled engines. The lower oil consumption for the monograde oils in the 4 cylinder air cooled engines is simple to explain — lower volatility, but the better performance for the multigrades in the 4 cylinder water cooled engines is an enigma. The Cummins NTC 400 results shown in Fig. 4.7.3 emphasise the point that in certain engines multigrade oils of lower volatility and viscosity can give lower oil consumption than monograde oils of higher viscosity and lower volatility.

531

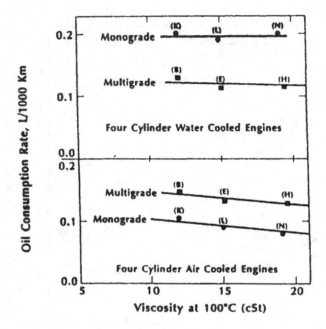

Figure 4.7.14: Multigrade Oils give Lower Oil Consumption than Corresponding Monogrades in Water Cooled Engines, Higher in Air Cooled.

4.7.4 Conclusions

Engine oil is consumed in a number of different ways and the mechanism is very complex. Many factors affect oil consumption and some have a primary influence and others secondary, particularly those affecting the long term changes in oil consumption as the engine ages. These many factors fall into three main categories:

— Engine Design
— Engine Operating Conditions
— Engine Oil Technology

With good engine design low levels of oil consumption can be achieved, even under arduous operating conditions, but the key to this is engine oil technology. This is particularly so if low levels of oil consumption are to be maintained as the engine ages.

High viscosity oils were long held to be the cure for engines with inherently high oil consumption, but now it is seen that it is low volatility rather than high viscosity what is required. Findings regarding oil consumption are also confusing with what holds for one engine type is the opposite to what holds for another. This is demonstrated in Fig. 4.7.14 and by the fact that whereas European turbocharged HD diesels rely on high ash SHPD oils to control bore polishing/oil consumption their North American counterparts rely on medium to low ash oils.

Engine oil additive chemistry can significantly affect long term increases in oil consumption as the engine ages. Because of growing concerns re oil seal performance, the European vehicle makers have recently introduced tough requirements which severely restrict the choice dispersants and other additives used in today's engine oil formulations.

To meet severe European volatility/oil consumption requirements with 5W or 10W multigrade oils, it is necessary to incorporate non-conventional basestocks on a partial basis.

An extensive field test involving 52 cars and four gasoline engine types compared the oil consumption performance of 9 10W30, 10W40 and 10W50 oils of high, medium, and low volatility. Analysis of the results showed the following for the range of oils and cars covered:

- For all four engine types there is a linear relation between viscosity and oil consumption, which decreases as viscosity increases.

- For 4 cylinder air cooled, 4 cylinder water cooled and 8 cylinder water cooled engines the viscosity effect is small, but it has a large effect in the 6 cylinder water cooled engines tested.

- For the 52 cars tested the weighted fleet average figure calculated for percent change in oil consumption for a one centistoke change in kinematic viscosity at 100°C was 2.8 %.

- For all four engine types there is a linear relationship between volatility and oil consumption, which increases as volatility increases.

- Volatility was a major factor controlling oil consumption for all four engine types, and was particularly large for the 8 cylinder water cooled engines.

- The weighted fleet average figure calculated for percent change in oil consumption for a 1 % change in ASTM D2887 375°C volatility was 3.06 %. This and the result obtained for viscosity change compared very favourably with the figures quoted from another test.

— The oil consumption results for the 9 multigrade oils were best correlated with an oil consumption model of the form:

Oil Cons. = B_0 + B_1. KV100 3 B_2. VOL375 (1/1000 km).

where B_0. B_1 and B_2 are constants depending on engine type.

— Because of variations between engine types it was not possible to develop one equation covering all 4. The following 4 equations best covered the oil consumption (OC) data from the field test:

— 4 cylinder air-cooled engines
OC = 0.117—0.00124 KV100 + 0.00369 VOL 375

— 4 cylinder water-cooled engines
OC = 0.191—0.00324 KV100 + 0.0049 VOL 375

— 6 cylinder water-cooled engines
OC + 0.333—0.016 KV100 + 0.00412 VOL 375

— 8 cylinder water-cooled engines
OC = 0.196—0.00498 KV100 + 0.00712 VOL 375

— In a comparison of the multigrade oils against low volatility monograde oils of equal viscosity, the multigrade oils gave higher oil consumption in 4 cylinder air cooled engines, but lower oil consumption in 4 cylinder water cooled engines.

4.7.5 References

(1)   Guertier, R.W.: — Mahle — "Excessive Cylinder Wear and Bore Polishing in Heavy Duty Diesel Engines: Causes and Proposed Remedies", SAE Paper 860165.
(2)   Roberts, D.C.: — Esso — "Why Synthetics in Modern Engine Oils?" Stichting Neder-lands National Cimite Motorporoven Smeerolie En Brandstoffen Symposium Fuels and Lubricants For Future Cars, Rotterdam, April 1987.
(3)   Oronite — "Diesel Engine Oil Technology", February 1988.
(4)   Carey, L.R.; Roberts, D.C.; Shaub, H.: — Esso/Exxon — "Factors Influencing Engine Oil Consumption in Today's Automotive Engines", SAE Paper 890726.
(5)   Didot, F.E.; Green, E.; Johnson, R.H.: — Sun Oil — ,,Volatility and Oil Consumption of SAE 5W-30 Oils, SAE Paper 872126.

4.8 The Contribution of the Lube Oil to Particulate Emissions of Heavy Duty Diesel Engines

P. Tritthart, F. Ruhri and W. Cartellieri
AVL-List Ges.m.b.H, Graz, Austria

4.8.1 Summary

First the analysis of particulates as practiced in the engine development department is introduced, especially to define the quantity of soluble particulates caused by lube oil. Then, the significance of lube oil particulates is demonstrated by means of some peculiarities. The major formation mechanisms and the correlation between lube oil particulate emission and lube oil consumption is demonstrated, from which a strategie for oil particulate reduction is derived.

4.8.2 Introduction

Different studies performed by AVL (1—4) and others (5—10) already showed the importance of lube oil as a significant contributor to particulate emissions.

The first studies concentrated on particulate emissions of diesel engines for light vehicles (passenger cars and light trucks) but in the last years the attention was put to heavy duty diesel engines because of the particulate reductions especially required in the U.S. for this engine category, Table 4.8.1.

After all the experience collected at AVL during the development of diesel engines to meet future particulate limits, we have to state that the particulate fraction caused by lube oil plays a key role, which finally determines whether development work becomes successful or not.

The essential lube oil particulate sources are:
— lube oil from the cylinder wall and
— lube oil from valve stem lubrication

Therfore, these two aspects deserve major attention in engine development.

In the following it is demonstrated which influences on lube oil particulate emissions are possible and in what direction and magnitude future developments will have to be led.

Table 4.8.1: Exhaust emission standards for Heavy Duty Diesel engines in Europe and USA

ECE	HC	CO	NO$_x$	Part.	
Basis	3.5	14.0	18.0	–	
1988	2.4	11.2	14.4	–	
	(–30 %)	(–20 %)	(–20 %)		g/kWh ECE R49
Switzerland					13 Mode-Test
1991	1.23	4.9	9.0	0.7	
USA					
1991	1.3	15.5	5.0	0.25	g/HP-hr Transient Test
1994	1.3	15.5	5.0	0.10	cold 1/7 hot 6/7

4.8.3 Particulate Analysis

All particulate fractions mentioned in the following were derived from the scheme in Fig. 4.8.1. The lube oil particulate fraction is the result of a gas chromatografic analysis which is described in detail in (1). This method is based on the comparison of the 50 %-boiling point of organic soluble particulates and the 50 %-boiling point of various fuel/lube oil-calibrating blends. For the latter only those fuel fractions that boil above 320°C are used, i.e. light fuel fractions are distilled off before. This method implies that the molecular distribution of the defined fuel/lube oil calibrating blend stays the same during its change to the particulate phase. Hence, at a given 50 %-boiling point the molecular distribution of the organic soluble particulate phase should coincide with that of the calibrating blend. Of course, this is a simplistic and not always correct assumption, since even at lowest loads or during motoring – depending on the absolute size of the oil particulate portion – remarkable crack reactions may occur. Thus, originally higher boiling lube oil fractions are split into lighter boiling components. In the gas chromotogram these lighter components are then wrongly identified as fuel portions. Therefore, this method leads to an underestimation of the lube oil portion on particulates. Furthermore, the contribution of lube oil to organic insoluble particulate fractions cannot be defined by this method. Despite these shortcomings the fuel/lube oil separation method has successfully detected lube oil particulate problems in engines since its introduction 8 years ago. Due to automation of gas chromatography laboratory technical analysis can also be kept within reasonable limits. Therefore, this method is conveniently applicable for the daily routine work.

536

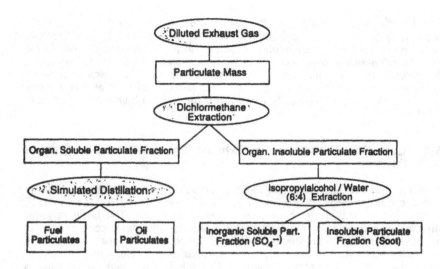

Figure 4.8.1: Analysis of Diesel Particulates (Schematic)

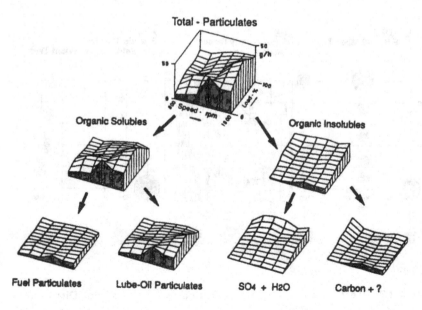

Figure 4.8.2: Topography of Particulate Components from 12 l DI/TCI Heavy Duty Diesel Engine, US-2D Fuel (0.3 % S by Weight)

Fig. 4.8.2 shows the topography of particulate components of a heavy-duty diesel engine (according to the analysis scheme of Fig. 4.8.1). This example presents a typical lube oil particulate problem. In a wide load and speed range lube oil particulates exceed more than half of the entire particulate emission, while carbon portions (dry soot) play a minor role for this engine because of prior combustion optimization. Fig. 4.8.2 also shows that sulphates and water present a considerable magnitude when using fuel with a high sulphur content (0.3 %).

4.8.4 Lube Oil Particulates in Exhaust Emission Tests — Current Position

The test procedure prescribed for exhaust emission tests is primarily essential for engine development. In Europe it is the 13-mode-steady test acc. to ECE R49 and in the U.S. it is the transient test (Fig. 4.8.3). The latter requires significant investments for electronic engine and dynamometer control and for exhaust measurement equipment. To reduce the costs for engine development, an 8-mode-steady state test is helpful to simulate the transient test. A comparison of the weighting factors of points in Fig. 4.8.3 shows that the ECE-test is primarily settled in the medium speed and the transient test in the upper speed range.

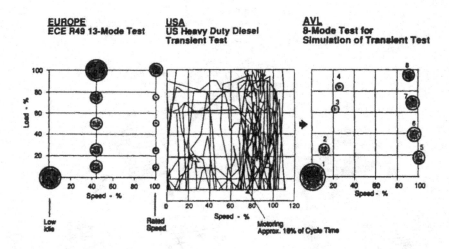

Figure 4.8.3: Test Procedures for Heavy Duty Diesel Engines Time Distribution in Load-Speed Zones

When testing the engine according to ECE R49 and the US-Transient test — depending on the engines type-different results on emissions may occur. Fig. 4.8.4 shows on three combustion-optimized engines of different sizes that in the US-test lube oil particulate emissions are usually higher and therefore, the entire particulate level is increased. This behaviour can also be seen at extremely low oil particulate levels (engine C in Fig. 4.8.4). The increase of oil particulates is especially distinct on engine A, where the lube oil portion increased from 35 to 43 % of the total particulate emission. The reason for this characteristic lies in the engine's load and speed behaviour (Fig. 4.8.5). As this figure shows, lube oil particulates first increase with load to a maximum and decrease then to higher loads to a minimum and then again increase slightly on each engine. With increased speed the maximum generally lies higher and is shifted to lower loads. On engine A the increase with speed is especially distinct which explains its high transient test emission, as this test primarily runs at high speed. These characteristics can be seen even better when comparing the modal lube oil particulate emissions in the 8-mode-test, Fig. 4.8.6. This figure shows that the high speed modes 5 and 6 are primarily decisive for lube oil particulates in the transient test, which are remarkably higher on engine A.

This example already indicates that a potential lube oil particulate problem is primarily a high speed/light load problem, which is manifest only on the transient test because of its high speed weightings. Therefore, all further investigations primarily concentrate on the transient test, as most of the lube oil particulate problems occur during this test procedure.

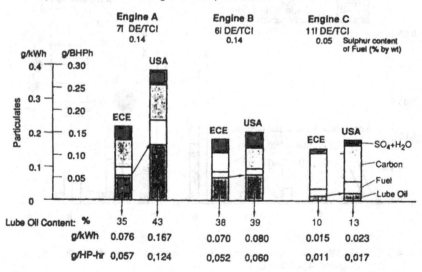

Figure 4.8.4: Contribution of Lube Oil to Particulate Emissions ECE = ECE R49 Test; USA = US Diesel Transient Test (AVL 8-Mode Simulation)

539

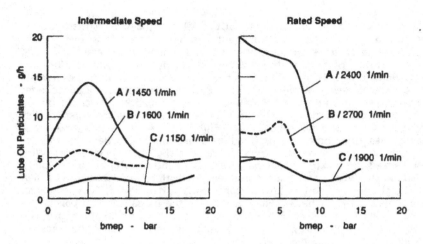

Figure 4.8.5: Effect of Speed and Load on Lube Oil Particulate Emissions from
DI/TCI Diesel Engines
(A ca. 7 l; B ca. 6 l; C ca. 11 l; all 6 Cyl.)

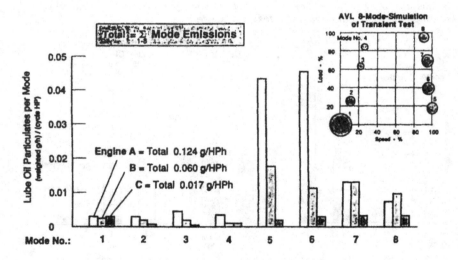

Figure 4.8.6: Modal Lube Oil Particulate Emissions from 8-Mode Test for
Engines A, B and C

4.8.5 Peculiarities of Lube Oil Particulates

4.8.5.1 The Influence of Cooling Water Temperature

Fig. 4.8.7 shows the influence of cooling water temperature on lube oil particulate emission in the transient test on two engines. Two variants are shown for
engine A — variant 1 at a high level of lube oil particulates and variant 2 at a very
reduced level. It can be seen that on engines with a high oil particulate level a
reduction in cooling water temperature by 30°C causes lube oil particulates to
increase by 35 %. This result is backed-up by the experience that oil consumption increases with reduced cylinder wall temperature (higher viscosity, i.e.
thicker lubricating film at the cylinder wall) (11). Surprising is the result on the
lube oil particulate optimized engine A/variant 2, where no influence of temperature exists.

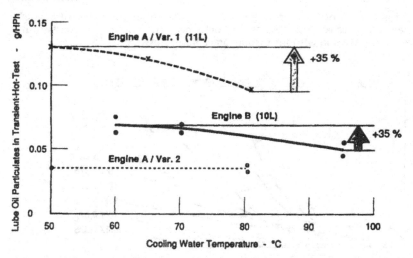

Figure 4.8.7: Effect of Cooling Water Temperature on Lube Oil Particulates in
Transient Test from DI/TCI Diesel Engines

4.8.5.2 Particulate Emission During Motoring

Fig. 4.8.8 shows particulate emission during motoring on various engines. The
corresponding piston displacement as well as the rate of lube oil particulates in
transient test is entered for each curve. The particulate rate during motoring
corresponds to the lube oil consumption (after ensuring that fuel injection is out
of order, as was the case for these tests) during this operation mode. For the
first it is striking, that motored particulate emission of all engines increases with
speed. The steeper the curve, the higher the lube oil particulate emission in the
transient test. This data suggests a correlation of lube oil particulate emissions

541

with the rate of motoring particulates at high speed, as the latter is important in the transient test, Fig. 4.8.9. For better comparison of engines, the particulates rates in Fig. 4.8.9 were normalized by the engine displacement, and the motored particulate value was taken as that of the engine at rated speed.

Fig. 4.8.9 shows that the curves of high quality engines are close together and the transient test oil particulates increase in a linear manner with motoring particulates to a certain level and then go apart. During motoring we found that the exhaust pipes are practically dry in good engines, whilst in engines with high motoring particulate rates the exhaust pipes are oil wetted, and in extreme cases oil drips out. It can be concluded that high oil particulate rates derive from a high oil availability on the cylinder wall, i.e. oil is fed into the exhaust system, partly stored on the walls and then evaporated at a higher exhaust temperature and fed into the particulate measuring system.

It should be mentioned that during motoring, substantial quantities of particulates may also derive from leaky turbocharger shaft seals, as comparative measurements with and without turbocharger have shown.

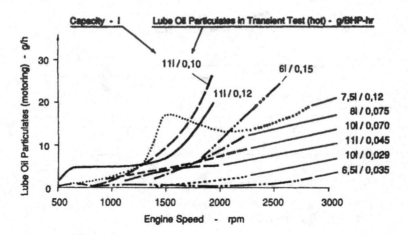

Figure 4.8.8: Lube Oil Particulates from Motored Engine Conditions Compared with Transient Test Oil Particulates

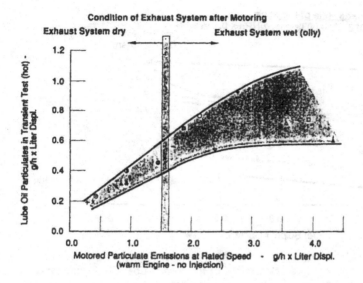

Figure 4.8.9: Lube Oil Particulates in Transient Test as Function of Oil Particulates from Motoring Test at Rated Speed
(1 — 2 l/Cyl.; Rated Speed 1800 — 3200 rpm)

4.8.5.3 Effect of Sulphur Content of Fuel on Lube Oil Particulates

Tests with fuel of different sulphur content showed that particulate emission is
reduced by 0.020 +/- 0.004 g/HP-hr in the transient test on an average of
various engines for each 0.1 weight % sulphur reduction (12). This influence
does not only derive from reduced sulfate formation but also, as Fig. 4.8.10
shows, from the reduction of lube oil particulates. The average lube oil particulate gradient of 0.0062 +/- 0.0033 g/HP-hr and 0.1 wt % fuel sulphur (Fig.
4.8.10) seems to be low, but is still important, if all particulate components
ought to be minimised in order to arrive at total particulates of 0.1 g/HP-hr
without exhaust after treatment.

The property of sulfates to attract hydrocarbons from the gaseous phase and
hence, to increase particulate emissions has already been mentioned (13). As
our tests show, this effect is primarily attributable to lube oil particulates.

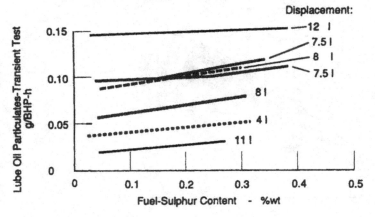

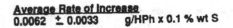

Figure 4.8.10: Effect of Fuel Sulphur Content on Lube Oil Particulates in Transient Test; DI/TCI Diesel Engines

4.8.5.4 Influence of Lube Oil Formulation on Particulate Emission

In earlier investigations (3) we found a remarkable influence of the lube oil evaporation characteristics as expressed by the Noack-value on particulate emissions, especially regarding the organis soluble lube oil fraction. These tests have been effected at a relatively high level of oil particulates. Therefore, further test with a lube oil particulate-optimized engine have been effected to see whether the same effect then occurs also. Fig. 4.8.11 shows the result of these tests with 4 different lube oils. It can be seen that the lube oil particulate fraction is influenced only a little whilst the biggest differences in significant magnitudes occur at the organic insolube particulates. In future more attention will have to be drawn to this fact.

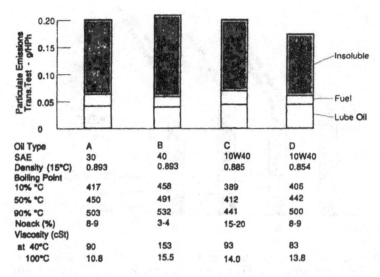

Oil Type	A	B	C	D
SAE	30	40	10W40	10W40
Density (15°C)	0.893	0.893	0.885	0.854
Boiling Point				
10% °C	417	458	389	405
50% °C	450	491	412	442
90% °C	503	532	441	500
Noack (%)	8-9	3-4	15-20	8-9
Viscosity (cSt)				
at 40°C	90	153	93	83
100°C	10.8	15.5	14.0	13.8

Figure 4.8.11: Effect of Lube Oils on Particulate Emissions 11 l DI/TCI Diesel Engine, US–2D Fuel

4.8.5.5 Effect of Valve Stem Sealing on Particulate Emissions

Until now this report is based on the assumption that only lube oil of the cylinder lubrication causes the major portion of oil particulates. But it is also known of various tests (1, 6, 14) that valve stem sealing can present a significant source for lube oil particulates especially at an increased level of lube oil availability on the valve stems. In case of a reduced level (low splash oil, good oil removal) this influence is dramatically diminished.

To define the influence of valve stem sealings at an extremely low level of lube oil particulates from the cylinder, these sealings were gradually removed in the inlet and outlet area, Fig. 4.8.12. As expected, the component of lube oil particulates increased, but proportionally also the carbon fraction, especially when there was no valve stem sealing mounted at the outlet. Thus, it can be concluded that already small quantities of lube oil which get along the valve stems into the induced air or into the exhaust significantly influence the particulate emission. Experience showed that this influence is higher at four-valve than at two-valve-engines due to the double number of sealings.

Similar effects of oil admission at inlet and outlet valves are described in (6), but the levels of oil consumption and oil admission were remarkably higher. Absolutely measured, the effects which are described here are rather low in comparison to those in (6).

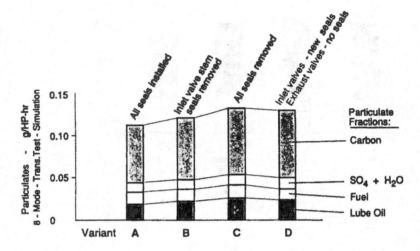

Figure 4.8.12: Effect of Valve Stem Seals on Particulate Emission 8-Mode Trans. Test Simulation; DI/TCI Diesel Engine, 2 Valves/Cyl., Exhaust Emissions Optimized, Fuel Sulphur Content: 0.05 % by wt.

4.8.6 Oil Consumption and Lube Oil Particulates

After all the experiences presented so far, the question arises what connection exists between lube oil consumption and lube oil particulates, especially regarding future targets for lube oil consumption. To follow this question, the oil ring tension has been increased by about 40 % at a lube oil optimized engine. Fig. 4.8.14 shows the results regarding oil consumption, oil particulates and insoluble particulates in significant points of the 8-mode-test (mode 1, 5, 6, 8), in a 4-mode-cycle (consisting of these 4 modes) and finally in the transient test.

The 4-mode-test should show the accuracy of projection from the single modes at a cylcic test driven in a 5 minute rhythm. The weighting and sequence of modes were adjusted to approximately reflect the operating mode of the transient test, i.e. low idling (mode 1) is fixed with 2 minutes duration and all other modes with 1 minute, and mode 8 (high load, high torque) follows immediately after idling (mode 1).

The percental reduction of all components due to the increase of oil ring tension is shown in Fig. 4.8.14.

The following statements can be derived from the results presented in Figs. 4.8.13 and 4.8.14:

— Depending on the load point or operation mode the ratio of oil consumption to lube oil particulates fluctuates between 2:1 and 15:1. No rule can be derived from these ratios. Tendenciously the ratio diminishes with reduced oil consumption, which means that the magnitude of reduction in oil consumption does not reflect completely in reduced lube oil particulate emissions. Mode 5 presents an exception.

— In mode 5 (high speed, low load) the reduction of oil consumption results in an overproportional reduction of lube oil particulates.

— At high loads the reduction of oil consumption primarily affects the reduction of insoluble particulates.

— The projected 4-mode results derived from the single mode results show a relatively good agreement with the dynamic 4-mode-results.

— In the transient test and also in the 4-mode-test the reduction of oil consumption is reflected primarily in the reduction of insoluble particulates, whilst lube oil particulate fractions practically remain unchanged.

— Therefore, it can be concluded that there are no longer any significant oil evaporations and oil feed processes below a certain oil quantity on the cylinder wall, but oil combustion processes forming insoluble particulates (soot, ash) become the dominant mechanism. This experience is in line with the earlier mentioned (Fig. 4.8.7) influence of cooling water temperature on lube oil particulate fractions, which no longer exists at a low level of oil particulates.

The question arises which oil consumption target is necessary to achieve this unsensitive threshold of oil particulates. For that purpose in Fig. 4.8.15 the lube oil particulate emission in the transient test of various engines is plotted versus their oil consumption at mode 8. If it is assumed that based on our tests this threshold is at a lube oil particulate value of 0.03 g/HP-hr, Fig. 4.8.15 shows that an oil consumption of max. 0.19 g/kWh (0.14 g/HP-hr) should be achieved in mode 8. Oil consumption figures of similar magnitude have been recommended in (9).

As the lube oil particulate problem has been identified as one at high speed/low load which is also becoming manifest during motoring, a further criteria, namely motoring particulate emission at rated speed, should be considered. In Fig. 4.8.16 this has been put into relation for various engines to their lube oil particulate emission in the transient test. If again the threshold value of 0.03 g/HP-hr for lube oil particulate emission is required, according to Fig. 4.8.16 the motoring particulate value at rated speed should not exceed 0.5 g/h x 1 piston displacement.

547

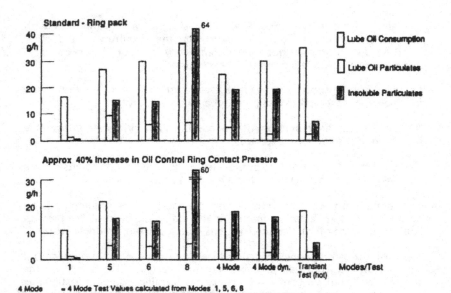

Figure 4.8.13: Effect of Oil Control Ring Contact Pressure on Oil Consumption and Lube Oil Particulate Emissions 11 l DI/TCI Diesel Engine, Optimized Oil Consumption, 230 kW

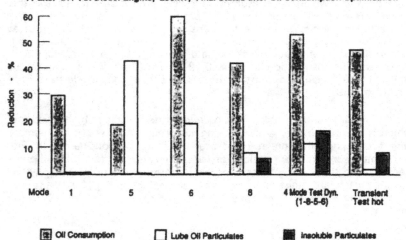

Figure 4.8.14: Reduction of Oil Consumption, Lube Oil Particulates and Insoluble Particulate Fraction by a 40 % Increase in Oil Control Ring Contact Pressure

548

DI/TCI - Diesel Engines, 1-2 l/Cyl., Rated Speed Range 1800 - 3200 rpm

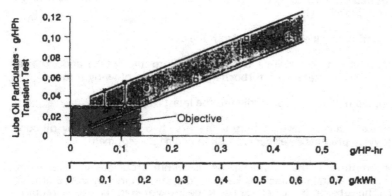

Figure 4.8.15: Lube Oil Consumption, Mode 8 (90 % Speed, 95 % Load). Dependency of (Transient Test) Lube Oil Particulates on Oil Consumption in Mode 8

DI/TCI - Diesel Engines, 1-2 l/Cyl., Rated Speed Range 1800-3200 rpm

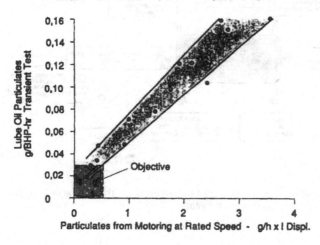

Figure 4.8.16: Dependency of Transient Test Lube Oil Particulates on Particulate Levels from Motoring at Rated Speed

4.8.7 Strategy for Lube Oil Particulate Reduction

In view of gained experience a strategy for oil particulate reduction ought to be put in place on 3 levels:

1) Minimization of oil consumption from the cylinder wall

2) Minimization of all possible oil entrances into the intake and exhaust system like valve stem seals, turbocharger shaft seals, blow-by return

3) Development of lube oil according to the latest requirements.

Engine producers mainly have to pursue the first two tasks, whereby oil consumption anyway presents a permanent task in engine development.

The oil consumption in the cylinder depends on numerous design parameters, which should have certain characteristics to secure the required oil consumption, Fig. 4.8.17. The lube oil performance has to be integrated into all these requirements.

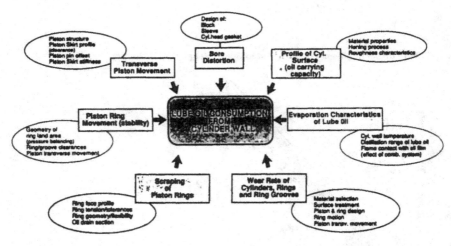

Figure 4.8.17: Criteria and Design Parameters Affecting the Lube Oil Consumption form the Cylinder Wall

4.8.8 Summary and Conclusions

These tests have shown that the organic soluble particulate portion caused by lube oil may contribute with more than 40 % to the entire particulate emission.

On the very same engine, lube oil particulates are generally higher in the US-transient test than in the ECE R49 test, as the US-test puts more weight to higher speeds and is based on a lower load factor, and the lube oil particulate emission generally increases with speed. Thus, work for optimization has to be done especially for the transient test.

The lube oil particulate emission has been primarily identified as a high speed/low load problem, which is also manifest during motoring.

The lubrication of valve stems presents a further essential source for lube oil particulates, but sometimes also the shaft sealing of the turbo charger.

The lube oil particulate emission correlates well with the particulate emission during motoring at rate speed. This value should be $\leqslant 0.5$ g/h x liter piston displacement to achieve a target of $\leqslant 0.03$ g/HP-hr lube oil particulates in the transient test.

The lube oil consumption at high load necessary to achieve the target of lube oil particulates in the transient test ($\leqslant 0.03$ g/HP-hr) should be $\leqslant 0.19$ g/KLh (0.14 g/HP-hr).

At this low level of oil consumption a further reduction in oil consumption hardly influences the organic solube particulate phase. But a reduction in the insolube particulate phase can be stated. Also the oil formulation at a low level of oil consumption primarily influences the insoluble rather than the soluble particulate phase. Therefore, in future the major emphasis has to be put on the insoluble particulate portion deriving from lube oil.

In general it can be stated that secondary influences to the lube oil particulate emission, like cooling water temperature or sulphur content of fuel, are the less distinct the lower the level of oil consumption.

A strategy for oil particulate reduction should be put in place on 3 levels:

- Minimization of oil consumption from the cylinder wall

- Minimization of all oil entrances into the intake and exhaust system

- Development of lube oils according to the latest requirements.

551

4.8.9 References

(1)   Cartellieri, W.; Tritthart, P.: "Particulate Analysis of Light Duty Diesel Engines (IDI & DI) with Particular Reference to the Lube Oil Particulate Fraction", SAE Paper 840418.

(2)   Cartellieri, W.; Wachter, W.F.: "Status Report on a Preliminary Survey of Strategies to Meet US-1991 HD Diesel Emission Standards Without Exhaust Gas Aftertreatment", SAE Paper 870342.

(3)   Cartellieri, W.; Herzog, P.L.: "Swirl Supported of Quiescent Combustion for 1990's Heavy-Duty DI Diesel Engines — An Analysis", SAE Paper 880342.

(4)   Cartellieri, W.; Ospelt, W.M.; Landfahrer, K.: "Erfüllung der Abgasgrenzwerte von Nutzfahrzueg-Dieselmotoren der 90er Jahre", Motortechn. Zeitschrift 50 (89) 9.

(5)   Mayer, W.J.; Lechmann, D.C.; Hilden, D.C.: "The Contribution of Engine Oil to Diesel Exhaust Particulate Emissions", SAE Paper 800256, also in SAE Proceedings, P-86 "Diesel Combustion and Emissions", 1980.

(6)   Maurer, M.: "Beeinflussung der Partikelemission eines Dieselmotors durch das Schmieröl", Dissertation (1986), Fakultät für Maschinenwesen der Rheinisch-Westfälischen Technischen Hochschule Aachen.

(7)   Springer, K.J.: "Diesel Lube Oils — 4th Dimension of Diesel Particulate Control", ASME Conference on Engine Emissions Technology for the 1990's, San Antonio, Texas, October 1988.

(8)   Shore, P.R.: "Advances in the Use of Tritium as a Radiotracer for Oil Consumption Measurement", SAE Paper 881583.

(9)   Moser, F.X.; Haas, E.; Schlögl, H.: "Die Partikel-Hürde U.S. 1991: Vergleich der Testverfahren für Nutzfahrzeugmotoren-Bewältigung mittels innermotorischer Maßnahmen", 10. Internationales Wiener Motorensymposium, April 1989.

(10)   Lewinsky, P.M.; Cooke, V.B.; Andrews, C.A.: "Lubrication Oil Requirements for Low Emission Diesel Engines", AVL-Conference "Engine and Environment", Graz, Aug. 1.—2. 1989.

(11)   Rulfs, H.W.: "Untersuchung des Schmierölverbrauchs eines aufladbaren Dieselmotors unter Anwendung eines Radioisotops", VDI-Forschungsheft Nr. 601 (1980).

(12)   Cartellieri, W.: "Die Partikelproblematik aus der Sicht der Motorenentwicklung von Nutzfahrzeug-Dieselmotoren", AVL-Conference "Motor und Umwelt", Graz, 1.—2. August 1989.

(13)   Wall, J.C.; Hoekman, S.K.: "Fuel Composition Effects on Heavy-Duty Diesel Particulate Emissions", SAE Paper 841364.

(14)   Amano, M.; Sami, H.; Nakogawa, S.; Yoshizaki, H.: "Approaches to Low Emission Levels for Light-Duty Diesel Vehicles", SAE Paper 760211.

4.9 Gasoline Engine Camshaft Wear: The Culprit is Blow-By

J. A. McGeehan and E. S. Yamaguchi
Chevron Research Company, Richmond, USA

4.9.1 Abstract

We were able to identify engine blow-by as a primary factor affecting camshaft wear in gasoline engines. Using a 2.3-liter overhead-camshaft engine, we isolated the valve-train oil from the crankcase oil and its blow-by using a separated oil sump. We find that:

— with engine blow-by, the camshaft wear was high.
— without blow-by, the camshaft wear was low.
— with blow-by piped into the isolated camshaft sump, the wear was high again.

Later studies identified nitric acid as a primary cause of camshaft wear. It is derived from nitrogen oxides reacting with water in the blow-by. But even in the presence of blow-by, camshaft wear can be controlled by the proper selection of zinc dithiophosphates (ZnDTP) and detergent type.

4.9.2 Introduction

Today's multi-valve gasoline engines rely on two or four overhead camshafts for their high performance and fuel economy. One of the keys to their success is camshaft wear control. But what are the causes of camshaft wear?

An overhead camshaft design allowed us to study the effects of engine blow-by on wear. Simply by isolating the valve-train oil from the crankcase oil and its blow-by, using a separated sump, we could run a fire engine with or without blow-by entering the valve-train area.

Operating with this configuration, we found that:

1. Blow-by affects camshaft wear and ZnDTP structural changes — for example, in its depletion rates.

2. Nitric acid is a primary cause of this wear. It is derived from the nitrogen oxides reacting with water in the blow-by.

3. The key to low wear — in the presence of blow-by — is the proper selection of ZnDTP and detergent type. Phosphorus, zinc, and sulfur from the ZnDTP

553

must adsorb, react, and remain as an intact film on the wear surface, in spite of the continuous production of nitric acid.

These findings were derived from a sequence of tests using a 2.3-liter overhead-camshaft engine, operating at Sequence V-D conditions. This paper is organized in the sequence of test events which led us to the final conclusions shown above.

4.9.3 Blow-by Caused Engine Deposits in Gasoline-Engines

We began this study because of the lack of correlation between bench test and engine test results on camshaft wear. We already knew that blow-by was a critical element in deposit formation. As early as 1932, researchers identified nitric acid and "nitro-hydrocarbons" in used engine oils, and associated them with resin formation (1—3).*

However, Diamond et al. (4), could not produce varnish deposits by introducing nitrogen dioxide into a non-fueled, motored engine. Later, Spindt et al. (5), pin-pointed the fact that deposits are caused by the reactions between oxides of nitrogen and unsaturated fuel constituents. They also found that high nitrogen content resulting from lean combustion mixtures, combined with low operating temperatures, caused major varnish deposits.

Spindt and Wolf (6) and Dimitroff et al. (7) provided evidence that related engine varnish to certain types of fuel constituents or their combinations, specifically diolefins, heavy aromatics, and certain naphtenes.

Rogers et al. found that low temperature sludge was derived almost exclusively from fuel degradation products. These products arrive in the crankcase lubricant mainly as low molecular weight oil solubles, which then undergo further reactions and are converted into an oil-insoluble form. Rogers also showed that the idling period was significant in increasing sludge deposits (8).

Following this research, Quillian et al. (9) demonstrated that varnish and sludge deposits could be significantly reduced by diverting the blow-by from the crankcase oil. Vineyard and Coran (10) also used blow-by diversion to collect the liquid products of blow-by, and to implicate these materials as deposit precursors. Their analysis showed that α,β-nitro-nitrates were present in the blow-by. Dimitroff et al. (11) measured NO_x (NO, NO_2) in the blow-by and found that its level was dependent on spark-timing.

As a result of this earlier research, later ASTM tests for sludge and varnish deposits — VB, VC, V-D, and VE — have operated with high blow-by rates, high NO_x conditions, low coolant temperatures, and idling cycle, and fuels prone to deposit formation (12—14).

---

* Numbers in parentheses designate reference at end of paper.

4.9.4   Diverting the Blow-by from the Camshaft

The Sequence V-D test introduced in 1980 used a 2.3-liter single overhead-camshaft engine (see Figure 4.9.1). In this test, the blow-by rises through the oil drain holes in the cylinder head and exits out the rear end of the rocker cover. The blow-by then passed through an external heat exchanger — maintained at engine coolant temperature — and finally exits through the positive crankcase pressure valve (PCV). This valve regulates the amount of blow-by being drawn from the crankcase to the vacuum side of the inlet manifold.

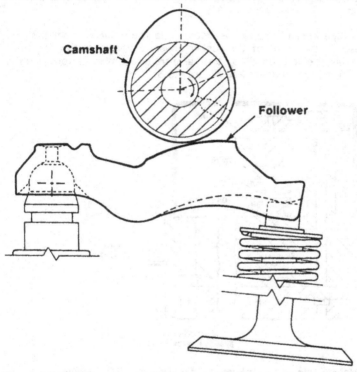

Figure 4.9.1:   Sequence V-D Overhead Camshaft Engine. Camshaft and Cam Follower

This engine configuration made it easy to divert the engine blow-by from the camshaft. Camshaft oil was isolated from crankcase oil by a few simple modifications:

1. Blocking off the normal oil drain holes in the cylinder head with tapered, threaded plugs, and providing alternative drain holes at the side of the cylinder head. Unblocked oil drain holes normally allow blow-by to rise through the valve-train area.

555

2. Blocking off the normal oil supply gallery to the camshaft from the crank-case oil pump with a Lee plug and substituting an external oil pump to supply the oil gallery in the cylinder head.

3. Providing a separate external oil sump and pump for the valve train.

With these changes, we could maintain the valve train oil at normal V-D test temperatures. The combination of frictional heating and cylinder deck heating raised the oil temperature, but we could use a heat exchanger to cool and regulate it (see Figure 4.9.2 and Table 4.9.1).

When using the separated oil sump, we could pipe the blow-by from the side of the crankcase to the front of the rocker cover. Entering at this location, the blow-by travels the full length of the valve-train assembly before it exits through the cooling heat exchanger (see Figure 4.9.3).

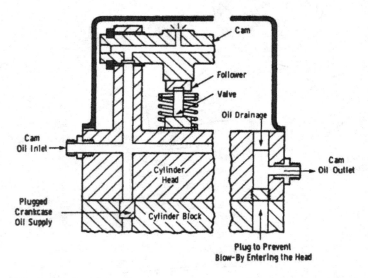

Figure 4.9.2: Valve-Train and Crankcase Oil Separated in 2.3 L Engine

556

Table 4.9.1: V-D Operating Conditions — Basic Information

| Cycle | rpm | bhp (kW) | Temperature | | | Inlet Air Humidity, % | Blow-by, cfm | Blow-by Condenser Temperature, °F (°C) | Spark Timing, °BTDC |
			H₂O Out, °F (°C)	Oil Gallery, °F (°C)	Inlet Air, °F (°C)				
I	2500	33.5 (25)	135 (57)	175 (79)	80 (27)	1.14	1.8-1.4	135 (57)	10
II	2500	33.5 (25)	155 (68)	187 (86)	80 (27)		—	155 (68)	46
III	750	1.0 (0.75)	120 (49)	120 (49)	80 (27)	—	—		46

Test Length: 192 Hours

No Follower Limits

Piston Ring End Gaps Increased to 0.051 Inches (1.295 mm)

Camshaft Wear Limits:
Average: 0.0010 Inches (0.0254 mm)
Maximum: 0.0025 Inches (0.0635 mm)

557

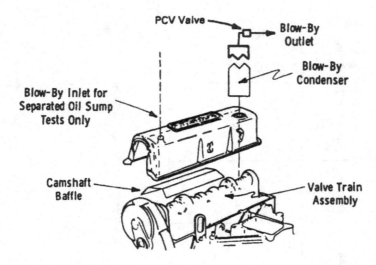

Figure 4.9.3:   Blow-by Inlet and Outlet in Separated Oil Sump System

4.9.5   Low Wear with no Blow-by

In our earlier research we developed a formulation which consistently produced high wear in a normal V-D engine configuration. This oil contained primary ZnDTP at a 0.05 % phosphorus level, in an all-calcium detergent package. Its sulfated ash was 0.76 %, and its viscosity grade was 10W-30 (we refer to this as Formulation A).

Using this oil in these tests, we found that:

1. With blow-by in the normal V-D engine configuration, valve-train wear was high in two separate tests.

2. Without blow-by in the separated-oil-sump configuration, valve-train wear was low in two separate tests.

3. With blow-by piped in to the separated-oil-sump, valve-train wear was high again (see Figures 4.9.4 – 4.9.6).

With the blow-by piped in to the front of the rocker cover and exiting at the rear, the cam lobe wear became progressively worse toward the rear of the engine. We assumed that increased wear at the rear was associated with increased condensation of the gas as it traveled from the front entry to the rear

558

exit. The rear lobes 7 and 8, which were not covered by the camshaft baffle*
and were directly below the cooling condenser, exhibited the greatest wear
(see Figure 4.9.6). Although we conducted only one test at this condition,
the wear results were the same as those of a normal V-D test with this oil.

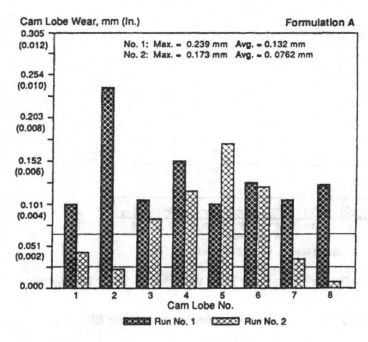

Figure 4.9.4: Normal V-D Cam Lobe Wear

---

* The camshaft baffle is a part added to the normal head assembly. First, the baffle prevented oil normally sprayed through the camshaft feed holes from washing sludge off
  the rocker arm cover. Second, it provided an additional part to rate for varnish (14).

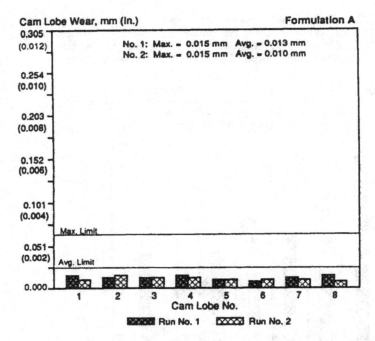

Figure 4.9.5:   No Blow-by in the Valve Train. Separated Oil Sump Configuration

4.9.6   Search for the Wear-Causing Component in the Blow-by

Initially, we ran two experiments to find out what blow-by component causes the wear. In the first experiment, we separated the liquid and gas phases of the blow-by, and returned only the gas phase to the engine. In the second experiment, we returned only the liquid phase to the engine.

First Experiment: (Only the gaseous phase returned to the engine.) It is difficult to separate the gas phase from the liquid phase due to the small particle sizes. So we decided to try blow-by cooling followed by a series of filtrations. Consequently, the crankcase blow-by passed sequentially through the following separators:

1. water-cooled heat exchanger.
2. low speed cyclone filter.
3. high speed cyclone filter.
4. Balston paper filter, which removed particles larger than 0.6 micron and is 99 % effective. The filter element was changed every 24 hours (see Figure 4.9.16).

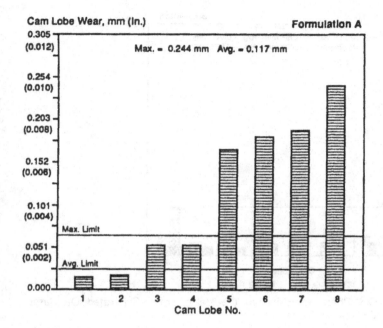

Figure 4.9.6: All the Blow-by Returned to the Valve Train. Separated Oil Sump Configuration

These separators removed 14 quarts (13.2 liters) of liquid from the blow-by in the 192-hour test. Separation was not complete, however, as evidenced by the lacquer that formed on the inside of the clear plastic tube connecting the paper filter to the rocker cover inlet. When only the "gas phase" was returned to the engine, the valve-train wear was low (see Figure 4.9.7).

Second Experiment: (Only the liquid phase was returned to the engine.) In this experiment, too, the wear was low. First, we injected the liquid phase into the separated sump at the same rate that it was extracted from the blow-by — 60 mL/hr. The camshaft seized in its bearing after 48 hours at test conditions, probably due to the fact there was 32 % water in the oil. However, the cam lobe wear was low. On the next test we decided to inject the liquid phase at a slower rate of 6 mL/hr. At this rate of injection, the test was successfully run for 192 hours with low wear (see Figure 4.9.8 and Table 4.9.2). We achieved this result in spite of the fact there was 2 % water in the oil at the end of the test, which is 10 times the normal water content.

Neither of these two experiments provided the key to the failure mechanism. But, along with the previously completed tests, they provided important information on ZnDTP depletion rates in the used oil.

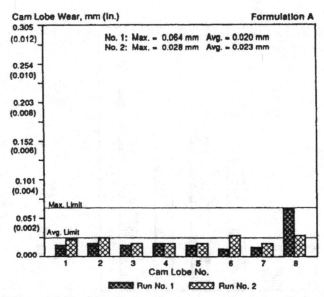

Figure 4.9.7:   Gas Phase Returned to the Valve Train. Separated Oil Sump
Configuration

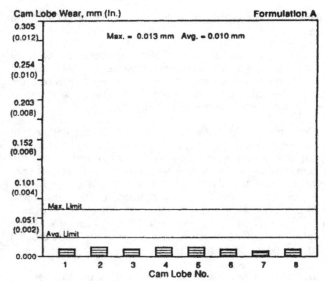

Figure 4.9.8:   Liquid Phase Returned to Valve Train. Separated Oil Sump
Configuration

4.9.7 Blow-by Causes ZnDTP to Deplete

It is known that the chemical structure of fresh ZnDTP changes in the engine environment. This change can be monitored by:

— Thin Layer Chromatography (TLC)
— Fourier Transform-Infrared Spectroscopy (FT-IR)
— ^{31}Phosphorus Nuclear-Magnetic Resonance Spectroscopy (^{31}P-NMR)

ZnDTP's are salts of esters of dithiophosphoric acids. TLC indicates only the presence or absence of the intact dithiophosphate anion. In contrast, the FT-IR indicates the changes in the P-O-C and the P=S bonds. ^{31}P-NMR monitors the ZnDTP conversion to other phosphorus-containing compounds.

We used all three of these techniques. Although these results agreed, the ^{31}P-NMR was the most specific (see Appendix Table 4.9.2). ^{31}P-NMR allows one to look at the fate of the ZnDTP during the engine test. The fresh or intact ZnDTP is composed of two distinct phosphorus-containing species — neutral and basic — which are clearly identified in the ^{31}P-NMR spectrum (see Figure 4.9.15). Any changes in these two signals, or any new signals which appear in the spectrum, indicate that the ZnDTP is being converted to other P-containing compounds.

Our ^{31}P-NMR results clearly demonstrated that only total blow-by causes complete ZnDTP depletion, while its components cause only moderate depletion rates and low cam lobe wear, as summarized below:

1. ZnDTP was fully depleted within 24 hours in the presence of total blow-by, and the wear was high.
2. ZnDTP remained intact — with no structural changes — in the absence of blow-by, and the wear was low.
3. 65 % of the ZnDTP was depleted after 192 hours in the presence of the liquid phase of the blow-by, and the wear was low.
4. 60 % of the ZnDTP was depleted after 100 hours in the presence of the gaseous phase of the blow-by, and the wear was low.

Blow-by affects both ZnDTP depletion rates and valve-train wear. We therefore analysed the composition of the blow-by to help elucidate the mechanism of ZnDTP depletion on the resulting wear rates.

4.9.8 Analysis of the Blow-by

The gaseous phase of the blow-by was collected and analysed by high resolution mass spectrometry (HRMS). The HRMS showed NO_x (NO and NO_2) in the sample. It also showed CO_2, O_2, and N_2.

The condensed-liquid phase, which we collected at room temperature, contained approximately 20 % hydrocarbons. Hydrocarbons in the same carbon number range as the fuel accounted for 90 % of this layer, while hydrocarbons (oil) greater than C_{14} were responsible for the remaining 10 %. FT-IR analysis showed organic nitrates, but no peroxides.

The remainder (80 %) of the liquid phase was analysed by Dionex Ion Chromatograph. The identifiable anions were:

Cl^{-1}	(10 ppm)
SO_4^{-2}	(310 ppm)
NO_3^{-1}	(1300 ppm)

Clearly the data shows that NO_3^{-1} is the predominant ionic species and suggests that it was derived from the nitrogen oxides reacting with water in the blow-by. The aqueous phase was very acidic (pH 2), and we attribute this acidity to nitric acid (HNO_3).

4.9.9 Nitric Acid Causes Valve-Train Wear

From all these findings, three facts were apparent:

1. Blow-by causes high camshaft wear with the specific formulation used throughout our studies.
2. Only the total blow-by causes ZnDTP to deplete within 24 hours. Its components — gas or liquid phases — showed significantly decreased depletion rates, and produced low wear.
3. The predominant anion in the condensed liquid phase of the blow-by was NO_3^{-1}.

This, combined with the strong acidity of the liquid phase (pH 2), led us to conclude that nitric acid (HNO_3) had formed. This strong acid probably gives rise to the rapid change in the ZnDTP structure, and may cause the wear inhibitors to become inactive.

Our experiment in which we injected the liquid phase only contained 2 to 9 millimoles of strong acid and caused only moderate ZnDTP depletion (see Appendix). These findings led us to the following premise: that it is the continuous production of NO_x, and its reaction with water produced in combustion, that forms enough nitric acid to cause rapid ZnDTP depletion.

Consequently, we estimated the total production of nitric acid in the blow-by during the 192-hour test. These calcuations, shown in the Appendix, suggest a total production of 82 millimoles of nitric acid. We actually observed 110 millimoles, based on the collection of 14 quarts of liquid blow-by. This small amount of nictric acid may cause ZnDTP depletion by:

— Acid-catalyzed hydrolysis and/or
— Non-catalyzed hydrolysis.

To test our hypothesis that HNO_3 is the harmful component in the blow-by, we vaporized solutions of nitric acid directly into the valve train and camshaft compartment of the separated-oil-sump configuration. We did one engine test with 10 times the 110 millimoles of nitric acid referred to above. In this test, we observed high wear, in the range for this oil in a conventional Sequence V-D test (see Figure 4.9.9).

4.9.10 Wear can be Controlled in the Presence of Nitric Acid

Our previous V-D wear results indicated that 0.05 % phosphorus oils can be developed to provide low wear — even in the presence of blow-by. These results showed that either magnesium or calcium detergents can be used in a low wear formulation provided the right type of ZnDTP and detergent is selected (15) (see Figure 4.9.10).

With calcium detergent low wear was achieved with either secondary or mixtures of secondary and primary ZnDTP's. In contrast, with magnesium detergent, low wear was achieved with primary ZnDTP.

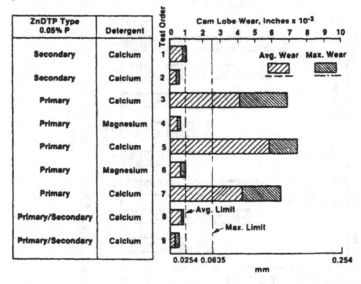

Figure 4.9.10: Sequence V-D Study at 0.05 % P in 0.76 % Ash Formulation. Effect of Detergents and ZnDTP Types on Wear (15)

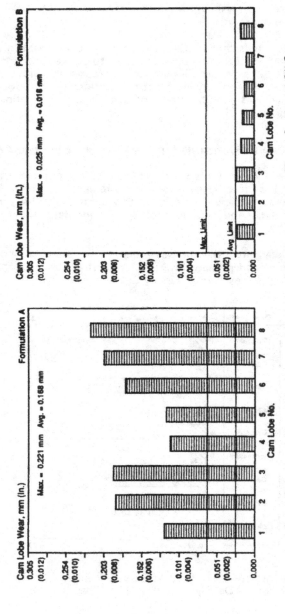

Figure 4.9.9 and 4.9.11: Nitric Acid (10 x 110 Millimoles) Returned to Valve Train. Separated Oil Sump Configuration

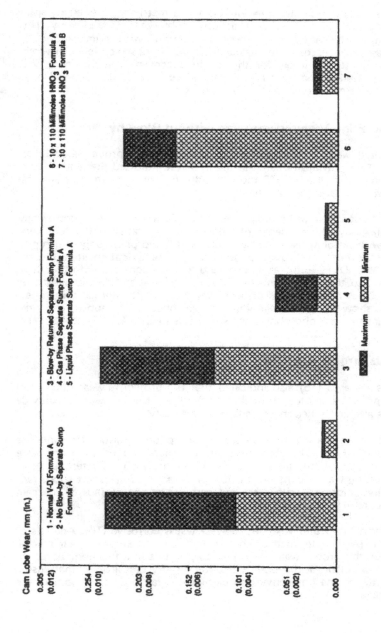

Figure 4.9.12: Summary of All Engine Test Results

So we selected a low wear formulation from this matrix and tested it, again using 10 times the 110 millimoles of nitric acid observed in the blow-by. This oil (Formulation B) contained magnesium detergent with primary ZnDTP, equal TBN, ash, and viscosity to Formulation A. We observed low valve-train wear comparable to the levels for this oil in a conventional Sequence V-D test (see Figures 4.9.11 and 4.9.12). We therefore concluded that the HNO_3 is definitely the harmful component in the blow-by.

4.9.11 Wear Film Analysed with and without Blow-by

The key to low wear in the presence of blow-by is the proper selection of ZnDTP type. Our previous research in the V-D test indicated that sulfur, zinc, and phosphorus from the ZnDTP must be adsorbed on the camshaft and follower to provide a protective wear film.

To confirm these previous findings, we used ESCA (Electron Spectroscopy for Chemical Analysis) and depth profiling on the surfaces of the followers. In our tests with no blow-by, gas phase only, and liquid phase only which produced low wear films, we found phosphorus in the oxidation state present in the original ZnDTP, sulfur as sulfide, and zinc. In contrast, with the blow-by piped back to the valve train, there was no sulfur as sulfide and almost no phosphorus. Zinc, however, was present. This agrees with normal engine test results which show that low wear films contain phosphorus, sulfur, and zinc, and high wear films lack phosphorus and sulfur (see Figures 4.9.13 and 4.9.17).

4.9.12 Metallurgical Analysis

Our previous research (15) also indicated that if a protective wear film is formed, it will protect the integrity of the camshaft and the follower and provide low wear. We also confirmed these findings in this study.

In all the low wear tests, a film containing phosphorus, sulfur, and zinc was present on the surfaces. In these tests the camshaft and its followers were completely intact, exhibiting very low wear. In the high wear test, abrasive wear and scuffing was evident on the followers; and the camshaft subsurface was plastically deformed, with fractured carbides at the surface of the cam lobes (see Figures 4.9.14 and 4.9.18).

West et al. postulated that wear in the V-D test is corrosive (16). Analysis of our high wear parts at the end of the test indicates an abrasive wear mechanism. Lack of the proper wear films probably results in a two-body abrasion. We suggest that nitric acid caused corrosion of the cast iron — martensitic matrix — allowing the unreactive iron carbides to fall out of, or abrade, the follower surface.

568

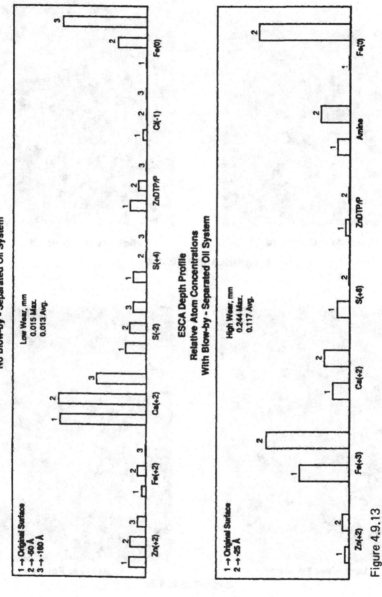

Figure 4.9.13

569

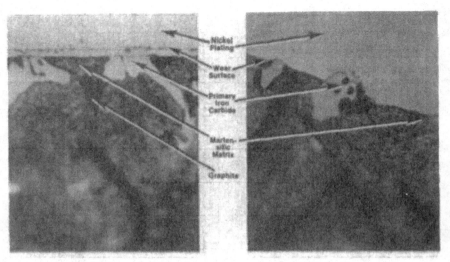

Low Wear Camshaft　　　　　　　High Wear Camshaft

Magnification 1500x

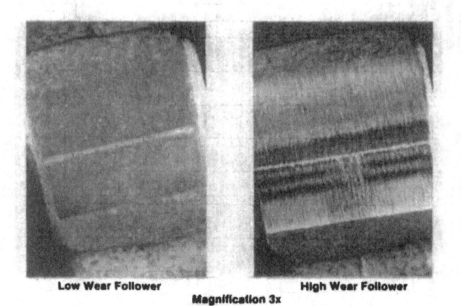

Low Wear Follower　　　　　　　High Wear Follower

Magnification 3x

Figure 4.9.14: Camshaft and Followers from Low and High Wear Tests

570

4.9.13 Other Studies Relevant to these Findings

C. T. West et al. (16) have also examined various factors, including blow-by, associated with wear in *motored*-engine tests. Their results were consistent with the hypothesis that blow-by is an important contributor to wear in moderate temperature gasoline service. In addition, they attempted to simulate the blow-by by making up a solution of water, nitric acid, sulfuric acid, and acetic acid.

Testing in an *electrically driven motored* engine produced significant wear; however, it was less than that observed in their fired engine tests. They therefore concluded that factors other than simulated blow-by are important. They proposed oxidative and thermal degradation as contributors to wear based on motored-engine tests. However, there was substantial variability in their tests. In contrast, our study puts the blame for valve-train wear squarely on blow-by, and specifically on the HNO_3 produced during the combustion process.

In gasoline engine studies, Murakami et al. (17) found that the concentration of nitrate ion in the oil was five times greater than that of the nitrous ion (50 ppm versus 10 ppm). However, below the dew point of water vapor, the nitrous ion concentration rapidly increased and exceeded the nitrate ion concentration. We believe that the rapid increase in the concentration of ion, as the engine temperature drops below the dew point, occurs according to the following reaction in the gaseous phase (18):

$$NO + NO_2 + H_2O = 2HNO_2$$

It is known that aqueous solutions of nitrous acid are unstable and decompose on heating as shown in the following reaction:

$$3 HNO_2 = HNO_3 + H_2O + 2 NO \text{ (19)}$$

In fact, with Sequence V-D operating temperatures of 125°F (52°C), the rate constant favors the production of nitric acid in the liquid phase (19). This would explain our observation of nitrate ion in the aqueous phase of the Sequence V-D blow-by. It is present at 26 times the concentration observed by Murakami in his oil analysis.

Our analysis of the aqueous phase of the blow-by and our subsequent test results indicate that it is the strong nitric acid that causes the camshaft wear. However, the weak nitrous acid may be involved in the wear mechanism when the engine is operating below the dew point of water vapor.

In the new Sequence V-E test, the valve-train temperature is 7°F above the dew point of water to 5°F below the dew point (20). At these conditions both nitric and nitrous ions should be present and they may both contribute to the wear. Even in this environment, however, wear can be controlled by the proper selection of ZnDTP and detergent type. With the proper selection, a protective wear film containing zinc, phosphorus and sulfur will form and remain intact despite the acid environment.

This acid may be transported to the oil by two sources — the blow-by and oil on cylinder wall. According to Murakami et al. (17), the NO_x which transforms into nitric acid in the oil mixes primarily through the oil film on the cylinder wall and through oil in the piston ring grooves. Their finding suggests twice the acount is transported via cylinder walls versus the blow-by.

Recently, Habeeb et al. (21) have suggested a different mechanism, not based on acid corrosion. They suggest that hydroperoxides, produced by oil oxidation, cause camshaft and follower wear by direct attack on rubbing surfaces. The authors use about an 18-fold excess of hydroperoxide to ZnDTP to obtain their relatively high (90) HPN (hydroperoxide number = millimoles hydroperoxide/1000 g oil). The estimate of this excess assumes an engine oil formulation of about 0.1 % P as ZnDTP, standard stoichiometry of organic hydroperoxides in iodometric titration procedures (22), and the chemistry outlined in Equations 1 and 2 shown below. In contrast, in their controlled field test study, three used oils that had previously passed the Sequence V-D test gave HPN's averaging less than 45 at the end of 36.000 km.

Their study suggests to us that all the ZnDTP in an average engine oil formulation would be consumed by the hydroperoxide when it is added at the concentration required to have HPN of 90. We believe that Habeeb et al. observed such high wear because there was no ZnDTP available to form an effective antiwear film. Instead, a set of ZnDTP depletion reactions occurred in the contact area where the antiwear performance is needed (23, 24). The chemistry involved was probably:

Equation 1

$$4[(RO)_2 PS_2]_2 Zn + ROOH \rightarrow [(RO)_2 PS_2]_6 Zn_4 O + [(RO)_2 PS_2]_2 + ROH$$

Equation 2

$$[(RO)_2 PS_2]_6 Zn_4 O \leftrightarrow 3[(RO)_2 PS_2]_2 Zn + ZnO$$

Willermet and Kandah (25) have found that the disulfide species is less effective than the ZnDTP as an antiwear agent. Our own work on Sequence V-D camshaft oil film temperatures is consistent with Equation (2), becoming important at temperatures between $100 - 130°C$. In this range, the basic ZnDTP species formed in Equation (1) decomposes to the corresponding neutral ZnDTP species and zinc oxide. Eventually, both the neutral and basic ZnDTP species will be totally consumed by the hydroperoxide, leading to the observation of high valve-train wear.

572

4.9.14 Conclusions

Using a 2.3-liter overhead-camshaft engine, we isolated the valve-train oil from the crankcase oil and its blow-by using a separated oil sump. Operating this engine at low to moderate coolant temperatures — Sequence V-D conditions — we reached the following conclusions:

1. Blow-by has a dramatic effect on ZnDTP depletion rates and camshaft wear.

2. Nitric acid is a primary cause of valve-train wear. It is derived from the nitrogen oxides reacting with water in the blow-by.

3. In the presence of blow-by, the key to low valve-train wear is the proper selection of ZnDTP and detergent types. Phosphorus, sulfur, and zinc from the ZnDTP must *adsorb, react, and remain as an intact film* on the wear surfaces in spite of the continuous production of nitric acid.

4. High wear parts indicate that there is an abrasive wear mechanism. The lack of the proper wear films probably results in two-body abrasion. The abrading particles are most likely broken pieces of iron carbides. We suggest that the nitric acid causes corrosion of the cast iron — martensitic matrix — allowing the unreactive iron carbides to fall out and abrade the follower surfaces.

4.9.15 Acknowledgements

The authors would like to thank Chevron Research Company and the Oronite Technology Division for their permission to publish this paper, and express their appreciation to:

— J. Q. Adams for his ESCA analysis of the camshaft followers.

— K. K. Kirkman for his metallurgical analysis of the valve-train parts.

— S. Kwok who was responsible for all the engine tests and assembly of the separated sump.

— P. R. Ryason for his review of this paper and his suggestions.

— T. V. Liston for all his support and guidance in these studies.

— And finally, we would also like to thank S. Korcek and M. Johnson of Ford Motor Company for their review of this paper and making us aware of references on blow-by composition.

APPENDIX

BLOW-BY CAUSES ZnDTP TO DEPLETE

ZnDTP depletion was determined by TLC, FT-IR, and ^{31}P-NMR. Although the results from each of these techniques were in agreement, the discussion will focus on the most specific among them — ^{31}P-NMR spectra. ^{31}P-NMR allows one to follow the fate of the ZnDTP during the course of the engine test, since only the ZnDTP contains P in the engine oil formulation. Any new signals that appear in a ^{31}P-NMR spectrum are indicative of ZnDTP depletion and conversion to P-containing compounds whose structures differ from that of the neutral and basic species of the ZnDTP, and which are less (or not) effective wear inhibitors.

The intact ZnDTP is actually composed of two distinct chemical structures, the neutral and the basic species (see Figure 4.9.15). Engine test conditions may result in reactions of these species. For the purpose of this discussion, a loss of neutral or basic species will be termed ZnDTP depletion.

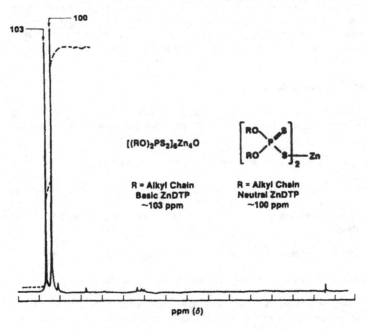

$$[(RO)_2PS_2]_6Zn_4O$$

R = Alkyl Chain
Basic ZnDTP
~103 ppm

R = Alkyl Chain
Neutral ZnDTP
~100 ppm

ppm (δ)

Figure 4.9.15: ^{31}P-NMR Spectrum

Figure 4.9.16: Blow-by gaseous phase returned to engine. Crankcase blow-by passed through the following separations: Water-cooled heat exchanger, low speed cyclone filter, high speed cyclone filter, and Balston paper filter. Those separations removed 14 quarts (13.2 liters) of liquid

575

NO BLOW-BY — ^{31}P-NMR studies were obtained on the used oil (192 hours) from the separate oil system. This spectrum showed that the ZnDTP remained intact [basic species (103 ppm); neutral species (100 ppm)]. No ZnDTP depletion was observed. This finding was supported by both thin layer chromatography and FT-IR. The used oil did not appear "used" (dark).

GASEOUS BLOW-BY — ^{31}P-NMR spectra were obtained of the used oils after both 100 hours and 192 hours of gaseous only blow-by introduction. We found that approximately 50 % of the alkyl ZnDTP remained intact after 100 hours of engine operation. This is much longer than the 24 hours normally required to deplete the ZnDTP when the full engine blow-by is used. We observed total ZnDTP depletion in the spectrum from the 192-hour sample. No ZnDTP depletion in the spectrum from the 192-hour sample. No signals at 103 ppm and 100 ppm were observed. All the ^{31}P-NMR signals were observed at less than 50 ppm, which indicate structures in which the phosphorus is bonded directly to oxygen rather than sulfur. We therefore concluded that the introduction of the gaseous phase of the blow-by decreased the rate of ZnDTP depletion.

LIQUID BLOW-BY — ^{31}P-NMR spectra were obtained of the end-of-test oils that had been treated with various percentages of liquid blow-by. We did not observe complete ZnDTP depletion by ^{31}P-NMR analysis in any of the end-of-test oils, whether the oil was exposed to 2 % or 32 % liquid blow-by. The fresh oil showed basic and neutral species at 103 and 100 ppm, respectively as in Figure 4.9.15. The used oil spectrum (Table 4.9.2) showed that 35 % of the ZnDTP remained intact at the end of the test from the 2 % liquid blow-by addition experiment. The acid species (86 ppm), the monothiophosphate (69 ppm), and phosphate species (< 50 ppm) were also present. This partial depletion of the ZnDTP was also indicated by our supporting FT-IR and thin layer chromatography studies. We concluded that the introduction of the liquid phase of the blow-by showed a significantly decreased rate of ZnDTP depletion.

In summary, our ^{31}P-NMR results clearly demonstrate that only the total blow-by causes complete ZnDTP depletion, and experiments with its components cause only moderate depletion rates.

Table 4.9.2: Analysis of Used Oils from V-D Separate Oil, System Blow-by Effects on Engine Test Oils

| | ZnDTP Depletion | |
	No Blow-By	With Blow-By
Test Hr	192	192
TLC, %	Little	100
FT-IR, %	5	100
^{31}P-NMR, %	5	100

576

Effect of Gaseous Phase of Blow-by on Engine Test Oils

	ZnDTP Depletion
Test Hr	192
TLC, %	100
FT-IR, %	100
^{31}P-NMR, %	100

Effect of Liquid Phase of Blow-by on Engine Test Oils

Run	ZnDTP Depletion			
	With Whole Blow-By, 24 Hr	With Liquid Blow-By EOT Water Content of Oil		
		20 % 24 Hr	32 % 48 Hr	2 % 192 Hr
TLC, %	100	Some	Some	Some
^{31}P-NMR, %	100	35	85	65
FT-IR, %	100	Some	Some	Some

NITRIC ACID CAUSES CAMSHAFT WEAR

The experiments with only the gaseous phase of the blow-by show that this phase alone is not responsible for the high wear normally observed from our 0.05 % phosphorus-containing oil.

Similarly, the liquid phase of the blow-by is not responsible for the normally observed high wear condition from this oil. Total ZnDTP depletion was not observed after 192 hours of engine testing.

The explanation for this phenomenon may be understood by considering the following. Addition of approximately 2.0 % of the liquid blow-by (pH 2) is equivalent to 2.2 millimoles of strong acid (probably nitric acid based on the blow-by analysis). [This 2 % is based on the end-of-test (EOT) water content of the oil (see Table 4.9.2)]. The all-calcium detergent (120 millimoles) in the formulation provides more than enough base to neutralize this small amount of acid, based on the assumption of a 3-quart valve-train compartment. No ZnDTP depletion would be expected due to the acid-catalyzed hydrolysis. Thus, a substantial amount of the primary ZnDTP would be expected to remain intact, even after 192 hours of engine operation under these conditions.

In the case of total blow-by, one must consider how much NO_x is produced during the 192-hour test. We calculated this quantity using:

1. The Sequence V-D operating conditions.
2. Theoretical air/fuel ratio on the Phillips "J" reference fuel.

3. An estimated airflow and NO_x concentrations for each of the three stages of the test.
4. An estimate that blow-by gases contain approximately 10 — 15 % of exhaust gases.
5. An estimate that increasing the end gaps from 0.015 — 0.051 in. should approximately double the blow-by produced.

This is shown below:

CALCULATION OF NO_x PRODUCTION DURING SEQUENCE V-D TEST

To Find the Moles of N_2,
O_2, CO_2, and H_2O in
the Exhaust Gas
_____

It is given that

Fuel: $CH_{1.69}$
Molecular Weight (MW) = 13.71 g/Mole
Air/Fuel (A/F) = 15/1
By Stoichiometry,
$CH_{1.69} + 1.422\ O_2 \rightarrow CO_2 + 0.845\ H_2O$
and 1 g of Fuel Corresponds to 0.073 Moles

15 g Air Corresponds to 0.519 Moles
From $(0.79 \times 28.01) + (0.21 \times 32.00) = 22.13 + 6.72 = 28.85$ g/Mole = MW of Air

This Means there is $0.21 \times 0.519 = 0.109$ Moles O_2
and $0.79 \times 0.519 = 0.410$ Moles N_2

But Combustion of 1 g of Fuel Uses up O_2, so the Moles of O_2 in the Exhaust Gas is: $0.109 - 0.104 = 0.005$ Moles O_2 from $(0.073$ Moles $\times 1.4225)$

And 0.062 Moles H_2O from $(0.073$ Molex $\times 0.845)$
and 0.073 Moles CO_2 from $(0.073$ Molex $\times 1)$
To Find the MW of the Exhaust Gas

We calculate the mole fraction (x) shown below:

Mole Fraction	Moles of Each Constutuent	MW	MW of Exhaust
0.410/0.550 = 0.745	0.410 N_2	28.01	20.88
0.005/0.550 = 0.009	0.005 O_2	32.00	0.29
0.073/0.550 = 0.132	0.073 CO_2	44.01	5.81
0.062/0.550 = 0.113	0.062 H_2O	18.02	2.04
Check = 0.999	0.550 = Total		29.02 g/Mole

To Find Moles of NO_x
Produced During the Entire Test

	Exhaust Flow Rate (Estimated)	Theoretical NO_x, ppm
Stage I	243 Lb/Hr = 3801 Moles/Hr	3200
Stage II	243 Lb/Hr = 3801 Moles/Hr	3200
Stage III	40.9 Lb/Hr = 641 Moles/Hr	Na[1]

[1] No data available

So,

	Moles NO_x Produced/Hr
Stage I	12.16
Stage II	12.16
Stage III	2.05[2]

[2] (Assumes 3200 ppm NO_x Level)

In the entire test, we calculate that 1.97×10^3 Moles of NO_x is in the exhaust produced from:

$$12.16 \times 156 \text{ Hr} = 1.896 \times 10^3$$
$$2.05 \times 36 \text{ Hr} = 0.074 \times 10^3$$

$$1.97 \times 10^3$$

P. A. Bennett (26) et al. and J. A. Spearot and N. E. Gallopoulos (27) have concluded that blow-by gases contain approximately 10 – 15 % exhaust gases, using engines very different from the 2.3 L Ford engine. If we assume the minimum value of 10 %, then, in the entire test, we calculate that at least 1.97×10^2 of NO_x is produced in the blow-by based on the quantities produced in the exhaust.

However, for the Sequence V-D engine test, the end gaps have been increased from 0.015 to 0.051 in. This increase is estimated to give a doubling of the blow-by constituents. Thus, we estimate that 3.94×10^2 moles of blow-by NO_x is produced in the Sequence V-D engine test.

Calculation of the Total
Amount of Nitric Acid (HNO_3)
in the Blow-By

From the mole fraction of water and the blow-by rate, we calculated the total production of water, assuming a temperature of 70°C (150°F), which is well within the range of oil gallery temperature for the Sequence V-D test.

579

To Find the Moles of H_2O in the Blow-By:

We can get an estimate of the H_2O present in the blow-by by: 0.113 (mole fraction H_2O) x 3.02 m^3/hr = 0.34 m^3/hr of H_2O.

If we assume 70°C (343°K),

Then (0.341) 273/343 = 0.271 m^3/hr at STP.

For the entire test, this is: 0.271 m^3/hr x 192 Hr x 1000 L/m^3 = 52 x 10^3 L of Water Vapor at STP

So,

$$\frac{52 \times 10^3 \text{ L}}{22.4 \text{ L/Mole}} = 2.32 \times 10^3 \text{ Moles } H_2O \text{ in the Blow-By} =$$

$$41.8 \times 10^3 \text{ g} = 41.8 \text{ L} = 11 \text{ Gal.}$$

If we again assume that the blow-by contains 10 % exhaust, and estimate that increasing the end gaps will double the amount of blow-by produced; then 8.36 L (2.2 Gal.) of H_2O would be produced in the blow-by.

However, we, in fact, collected 14 quarts (13.2 L) of liquid blow-by, 80 % of which was the aqueous phase. This means that 10.6 L of H_2O was collected. Our collection was known to be incomplete (see Figure 4.9.16), because it was done at an elevated temperature (75°F). An elaborate, cooled-trap system would be necessary to collect the full amount. Nevertheless, the collected amount is well above the calculated amount.

So, based on the 10.6 L (2.8 Gal.) of collected aqueous phase, we find that:

Since pH $= -\log[H+]$
  2 $= -\log[H+]$
  [H+] $= 0.010$ M

and assuming complete dissociation of HNO_3 to ions,* H+ and NO_3^-, there is at least 0.106 moles HNO_3 in the blow-by.

The fact that 10.6 L of water was collected, and (10.6 liters x 0.01 moles per liter), suggests that 106 millimoles of HNO_3 was produced in the blow-by during the entire Sequence V-D test. To get an appreciation for this value, we note that 106 millimoles is only 0.027 % of the N in the NO_x in the blow-by, assuming the blow-by contains 10 % exhaust and estimating that increasing the end gaps will double the amount of blow-by produced. Obviously, a very small amount goes a long way to produce high wear. Under these conditions, ZnDTP depletion would be expected, perhaps by:

---

* Ion — An atom or group of atoms that carries a positive or negative charge.

1. Acid-catalyzed hydrolysis and/or
2. Non-catalyzed hydrolysis, in spite of the presence of 120 millimoles of base in a 3-quart valve-train compartment.

If we disregard the fact that blow-by gases contain a percentage of the exhaust gases and the effects of end gap increases on blow-by, then we derive a number of 0.418 moles HNO_3 in the blow-by with 41.8 L (11 Gal.) of water produced. Obviously, these are maximum values. Our experimentally observed level of the total amount of aqueous blow-by is well within the minimum and maximum values.

This discussion is summarized in Table 4.9.3.

Table 4.9.3: Comparison of Minimum, Maximum, End Gaps Increased, and Observed Values of NO_x, HNO_3, and H_2O in the Blow-by of Sequence V-D Test

	Minimum Value (Assuming 10 % Exhaust Gas in Blow-By)	End Gaps Increased in Sequence V-D From 0.015 to 0.051 In.	Observed	Maximum Value (Assuming Blow-by is 100 % Exhaust)
NO_x Produced in Blow-By (Moles)	1.97×10^2	3.94×10^2	Not Measured	1.97×10^3
Total Amount of H_2O in the Blow-By (Gal.)	1.1	2.2	2.8	11
HNO_3 in the Blow-By (Moles)	0.041	0.082	0.106	0.418

When we injected the liquid phase into the engine, we used a rate of 6 mL/Hr. This suggests, based on the 106 millimoles HNO_3 from the collected blow-by and the fact that 80 % of the liquid blow-by is the aqueous phase, that about 9 millimoles of acid was initially injected. At the end of test, the oil was found to have about 2.2 moles.

In our study, we did not directly measure the concentrations of the oxides of nitrogen (NO_x = NO + NO_2) in the blow-by. However, in a previous study, our colleagues (28) have determined values for crankcase NO and NO_2 in a Ford gasoline engine, which agree with the level and direction indicated by Messrs. Korcek and Johnson (29).

It is now apparent why we saw less ZnDTP depletion and low wear in the case of the gaseous phase only or the liquid phase only experiments. It takes both NO_x and H_2O to form the deleterious HNO_3. With no blow-by added, neither of the two necessary reactants was available for HNO_3 productions, and low wear resulted. The depletion of the ZnDTP, and consequently the valve-train wear, are dramatically affected by blow-by. Our analysis indicates that ZnDTP depletion is caused by excess nitric acid, which results from the interaction of water and NO_x produced by combustion. This excess nitric acid is also responsible for the production of inactive or less active wear inhibitors.

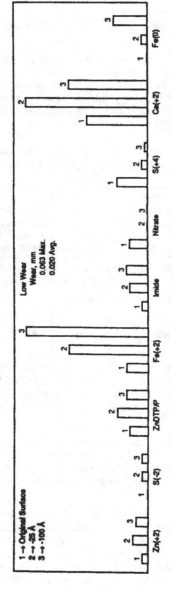

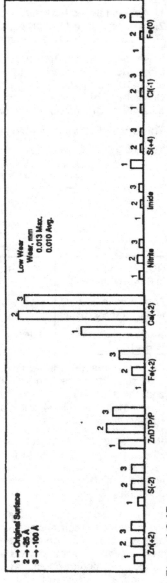

Figure 4.9.17

Low Wear

High Wear

**Mild Abrasion and Numerous Pits
on Surface of Cam Lobes (500x)**

**Abraded Wear Surface on
the Cam Lobe (500x)**

Figure 4.9.18: Cam Lobe and Cam Follower with High and Low Wear

4.9.16 References

(1) Hanson, T.K., Egerton, A.C.: "Nitrogen Oxides in International Combustion Engine Gases", Proc. of the Royal Society of London, Series A, Vol. 163. November 1937.

(2) Graefe, E.: "Petroleum (Berlin)", Vo. 27, No. 23. Motorenbetrieb u. Maschinen-Schmierung, Vol. 5, No. 6, pp 5 and 6, 1932.

(3) Ehler, K.: "Petroleum (Berlin)", Vol. 28, No. 41. Motorenbetrieb und Maschinen-Schmierung, Vol. 5, No. 10, p 8, 1932.

(4) Diamond, H.; Kennedy, H.C.; Larson, R.G.: "Investigation of Deposits and Oil Deterioration Phenomena in Motored Engines", American Chemical Society, Petroleum Division, p 147, September 1947.

(5) Spindt, R.S.; Wolfe, C.L.; Stevens, D.R.: "Nitrogen Oxides, Combustion and Engine Deposits", SAE Transactions 64, 797, 1956.

(6) Spindt, R.S.; Wolfe, C.L.: "The Where and How of Engine Deposit", SAE Journal 60, 39 (1952); Paper No. 636, SAE Meeting, August 14, 1951.

(7) Dimitroff, E.; Moffitt, J.; Quillian, R.D.: "Aromatic Compounds in Fuels Identified as Main Precursors of Engine Varnish", SAE Journal, Vol. 77, p 52—58, 1969.

(8) Rogers, D.T.; Rice, W.W.; Jonach, F.L.: "Mechanism of Engine Sludge Formation and Additive Action", SAE Transactions Vol. 64, pp 783—796, 1956.

(9) Quillian, R.D.; Meckel Jr., N.T.; Moffitt, J.V.: "Cleaner Crankcase With Blow-by Diversion", SAE Paper No. 801B, 1964.

(10) Vineyard, B.D.; Coran, A.Y.: "Gasoline Engine Deposition: Blow-by Collection and the Identification of Deposit Precursors", American Chemical Society, Division of Petroleum Chemistry, Ind., pp A25—A32, 1969.

(11) Dimitroff, E.; Moffitt, J.V.; Quillian Jr., R.D.: "Why, What, and How: Engine Varnish Formation", Preprint, ASME and ASLE Lubrication Conference: October 8—10, 1968.

(12) Bardy, D.C.; Franklin, T.M.; Roberts, C.E.: "A 2.3-L Engine Deposit and Wear Test — An ASTM Task Force Report". SAE Paper 780260, 1978.

(13) Harris, S.W.; Eggerding, D.W.; Udelhofen, J.H.: "Analysis of the PV-1 Test Via Used Oils", "Lubrication Engineering". Vol. 38, 8 487—496, 1980.

(14) Nahumck, W.M.; Hyndman, C.W.; Cryvoff, S.A.: "Development of the PV2 Engine Deposit and Wear Test — An ASTM Task Force Progress Report", SAE Paper 872123, 1987.

(15) McGeehan, J.A.; Yamaguchi, E.S.; Adams, J.Q.: "Some Effects of Zinc-Dithiophosphates and Detergents on Controlling Engine Wear", SAE Paper No. 852133, 1985.

(16) West, C.T.; Passut, C.A.; Chamot, E.: "Wear Mechanisms in Moderate Temperature Gasoline Engine Service", SAE Paper No. 860374, 1986.

(17) Murakami, Y.; Aihara, H.; Kuniya, J.: "Analysis of Mechanism Intermixing Combustion Products in Engine Oil", Nippon Kikai Gakkai Ronbunshu, B Hen, Vol. 53, No. 496, pp 3814—3821, December 1987.

(18) Cotton, F.A.; Wilkinson, G.: "Advanced Inorganic Chemistry", John Wiley and Sons Inc., pp 349, 1966.

(19) Yost, D.M.; Russel Jr., H.: "Systematic Inorganic Chemistry", Prentice Hall Inc., p 60, 1944.

(20) Hanson, J.B.; Harris, S.W.; West, C.T.: "Factors Influencing Lubricant Performance in Sequence V-E, Test", SAE Paper No. 881581, 1988.

(21) Habeeb, J.J.; Rogers, W.W.; May, C.J.: "The Role of Hydrogen Peroxide in Engine Wear and the Effect of ZnDTP-Dispersant-Detergent Interactions", SAE Paper No. 872157, 1987.

(22) Mair, R.D.; Graupner, A.J.: "Determination of Organic Peroxides by Iodine Liberation Procedures", Analytical Chemistry, Vol. 36, No. 1, pp 194–204, 1964.

(23) Korcek, S.; Mahoney, L.R.; Johnson, M.D.; Siegel, W.O.: "mechanism of Antioxidant Decay in Gasoline Engines Investigation of Zinc Dialkyldithiophosphate Additives", SAE Paper No. 810014, 1981.

(24) Bridgewater, A.J.; Dever, J.R.; Sexton, M.D.: "Mechanisms of Antioxidant Action, Part 2, Reactions of Zinc Bis-O, O-Dialkyl (aryl) Phosphorodithioates and Related Compounds With Hydroperoxides", J.C.S. Perkin II, pp 1006-1016, 1980.

(25) Willermet, P.A.; Kandah, S.K.: "Lubricant Degradation and Wear, Reaction Products of a Zinc Dialkyldithiophosphate and Peroxy Radicals", ASLE Transactions, Vol. 27, No. 1, p 67–72, 1984.

(26) Bennett, P.A.; Murphy, C.K.; Jackson, M.W.; Randall, R.A.: „Reduction of Air Pollution by Control of Emission from Automotive Crankcases", SAE Transactions, Vol. 68, 1960.

(27) Spearot, J.A.; Gallopoulos, N.E.: "Concentrations of Nitrogen Oxides in Crankcase Gases", SAE Paper No. 760563, 1976.

(28) Erdman, T.R.; Alexander III, W.: "The Effects of Operating Conditions and NO_x on Engine Deposits and Oil Degradation", Chevron Research Company Report, 1976.

(29) Korcek, S.; Johnson, M.: personal communication based on work presented at Gordon Research Conference on Tribology, 1988.

585

4.10 Lubrication Technique of Not Continuously Operating Vehicles

A. Zakar
Hungarian Hydrocarbon Institue, Szazhalombatta, Hungary
G. Borsa
Danube Refinery, Szazhalombatta, Hungary

A significant proportion of vehicles are not operated continuously. During the period when the vehicles are not in use corrosion causes significant damages both in the engine space and in the equalising gears. We shall briefly describe the different corrosion phenomena taking place in the intermittently operated vehicles. The most effective protection against corrosion is the use of protective lubricants. We shall sum up the results obtained by product development, with special regard to the interaction between the lubricating and anti-corrosion additives.

During the past few years there has been an increase of the number of machines and vehicles which are not in continuous use but are left to stand idle for shorter or longer periods. Experiences show that the lubricants used for those machines are not sufficient to prevent corrosion, and both the engine compartment and the equalising gear may become significantly damaged by corrosion. When left standing idle even in garages for comparatively brief periods (3—6 months) the engines of vehicles operated with popular motor oil the damages caused by electrochemical corrosion become visible in the form of dots and spots on cylinder liners, valves, springs and bearings as well as on other important parts (1—5).

Detailed surveys have led to the conclusion that seasonally used engines abrade faster than those in continuous use due to corrosive wear. For example seasonally employed self-propelled combines wear out 3 to 5 times faster than trucks built of identical main parts and components but used throughout the year. Consequently the maintenance of combines requires 4 to 5 times as much labour and assets than that of trucks. During a one year storage of agricultural machines seizure marks of 0.1—0.2 mm depth develop on main parts of engines of agricultural machines. Even when left to stand for a shorter period (about 1 month) electrochemical corrosive wear takes place which corresponds to the mechanical wear of one year operation. Abrasion is greatest when mechanical and corrosive wear alternate (6—10).

Figure 4.10.1 presents the depreciation of vehicles in different climatic zones in cases of continuous and cyclical operations. In the first year depreciation is greater for continuously operated vehicles, but after two years depreciation is much greater for vehicles with cyclical operation (7).

586

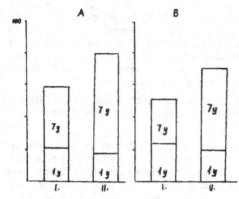

Figure 4.10.1:
Depreciation of Vehicles
I. Under continuous use
II. When used seasonally
A In temperate climate
B In subtropical climate

In internal-combustion engines and different grease depending on operation conditions three types of different corrosive damages may occur (11—15).

— During the period of transportation and storage atmospheric corrosion may occur with vehicles in the pre-operational stage.

— When left idle after operation, in addition to atmospheric corrosion the corrosive effect of the used oil in the vehicle has also to be reckoned with.

— During operation the inner surfaces of the engine and differential gear are subjected to corrosion caused by the reaction products of fuel and different additives.

Taking all that into consideration the methods of protection can be as follows:

— Protection during storage and transportation, using storage-engine-protecting oil. This oil is to have only anticorrosion effect and cannot be used as lubricant.

— Protection during operation, which can be achieved with engine oil and oil with adequate additives. However, these lubricants do not provide for protection during storage, during periods when the vehicle is not operated.

— Protection in the state of ready for use can only be achieved by applying protective-lubrication oil both as engine oils as well as for the transmission.

However, protective-lubricating oils are suited not only for keeping the vehicles in the state of ready to use but also for protection against corrosion during storage, transportation and operation.

587

Protection in the state of ready for use is the type of anticorrosion treatment we apply for vehicles which are parked for shorter or longer periods but may be available for use at any time. The possibility for immediate start may be of special importance with certain groups of vehicles, such as defence equipment or facilities required for flood protection, because there may be cases when there is no time for lengthy preparation before starting the engine-which frequently occurs with vehicles filled with oils of several functions.

In a wider sense of the word we speak of protection in the state of ready for use whenever protective lubricants are used for the engine and transmission of the vehicle.

Developing on the types of intermittently operating vehicles a widge range of corrosion phenomena can take place. In the course of tribocorrosion processes taking place in internal combustion engines the products of complete and partial combustion reach metal surfaces mainly via the circulating lubricant and induce corrosion. This process can be observed both with Diesel and Otto engines, but the causes and the processes themselves are highly different (16-18).

The corrosion of the lubricating systems of vehicles may additionally also be caused by aggressive corrosive substances entering from outside. Their entrance depends on the environment, on the construction of the vehicle but also on several other factors.

Of the pollutants entering from ouside the most important ones include steam, condense water, sulphur-dioxide, sulphur-trioxide and other air pollutants the character of which cannot be a priori assessed. Water, SO_2 and S_2O_3 threaten more or less all the places to be lubricated. In the case of Otto engines HBr may be a specific pollutant coming from an anti-wear additive of fuel oil.

The most important and most dangerous corrosive agent in lubrication technology is water. There is practically no vehicle lubrication system which is free of water. In these cases water is always present in condensed fluid phase (lubrucant) present in relatively nonpolar. Consequently water can always replace the lubricant on the lubricated surface and create corrosion.

One may ask the question: "How can water enter from outside a properly closed lubricating system (e.g. engine oil, gear oil)?" The reason is the well-known "breathing phenomenon". For example water enters during night, when the air cools off, as water gets condensed from the atmosphere. When the atmosphere is heated during the day the condensed water does not escape into the atmosphere again. During the subsequent cooling down period additional water gets condensed, and the process is being repeated, the engine keeps on "breathing" so that finally there is a significant quantity of water in the engine space. As the system is closed water remains and becomes accumulated in the bottom of cavities. But water can enter the engine space also when for example there is a defect in the packing between the engine block and the water space. In intermittently operated closed lubrication systems the temperature difference between the state of operation and standing idle is by itself enough to have "breathing condensation" to occur.

With internal combustion engines the reaction water developing by the oxidation of hydrocarbons gets also collected in the oil sump. Additionally, water can also transmit atmospheric pollution to the lubricated surfaces. The most significant pollutants of this type and at the same time the most damaging aspects of lubrication corrosion are compounds with S content: H_2S and SO_2. Lubricating systems open to the atmosphere absorb H_2S and and/or SO_2 and transmit them to the lubricated surface. Sulphuric corrosion of this type often causes significant seizures. The corrosive effect is further increased by the catalytic effect of the metal surface which helps the compounds with sulphur content present in the oil to oxidise into S_2O_3 through several reactions; this compound forms sulphuric acid with the water present causing great damages to the metal surface.

Earlier we have described the most corrosion phenomena taking place in intermittently operated vehicles.

Operation conditions pose extremely complex demands on protective lubricants of the vehicle industry. The lubricant must be of adequate viscosity, of high efficiency during operation, while in idle it should have anticorrosive effect. At first sight it is easy to meet these requirements: the lubricant should have additives which make it suitable from the aspects of lubrication technology, and should also be a certain corrosion inhibitor. However, in practice this is a most complex task due to prevailing synergistic-antagonistic effects, as it is a mixture of any components. Of the anti-corrosion additives literature recommends the use mainly of alkyl-zinc-dithiophosphate, glycerine-distearate, metylchinlone, calcium-alkyl-phenolate, aluminium-isopropylate. However, these additives provide protection only for brief periods (19—20).

Nitrate lubricants and their mixtures with stearine as well as their more advanced modified versions are widely used, but their efficiencies do not fully meet expectations as far as their protective qualities over a range of time are concerned (21 — 23).

In addition to corrosion inhibitors usually promoters are also employed such as the calcium salts of acids produced by the oxidation of paraffine, calcium-petrosulphonates, alkyl-phenols, Ba and Mg salts of partial esters of synthetic dicarbonic acids as well as superbasic detergents. A primary requirement with each inhibitor is that they have to be compatible with the other additives of the lubricant and should not reduce their active life. Thus the synergistic and antigonistic effects of the additives are of utmost importance. Synergism is the rule, that the molecular interactions cause the characteristics of the components of mixtures not to change additively.

The molecular association of the additive plays an important role in these effects (24—25).

First the interrelations of anticorrosive additives of different types were investigated in the protective lubricants. This is also a very complex field as most of the additives employed for the purpose have several functions: they also play detergent-disperging roles and inhibit oxidation.

The importance of the problem is underlined by the fact that in practice single anti-corrosive agents are very seldom used, but rather combinations of 2 – 4 inhibitors.

The oil base by itself has an important defining role from the aspects of anti-corrosive agents (refinement grade, aromatic, naphthene type, synthetic components, etc.).

Here synergistic effects depend on the steric structure and polarity combined of the oil molecules (26–27). Synergism may occur between inhibitors of the same type also, for example when two hydroperoxide decomposing agents are used together. This may be due to mutual regeneration through H transfer. Synergism may play a significant role when zinc-dialkyl-dithiophosphate, the antioxidant used in greatest quality by industry is employed. It acts by reducing hydroperoxides and it also decomposes radically (28).

Surfaces treated with zinc-dialkyldithiophosphate containing protective lubricating oil have surface layers formed which have Zn and P in greater concentration. The decomposition products of the primary additive react with the metal components of the structural compounds and in the meantime — when suitable additives are present — the surface is sulfidated and phosphated, becoming more resistant to wear and corrosion. The metal cation is also built in into these layers. In this aspect antioxidants of this type inhibit aging as well as they protect the surface.

With the protective lubricating materials used as auxiliary agents for machining metals it is worth noting that inorganic acids formed during the thermal decomposition of additives also take part in the development of the layers with EP effect formed from the antioxidants (29).

Considering the fact that corrosion inhibitors are usually employed not as single additives but in combination, and that in most cases antioxidants are also present in the system, it becomes obvious that it is a complex task to select the optimum anticorrosion additive for the protective lubricant. No generally valid guidelines can be offered; one must always consider the requirements of the given user, and target oriented additives are to be chosen based on detailed laboratory experiments. This is illustrated by Figures 4.10.2 and 4.10.3 presenting the efficiency of the characteristic anticorrosion additives of protective lubricating oils. Of the single additives, those of the alkyl-aryl sulfonic acid type prove to be most effective, but the best results are obtained with the basic alkali-earth metal salt of alkyl-aryl sulfonic acide of different composition because of the synergistic effect (30).

Synergistic effect can usually be observed when alkali-earth metal sulfonates and phenolates are employed, or when succinimide additives are used. The explanation is among others that alkali earth metal compounds protect metal surfaces mainly at high temperature, while succinimides are effective at low and medium temperatures (31).

590

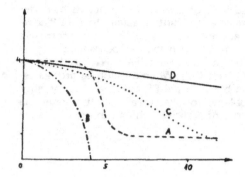

Figure 4.10.2: Effectivity of the Corrosion Inhibitor Additives for Protective
Lubricating Motor Oil

↑ Corrosion inhibition effect, corrosion grade
→ Corrosion additive, %
A Alkanol-amine-ester type additive
B Alkyl-benzene-sulfonic acid type additive
C Alkyl-naphtalane-sulfonic acid type additive
D Nitrated lubricating oil type additive

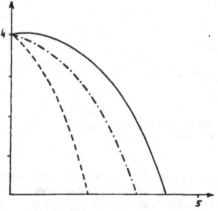

Figure 4.10.3: Synergism of Alkyl-benzene-sulfonic Acid Type Additive in
Protective Lubricating Motor Oil

↑   Corrosion inhibition effect, corrosion grade
→   Corrosion additive, %
AB Several compositions of alkyl-benzene-sulfonic acid earth
metal salts.

591

Certain types of non-anticorrosive additives of protective lubricants have decisive effect on corrosion inhibitors. We have to point especially to the EP additives role on gear oils which usually act antagonistically to the anticorrosive additives (Figure 4.10.4). The inhibitor reacts with the democomposition products of the EP additives inhibiting the acidic reagent (HCl, H_3PO_4, H_2S) to react to the metal, preventing the EP effect (32). When developing protective-lubricating gear oil, it is the interaction between the EP additives and the anticorrosion additives which creates problems for the researches. It is only the joint, complex analysis of the two types of additives which can help to select the best solution.

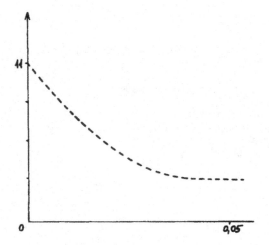

Figure 4.10.4: Antagonistic Effects of Corrosion Inhibitors and EP Additives

↑ FZG test
→ Corrosion additive, %
Corrosion inhibitor: dodecyl-succinic acid semi-ester
EP additive: ashfree dithiophosphate (0.5 %)

Only compounds which contain no Cl can be used as EP additive to protective-lubricating gear oil. Figure 4.10.5 presents the effect of different types of EP additives on the anticorrosion efficiency of corrosion-lubricating gear oil. The figure shows that efficient corrosion preventing effect can be obtained only when EP additive-combination is employed in the gear oil, for in such cases the antagonistic effect with the anti-corrosion inhibitor observed with single EP additives cannot be observed. The oil-change period of the first protective-lubricating engine oil types was very short: 100—1500 km, as against that of the high performance engine oils which could be used for 10000—15000 km.

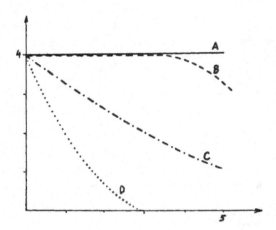

Figure 4.10.5: Effect of Different EP Additives on the Corrosion Inhibition Properties of Protective Lubricant Gear Oil

↑ Corrosion inhibition effect, corrosion grade
→ Corrosion inhibitor concentration, %
A S-P type EP additive (a)
B S-P type EP additive (b)
C 1:2 mixture of (a + b) type EP additives
D 1:1 mixture of (a + b) type EP additives

Meanwhile the corrosion-inhibiting effect has also been reduced during the use of the oil. The break-in engine oils also come into this group. The protective-lubricating oils of this type are used to lubricate new engines, to run them on block testing pad and during storage and transportation. Their oil change period is short: 500 to 1000 km. With the up-to-date protective lubricating engine oils, well selected combinations of detergent-dispergent additives and chemically stable, high temperature resistant anti-corrosion inhibitors have helped to increase the oil-change period to 10000 — 15000 km without having the anti-corrosive effect reduced during use (Figure 4.10.6), while with the gear oils used during the "ready to use" period the oilchange period is now around 60—100 thousand km (33).

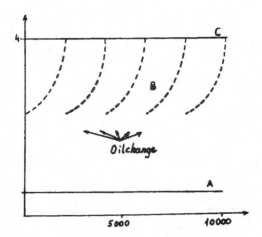

Figure 4.10.6: Changes of the Corrosion Inhibiting Performance of the Protective-lubricating Engine Oil during a Run Test

↑ Corrosion inhibition effect, corrosion grade
→ Run test (km)
A Up-to-date protective-lubricating engine oil
B Nitrated lubricating oil type protective-lubricating oil
C General engine-oil

Demand in protective-lubricating engine oils and gear oils has been growing lately and the range of users is also on the increase. There are several types of operations for intermittently used vehicles, the most typical of which are the following:

— Following several months of intensive use the vehicle is left unused for a predetermined period. This is the case with agricultural single purpose machines, harvesters, single-purpose machines of the building industry. These are very expensive machines the conservation of which is most important.

— The vehicles are left unused for longer periods but must be in a state when they can be started immediately. This is the case for the technical equipment of the armed forces, equipment used for flood prevention, average vehicles in industry and transportation. The use of non-suitable lubricants may risk state-, economic interests, the life and wealth of people.

— Cyclically operating vehicles the use of which can be foreseen during the year: e.g. single purpose machines for the construction of roads, those used for cleaning the roads (machines spraying salt during winter, snow-plough), boatengines, etc.

— It is an important area: corrosion prevention of new and regenerated engines and gears during transportation and storage.

Growing fuel costs induce an increasing number of private motorists in several countries not to use their cars during winter. Consumers of this type should also be encouraged to use protective lubricating oils, because the engines and gears may corrode during the winter months when the cars are left unused.

Table 4.10.1 sums up the characteristic values of some protective lubricating oils used in the vehicle industry. The table also shows the analysis of the usual engine and gear oils for the sake of comparison.

Table 4.10.1: Protective Lubricants of the Vehicle Industry

| Parameters | Engine oil | | Engine oil | |
	Protective lubricating engine oil	General engine oil	Protective lubricating engine oil	General gear oil
Performance level according to API	CC/SE	CC	GL-5	GI-5
Viscosity grade according to SAE	20W/40	30	80 W/90	90
Viscosity at 100°C mm²/s	15	12	14	15
Flash point, °C	220	210	220	210
Cold point, °C	−25	−17	−30	−18
Humidity chamber resistance 672 h, corr. grade (DIN 51359)	0	4	0	4
Anti-corrosion effect in salt water, corr. grade (DIN 51358)	0	4	0	4
Change period in '000 km	10-15	10-15	60-100	60-100

Protective lubricating engine oils provide protection against corrosion for the minimum of 2 years, protective-lubricating gear oils for the minimum of 4 years for engines, and for mechanical gears, differential gears, equalizing gears whether they are stored in the open or in closed, covered premises (garages). The protective lubricants of the vehicle industry have been developed so that they should be compatible with the engine oils and gear oils of the known brands available on the international market, and if necessary they can also be used for replenishment. Naturally in the latter case one must expect the anti-corrosion effect to be reduced.

References

(1)   Krejna, Sz.E.: Ingibirovannüe maszla i topliva CNIIT Enegtehim 88. (1964).
(2)   Bilenkina, A.V.: Ingibitorü Korrozii i mechanizm ih gejsztvija. M. Izd. Doma Naucsno-technicseszkoj Pron. F.E. Dzer. 157 (1971).
(3)   Lavrusko, L.: Ingibitorü, zsacsitnüe szarnzki is Konszervanija metallicseszkih izdelij, CNIIT Eneftehim T.1.114. T.2.120 (1969).
(4)   Bilenkina, A.B.: Ingibirovanüe neftejanüe pokrütija. Izd. Doma Naucsnotechnicseszkoj Pron. F.E. Dzer. 124 (1973).
(5)   Gilelah: Plaszticsnüe szmazki, Kijev, Naukova Dumka (1975).
(6)   Sehter, Ju.N.; Krejn, Sz.E.: Poverhnosztno-aktivnüe vescsesztva iz nefjanogo szürja. Himija, Moszkva (1971).
(7)   Dolberg, A.L.; Sehter, Ju.N.; Evsztratova, N.I.: Avtomobilnaja promüslennoszt 41—44 (1970).
(8)   Antipenki, A.M.; Bilinkin, A.V.; Krjen, Sz.E.: Vesztnik Masinosztronenija. 9, 15—16 (1970).
(9)   Dolberg, A.L.; Sehter, Ju.N.; Evsztratova, N.I.: Sztanki i insztrument, 12, 36—38 (1970).
(10) Dolberg, A.L.; Sehter, Ju.N.; Evsztratova, N.I.: Avtomobilnaja promüslennoszt. 5, 43—44 (1971).
(11) Vámos, E.: Krróziós Figyelö 11, 1. 29—32 (1971).
(12) Temporart rostkydd.: IVA. S. Korrosionsnamd Stockholm (1963).
(13) Zakar, A.: Kórrózios Figyelö, 26, 1. 16—20 (1986).
(14) Vámos, E.; Rónai, D.: II. Gépjármü és motortechnikal Konferencia. 155—162. Sopron (1971).
(15) Zakar, A.; Gyöngyössy, L.; Pintér, Gy.: Mezőgazdaság kemizálása konferencia T. 193—195. NEVIKI-DATE, Keszthely (1981).
(16) Vámos, E.: A kenőolajadalékok hatásmechanizmusa, Budapest, INFORMATIK (1985).
(17) Vámos, E.: Tribológia, Budapest Műszaki Könyvkiadó (1983).
(18) Jäger, G.: Schmierstoffe und ihre Prüfung im Labor, Licspe, Verlag für Grundstoffindustrie (1984).
(19) Sehtyer, Ju.N.: Zascsita metallov ot Korrozii, Moszkva, Himija (1966).
(20) Vámos, E.: Átmeneti Korrózióvédelem, Budapest, Műszaki Könyvkiadó (1978).
(21) Ermolov, F.N.; Englin, A.B.: Himija i technologia topliv i maszel 3. 25—27 (1982).
(22) Kopylov, L.I.: Himija i technologija topliv i maszel 7. 33—24 (1981).
(23) Prokopjev, I.A.: Himija i technologija topliv i maszel 7. 36—37 (1982).

(24)   Vámos, E.; Pintér, J.: Characterization of mutual effect of lubricating oil additives LAWPSP Symp. A/4.1—4.11 Bombay (1985).
(25)   Vámos, E.; Pintér, J.; Zakar, A.: Synergism of EP additives. Jugoma '86. Portoroc (1986).
(26)   Fuksz, G.I.: Himija technologija topliv i maszel 20. 3. 41—43 (1975).
(27)   Kennedy, P.I.: ASLE Trans 23, 370—378 (1980).
(28)   Studt, P.: Erdöl und Kohle 31, 439—443 (1978).
(29)   Vámos, E.: Ropa a Uhlie 21. 5. 289—291 (1979).
(30)   Zakar, A.; Vámos, E.: 5th Int. Colloquium. Additives for Lubricants and Operational Fluids. Esslingen, TAE. I.7.1.1—7.1.10 (1986).
(31)   Knispel, B.; Hötzel, D.; Weber, K.: Schmierungstechnik 12. 179—183 (1981).
(32)   Bartz, W.I.: Additive für Schmierstoffe, Hannover, Curt R. Vincentz Verlag (1985).
(33)   Zakar, A.; Vámos, E.: Védo-kenőolajok fejlesztési eredméyei III. Tribológiai Konferencia, Budapest (1983).

597

4.11 Environmental Effects of Crankcase and Mixed Lubrication

P. van Donkelaar
Greentech Research sprl, Essen, Belgium

Summary

The environmental effects of crankcase- and mixed lubrication are evaluated by following the lubricant in its path through the engine and into the environment. In a *crankcase-lubricated* engine, the oil is subjected to changes: it absorbs other substances, it also supplies substances to particulate emissions in the exhaust. The contaminated oil then reaches the environment in various ways of which many are uncontrolled. Often, these uncontrolled immissions lead to high concentration of dangerous substances in the environment. In *mixed-lubricated* engines, there are no significant changes and no cumulative effects in the engine. The oil returns to the environment in a well-defined and finely dispersed way and in extremely low concentrations. In addition, most of the oils used in mixed-lubricated engines are biologically degradable (some are designed to be very well degradable) and no negative environmental effects have been found.

The presentation concludes with a recommendation for engine builders to reconsider the application of mixed-lubrication in all types of engines.

4.11.1 Introduction

In industrial nations, approximately 4.5 % of the gross national product (GNP) is consumed in friction, wear and corrosion of moving components in technical products (1). This means a loss of raw materials and energy with a value of several hundred billion dollars per year world-wide. These losses could be reduced considerably with optimum lubrication. Energy could be saved and the life of products lengthened. These are basic elements of effective environmental protection. When we discuss the environmental disadvantages of lubricants and lubricating systems in this article, we should not lose sight of these basic positive elements, which give lubricants a significant environmental credit to start with.

Lubricants represent only 1 % of the total consumption of mineral oil products. But because they are not fully consumed, used lubricants present a potential environmental problem. Consumption varies between less than 20 % for oil used in turbines and transmissions to 100 % for oil used in mixed-lubricated engines.

598

In the EEC, total sales of lubricants amounts to about 4.5 million tons per year, half of which is totally consumed. This results in a quantity of over 2 million tons of used oil, which is disposed as follows: 750,000 tons are used as fuel, up to 700,000 tons are reprocessed and 600,000 tons "disappear" uncontrolled in the environment.

It is exactly this last quantity of 600,000 tons that is causing increasing concern. It represents a potential threat to the environment (23) and a loss of natural resources (2).

In this article, we will follow the lubricant in its path through the engine and into the environment. In a *crankcase-lubricated* engine, the oil is subjected to changes: it absorbs other substances, it also supplies substances to deposits in the engine and to particulate emissions in the exhaust, it is partly emitted with the exhaust, a further part is lost in leakages and the rest is finally exchanged for fresh oil. Used oil then reaches the environment in various ways, of which many are uncontrolled. Often these immissions lead to high concentrations of dangerous substances in the environment.

In *mixed-lubricated* engines there are no significant changes and no cumulative effects in the engine. The oil returns to the environment in a well-defined and finely dispersed way and in extremely low concentrations. In addition, most of the oils used in mixed-lubricated engines are biologically degradable (some are designed to be very well degradable) and no negative environmental effects have been found (3).

In our conclusion we recommend engine builders to reconsider the application of mixed-lubrication, also for those engines which are traditionally equipped with crankcase-lubrication. This could make a significant contribution to the solution of one of the most urgent environmental problems related to the use of internal combustion engines.

Because of lack of effective control over waste oil disposal, this problem might well be more serious than the problems supposed to be caused by exhaust emissions and on which environmentalists and governments have been concentrating until now.

4.11.2 Basic Requirements of Engine Lubrication Systems

The basic task of a lubricating system is to transport the required quantity of lubricant should protect against wear, reduce friction, keep the engine clean (detergence), protect against corrosion, help cool the engine and seal the pistons. We will see that it is especially the detergent function that has environmental significance.

In internal combustion engines, two different lubrication systems have emerged:

With *crankcase-lubrication*, the same oil remains in the engine for a considerable period of time. From the sump, the oil is pumped into the engine and it then returns to the sump, from where it is ultimately removed to be exchanged for fresh oil. Oil change intervals have significantly increased over time.

With *mixed-lubrication,* the engine is continuously supplied with fresh oil, which is usually accomplished by pre-mixing the oil with the fuel. Modern engines, however, have automatic and variable ratio mixing and the engine is supplied with just the amount of oil it requires for each operating condition. Mixed ratios have tended to become leaner over time.

4.11.3 Development of Engine Lubrication Systems

Improvements in engine design and in engine lubricants have resulted in increased lubricant loads over time.

For crankcase-lubricated engines, the following lubricant load factor is a good indicator for this (4):

$$\frac{\text{engine power}}{(\text{oil volume} + \text{oil suppletion}) / 1000 \text{ km}}$$

For spark-ignited engines, this index number was typically increased from 10 to 340 HP/1/1000 km in the period from 1949 to 1986. In the same period, oil change intervals increased 10-fold from 1500 km to 15.000 km (Table 4.11.1).

For mixed-lubricated engines a similar load factor may be used, which is inversely related to the amount of lubricant available per cc displacement to produce one kW (5):

$$\frac{\text{mixing ratio}}{\text{displacement for 1 kW}}$$

For outboard motors, this index number increased from 0.73 to 6.94 in the period from 1958 to 1985 (Figure 4.11.1). In the same period, the mixing ratio went from 24:1 to 125:1 (Figure 4.11.2). More and more mixed-lubricated engines use automatic lubrication and variable mixing ratios. Direct oil injection is also used.

Table 4.11.1: *Development of Engine Lubricant Load*
in crankcase-lubricated engines (according to Eberan)

	Ford-Taunus 1949	*Audi 80 Quattro 1987*
Engine:	4 cyl., 4-stroke, carburetor	4 cyl., 4-stroke, injection, catalyst, lambda regulated, Electronic ignition, knock regulated
Displacement:	1172 cc	1847 cc
Compression:	6.6 : 1	10.5 : 1
Power:	34 HP at 4200 rpm	113 HP at 5600 rpm
Specific power:	29 HP/1	61 HP/1
Sump contents:	3 l	3.5 l
Lubricant load-factor:	10 HP/1/1000 km	340 HP/1/1000 km
Oil consumption:	check every 500 km	max: 1.5 l/1000 km normal: 0.1 l/1000 km
Oil changes 1st: 2nd: 3d:	500 km 1500 km 3000 km, etc.	15000 km 30000 km, etc.
Oil recommendation:	— mix oil with fuel for first 1000 km — special oil for first 3000 km — small lubrication service at 1500 km — large lubrication service with gearcase oil change every 7500 km	— high performance oil according to VW—standard 500.00 and 505.00 — no oil changes for gearcase (filled for life)

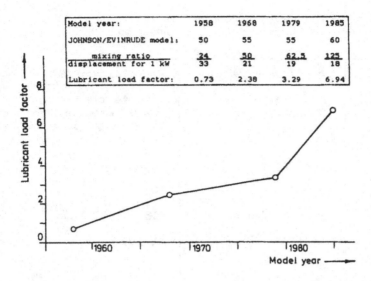

Model year:	1958	1968	1979	1985
JOHNSON/EVINRUDE model:	50	55	55	60
mixing ratio / displacement for 1 kW	24/33	50/21	62.5/19	125/18
Lubricant load factor:	0.73	2.38	3.29	6.94

Figure 4.11.1: Development of Engine Lubricant Load in Mixed-lubricated Outboard Motors

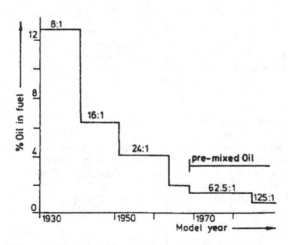

Figure 4.11.2: Mixing Ratio for Outboard Motors

Table 4.11.2: Requirements for Oil for Crankcase-Lubricated Engines
(according to Gairing)

Friction and wear
1. Friction reduction
2. Fuel consumption reduction
3. Wear protection
4. Oil film stability
5. Surface damage prevention
6. Retain honing pattern
7. Harmonized reaction temperatures of additives
8. Neutralizing capability
9. Adhesion properties

Temperature and viscosity
10. Thermal stability
11. Oxidation resistance
12. Nitration resistance
13. Viscosity at high temperatures
14. Viscosity at low temperatures
15. Viscosity stability
16. Shear stability

Cleanliness
17. Dispersive capacity
18. Detergence capacity
19. Prevention of ring sticking
20. Prevention of hot sludge
21. Prevention of cold sludge
22. Prevention of lacquer
23. Water and coolant resistant
24. Anti-rust and anti-freeze resistant

Deposits
25. No deposits in the inlet valves
26. No deposits in the combustion chamber
27. No pre-ignition
28. No deposits in the bearings of turbo chargers

Oil/engine components
29. Corrosion protection
30. Compatibility with metals and paint
31. Compatibility with elastomers (seals)
32. Compatibility with filter materials
33. Prevention of filter clogging
34. Heat transfer capability/cooling capacity
35. Sealing capability

Base stock/additives

36. Additive solubility
37. Homogeneity
38. No filtering out of additives
39. No heat generation
40. Prevention of foaming
41. Air release capability
42. Low evaporation

Application

43. Compatibility with fuels
44. Mixing capability/compatibility with other oils
45. Running-in capability
46. Long oil change intervals
47. Application in various types of engines
48. Consistent quality
49. Economic production
50. Availability
51. Storage stability

Environment

52. No negative influences on health and the environment
53. No negative influences on catalysts and sensor
54. No contribution to particulate emissions
55. No offensive smell
56. No negative effects during recycling

4.11.4 Development of Engine Lubricants

Because of ever increasing lubricant loads in all types of engines, the requirements for these oils have also increased in severity.

Table 4.11.2 gives a summary of 56 requirements for oils used in crankcase-lubricated engines (4 and 6). This list includes the essential requirement that used engine oil should not harm the environment.

Requirements for oils for mixed-lubricated engines are more modest (7) because some of the potential problems with oils in crankcase-lubricated engines simply do not exist (foaming, oxidation, black sludge, etc.). In the meantime there are now readily degradable lubricants for mixed-lubricated engines and there is also a good, standardized method (CEC-L-33-T-82) for evaluating and comparing the biodegradability of these oils so that this favorable characteristics can be measured and controlled.

4.11.5 Environmentally Relevant Engine Operating Conditions

Engine oil must be capable of washing off deposits and dirt which then accumulate in the oil. Non-soluble deposits must be kept in suspension. These cleaning and dispersing capabilities cause increasing quantities of contaminants to accumulate in the oil of a *crankcase-lubricated* engine. Amongst these contaminants, the polycyclic aromatic hydrocarbons (PAH) are of particular environmental relevance because of their carcinogenic (mutagenic) properties.

140 different kinds of PAH have been found in used oil of crankcase-lubricated engines. These PAH are also present in fresh oil but in much lower quantities. They stem mostly from the fuel (21) and also from the combustion process. When fuels with a low PAH content are used, the PAH concentration in the oil also tends to be lower.

The mutagenic effects of used engine oil have been determined by means of Ames testing (8). It was found that 70 % of these effects were caused by PAH with more than three rings although this fraction represents not more than 1 % of the volume of used oil. Of this highly mutagenic fraction, 18 % of the effects were caused by Benz(a)pyrene. There were hardly any mutagenic effects caused by the PAH-free portion of the oil, which represents 92 % of its volume. The remaining 7 %, consisting of PAH with 2 and 3 rings, cause about 10 % of the mutagenic effects.

The concentration of Benz(a)pyrene in the lubricating oil in crankcase-lubricated engines of passenger cars increases at a rate of about 5 to 10 mg/l per 1000 km. For crankcase-lubricated Diesel engines in passenger cars, the increase is about 10 times lower at 0.1 to 1 mg/l per 1000 km. For trucks and busses no increase in the PAH concentration of lubricating oil was determined.

The average values of Table 4.11.3 are graphically shown in Figure 4.11.3. From this graph, it can be seen that there is an average increase of PAH with 5–7 rings of 26.8 mg/kg per 1000 km in the oil of *crankcase-lubricated spark-ignited* engines. With oil change intervals of 10.000 to 15.000 km, which are now common practice, this would result in a concentration of *270–400 mg/kg* PAH with 4–7 rings in the used oil from these engines.

For passenger cars with *Diesel* engines, the accumulation will be about 10 times lower and the oil change intervals are shorter. This would result in a concentration of *20–30 mg/kg* of PAH with 4–7 rings in the used oil from these Diesel engines. For Diesel engines in busses and trucks, this value is about *4 mg/kg*.

The average concentration of PAH with 4–7 rings in oil for *mixed-lubricated* engines is *0.345 mg/kg*. This is very low indeed and about 1000 times less than the concentration in used oil from crankcase-lubricated spark-ignited engines. The basic reason is that PAH's do not accumulate in mixed-lubricated engines.

Table 4.11.3: *Carcinogenic PAH in Engine Oils* (according to Grimmer)

| | PAH with 4–7 rings | | | | | |
| | mg/kg oil | | | | | |
	chrysene & triphenyls	benzofluor anthenes	indeno pyrene	anthan threne	benzo(A) pyrene	sum
Fresh oils						
Best 2 T	0.207	0.027	0.001	0.008	0.008	0.251
2T 3 oils	0.288	0.031	0.006	0.005	0.015	0.345
4T 16 oils	1.705	0.151	0.002	0.012	0.070	1.940
4 T Diesel	2.860	0.085	0.009	0.019	0.051	3.024

After ...... km in 4T spark-ignited engines in various passenger cars and using various fuels.

	chrysene & triphenyls	benzofluor anthenes	indeno pyrene	anthan threne	benzo(A) pyrene	sum
1000 km	11.0	5.7	2.1	1.6	5.2	25.6
1300 km	8.7	14.4	4.9	3.7	9.7	41.4
1600 km	17.3	14.3	2.9	3.5	10.9	48.9
2300 km	39.6	27.2	8.5	8.8	22.6	106.7
4000 km(2)	31.5	33.6	9.9	6.0	21.9	102.9
4900 km	18.1	16.6	5.5	3.0	12.5	55.7
5000 km	54.0	28.2	10.9	9.4	24.2	126.7
6000 km	71.4	43.9	10.6	7.8	29.8	163.5

Taxis with Diesel engines and % city traffic

	chrysene & triphenyls	benzofluor anthenes	indeno pyrene	anthan threne	benzo(A) pyrene	sum
20 %	7.5	1.8	0.8	0.5	0.7	11.3
25 %	9.1	4.5	2.4	1.1	3.5	20.6
50 %	7.6	2.0	1.0	0.7	1.1	12.4
90 %	34.0	3.6	1.2	1.0	1.9	41.7
99 %	40.0	16.8	9.0	4.4	11.9	82.1
				average:		33.6

Average of 12 trucks

	chrysene & triphenyls	benzofluor anthenes	indeno pyrene	anthan threne	benzo(A) pyrene	sum
	2.67	0.56	0.13	0.05	0.24	3.65

Average of busses

	3.2	0.57	0.15	0.08	0.22	4.22

Average of re-processed oils

	14.6	6.0	0.34	–	0.51	21.45

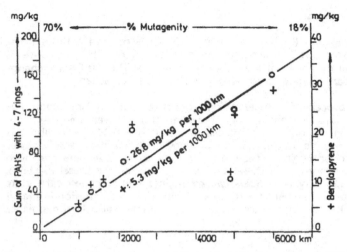

Figure 4.11.3: Carcinogenic PAH in Lubricating Oil of Crankcase-lubricated Spark-ignited Engines (according to Grimmer)

4.11.6 Lubricant Emissions*

There are considerable differences between the oil emissions of crankcase- and mixed-lubricated engines:

4.11.6.1 Lubricant emissions by Crankcase-Lubricated Engines

Oil from crankcase-lubricated engines is emitted in various ways:

— direct with the exhaust
— together with particulate emissions in the exhaust
— through leakage
— during and after oil changes.

Exhaust

For crankcase-lubricated engines in perfect technical condition, the direct oil emissions in the exhaust are very low indeed: 0.1—0.25 l/1000 km (22). However, as the engine ages, the piston rings do not seal as well and more and more oil is directly emitted. Older engines can reach oil emissions which are not much less than those of modern mixed-lubricated engines with automatic lubrication.

607

Particulate in the exhaust

An English investigation (9) has shown that 40—70 % of the PAH in exhaust particulate stem from the PAH accumulated in the engine oil. The other 3—60 % originate from the fuel or from the combustion process. There is PAH exchange between oil and particulate whereby an equilibrium level is reached.

The combustion process of a *Diesel* engine produces a relatively large amount of particulate. Therefore, the equilibrium is reached at a relatively low PAH concentration both in the oil and in the particulate (Table 4.11.6). Despite this lower concentration, the total mutagenic potency per km is considerably higher than for spark-ignited engines (Table 4.11.4). This is because there are much higher particulate emissions in Diesel exhaust and because the Diesel PAH-mix has a higher mutagenicity (10). *Spark-ignited* engines emit much less particulate and the equilibrium is reached at a 10 times higher PAH concentration in the oil (Table 4.11.5). The PAH concentration in the particulate is also 10 times higher (Table 4.11.6).

Table 4.11.4: Mutagenicity of Exhaust Particulate per Mile in the Ames Test (according to Shore)

		Vehicle with:			
		Spark-ign. engine normal	Spark-ign. engine lean	U.S.Diesel	Euro Diesel
Test animal	with (+) without (—) S9				
TA 98	+	3456	1376	64440	73032
	—	3915	2896	112680	96694
TA 98NR	+	2454	509	26914	34860
	—	2741	1608	58594	50282
TA 98/1 8 DNP	+	1210	179	3845	26145
	—	1611	927	47326	37711
CM 891	+	2781	856	28440	36312
	—	4131	2264	106560	116688

Table 4.11.5: Concentration of Carcinogenic PAH with 4–7 Rings in Used Oil of Internal Combustion Engines (according to Grimmer)

	PAH with 4–7 rings mg/kg
Crankcase-lubricated spark-ignited engines	270 – 400
Crankcase-lubricated Diesel engines	20 – 30
Mixed lubricated spark-ignited engines	0.2 – 0.4

Table 4.11.6: PAH Concentrations in Particulate (according to Shore)

PAH	g PAH/g particulate	
	Spark-Ignited Engine	Diesel Engine
Fluoranthene	320	44
Pyrene	694	91
Chrysene	228	49
Benz(a) anthracene	145	13
Benz(b) fluoranthene	100	7
Benz(k) fluoranthene	26	2
Benz(a)pyrene	109	3
Benz(ghi) perylene	451	9
Coronene	199	1
TOTAL (nine PAH's)	2272	279

Leaks

There are many seals in the oil circuit of crankcase-lubricated engines which are prone to leaking. Although substantial progress has been made in testing the compatibility of these seals with various lubricants and fuels (and vice versa), many older engines in the real world do leak. Few cars are equipped with oil level indicators so that most car drivers are not aware of this.

Therefore, there is no adequate control of oil emissions caused by leaks. Neither is there much control over the quality of service in garages.

Oil Changes

If well-trained mechanics in a well-equipped garage change oil, very little oil will end up in the environment. However, it is estimated that 10–15 % of all oil changes are done by do-it-yourself amateurs (11). Many of those do not bring the used oil back to authorized stations but — disregarding warnings on the packs — simply dispose of the used oil in household garbage, in sewage or just anywhere. These emissions cause very high concentrations of used oil in the environment.

4.11.6.2 Lubricant Emissions by Mixed-Lubricated Engines

Mixed-lubricated engines emit their oil almost exclusively through the exhaust. These emissions remain constant over the life of the engine and do not increase with engine wear.
In a Japanese study (12) it was found that 95 % of the matter caught in measuring filters in the exhaust of mixed-lubricated engines consisted of slightly oxidized or unburned oil (Figure 4.11.4).
The extremely low mutagenicity of the fresh oil therefore remains low for the oil components in the exhaust because there is no PAH accumulation in mixed-lubricated engines. These oil emissions have been getting steadily lower over the years with the introduction of leaner mixing ratios and automatic oiling.

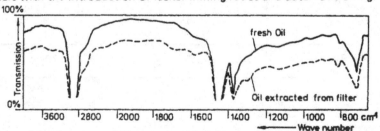

Figure 4.11.4: Infra-red Spectrum of Oil Components in Exhaust Particulate of Mixed-lubricated Engines in Comparison to Fresh Oil
according to Segiura

610

4.11.7 Lubricant Immissions *

In general, oil emitted with the exhaust arrives in the environment in very low immission concentrations. If the concentrations of contaminants are also low, as in *mixed-lubricated* engines, then the environment risk is negligible. The situation is even better for outboard engines, where oil immissions in richly aerated water are immediately subjected to various degradation processes (3).

The situation is much less favorable for immissions which do not reach the environment via the exhaust, such as oil leaks and used oil from *crankcase-lubricated* engines. Oil *leaks* have a tendency to accumulate in the environment. This can be seen quite clearly in parking lots, in garages and in the central part of highway lanes, especially where there is a dip in the road. When it rains, these PAH containing oils then reach the sewers, rivers and lakes and finally the sediments.

As mentioned earlier, *600,000 tons of used oil* are entering the environment in the EEC in an uncontrolled way every year. Although there are strict European and national environmental laws that should ensure the proper disposal of used engine oils. investigations and inquiries (13) (14) (15) have shown that it is impossible for these laws to be enforced effectively. With this lack of adequate control, one has to rely on the motivation of each individual. Motivation is difficult to achieve when authorized stations have high recycling costs. The do-it-yourself customer with his container of used oil is simply sent home again. This is a typical example of strict legislation (20) having the opposite effect of what was intended to achieve.

A very dangerous application of used oil is to use it as chain lubricant, which is the case for about 30 % of all chainsaws (16).

With a consumption of 1 liter used oil for 2–3 liters of fuel, the mutagenicity of these immissions is about 4 orders to magnitude higher than that of the exhaust emissions of the chainsaw engine **. This example shows that these types of immissions can be far more dangerous than those caused by the engine exhaust.

As mentioned in the previous paragraph, a substantial portion of PAH, accumulated in the oil, is emitted via *particulates.* In the environment, particulates concentrate in pockets, where they tend to accumulate preferentially due to topographic features, air flow patterns and other meteorological factors (24). These accumulation effects are quite similar to snowdrift patterns and also

---

* We are using the terms "emissions" and "immissions" in the same sense as in other languages: emissions are polluting substances as they come out of the point of origin (e.g. exhaust), immmissions are the substances at the point where they cause environmental effects (18).

** In the meantime, there are now chain lubricants on the market, based on vegetable oils, which are not toxic, not mutagenic and very well degradable.

4.11.8 Fate of Engine Lubricants in the Environment

correlate well with environmental lead concentrations, but do not occur with gaseous immissions.

Readily degradable lubricants from *mixed-lubricated* engines enter the environment in very low concentrations and without accumulating effects of dangerous substances in the engine or in the environment. These immissions are quickly dispersed and degraded, causing no environmental load of any significance.

In contrast, used engine oil from *crankcase-lubricated* engines with their anti-oxidants and much higher levels of contaminants often enter the environment in high concentrations. As we have mentioned before, the persistent components of these oils are transported via sewers and rivers into lakes where they finally sink into the sediments.

A recent study of Bodensee sediments (17) has shown clearly that these types of hydrocarbons are accumulating near river exits (Figure 4.11.5). There is also a diffuse contribution via rainfall over the whole surface area of this lake.

The more easily degradable lubricants used in mixed-lubricated engines were not found in this study, not even in the entrances of busy marinas where they could have been expected in measurable concentrations. A qualitative comparison of the C-number distribution from gas chromatograms also indicated used oil from crankcase-lubricated engines as a possible source for hydrocarbons in the sediments (Figure 4.11.6).

In another study (19), a bio-assay of the Bodensee sediments confirmed that the environmental load with organic compounds was mainly coming from the rivers. No such loads were found near pleasure boat harbours where there was considerable boat traffic, mainly using mixed-lubricated marine engines.

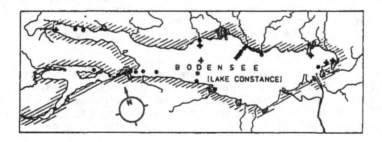

Figure 4.11.5: +: Particularly High Hydrocarbon Concentrations in the Sediments Near River Mouths (according to Howells)

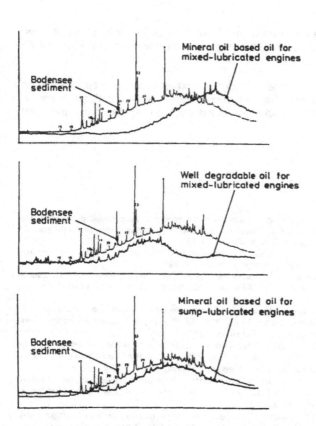

Figure 4.11.6: C-number Distribution in Bodensee Sediment Compared to C-number Distribution of Lube Oils (according to Howells).

4.11.9 Conclusions

We have seen that there are some basic differences between crankcase- and mixed lubrication with respect to their environmental effects. This was demonstrated in particular with the example of the polycyclic aromatic hydrocarbons (PAH). The PAH concentrations in used oil from crankcase-lubricated spark-ignited engines is about 1000 times higher than in lubricants used in mixed-lubricated engines. There are also considerable differences in the concentrations of the emissions and immissions of the lubricants from both types of engines into the environment.

We therefore arrive at the conclusion that from an environmental point of view, mixed lubrication is many orders of magnitude better than crankcase lubrication.

In the development of new internal combustion engines, the application of mixed lubrication should therefore be reconsidered, also for those types of engines that have been traditionally equipped with crankcase-lubrication until now.

4.11.10 References

(1)   Güsmer, S.F.: Werkstattplanung, Ölabgabe-Einrichtungen, Altölbeseitigung und Alt-öl-Aufbereitung. Presentation in course 11211/68.183 of the Technische Akademie Esslingen, Ostfildern, West-Germany, February 1989.

(2)   Concawe: The collection, disposal and regeneration of waste oils and related materials. Report 85/53, The Hague, the Netherlands, 1985.

(3)   Donkellar, P. van: Außenbordimmissionen in der Umwelt. Presentation in course 6543/14005 of the Technische Akademie Esslingen, Ostfildern, West-Germany, October 1983.

(4)   Eberan-Eberhorst, C. von: Motoren-Schmierstoffe als Partner der Motoren-Entwicklung. Presentation in course 11211/68.183 of the Technische Akademie Esslingen, Ostfildern, West-Germany, February 1989.

(5)   Donkellar, P. van: CEC-Öltestverfahren in wassergekühlten Außenbordern — Klasse TSC-4 (2T-W). Presentation in course 7367/68.095 of the Technische Akademie Esslingen, Ostfildern, West-Germany, December 1984.

(6)   Gairing, M.: Anforderungen an Schmierstoffe für Kraftfahrzeugmotoren. Presentation in course 11211/68.183 of the Technische Akademie Esslingen, Ostfildern, West-Germany, February 1989.

(7)   Heinz, H.: Weltweite Zweitaktöl Klassifizierung — Vorgeschichte und heutige Lage. Presentation in course 10834/68.163 of the Technische Akademie Esslingen, Ostfildern, West-Germany, November 1988.

(8)   Grimmer, G. et al: Untersuchungen über die carcionogene Wirkung von gebrauchtem Motorenschmieröl aus Kraftfahrzeugen. Erdöl und Kohle Bd. 35, Heft 10, pp 466–472, West-Germany, October 1982.

(9)   Abbass, M.K. et al: The aging of lubrication oil, the influence of unburnt fuel and particulate SOF contamination. SAE paper 872085, November 1987.

(10)   Shore, P.R. et al: Application of short-term bio-assays to the assessment of engine exhaust emissions. SAE paper 870627, February 1987.

(11)   Visser, K.: Afvalolie, sluipende aanslag op het milieu. Autokampioen, pp 2873–2878, the Hague, the Netherlands, December 1973.

(12)   Sugiura, K.; Kakaya, M.: A study of visible smoke reduction from a small two-stroke engine using various engine lubricants. SAE paper 770623, June 1977.

(13)   Vanhoutte, P.: Arvalolie — een onderschat probleem. Milieurame 7/8 pp 31–33, Brussels, Belgium, 1988.

(14)   Dunker, R. von: Von sauberen Vorschriften, deutscher Gründlichkeit und den Problemen, sie zu verwirklichen!. Internationale Bodensee Nachrichten 25 p 3, D 7460

Balingen, West-Germany, January 1988.

(15) NRC Handelsblad: Garages erg slordig met het milieu. Newspaper article in NRC-Handelsblad, Rotterdam, the Netherlands of 25 February 1988.

(16) Siede, R.: Schadstoffbelastung durch Betriebsstoffe der Motorsäge. Presentation in colloquium of the Kuratorium für Waldarbeit und Forsttechnik: Gefahrstoffe beim Einsatz der Motorsäge, pp 143–166, Groß-Umstadt, West-Germany, February 1988.

(17) Howells, S.E. et al: Analysis of hydrocarbons in sediments of the Bodensee (Lake Constance). FSC/OPRU/3/86, Oil Pollution Research Unit, Pembroke Dyfed, SA71 5EZ U.K.; March 1986.

(18) Umweltschutzgesetz 814.01 of the Schweizerische Eidgenossenschaft, Bern, Switzerland, April 1985.

(19) Internationale Gewässerschutzkommission für den Bodensee: Die Oligochaeten im Bodensee als Indikatoren für die Belastung des Seebodens (1972 bis 1978). Bericht 38, Bodensee, West-Germany, 1988.

(20) Möller, U.J.: Umweltschutzregelungen für Frisch- und Altöl. Presentation in course 11211/68.183 of the Technische Akademie Esslingen, Ostfildern, West-Germany, February 1989.

(21) Westerholm, R.N. et al: Effect of fuel polycyclic aromatic hydrocarbon content on the emissions of polycyclic aromatic hydrocarbons and other mutagenic substances from a gasolie-fueled automobile. Environ. Sci. Technol., Vol. 22, No. 8, 1988.

(22) Reinhardt, G.P.: Anforderungen an Schmierstoffe durch den modernen Furhpart. Presentation in course 11211/68.183 of the Technische Akademie Esslingen, Ostfildern, West-Germany, February 1989.

(23) Vazquez-Duhalt, R.: Environmental Impact of Used Motor Oil. The Science of the Total Environment, 79 1–23, 1989.

(24) Orsi, E.V. et al: Vehicular traffic and airborne particulate patterns in urban and mountain areas in northeastern U.S.A. Paper presented at the Second European Meeting of Environmental Hygiene, Düsseldorf, May 31 – June 2, 1989.

615

4.12 Development in Synthetic Lubrication for Air Cooled Two Cycle Engine Oils: Effect of Esters on Lubrication and Tribological Properties

D. Moura and J.-P. Legeron
Cofran Research Sarl, La Rochelle, France

Summary

This paper reports the development of an air-cooled two-stroke engine lubricant using esters as synthetic base stocks and performance additives.
The viscosimetric and tribological properties as well as the performances in two cycle gasoline engines will be described in comparison with standard lubricants. It is shown that the use of suitable esters make possible the formulation of high performance lubricants whatever the severity of use.

Keywords: Two-stroke, ester, boundary lubrication, tribometry.

4.12.1 Introduction

Two-stroke gasoline engines offer a combination of characteristics which differentiate them in many applications.
The most advantageous characteristics of the two-stroke engines are their low initial cost, their simplicity combined with low weight and relatively high power output, and their quick low temperature starts.
However, they are also considered as severe devices generating lubricating problems which cannot be always easily solved.
In this paper, the development of such synthetic lubricants will be described. It is shown that well chosen esters are good candidates for two-stroke engine lubrication whatever the severity of use.

4.12.2 Two-Stroke Engines and their Conditions of Use

When two surfaces moving relative to each other are completely separated by a thick film of oil, they are considered as being in a hydrodynamic lubrication regime. In theory:
- moving surfaces do not touch each other
- no surface wear occurs
- main lubricant property required is viscosity

If the oil film is so thin that the surface asperities come in contact, it can be considered as being in a boundary lubrication regime:

616

— surface wear can occur
— lubricant properties are required to reduce wear to an acceptable rate

Some parts of an engine, such as the bearings, operate in a hydrodynamic lubrication regime. On the other hand, valve gears operate in a boundary lubrication regime.
There are also some areas in the engine, such as pistons and cylinder walls, that alternate back and forth between boundary and hydrodynamic lubrication regimes.

Generally, auto-lubrication of two-stroke engines presents some disadvantages that it is necessary to describe shortly:

— the ratio of consumed oil required for lubrication of connecting rod bearings and cylinder walls is very low. The main part is carried off through the transfer line to the combustion chamber. Oil is not completely burned and then gives carbon deposits either in the discharge port or in the engine exhaust. Consequences are decrease of output power and clogging of the upper part of the engine.

— polarized ignition plugs attract ash which then causes short-circuit and ignition failure.

For these reasons, the lubricant must have a specific grade to allow good lubrication in two-stroke engines.

4.12.3 Lubricant Choice and Performance Considerations

Two-stroke engine oils generally need the following components:

— Mineral or synthetic base fluids;
— Ashless dispersant additives, added to meet appropriate performance of blended oils:
— High boiling point solvent, to improve self miscibility properties.

Three kinds of lubricants were tested (Table 4.12.1).

The results show that it is quite possible to establish a correlation between the structure of a synthetic ester based oil and the antiwear and friction properties.

The influence of the ester group has a direct effect on the physical properties of finished oils:

— Ester A: It is a non-hindered dimer ester. Generally, deacids used for such products are oleic acid dimers (C36). They are then combined with linear monoalcohols.

617

The structure of these esters is complex, and occasionally contains C4 chains and C6 insaturated cycles coming from the Diels and Alder reaction. Double bonds stay in these molecules but can be saturated by hydrogenation.

— Ester B: Resulting from linear monoalcohol and polycarboxylic acid reaction, it presents a comb-like structure with the main chain consisting of carbon atoms only.

Generally, these esters with straight chain have higher VI values, flash points and pourpoints than the branched chain esters (Table 4.12.2).

This kind of esters are already used in several applications as anti-wear additives and friction reducers in boundary lubrication.

— Ester C: Because of their branched chain, the viscosity index of this pentaerythritol neopolyol ester (NPE) is low.

However, the structure of the hydrocarbon chain allows to NEP to be the most thermally stable molecules among esters.

Related to the absence of hydrogen on carbon of alcohol function, thermal degradation requires a radical reaction with greater energy than in the decomposition of diester.

Consequently, it is not easy with such esters to obtain high viscosity index. The higher the acid chain is, the higher is the pourpoint.

Thus, the use of neopentylpolyol esters as lubricants is restricted in this case.

Table 4.12.1: Summary of Oil Characteristics

Oil reference		A	B	C
Ester E1	%wt	76		
Ester E2	%wt	5		
Ester E3	%wt		75.4	
Mineral base	%wt			79
Polybutene	%wt		5	
Dispersant additive	%wt	8	8	8
Solvent	%wt	11	11.6	13
Viscosity at 100°C	mm²/s	11.4	10.6	10.6
Viscosity at 40°C	mm²/s	72.5	82	88.6
Viscosity Index		150	114	103
Pourpoint	°C	− 36	− 33	− 9
Flashpoint	°C	192	190	191
Acid value	mg KOH/g	1.4	1.2	1.2
Sulfated ash	%wt	0.17	0.16	0.24
Density at 20°C		0.897	0.94	0.882

Table 4.12.2: Ester Properties

Ester reference		A	B	C
Viscosity at 100°C	mm²/s	12.8	35	10
Viscosity at 40°C	mm²/s	88	357	87
Viscosity Index		144	141	94
Flashpoint	°C	300	255	270
Pourpoint	°C	− 45	− 30	− 36
Noack volatility	%wt loss	1.1	−−	1.3
Acid value	mg KOH/g	0.05	−−	0.1

4.12.4 Friction Test Procedure

The friction and wear experiments were carried out on two types of testing apparatus: PLINT friction machine and Cylinder/Cylinder Tribometer.

4.12.4.1 PLINT Friction Machine

An outline of the test procedure is as follows:
This apparatus allows to measure continuously coefficient of friction and electrical resistance between ball and plate (Figure 4.12.1).

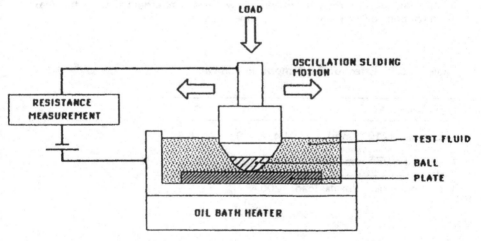

Figure 4.12.1: PLINT Friction Test Rig Schema

At the end of the test, the specific wear volume is rated from the following equation:

$$K' = \frac{\text{volume of wear} \cdot 100}{4 \cdot \text{amplitude} \cdot \text{frequency} \cdot \text{time} \cdot \text{load}}$$

The results are shown in Figure 4.12.2 and Table 4.12.3.
Fluid A indicates a better response as well in terms of friction coefficient as in wear behaviour than the other tested lubricants.
Because of the two-ester mixture, it has an active head which attaches itself by electrostatic forces to the metal surface while the backbone is a hydrocarbon chain. Then, the hydrocarbon chain forms a transition surface on the metal which converts it into a genuine hydrocarbon surface.
In the present case, metal and surfactant are reactive, and this form a chemical compound which sticks to the surface until its decomposition.
The hydrocarbon tail has a strong affinity for oil, and hence, the surface is easily lubricated.

The behaviour of oil C is better than that of the ester based oil B in terms of wear level. It does not contain synthetic bases but includes in its formulation an important level of Bright Stock Solvent (approximately 10 %) to provide satisfactory wear protection.
However, the friction coefficient value varies during the time. Film oil formation and hydrodynamic lubrication has no good continuity. Boundary lubrication (high wear) and hydrodaynamic one (lower wear) alternate even if the maximum friction coefficient value is the same in any case.

Table 4.12.3: Effect of Oil Composition on Wear

Oil reference		A	B	C
Wear scar	$(\times 10^{-3}\,mm)$	4.4	9.5	6.8
K'	$(\times 10^{-9}\,mm^3/Nm)$	56.4	129	92.5

Test conditions: Normal load: 200 N
Time   :   60mn
Temperature: 228°C

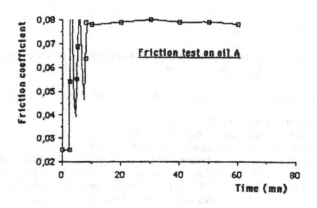

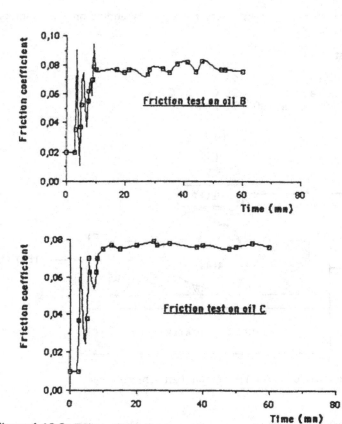

Figure 4.12.2: Effect of Oil Composition on Friction Coefficient

4.12.4.2 Cylinder/Cylinder Tribometer

In order to assess the influence of ester type on the wear rate, a series of tests were conducted on a Cylinder/Cylinder machine.

On this test, a cylindrical roller, 10 mm diameter, is loaded against a rotating ring.
The device is shown schematically in Figure 4.12.3.
On the machine, rotating speed can be settled up to 1100 rpm; load of up to 1900 N can be achieved and temperature of oil bath ranged from 20°C up to 300°C. The measure of the electrical resistance of the oil film between roller and ring gives information on film formation and thus, the lower level of wear reached. Values of resistance have to be measured under conditions of hydro-dynamic lubrication. In this test, we have not tried to correlate the value of the resistance with the film thickness.
Electric flow through thin oil film is very complex, depending on many factors other than film thickness.

Conditions of test:
- Load . . . .   270 N
- Rotating speed . . . .   260 rpm
- Oil Temperature . . . .   20°C – 80°C
- Ratio of oil dilution in 100 NS.. 5 %wt

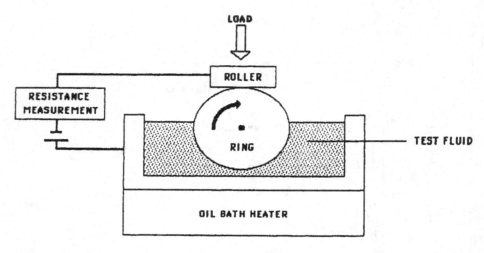

Figure 4.12.3: Cylinder/Cylinder Tribometer Test Schema

4.12.4.3 Results

For both oils A and C, as shown in Figure 4.12.3, there is a similar response and the wear rates were reduced by about a factor of 3 with oil B.
In all cases, the wear rates increase with increasing temperature, typically by a factor of 4 at 80°C compared with the room temperature ones (Figure 4.12.5). In conclusion of these two tests, it appears that oil B, based on NPE, cannot be selected, protection against wear being not sufficient in case if boundary lubrication which occurs during each engine revolution, very accentuated when sliding speed of the piston is high and clearances are low.

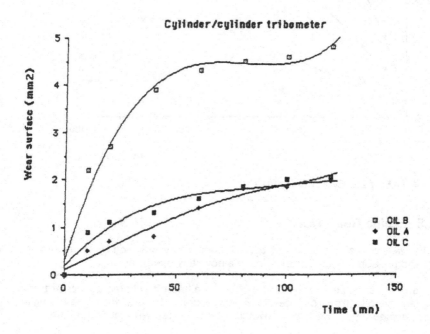

Figure 4.12.4: Wear Evolution at Room Temperature

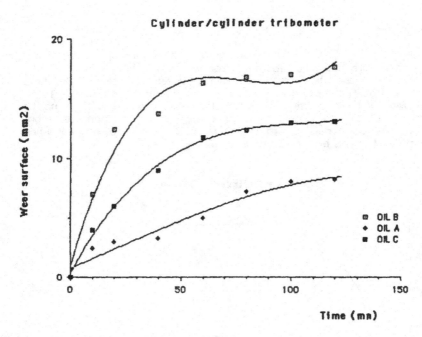

Figure 4.12.5: Wear Evolution at 80°C

4.12.5 Engine Road Tests

An advantage of field tests is that they duplicate very well ring belt area temperature and pressure conditions existing in a normally operating engine.

A single cylinder engine was used to evaluate the lubricity characteristics of two-stroke engine oils from our development program. Results from these single cylinder engine tests have been shown to correlate well with laboratory friction tests.

Piston head, cylinder walls, combustion chamber and free exhaust aspect could be well controlled. Piston skirt clearances were measured before and after road test.
Table 4.12.4 shows the main characteristics of the engine used.

The performance of two-stroke cycle engine premix lubricants were evaluated on a go-kart chassis operated by a go-kart racer. Three field tests were conducted in race-type service to be able to differenciate between ester and mineral base stocks effects on two-cycle oils:

— evaluation of lubricity:
it was measured by piston tightness-power loss through obstruction of the exhaust system (significant drop in the measured temperature between beginning and ending of the test)- and ring sticking, piston clearance and piston skirt wear tracks.

— deposit formation:
deposit control on piston, piston underhead and combustion chamber as well as observation of spark plug fouling were conducted.

The engine ratings with oil A were excellent, particularly in the critical areas of piston scuffing, bearing lubrication, piston ring sticking and exhaust port plugging.

As indicated in Figure 4.12.6, after 60 minutes of service, the engine was essentially clean and no ring sticking was encountered.

Figure 4.12.6: Typical End of Test Pistons

After eight laps, test with oil B was stopped after engine tightening. Inspection revealed deficient lubrication at this fuel to oil ratio even in the lower part of the engine.

After the test, the oil C results for piston deposits, scuffing and ring sticking were considerably worse than those obtained with oil A (Figure 4.12.6).
More deposits appeared after combustion due to Bright Stock Solvent presence which have tendency to clogg the exhaust system and gum up the piston rings.

In all tests, oil blended with ester mixture performed better than oil blended with NPE and mineral base stocks (Table 4.12.5).
In the same way, reduction of smoke was very sensitive using oils A and B. However, it depends of fuel to oil ratio and presently, to get significant results, it should have been necessary to increase this ratio.

Table 4.12.4: YAMAHA lubrication field test conditions

Engine description

Type	YAMAHA KT 100S,2-cylinder air-cooled
Swept volume	100 cc
Rated output	15 HP (12 000 rpm)
Compression ratio	7.9:1
Bore	50 mm
Stroke	46 mm
Fuel to oil ratio	33:1 (premix)
Fuel	Leaded gasoline

Operating conditions

Speed	from 10 000 to 15 000 rpm, max. load
Intake air temp.	~ 20°C
Exhaust temp.	~ 500°C
Piston clearance	0.02 mm
Test duration	60 mn (40 laps)

Table 4.12.5: YAMAHA go-kart type field test results

Oil reference		A	B(*)	C
Test conditions				
Test duration	mn	60	24	60
Laps		40	16	40
Miles		44	18	44
Initial piston clearance	mm	0.02	0.02	0.02
Inspection results				
Terminal piston clearance	mm	0.03	——	0.08
Exhaust port aspect		brown	grey	oily brown
Piston skirt varnish		none	none	light
Piston scuffing		none	——	severe
Piston ring		free	free	stuck
Piston undercrown		clean	clean	coked
Piston top deposits		normal	normal	high
Spark plug aspect		good	good	fouled
Exhaust temperature				
—at the beginning	°C	~ 510	~ 530	~ 530
—at the end	°C	~ 490	——	~ 470

(*) The test on oil B was stopped because engine could not be restarted.

4.12.6 General Conclusion

Oils blended with two ester mixtures have demonstrated superior lubricant performance in laboratory friction tests and in actual race-type service field tests when compared to the other based lubricants at equal fuel to oil ratio.

Laboratory friction test performance

Improved bearing and especially piston lubrication occurred when:
- sliding speed around the midpoint of the piston course promotes hydrodynamic lubrication,
- sliding speed around piston top and bottom dead center induces boundary lubrication.

Field test performance

In very severe conditions of use, and a relatively low treat level, it shows:
— good varnish control and ring sticking protection,
— better scuff and ring wear protection.

The number of synthetic lubricants are growing fast and will continue to develop during the next years.
Application studies must be used to identify needs, select the proper synthetic components and determine cost savings.

In applications involving two-cycle race lubrication, it is quite important to select the better low oil feed rate. It will contribute to increase the output power, reduce environmental and maintenance problems, and control varnishes and deposit formation.

Therefore, two-stroke engine life is increased through improved performances.

4.12.7 Acknowledgements

The authors wish to thank the AKZO Chemical Division for their assistance in conducting the Plint Friction test reported in this paper and the Gallo Racing team for their collaboration in the field tests procedure.

4.12.8 References

— Oliver, C.R.; Reuter, R.M.; Sendra, J.C.: "Fuel Efficient Gasoline Oils" Lubrication, Vol 67,1 (1981), p 2.
— Kenbeek, D.; Waal, G. van der: "Development of High Dilution, Low Pollution Outboard Oils", Journal of Synthetic Lubrication, Vol 5,3 (1988), p 216.
— Legeron, JP.: "Synthetics in Engine Lubricants", Tashkent, International Scientific Conference — Friction, wear and Lubricants, May 22nd —26th, 1985, Tashkent, USSR.
— Denis, J.: "Bases lubrifiantes Naturelles et de Synthèse". Publication IFP-ENSPM (1981), p 111—12, 111—26.
— Szydywar, J.: "Ester Base Stocks", Journal of Synthetic Lubrication, Vol 1,2 (1984), p 164—165.
— Venkataramani, P.S.; Kalra, S.L.; Raman, S.V.; Srivastava, H.C.: "Synthesis Evaluation and Application of Complex Esters as Lubricants: A Basic Study". Journal of Synthetic Lubrication, Vol 5,4 (1989), p 272.
— Beltzer, M.; Jahanmir, S.: "Effect of Additive Molecular Structure on Friction". Lubrication Science, Vol 1,1 (1988), p 3—26.
— Zakar, A.; Kovacs, A.; Vamos, E.; Olajos, E.: "Polyols Esters Oils for Industrial Purposes". Ostfildern, Industrial Lubricants-Properties, Application, Disposal. Technische Akademie Esslingen, January, 12th — 14th, 1988, Ostfildern, FRG.
— Robertson, W.S.: "Lubrication in Practive". Mac Millan Press, London, 1983.
— Kotvis, P.V.: "Overview of the Chemistry of Extreme-Pressure Additives". Lubrication Engineering, Vol. 42,6 (1986), p 363—366.

4.13 High Performance Ester-Based Two-Stroke Engine Oils

D. Kenbeek and G. van der Waal
Unichema International, Gouda, The Netherlands

4.13.1 Introduction

The two-stroke engine has always been a popular power source. It is used for a variety of applications ranging from small mopeds to big outboards and racing engines. This popularity has much to do with its simple design and construction resulting in:

— low weight
— compact engines
— high specific output
— relatively low costs

Another point of interest is the low NO_x emission. This will be of growing importance with the increasing concern about environmental issues. On the other hand, also a number of negative aspects have to be mentioned. Most well known are:

— high smoke emissions
— noise
— pollution of water (outboard engines)
— high emission of hydrocarbons

During the last years many technical improvements have been realized to reduce this. Examples of these are the introduction of electronic ignition systems, exhaust tuning, exhaust catalysts, automatic lubrication systems and improved fuel spray techniques (1). So far this has resulted in much more advanced engines. Further technological changes in design are foreseen as a result of:

— higher specific horsepower
— application of light alloy cylinders with special surface coatings
— low oil percentage in the feeding mixture
— more severe regulations concerning exhaust emissions and fuel consumption

These changes will definitely have an impact on engine performance and lubricant requirements (2).

Further improvements to the two-stroke engine may be achieved through the so-called Orbital Combustion Process, which is based on inventions of Australian engineer Ralph Sarich. This engine features an electronically controlled direct fuel injection system combined with other characteristics such as special combustion chamber geometry.

The Orbital Combustion Process is claimed to be extremely low regarding both hydrocarbon and NO_x emissions. A high potential for use in passenger cars is predicted. Compared with a conventional four-stroke engine following advantages are claimed (3):

- up to 30 % better fuel economy
- weight reduction by 50 %
- reduction of packaging volume by 70 %
- cost reduction by 25 %

So far this has attracted attention of some major OEM's. The impact on lubricant requirements is not yet known.

In summary it is concluded that the two-stroke engine is more alive than ever and, when everything comes through as predicted, might face a bright future.

4.13.2 Lubricant Requirements and Composition

At first sight the relatively simple two-stroke engine seems to require relatively simple lubricants. A more close look shows that compared with the four-stroke engine, the two-stroke engine is more critical towards (4):

- deposit formation in ports and exhaust
- formation of ashcontaining deposits in the combustion chamber, leading to pre-ignition
- piston wear and piston cleanliness
- detergency in piston ringzone
- miscibility of lubricant and fuel
- corrosion protection

Due to the requirements mentioned above the formulator has to balance carefully the lubricant composition in terms of:

- basefluid
- type of detergent/dispersant additive
- ash content
- type and amount of solvent

4.13.2.1 Basefluid

Both mineral oil and synthetic base fluids are used. Mineral oil based formulations normally contain brightstock (approximately 10 %), in order to provide proper wear protection. To avoid coke and deposit formation, the amount and quality of brightstock should be as optimal as possible. To diminish the dependency of good quality brightstock, the use of synthetic countertypes is considered as well. Both high viscosity esters and poly-isobutylenes are options, but from a lubricity point of view, esters seem to be the preferred route.

The growing demand for engine performance has resulted in a demand for high quality, fully synthetic products. Again esters and poly-isobutylenes are applied. In Europe esters are favoured especially with regard to lubricity and biodegradability.
Esters do provide excellent lubricity due to the presence of polar carboxylic groups, giving better adhesiveness to the metal surface. Therefore the use of brightstock can be omitted. Additionally the polar groups contribute to lubricant detergency.

A point of growing importance is biodegradability. It is well known that oleochemical esters, derived from natural oils and fats, respond well to biodegradation processes. The first application of esters for reasons of biodegradability has been for two-stroke outboard oils. This has led to the development of environmentally friendly high dilution, low pollution outboard oils (5). After a bit hesitant start in the nearly eighties, the market has responded well and currently esters have an established position within this market segment.

It has to be noted that not all esters show the same high degree of biodegradability. This is dependent on type of ester and choice of raw materials. Esters based on hindered polyols are preferred.

4.13.2.2 Additives

To ensure appropriate lubricant performance in terms of detergency/dispersancy it is necessary to use selected additives. Both ashless and ashcontaining additives are applied. Ashless types are mainly used for water-cooled engines, which are running at relatively low temperatures. However, it will be demonstrated further on that these additives show also potential for air-cooled applications. Chemically the ashless additives are often reaction products of polyamines and long chain fatty acids. The following generic structure is an example of the chemistry applied.

$$R_1 - C \begin{array}{c} N \\ \\ \end{array} \begin{array}{c} - CH_2 \\ | \\ CH_2 \end{array}$$
$$\begin{array}{c} N \\ | \\ R_2 \end{array}$$

R_1 = alkyl $(C_{16} - C_{18})$

$R_2 = R_1 (NH - CH_2 - CH_2)_3$

The ashcontaining additives are mainly used for air-cooled applications, where high temperatures are involved. Favoured chemistry are the calcium-sulphonates. Currently, the so-called low ash additives are preferred. These offer a balanced performance between detergency and precention of ashformation (pre-ignition).

4.13.3 Classification of Two-Stroke Lubricants

For four-stroke engine lubricants a worldwide accepted classification system does exist already for decades. In contrast to this, only a fragmented classification system for two-stroke lubricants did exist. This is now going to be changed. For many years, only a classification for two-stroke outboard oils was known, the so-called BIA TC-W approval.

However, since 1975 organisations both in Europe (CEC) and the United States (API, SAE, ASTM) have been working towards a comprehensive classification system for two-stroke oils. In 1988 this led to a classification for lubricants for motor-cycles. Other classifications are expected to follow soon.

The proposed and now partly in place being system has appeared under various designations, such as TSC-, ISO- and API-ones. Although the TSC-designation is probably most widely known, the API-one will be likely adopted as the official designation, TSC, ISO and API do compare as follows:

Table 4.13.1

ASTM	ISO	API
TSC-1	ISO-L-ETA	API TA
TSC-2	ISO-L-ETB	API TB
TSC-3	ISO-L-ETC	API TC
TSC-4	ISO-L-ETD	API TD
TSC-5	ISO-L-ETE	API TE

The five designations are covering five applications (see Appendix).

In addition to these five designations, two others are being considered:

The first one will be essentially API TE plus a biodegradation requirement. In particular in Europe a wish does exist to come to well defined high dilution, low pollution outboard oils, reflecting current market trends. With the new API TE specification (NMMA TC-WII) in place, this might be established soon.

The second designation is intended for chainsaw engine oils. Although this category should be covered by API TC, field problems have been noted both in Europe and the United States. Chainsaw engine manufacturers feel that a separate oil category is required.

4.13.3.1 API TC and API TD/TE Test Conditions

4.13.3.1.1 API TC (ASTM D4859-89)

Following oil performance characteristics are considered:

Table 4.13.2

—Ringsticking/cleanliness (ASTM D4857-89)

Test engine	Yamaha RD-350B, two-cylinders, 347 cm^3
Test procedure	Y-350M-2
Fuel/oil ratio	50:1 (unleaded)
Total running time	20 hours

—Tightening/lubricity (ASTM D4863—89)

Test engine	Yamaha CE50S, single cylinder, 49 cm^3
Fuel/oil ratio	150:1 (unleaded fuel)
Torque measurements	At 200°C and 350°C

—Preignition (ASTM D4858-89)

Test engine	Yamaha CE50S, single cylinder, 49 cm^3
Fuel/oil ratio	20:1 (leaded fuel)
Test duration	50 hrs at Wide Open Throttle

4.13.3.1.2 API TD/API TE

Since the end of 1988 the NMMA TC-WII (API TE) procedures have replaced NMMA TC-W (API TD). Current NMMA TC-W approvals will be valid till the end of the original life time, but will not be renewed. Herewith the oldest two-stroke engine oil classification is continuing its service life. The NMMA TC-W and NMMA TC-WII test conditions do compare as follows:

Table 4.13.3

NMMA TC-W/API TD	NMMA TC-WII/API TE
Engine tests OMC 90TLCOS	**Engine tests Yamaha CE50S**
- Lubricity Fuel/oil ratio; 150:1 (leaded) Test duration: 10 minutes	— Lubricity Fuel/oil ratio; 150:1 (unleaded) Torque measurements: At 200°C and 350°C
	OMC 40ECC
— General performance Fuel/oil ratio: 50:1 (leaded) Test duration: 21 hours	— General performance Fuel/oil ratio: 100:1 (unleaded) Running time: 98 hours
	Yamaha CE50S
— Preignition Fuel/oil ratio: 24:1 (leaded) Test duration: 98 hours	— Preignition Fuel/oil ratio: 20:1 (leaded) Test duration: 100 hours
Bench tests — Miscibility — Rust prevention	**Bench tests** — Miscibility/Fluidity (SAE J 1536) — Rust prevention — Filterability

The most pronounced difference between NMMA TC-W and NMMA TC-WII is found within the general performance test sequence, prescribing a leaner fuel/oil ratio for NMMA TC-WII. It suggests that NMMA TC-WII defines higher quality oils. However, this seems to be denied by the fact that essentially a similar reference oil is used for both TC-W and TC-WII (same basic additive chemistry, see chapter 2.2, but with a different anti-rust additive and same additive treat level). It is therefore concluded that one of the main characteristics of the OMC model 40 test engine is its ability to accept 100 : 1 fuel/oil ratios.

Furthermore it is concluded that one should be careful in using NMMA TC-WII oils at 100:1 in older engines. Due to lack of detergency (the engine "sees" only half the amount of additive compared to 50:1) problems may occur with regard to engine cleanliness and deposit built-up. It is therefore expected that OEMs will recommend to use NMMA TC-WII oils at 50 : 1 for older engines.

A new aspect of NMMA TC-WII is filterability. This is introduced to determine the tendency of an oil to form a gel precipitate in the presence of small amounts of water, which can plug oil filter screens in two-stroke engine oil injection systems.

4.13.4 Synthetic Two-stroke Oils: Engine Test Results

4.13.4.1 Air-cooled Performance According to TSC-1/API TA

As said before, esters show excellent lubricity due to the presence of polar carboxylic groups. A well-known test method to establish lubricity is the Motobecane tightening test. Although the Motobecane engine is going to be phased out due to lack of availability and will be replaced by the Yamaha CE 50S engine, it is still a valuable tool to test lubricity.

The test is based on the principle that during the operation of an engine under severe conditions the temperature will increase and the clearance between piston and liner will decrease. This will create conditions for metal-to-metal contact resulting in micro welds (scuffing). The formation of these micro welds will have a negative influence on engine output. It is the function of the lubricant to prevent or ristrict the formation of micro welds and to minimize power loss.

Outline of the Testmethod

A Motobecane AV 7L 50 cm^3 test engine is fitted with a continuous temperature reading device at the spark plug seating ring. The engine is run under high load at 4000 rpm and the power at the driving shaft is recorded.

After having obtained the normal operating conditions the cooling of the engine is cut off and the torque drop is measured between spark plug seating ring temperatures of 200°C and 300°C. The lower the torque drop, the better the lubricity.

The tests have been carried out with two ester base fluids and a blend of a 400SN paraffinic mineral oil with 10 % brightstock. Base fluid characteristics are:

Table 4.13.4

	Ester A	Ester B	400 SN mineral oil plus 10 % brightstock
viscosity 100°C (mm²/s)	6.5	13.5	13
viscosity 40°C (mm²/s)	31	94	128
VI-E	164	145	95
flashpoint COC (°C)	235	295	265
pourpoint (°C)	- 21	- 42	-12

635

The tightening tests have been carried out at different fuel/oil ratios. Following test results have been obtained:

Table 4.13.5

Fuel/oil ratio (leaded fuel)	50:1	100:1	200:1	400:1
Ester A	22	25	–	–
Ester B	–	21	25	33
400 SN min oil/ 10 % brightstock	25	35	–	–
Reference oil RL 09	–	25	–	–

Results are expressed as drop in torque (N.mm).

The test results demonstrate the outstanding lubricity of ester A and B. At a fuel/oil ratio of 100:1, low viscosity ester A has equal performance as the brightstock containing mineral oil at 50:1.

Ester B is even better, having a performance at 200:1, which equals those of the mineral oil based formulation at 50:1. The reference "pass" oil RL 09 is based on a blend of 40 % 100 SN mineral oil plus 60 % brightstock.

4.13.4.2 Air-cooled Performance According to TSC-3/API TC

The Yamaha RD-350B twin-cylinder engine is widely used to evaluate two-stroke lubricants for large air-cooled engines with a swept volume of 250 cm^3 and more. The engine provides valuable information about engine cleanliness and high temperature ringsticking as well as anti-scuffing properties.

Outline of the Test Method

In the first run one cylinder is supplied with the fuel/candidate oil mixture, whilst the other cylinder is supplied with the fuel/reference oil blend. Normally the test is run twice, exchanging the oils between the cylinders for the second run. In case the performance of the candidate oil exceeds that of the reference oil by a specified margin, the second run does not need to be made.
Total test period is 20 hours. After the test the engine is dismantled and examined for deposit and varnish formation, ring sticking and scuffing and wear.

The test has been carried out with ester B as the base fluid (see Table 4.13.4). An ashcontaining additive package has been used with following characteristics.

636

Table 4.13.6

Additive X		
Calcium	(% wt)	0.75
Sulphur	(% wt)	0.60
Viscosity 100°C	(mm²/s)	225

Test procedure has been Y-350M. A commercial super two-stroke oil was used as the reference.

Test results:

Table 4.13.7

	Ester B		Commercial super two-stroke oil
Fuel/oil ratio	50:1	100:1	50:1
Amount of additive	7	7	—
Piston skirt (average)	9.5	7.0	7.0
Piston undercrown	5.0	1.0	1.3
Piston top deposits	8.6	9.0	8.6
Cylinder head deposits	8.3	8.0	7.7
Ring land	9.0	2.8	2.8
Ring sticking:- top	10.0	10.0	10.0
- second	9.0	7.8	8.2
Exhaust port clogging	9.8	9.8	9.8
Piston scuffing	10.0	10.0	10.0
Piston ring wear (mg)			
- top	6.4	5.8	4.1
- second	8.8	14.3	9.7

As can be seen, the oil based on ester B shows superior performance compared to the reference oil, at a fuel to oil ratio of 50:1.

At 100:1 a satisfactory level of detergency is obtained despite the high fuel to oil ratio.

Detergency is about equal to those of the commercial mineral oil based two-stroke oil at 50:1.

As the Yamaha RD-350B test engine is susceptible to seizure, the protection offered by ester B at a fuel to oil ratio at 100:1, is considered as excellent.

Even better results are expected by a slight increase of the viscosity of the ester.

Overall conclusion is that ester B shows excellent performance regarding both engine cleanliness and wear protection.

4.13.4.3 Water-cooled Performance According to TSC-4/API TD(NMMA TC-W) with Air-cooled Performance According to API TC

The objective here was to develop an outboard engine oil with excellent engine performance and high biodegradability. Forthermore this oil should have potential for use in air-cooled engines as well. To achieve these goals both the base fluid and the additive package had to be selected carefully. The biodegradation requirement was leading to the choice of ester C. This ester is combining high biodegradability with excellent performance in outboard engines (5).

Seen the primary application, being for outboard engines, the additive package had to be ashless. To ensure adequate performance in both water-cooled and air-cooled engines, additive Y was selected on the basis of good field performance in conventional two-stroke oils.

Table 4.13.8: Characteristics of Ester C and Additive Y

Ester C	Additive Y
Viscosity 100°C (mm^2/s) 13	Nitrogen (% wt) 1.8
Viscosity 40°C (mm^2/s) 88	Viscosity (mm^2/s) 66
VI-E 146	
Flashpoint COC (°C) 300	
Pourpoint (°C) −25	
Biodegradability * (%) 95	

*CEC L-33-T-82 (6)

Ester C and additive Y have been combined in the following formulation:

Table 4.13.9: Oil Composition and Characteristics

Ester C	(% wt)	75
Additve Y	(% wt)	17
Solvent*	(% wt)	8
Viscosity 100°C	(mm^2/s)	11
Pourpoint	(°C)	− 27
Flashpoint COC	(°C)	87
Biodegradability (CEC L-33-T-82)	(%)	75

*High boiling solvent e.g. white spirit.

Yamaha CE50S Tightening Test Results

Table 4.13.12

	Candidate	Reference*
Torque drop between 200°C and $-$ 300°C (N.M.) $-$ 325°C $-$ 350°C	0.32 0.39 0.46	0.34 0.41 0.48

* Good quality mineral oil based two-stroke oil

The results of the tightening test show a significant difference between the candidate oil and the market reference. It proves the good lubricity of the ester compared to a conventional base fluid.

Yamaha RD-350B Test Result (Y-350M2 procedure)

Table 4.13.13

	Candidate oil	Reference oil ASTM VI D
Piston skirt	8.2	7.8
Piston undercrown	4.8	1.0
Piston top deposits	7.5	7.5
Cylinder head deposits	7.7	7.6
Ring land	3.6	2.3
Ring sticking -top	10.0	10.0
-second	7.9	6.6
Exhaust part clogging	9.8	9.8
Piston scuffing	10.0	10.0
Piston ring wear (mg) -top	8.4	5.6
-second	6.8	2.6

Conclusion: Using ester C and additive Y it is possible to formulate a two-stroke oil with excellent water-cooled performance at a 100:1 fuel to oil ratio and a high degree of biodegradability. Furthermore this oil has good potential for air cooled applications at a 50:1 fuel to oil ratio.

This formulation has a biodegradability of 75 % according to CEC L-33-T-82. As can be seen the biodegradation of the formulated oil is less than those of the ester. It is well known that two-stroke additives have a negative influence on biodegradability. However, the figure of 75 % is significantly exceeding the minimum level of 67 %, which is set by ICOMIA in its standard nr. 38—88 (7). (ICOMIA stands for International Council of Marine Industry Associations).

The formulation has been subjected to a number of appropriate engine tests to demonstrate its ability for water-cooled and air-cooled applications:

Table 4.13.10

	Test parameter	Fuel/oil ratio
Water-cooled,	Lubricity	150:1 (standard)
NMMA-TC-W	General performance	100:1 (special; standard is 50:1)
	Preignition	24:1 (standard)
Air-cooled, Yamaha CE50S	Lubricity	150:1
Air-cooled, Yamaha RD-350B, procedure Y-350M2	General performance	50:1

NMMA TC-W Test Results

Table 4.13.11

Fuel/oil ratio in phase II	Candidate 100:1	Reference 50:1	Pass limits
Average piston varnish (10 = clean)	8.8	9.1	0.5 max. below reference
Average top ring stick (10 = free)	8.6	8.3	1.0 max. below reference
Spark plus fauling	1	0	1 max.

So the candidate oil, based on ester C passed the NMMA TC-W test with the added challenge of using a fuel to oil ratio of 100:1 in phase II versus 50:1 for the reference oil. Combined with high biodegradability this results in a real "high dilution, low pollution outboard oil".

4.13.4.4 Air-cooled Performance in Chainsaw Engines

Chainsaw engines are running at very severe conditions, often more severe than those of motor cycles. This is illustrated by the following characteristics:

- small engines   : 20–65 cm^3
- high running speeds   : appr. 10.000 rpm, sometimes up to 15.000 rpm
- high ringzone temperatures : up to 300–310°C.

In addtion to this, running hours are high: 800–900 hours for a professionally used chainsaw, compared with 30 hrs for an outboard engine.

These conditions pose high requirements towards lubricity and high temperature properties, in order to avoid ringsticking.

During the last years field problems have been reported, both from Europe and the U.S.A., about ringsticking and crankcase deposits (leading to bearing failure). These problems are partly oil, partly fuel related. Especially the introduction of unleaded fuels as well as so-called environmentally friendly fuels (low in aromatics, olefins and sulphur) has led to a requirement for lubricants with higher performance.

To cope with the problems observed engine manufactures have developed in-house test methods, as they feel that current test methods and test engines do not cover adequately chainsaw engine requirements. In fact these should be covered by TSC-3/API TC. The main reason for not doing so, is that the TSC-3 reference oil is based on an ashless additive. It is however the general feeling that chainsaw engines need low ash additive technology. An example of such an in-house test is the Husqvarna cleanliness test.

Outline of the Test Method

A single cylinder two stroke, spark-ignition, Husqvarna 266 chainsaw engine, capacity 67 cm^3, is run for 5 hours, under cyclic conditions for 55 minutes full load, 5 minutes idle, with a spark plug temperature of 260°C at full load.

Power absorption is by means of a dynamometer. Two-stroke fuel mixture is fed to the engine carburetor from an overhead fuel tank via a pipette fuel measuring system.

At the end of the 5 hours an assessment is made of piston ring sticking and cleanliness allowing evaluation of performance of the lubricant in the selected fuel.

Using this test method, engine tests have been carried out with ester based and partly ester based lubricants. A good quality 600 SN mineral oil has been used for the partly synthetic oils.

Tests have been run with a regular grade unleaded gasoline with a 50:1 fuel to oil ratio.

Table 4.13.14: Base Fluid Characteristics

Ester Type of ester	B di-ester	C polyolester	D di-ester
Viscosity 100°C (mm²/s)	13.5	13.0	12.6
Viscosity 40°C (mm²/s)	94	88	88
VIE	145	146	144
Flashpoint COC (°C)	295	300	300
Pourpoint (°C)	− 42	− 25	− 45

Lubricant formulation:

Base fluid	:	82.5 %wt
Additve	:	7.5 %wt
Solvent	:	10.0 %wt

Table 4.13.15: Test Results

Base fluid ester Mineral oil	100 % B −	66 % B 34 % 600 SN	50 % B 50 % 600 SN	100 % D −
Fuel/oil ratio	50:1	50:1	50:1	50:1
Ring Stick	9.0	6.7	4.5	10.0
Ring Groove	7.0	0	0	9.1
Skirt Exterior	10.0 ·	9.4	9.6	9.8
Undercrown	10.0	10.0	10.0	9.8
Crownland	4.3	4.8	4.0	5.1

The test results clearly demonstrate the positive influence of esters on engine cleanliness, resulting in a high performance chainsaw engine oil when using a 100 % ester base fluid.

The difference between esters B and D is remarkable. Although based on the same chemistry, with a slight difference in raw material origin, ester D is slightly better with regard to ring stick and ring groove merits.

Considering biodegradability esters B and D are not the best choice, being moderately biodegradable. Ester C with high biodegradability is the best option in this respect. Unfortunately, chainsaw engine performance of ester C is not adequate. The solution can be found in working with a blend of both.

Table 4.13.16

Base fluid Ester	100 % B	100 % C	50 % B 50 % C
Fuel/oil ratio	50:1	50:1	50:1
Ring Stick	9.0	6.2	9.0
Ring Groove	7.0	0	8.1
Skirt Exterior	10.0	8.5	9.8
Undercrown	10.0	6.1	10.0
Crownland	4.3	4.8	5.2
Biodegradability (base fluid)	40 %	95 %	85 % *

*CEC L-33-T-82

As can be seen, ester C has a synergistic effect on biodegradability, whilst ester B has a synergistic effect on engine performance. So the use of the ester blend offers the possibility to arrive at a chainsaw engine oil with both good biodegradability and good engine performance.

4.13.5 Summary and Conclusions

In this paper a survey is given of benefits and future developments of two-stroke engines, as well as the state of affairs regarding a two-stroke oil classification system. Engine tests are discussed with fully synthetic ester based engine oils, according to TSC-1, TSC-3 and TSC-4 requirements. Doing so it is demonstrated that esters show superior properties with regard to engine cleanliness and lubricity. Using a particular ester it is possible to formulate a high dilution, low pollution outboard oil with good potential for air-cooled applications as well.

Finally, chainsaw engine test results are reported with partly and fully synthetic ester based formulations. In these small but critical engines esters show again a most promising performance.

4.13.6 References

(1)   Donkellar van, P.: The modern two-stroke engine as a superior power source. TAE course Modern Two-Stroke Lubrication, Esslingen, November 1985.
(2)   Minutes IGL-11 meeting, London, 13-10-1986.
(3)   Schlunke, K.: The Orbital combustion process of the two-stroke engine. 10th International Engine Symposium, Vienna, 27—28 April 1989.
(4)   Eberan von, C.: Worldwide two-stroke oil classification — History and current situation. TAE course Modern Two-Stroke Lubrication, Esslingen, November 1985.
(5)   Kenbeek, D.; Waal van der, G.: Development of high dilution, low pollution outboard oils. 6th International Colloquium, Esslingen, January 1988.
(6)   CEC L-33-T-92: Biodegradability of 2-stroke cycle outboard engine oils in water.
(7)   ICOMIA standard 38-88: Lubricating oil for 2-stroke cycle outboard motors — ecologically friendly.

Appendix

Designation	Critical Lubrication Requirements	Application examples	Test procedures
API TA/ TSC-1	Piston Scuffing Exhaust System Blocking	Mopeds Lawn Mowers Small Generators/Pumps	Yamaha CE50S — Power Loss — Tigthening
API TB/ TSC-2	Piston scuffing Deposit—induced Preignition Power loss due to Combustion Chamber Deposits	Motorscooters Small (<255cc) Motorcycles Higher Oil/Fuel Ratio Chainsaw	Vespa 125 T5 — Tightening — Power Loss — Preignition
API TC/ TSC-3	Piston Scuffing Deposit-Induced Preignition Ring Sticking	Lean Oil/Fuel Ratio Chainsaw ı i-Performance Motorcycles Snowmobiles	Yamaha Y-350M-2 — Cleanliness — Ring sticking Yamaha CE50S — Tightening — Preignition

API TD/ TSC-4	Piston Scuffing	Outboard Motors	NMMA TC-W* OMC model 90
	Ring Sticking		— Lubricity — Gen. Performance
	Deposit-Induced Preignition		— Preignition Rust Miscibility
API TE/ TSC-5	Piston Scuffing	Outboard Motors	NMMA TC-WII * Yamaha CE50S
	Ring Sticking		— Lubricity — Preignition
	Deposit-Induced Preignition		OMC model 40-Gen. performance, Rust Miscibility Filterability

* Note: BIA has been replaced by NMMA.

5. Tractor Lubrication

5.1 Tractor Lubrication

D. J. Neadle
Smallman Lubricants Ltd., West Bromwich, Great Britain

Summary

The modern tractor is a complex machine which places special demands upon lubricants.

The systems to be lubricated are engine, transmission and gearbox, rear axle and wet brakes and hydraulics.

There are two main approaches to tractor lubrication:

1) Super Universal Tractor Oil (SUTO/STOU)
2) Engine Lubricant
   Universal Tractor Transmission Oil (UTTO)

Certain conflicts can arise between the obvious need for the operator to mini-mise the number of lubricants in use and the specified requirements of the OEM.

The paper describes in detail lubricant specifications for farm tractors and discusses the technical aspects of the compromises which have to be considered in attempting rationalisation.

Finally current trends in farm tractor lubrication are described together with probable future developments.

Tractor Lubricants

5.1.1   Introduction

5.1.2   Functional Requirements Placed on Tractor Lubricants

        5.1.2.1   Engine Lubrication
        5.1.2.2   Transmission and Gear Box
        5.1.2.3   Rear Axle and Wet Brakes
        5.1.2.4   Ipto
        5.1.2.5   Hydraulics

5.1.3   Specification Requirements

5.1.1   Introduction

One can argue that life as we know it depends upon farm tractors because of the key roles they play in food production. It is a small step from this to realise the immense importance of correct lubrication of these machines.

The modern tractor is a complex machine which places special demands upon lubricants.

The mechanical systems to be lubricated are:

- Engine
- Transmission and Gearbox
- Rear Axle and Wet Brakes
- Independent Power Take off
- Hydraulics

There are different approaches to tractor lubrication which can be classified in the following way:

Monofunctional

- Engine Oil
- Gear Oil
- Hydraulic Oil

Multi-Functional/Universal

Tractor Oil Universal (TOU)

- Naturally Aspirated Engines
- Manual Transmissions
- Final Drives
- Hydraulics (Not Wet Brakes)

Super Tractor Oil Universal (STOU)

- Engines (naturally aspirated and turbo-charged)
- Manual Transmissions
- Final Drives
- Wet Brakes
- Independent Power Take Off
- Hydraulic Systems

Universal Tractor Transmission Oil (UTTO)
- Manual Transmissions
- Final Drives
- Wet Brakes
- Independent Power Take Off
- Hydraulic Systems

Farm tractor manufacturers approve particular lubricants as meeting their performance specifications. An increasing trend is the marketing of lubricants by OEMs, with equivalents being accepted provided they meet the appropriate specification.

European experience shows that farms prefer STOU because it is economical and simplifies servicing.

At the end of 1987, approximately 7.5 million farm tractors were in use in Western Europe. Lubricant consumption in total was 117,000 tonnes of which 68,000 tonnes was STOU.

Acceptance of STOU by North American manufacturers is very slow and this is a major market for hydraulic/transmission fluids.

Traditionally, farm tractors manufactured in Europe have been in the medium power range (40–90 hp), whilst those built in North America are generally greater than 100 hp.

Japanese tractors are relatively smaller and less powerful (typically 40 hp).

However, the trend towards larger scale agricultural work in Europe is resulting in the need for progressive introduction of larger tractors.

5.1.2 Functional Requirements Placed on Tractor Lubricants

5.1.2.1 Engine Lubrication

Required characteristics:
- Detergency
- High Temperature Viscosity
- Low Temperature Fluidity
- Thermal Stability and Oxidation Resistance
- Anti-wear
- Rust/corrosion Prevention
- Seal Compatibility
- Resistance fo Foaming.

The engine lubricant for a modern farm tractor should have comprehensive performance in diesel engines e.g. API CE, MIL-L-2104.D, DB 227.1, and CCMC D2. In addition the requirements of API SF, MIL-L-46152.B and CCMC G2 should be satisfied.

Engine tests used in the development of a modern Super Tractor Oil Universal:

Caterpillar 1G2
Mack T.6. and T.7.
NTC 400
Sequences IID, IIID, VD.
CRC L-38
GM 6V53T
DB OM616
DB OM364.A.
Petter AVB
Ford Cortina HT
FIAT 124 AC
MWM-B
VOLVO B.20

5.1.2.2 Transmission and Gearbox

Required characteristics:
— Load Carrying Performance
— Anti-wear Performance
— High Temperature Viscosity
— Rust Prevention
— Resistance to Foaming

Due to the importance of ZF as a supplier of tractor components, demonstrable performance in *synchromesh tests* using ZF equipment is a very desirable feature of tractor lubricants (see also Axle and FZG tests in Section 5.1.2.3).

5.1.2.3 Rear Axle and Wet Brakes

Required characteristics:
— Reduce Wet Brake Noise
— Oxidation Stability
— High Temperature Viscosity

The wet brakes are disc brakes which are cooled by immersion in the hydraulic/transmission oil. This type of brake has particular advantages for tractors because abrasive dust is excluded.

For the correct operation of oil immersed brakes, the hydraulic/transmission oil must not be so high in friction that stick/slip occurs. However, the friction must be sufficiently high to enable the braking force to be effective.

When stick/slip occurs, the noises produced are described as squawk, squeal, end-grunt and end-groan. Stick/slip may occur during tight circle manoeuvring, heavy braking or low speed braking. Tractors fitted with sintered bronze brake facings are particularly prone to this problem if fluid performance is inadequate.

The Ford 6610 tractor is used for assessing anti-squawk characteristics of tractor transmission oils. Sintered bronze brake plates are fitted. Weights are added to the tractor to increase test severity. During the tests, the brake squawk noise is measured in decibels:

In the *High Speed Deceleration Test* the tractor is driven at maximum speed, then the brakes are applied with pedal efforts up to 120 lbs. This test is repeated with the tractor towing a loaded trailer.

In the *Fade Test*, the tractor is towed for 1000 metres with the brakes applied to give first 1500 lb. drawbar pull and then 1700 lb. drawbar pull.

Spin turns are carried out, the tractor being driven on full lock with the inner brake applied.

Finally an assessment of noise is made during *typical service manoeuvres*.

The noise assessments are made in comparison with commercial oils of known performance.

Other squawk noise assessments have been made in the Valmet 850 and David Brown 1690 tractors. Massey Ferguson also make squawk assessments using a test ractor in their test centre.

The tractor hydraulic/transmission fluid must also provide satisfactory extreme pressure and anti-wear performance to protect differential and final drive gears. Performance should at least satisfy the requirements of API GL4, providing protection under conditions of low speed/high torque. Tractor manufacturers also specify protection against pitting, spalling and wear, defined by proprietary test procedures. Examples of these are the Ford 3000 Axle test called up by Ford Specification M2C 86A/B and the Final Drive Durability Test required by John Deere Specification J.20.A.

In the *Ford 3000 Tractor High Torque Axle Test* the ability of the lubricant to protect the teeth of the axle crownwheel and pinion is assessed under conditions of high torque/low speed. The tractor is run on a chassis dynamometer with the drawbar, fitted with a load sensor, firmly anchored. The tractor is then driven for 100 hours in 2nd gear with an engine speed of 2000 rpm and drawbar pull of 2273 kg. The hydraulic/transmission lubricant is maintained at 90°C.

Following the test, assessments are made based on (i) increase in backlash between crownwheel and pinion (indicating wear) and (ii) ridging, rippling, pitting or other surface distress on the teeth of the crownwheel and pinion.

An oil under assessment is compared with a reference oil of known performance.

Specific evaluations to confirm suitability for use in ZF axles are also essential. A front axle test is performed with a *ZF APL 335 fitted to a Ford 5610 Tractor.* The axle is run for an extended period under conditions of overload and the sun pinion is subsequently examined.

Although the stated requirement is for API GL5 performance, correctly developed STOU formulations have been proven to offer similar performance to the high reference API GL5/SAE 90 lubricant used for performance comparison in the test procedure.

Low speed FZG tests are also used to assess load carrying ability for transmission applications. In the *FZG (DIN 51354)* test, an SAE 10W/30 STOU can achieve Damage Load Stage 13.

5.1.2.4 IPTO

The Power Take Off is driven through a multiple plate clutch which can stall the tractor engine. This is readily demonstrated using a dynamometer under laboratory conditions and when an implement becomes jammed in service.

The hydraulic/transmission fluid must not be too high in lubricity. The Power Take Off clutch will only operate correctly if the coefficient of friction is sufficiently high.

In the *Ford 4600 Tractor Power Take Off Test,* a horizontal arm is attached to the PTO shaft and the free end is attached to a load cell. Therefore when the PTO clutch is engaged the tractor engine should stall. The lubricant must allow 100 engine stalls to be achieved (thereby demonstrating that the PTO friction plates are locking up) without damage to the plates. During the procedure, oil temperature is maintained at 84°C.

The time from PTO clutch engagement to engine stall is normally between 4 and 8 seconds. Long stall times may result in severe clutch plate deterioration. Three consecutive stall times of fifteen seconds would be considered as a lubricant failure.

5.1.2.5 Hydraulics

Required characteristics:

- Low Temperature Fluidity
- Oxidation Stability
- Good Filterability
- Resistance fo Foaming
- Seal Compatibility
- Rust Resistance
- Anti-wear Performance
- Copper Corrosion Resistance
- High Water Tolerance

The conditions of operation of farm tractors result in exposure to all types of weather. Dry conditions may result in contamination by airborne dust which can cause abrasive wear and clogging.

Under wet conditions, the tractor hydraulic oil may have to operate with much higher levels of water contamination than industrial hydraulic oils.

Both conditions place strict demands upon the oil filterability. This property is usually measured by the *Denison HF—O Filterability Test.*

Anti-wear properties of tractor oils for use in hydraulic circuits are usually assessed by weight loss of rings and vanes in the *Vickers 105.C Pump* and *35 VQ 25 Pump.*

5.1.3 Specification Requirements

5.1.3.1 Super Tractor Oil Universal

Tables 5.1.1 and 5.1.2 show examples of the specification requirements of tractor manufacturers for STOU. (Massey Ferguson M1139 and FORD ESN—M2C—159—B).

Table 5.1.1:   Super Tractor Oil Universal

SPECIFICATION: MASSEY FERGUSON M.1139	
Kinematic Viscosity at 100°C	: 10.2—11.9 mm²/s
Viscosity at—18°C (Brookfield)	: 8000 cP max.
Pour Point	: —30°C max.
FOUR BALL WEAR	
Scar Diameter (1hr, 65°C, 1500 rpm, 40 Kg)	: 0.40 mm max.
LOAD CARRYING CAPACITY (ASTM D.2733)	
Load Wear Index	: 55 kg min.
WET BRAKE CHATTER/SQUAWK TEST	: Pass
PTO CLUTCH TEST	: Pass
HIGH TORQUE AXLE TEST	: Pass
RUST PROTECTION	: Pass
COPPER STRIP CORROSION	
3 hours at 150°C	: 1a max.
FOAMING	
Sequence 1, ml	: 100/0 max.*
Sequence 11, ml	: 100/0 max.
Sequence 111, ml	: 100/0 max.
OXIDATION TEST (100 hrs at 149°C)	
Viscosity increase at 99°C	: 10 % max.
Separation/sludge	: None
SEAL TEST (168 hrs at 120°C)	
(RDR 008)	: Pass

* 1.0 % Volume Water Added to Foam Test

Table 5.1.2:   Super Tractor Oil Universal

SPECIFICATION: FORD ESN–M2C–159–B	
M2C–159–B1 SAE 10W/30	
M2C–159–B2 SAE 15W/30	
M2C–159–B3 SAE 20W/40	
FLASH POINT	: 190°C min.
FOUR BALL WEAR	
Scar Diam. (1 hr, 65°C, 1500 rpm, 40 kg)	: 0.40 mm. max.
ALLISON C3	
Powerglide Oxidation Test	: Pass
Power Steering Pump Test	: Pass
Seal Test	: Pass
Anti-foam	: Pass
FALEX PIN CORROSION	: Pass
COPPER STRIP CORROSION	
(3 hrs at 150°C)	: 1b max.
FOAMING	
Sequence 1, ml.	: 20/0
Sequence 11, ml.	: 50/10
Sequence 111, ml.	: 20/10
OXIDATION TEST (100 hrs at 149°C)	
Viscosity Increase at 99°C	: 10 % max.
SEAL TEST (70 hrs at 125°C)	
(100 Buna-N)	: 0–10
WATER TOLERANCE	
Water, Vol	: 0.5 %
Sediment, Vol	: 0.1 % max.
Water Separation	: Trace

5.1.3.2 Universal Tractor Transmission Oils

Table 5.1.3 shows the viscosity requirements of a number of Farm Tractor Hydraulic/Transmission Fluid specifications at 100°C and Viscosity Indices where specified. This clearly shows the division into high and low viscosity ranges which is discussed in Section 5.1.4.2.

Table 5.1.4 provides an example of a tractor manufacturers full specification for UTTO (John Deere J.20.A.).

Table 5.1.3: Viscometrics of Farm Tractors Hydraulic/Transmission Fluids

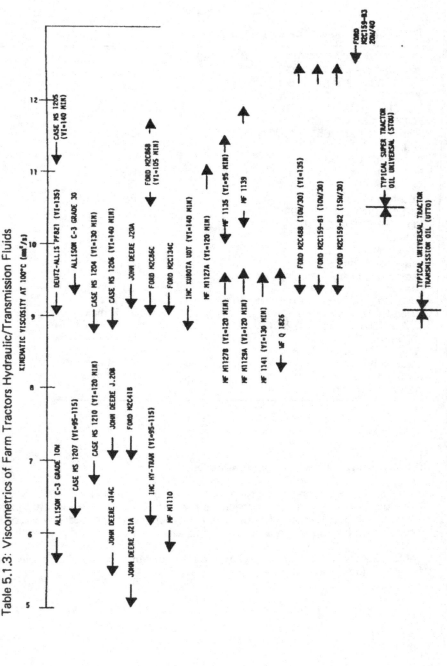

Table 5.1.4: Universal Tractor Transmission Oil

SPECIFICATION: JOHN DEERE J.20.A.
Kinematic Viscosity at 100°C   : 9.1 mm²/s min.
Viscosity at minus 18°C (Brookfield)   : 4,500 cP max.
Viscosity at minus 34°C (Brookfield)   : 70,000 cP max.
Pour Point   : minus 37°C max.
Flash Point   : 200°C min.
Timken EP Test (OK Load, 1bs)   : 10 min.
Timken Abrasion Test (6 hrs, 101bs)   : 1.5 mg. weight loss max.
Wet Brake Chatter/Squawk Test   : Pass
PTO Clutch Test   : Pass
High Torque Axle Test   : Pass

Allison C-3
Powerglide Oxidation Test   : Pass
Power Steering Pump Test   : Pass
Seal Test   : Pass
Anti-Foam   : Pass

Rust Protection   : Pass
Copper Strip Corrosion (3 hrs, 150°C)   : 1b max.
Foaming Sequence 1 (ml)   : 25/0 max.
Foaming Sequence 11 (ml)   : 50/0 max.
Foaming Sequence 111 (ml)   : 25/0 max.

Oxidation Test (100 hrs, 149°C)
Evaporation Loss   : 5 % max.
Viscosity Increase (99°C)   : 10 % max.
Separation/Sludge   : NIL

Seal Test (70 hrs, 100°C)
Volume change   : 0 to plus 5 %
Durometer change   : 0 to minus 5 %
Precipitation   : None

Water Tolerance
Water, Vol %   : 0.4
Sediment, Vol %   : 0.1 max.
Additive Loss, Mass %   : 15 max.

657

5.1.4 Tractor Lubricants

5.1.4.1 Super Tractor Oil Universal

Table 5.1.5 shows the composition, characteristics and performance levels for a representative SAE 10W/30 Super Tractor Oil Universal.

Table 5.1.5:   Example SAE 10/W30 Super Tractor Oil Universal

	% WEIGHT
DETERGENT—INHIBITOR ADDITIVE PACKAGE	13.0
VISCOSITY INDEX IMPROVER	7.5
POUR POINT DEPRESSANT	0.5
100 SOLVENT NEUTRAL BASE OIL	16.0
150 SOLVENT NEUTRAL BASE OIL	63.0

CHARACTERISTICS		
KINEMATIC VISCOSITY AT 100°C	:	10.53 mm² /s
CCS VISCOSITY AT MINUS 20°C	:	3500 mm² /s
POUR POINT	:	MINUS 37°C
TBN	:	10.2 mg KOH/g
SULPHATED ASH	:	1.5 % weight.
ZINC	:	0.14 % weight.
CALCIUM	:	0.38 % weight.
PHOSPHORUS	:	0.15 % weight.

PERFORMANCE
Ford M2C 1598
MASSEY FERGUSON M1139   } OEM
JOHN DEERE J.20.A   SPECIFICATIONS

API CE/SF
MIL-L-46152.B.
MIL-L-2104.D.   } ENGINES
CCMC G2/D2
DAIMLER—BENZ SHEET 227.1.

VICKERS HYDRAULIC PUMPS
DENISON HF—0   } HYDRAULICS/
AIR COMPRESSOR   OFF—HIGHWAY
PERFORMANCE

5.1.4.2 Universal Tractor Transmission Oil

Table 5.1.6 shows the composition, characteristics and specifications satisfied by a representative UTTO. The fluid selected for this example is intermediate viscosity and is intended to provide a universal approach to Tractor Hydraulic/Transmission Fluid requirements.

This "compromise universal" approach using a fluid of intermediate viscosity is anticipated to provide satisfactory *performance* (in all but the most severe cold temperature conditions) when used in equipment for which lower or higher viscosities are specified.

Examples of LOW VISCOSITY SPECIFICATIONS:
- ALLISON C—3 (10 GRADE)
- J. I. CASE 145
- INTERNATIONAL HARVESTER HY—TRAN

Examples of HIGH VISCOSITY SPECIFICATIONS:
- MASSEY FERGUSON M—1127.A.
- MASSEY FERGUSON M—1135
- FORD M2C-134-C
- FORD M2C-86-B

Table 5.1.6:   Example Intermediate Viscosity UTTO to Provide Universal Approach

COMPOSITION	% WEIGHT
PERFORMANCE ADDITIVE PACKAGE	: 7.2
VISCOSITY INDEX IMPROVER	: 3.5
HIGH VISCOSITY INDEX 170 NEUTRAL	: 89.3
SILICONE ANTI—FOAM	: 20 ppm.
CHARACTERISTICS	
KINEMATIC VISCOSITY AT 100°C	: 9.1 mm^2/s
BROOKFIELD VISCOSITY AT MINUS 18°C	: 3500 cP
CALCIUM	: 0.392 % weight.
PHOSPHORUS	: 0.112 % weight.
SULPHUR	: 0.299 % weight.
ZINC	: 0.079 % weight.
TBN	: 10.7 mg KOH/g
SPECIFICATIONS	
JOHN DEERE J.20.A.	
MASSEY FERGUSON M-1127.B.	
ALLIS CHALMERS PF821	

5.1.5 Current and Future Developments

In Europe the trend towards an increase in scale of agricultural operations is resulting in larger tractors with greater power, turbocharging, stronger transmissions and four wheel drive. The success of Super Universal Tractor Oil in Europe is likely to continue because of:
- good performance and user experience.
- simplification of servicing and inventories.
- very competitive fluid prices.

The largest market for Hydraulic/Transmission fluids will continue to be North America where fluid developments are likely to be targeted on:
- extreme pressure performance even in the presence of water.
- water tolerance and filterability.
- low temperature fluidity.
- wide operating temperature range.
- reduced viscosity.
- specialised testing defined by OEM.

Future developments in Tractor Lubricants generally will be aimed towards improvements in:
- gear protection.
- wet brake performance/frictional properties.
- all weather operation.

A trend is likely towards increased use of Contractors Plant Oil (CPO) which represents a widened concept of STOU to incorporate air compressor applications.

For marketing reasons, such products will usually be named and promoted separately to the agricultural community of Farm Tractors; however, CPO is an attempt to satisfy all off-highway plant lubrication with one product.

The general need to incorporate synthetic base stocks to satisfy performance requirements in the areas of low temperature viscosity, volatility and oxidation resistance is likely to influence Tractor Lubricant formulation.

The performance requirements of modern Farm Tractors have resulted in the development of lubricants of great complexity and versatility which arguably represent the current state of the art in lubricant technology.

The writer gratefully acknowledges the information provided by:

The Lubrizol Corporation
Exxon Chemical Ltd.
Ethyl Petroleum Additives
Watveare Ltd.

5.2 Use of Low Speed FZG Test Methods to Evaluate Tractor Hydraulic Fluids

B.M. O'Connor, The Lubrizol Corporation, Wickliffe, USA
H. Winter, Technical University Munich, Germany

5.2.1 Introduction

Assessing the wear performance of tractor hydraulic fluids is an expensive and time consuming process generally involving lengthy field service trials or full-scale tractor evaluations in the laboratory. These are not the most suitable or desirable methods when trying to develop lubricant products where many additve combinations must be tested.

The objective of this work was to investigate the possibility of simulating the type and relative amount of wear encountered in the planetary reduction units of large agricultural tractors in service as influenced by the fluid. In recent years considerable effort has been expended by the industry to develop a screen test for tractor gear wear. Since the start of that development effort, however, tractor original equipment manufacturers (OEM) have sought lubricant performance at still higher temperatures and loads than those initially addressed. Thus, this program was aimed toward the newer, more severe test requirements.

The test method of interest was a 50 hour full-scale tractor laboratory test used by the industry of measure wear performance of fluids under high temperature, high load, low speed conditions. The test is conducted in two stages and was designed by the OEM to represent severe field operation. The principal concern of the test is wear of the primary (spiral bevel) and secondary (planetary) reduction systems as a function of the fluid. The primary mode of failure observed with the spiral bevel gearing is scuffing, while the planetary units encounter normal abrasive wear. Since there are a number of methods to evaluate scuffing, this work concentrated on simulating the abrasive wear of the planetary gears.

Based on the information available about the equipment and operating conditions, calculations of critical parameters were made and translated to an FZG test rig for evaluation. Two ASTM tractor reference oils, TF-7 and TF-11, were chosen because of observed wear differences in a laboratory full-scale tractor test. A large difference in wear was also observed in the FZG test. Further, a modified test method using water as a contaminant showed large differences among three lubricants which had previously demonstrated similar wear characteristics in the non-contaminated test method. Since water is a common contaminant of tractor final drive systems in some geographical service areas, further improvements in water sensitivity of tractor formulations may be desirable.

661

5.2.2 Experimental

5.2.2.1 Full-Scale Tractor Test

The tractor final drive used in the laboratory test is driven by a 110 kW (150 HP) engine through a synchronized manual transmission. The final drive consists of a spiral bevel primary reduction (4:1) and a planetary final production (3:8:1); the overall reduction ratio is 15.2:1. Exact detailed information about the gear geometry used in this test equipment was not available. Estimates of these data based on measured values have been used for the purposes of calculating the Hertzian contact stresses and film thicknesses in the tractor test (Table 5.2.1). A summary of the test conditions is shown in Table 5.2.2 along with a further breakdown of speed and load for the sun and ring gears of the planetary unit.

Table 5.2.1: Data for Tractor Planetary Gearset

Parameter	Symbol	Dimension	Units
Number of teeth			
Sun gear	z_Z	15	––
Planets (3)	z_P	21	––
Ring gear	z_H	57	––
Module	m	6,0	mm
Tooth Width	b	38,1	mm
Pressure angle	a	20	deg
Helix angle	β	0	deg
Pitch diameter			
Sun	d_Z	90,5	mm
Planet	d_P	123.8	mm
Ring	d_H	– 344.5	mm
Base circle diameter			
Sun	d_{bZ}	84.6	mm
Planet	d_{bP}	118.4	mm
Tip circle diameter			
Sun	d_{aZ}	104.8	mm
Planet	d_{aP}	138.1	mm
Ring	d_{aH}	–341.3	mm
Center distance	a_{ZP}	107.15	mm

662

Table 5.2.2: Tractor Test Conditions

		Tractor Test Conditions		
Part	Torque [N·m]	Wheel Speed [rpm]	Oil Temp. [°C]	Time [h]
1	23454	24	121	20
2	23454	14	121	30

		Breakdown of Speeds			
Test Part	Rotational Ring (n_H) Sun (n_Z) [rpm]		Peripheral Ring (v_H) Sun (v_Z) [m/s]		Angular Velocity Ring (ω_H) Sun (ω_Z) [s^{-1}]
1	24	91.2	0.43 0.43		2.51 9.57
2	14	53.2	0.25 0.25		1.47 5.59

		Breakdown of Load			
Test Part	Time (t) [h]	Torque Ring (M_H) Sun (M_Z) [N·m]	Force Sun (F_{tZ}) [N]	Power (P) [kW]	Work (W) [kWh]
1	20	23454 2057	45468	59.0	1179
2	30	23454 2057	45468	34.5	1034

5.2.2.2 Calculation for the Full-Scale Tractor Test

5.2.2.2.1 Calculation of Contact Stress

Using the tractor gear dimensions and the full-scale rig test conditions shown in Tables 5.2.1 and 5.2.2, the contact pressures were calculated for the sun gear against the planet gears using the equations shown below [1, 2].

$$\sigma_{HO} = p_c Z_B Z_\epsilon Z_\beta \tag{1}$$

where

$$p_c = Z_H Z_E \left\{ (F_t \cdot (u+1)/d_1 \cdot b \cdot u) \right\}^{0.5} \tag{2}$$

$$Z_B = \tan \alpha_W \cdot \cos \beta / (\tan \alpha_{B1} \cdot \tan \alpha_{B2})^{0.5} \tag{3}$$

$$Z_\epsilon = \left\{ (4 - \epsilon_a)/3 \right\}^{0.5} \tag{4}$$

$$Z_\beta = (\cos \beta)^{0.5} \tag{5}$$

$$Z_H = \left\{ (2\cos\beta \cdot \cos\alpha_W)/(\cos^2 \alpha \cdot \sin\alpha_W) \right\}^{0.5} \tag{6}$$

$$Z_E = \left\{ 1/\pi \left[(1 - \nu_1^2/E_1) + (1 - \nu_2^2/E_2) \right] \right\}^{0.5} \tag{7}$$

Using the equations above and the information provided in Table 5.2.1 and 5.2.2 the estimated contact stress, σ_{HO}, between the sun and planet gear in the full-scale rig test is calculated to be 1954 N/mm^2.

5.2.2.2.2   Estimation of Film Thickness

Since the primary objective of this work is to determine the performance of the fluids being used in tractor applications, it is useful to know the film thickness and also the lubrication regime in which the equipment operates. The film thickness calculation is based on the Dowson and Higginson formula for isothermal elastohydrodynamic (EHD) line contacts [3]

$$h_{min}/R = 2.65 \cdot G^{0.54} \cdot U^{0.7} \cdot W^{-0.13} \tag{8}$$

where

$$G = \alpha^* \cdot E' \qquad \text{(material parameter)}$$

$$U = \nu_M (u_1 + u_2)/2 \cdot E' \cdot R \qquad \text{(speed parameter)}$$

$$W = w / E' \cdot R \cdot L \qquad \text{(load parameter)}$$

For the puposes of these calculations the pressure-viscosity coefficient, α^*, was estimated to be 16 GPa^{-1} for a mineral based oil at 120°C. The viscosity at the contact inlet, ν_M, was calculated using the known oil viscosities at 40 and 100°C and the calculated temperature of the gear contact at the operating condition according to equation 9 [4].

$$\log\log(\nu_M+0.8) \ = \ \log\log(\nu_{40}+0.8) \ + \ m[(\log 313)-\log(273+T_M)] \quad (9)$$
where
$$m = [\log\log(\nu_{40}+0.8) \ - \ \log\log \quad (\nu_{100}+0.8)]/0.07616$$

and
$$T_M \ = \ T_O + 7400 \ [P_{Vz}/(a \cdot b)]^{0.72} [X_S/(1.2 \cdot X_{Ca})]$$

The calculated film thickness, h_{min}, for the full-scale tractor planetary gears was $0.008\mu m$ for the second (slower speed) stage of the test. The lambda ratio, $\lambda = h/Ra$, for these gears is very low (<0.10) indicating a boundary lubrication regime.

5.2.2.3 Establishing FZG Test Conditions

5.2.2.3.1 Test Stand

The standard FZG test rig described in DIN 51 354 was modified to achieve low operating speeds (5). This was done by inserting a suitable reduction unit between the variable speed DC motor and the input drive gearbox as shown in Figure 5.2.1. In the work conducted here, the reduction unit consisted of a combination of a worm gear reducer and belt pulley system to achieve very low circumferential velocities if desired.

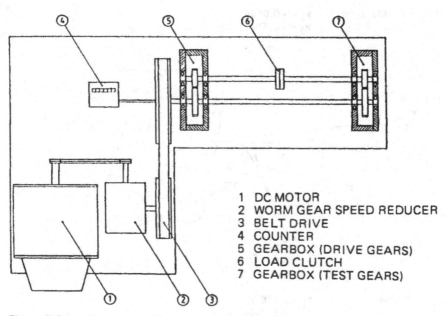

1 DC MOTOR
2 WORM GEAR SPEED REDUCER
3 BELT DRIVE
4 COUNTER
5 GEARBOX (DRIVE GEARS)
6 LOAD CLUTCH
7 GEARBOX (TEST GEARS)

Figure 5.2.1: Schematic of Low Speed FZG Test Stand

665

5.2.2.3.2 Test Gears

The test gears chosen for this investigation are commonly referred to as "C" profile gears. They are case carburizes steel (16 MnCr 5 E) gears with a typical hardness of 60 – 62 HRC. Ra values for the gears used in this study were generally between 0.5 – 0.7 μm. The geometry and profile characteristics are similar to planetary gears used in tractors and less likely to encounter scuffing compared to the standard "A" profile gears used in the FZG scuff test. A summary of the pertinent geometry characteristics is provided in Table 5.2.3; a schematic of both "C" and "A" profile gears is shown in Figure 5.2.2.

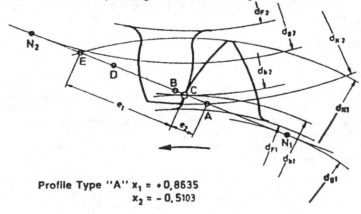

Profile Type "A" $x_1 = +0.8635$
$x_2 = -0.5103$

Scale 4 : 1

1 cm

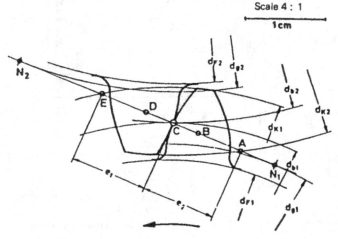

Profile Type "C" $x_1 = +0.1817$
$x_2 = +0.1715$

Figure 5.2.2: Schematic of FZG Test Gear Profiles

666

Table 5.2.3: Data for FZG Gears

Parameter	Symbol	Profile Type "A"	Profile Type "C"	Units
Number of teeth				
Pinion	z_1	16	16	
Gear	z_2	24	24	
Module	m	4.5	4.5	mm
Tooth Width	b	20.0	19.0	mm
Pressure angle	α	20	20	deg
Working pitch diameter				
Pinion	d_{w1}	73.2	73.2	mm
Gear	d_{w2}	109.8	109.8	mm
Base circle diameter				
Pinion	d_{b1}	67.7	67.7	mm
Gear	d_{b2}	101.5	101.5	mm
Tip circle diameter				
Pinion	d_{a1}	88.6	82.5	mm
Gear	d_{a2}	112.2	118.4	mm
Center distance	a	91.5	91.5	mm
Addendum Modification				
Pinion	x_1	0.86	0.18	——
Gear	x_2	−0.51	0.17	——
Max Relative Sliding Velocity	v_G/v	0.67	0.44	——

5.2.2.3.3 Operating Conditions

The operating conditions of speed, load, temperature, etc. are normally calculated for the full-scale equipment and then translated to appropriate conditions in the smaller test rig. In this case, however, it is not possible to directly simulate the contact stress of the tractor gears with the FZG test gears. The FZG test rig has a normal maximum torque capacity of 535 N•m, i.e. Load Stage (LS) 12, using the standard 23 mm diameter torsion shaft. This produces a contact stress of 1747 N/mm² on the test gears which is approximately 10 % less than the 1954 N/mm² calculated for the tractor. As an alternative to the use of contact stress directly, film thickness was evaluated as a criteria for developing the conditions in the FZG rig.

The film thickness was calculated for a variety of speed/load conditions in the FZG rig using equations 8 and 9. As expected from the Dowson-Higginson equation, load was not a significant factor in the film thickness developed for load stages 8 through 12. These data are plotted in Figure 5.2.3. These data suggest that film thicknesses equivalent to the tractor gear test can be achieved with the FZG rig using a more moderate load (LS 10; σ_{HO} = 1459 N/mm²) and slightly lower speeds (0.35 and 0.20 m/s). Details of all the test methods used in this work can be found in Table 5.2.4.

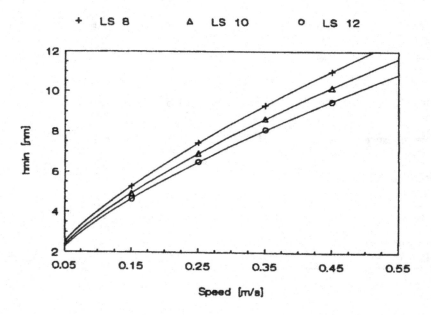

Figure 5.2.3: Film Thickness — FZG Test: Influence of Speed and Load

Table 5.2.4: Summary of Test Conditions – Low Speed FZG Test Methods

	Method A Phase 1	Phase 2	Method B Phase 1	Phase 2	Method C Phase 1	Phase 2
Load Stage	10	10	10	10	10	10
Gear Torque [Nm]	372.6	372.6	372.6	372.6	372.6	372.6
Circumferential Speed [m/s]	0.35	0.20	0.25	0.25	0.25	0.25
Rotational Speed [rpm] Pinion Gear	93 62	53 35	66 44	66 44	66 44	66 44
Oil Temperature [C]	120	120	120	120	80	80
Time [h]	20.0	30.0	28.1	24.1	28.1	24.1
Transmitted Work[kWh]	72.4	62.2	72.8	62.5	72.8	62.5
Contaminant	——None——		——None——		1 vol % Water	

The first method, Method A, essentially duplicates the tractor test method; two phases each having a different speed and time are used. Method B utilized a speed of 0.25 m/s which was derived as the time weighted average of the speeds in each phase of Method A. The test length is slightly longer to insure that the same amount of work is done on the gears as in Method A. Method C was developed to investigate the influence of water as a contaminant in the wear process. This method is the same as Method B except that the temperature is reduced from 120°C to 80°C to minimize the loss of entrained water during the test. In this procedure 1 % volume of water is mixed with the test fluid prior to the evaluation.

Another reason for selecting load stage 10 rather than a higher load was to minimize the possibility of scuffing (adhesive wear) during the test. The lubricants used in tractor final drives are low viscosity and generally not designed for high scuffing load capacity. Typical tractor hydraulic fluids have a scuff load capacity in the range of load stage 8 to 10 as measured by DIN 51 354 test method (5). Although the speeds are greatly reduced in these methods compared to the DIN 51 354 method the probability of adhesive wear still exists. Scuffing is normally not observed in tractor planetary gears but has been encountered with several fluids in a similar method proposed in ASTM using "A" profile gears.

Several possible factors which may have contributed to the scuffing are:

1) use of gears with a high sliding velocity profile
2) vibration or surging of the drive motor at low speeds; or
3) inadequate scuff load capacity of the fluids for the load concitions.

Attempts have been made to reduce the probability of scuffing in the screen test reported here. Type "C" gears were chosen here to reduce the sliding velocity. Compared to type "A" gears the tip sliding velocity is reduced by 35 %. The use of a reduction gear between the motor and the test unit also provides smoother operation during the test by eliminating low speed motor surging. The scuffing capacity of the fluid is primarily a function of the additive and cannot be changed. Thus, the test conditions must be adjusted to reproduce the abrasive wear observed in the full-scale test without scuffing.

Supporting the choice of using "C" profile gears and load stage 10 in the simulation are the scuffing safety factors calculated according to DIN 3990, Part 4 (2). Figure 5.2.4 is a summary of the scuffing safety factors as a function of the scuff capacity of the oil. The factors are calculated for the conditions to be used in the simulation with "C" profile gears (LS 10; v_t = 0.35 m/s; T_O = 121°C). The safety factors are much higher when "C" profile gears are used compared with the "A" profile gears operating under the proposed ASTM conditions (LS 10; v_t = 0.57 m/s; T_O = 121°C). The safety factor is only a gauge or probability about the occurrence of scuffing; the higher the value the less likely scuffing will occur. Experience has shown that safety factors above 1.6 are less subject to scuff damage while those below 1.4 are more likely to encounter it. Thus, even with oils that demonstrate load stage 8 capacity in the DIN 51354 test there is a fair chance they can successfully be evaluated for abrasive wear with the "C" profile gear test procedure.

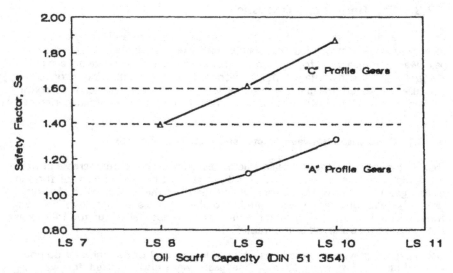

Figure 5.2.4:   Scuffing Safety Factor – Influence of Oil Scuff Capacity

5.2.2.4 Test Lubricants

Four different tractor hydraulic fluids were used during this investigation. The physical and chemical characteristics of these oils are shown in Table 5.2.5. Two of the fluids, A and B, are used as standard reference oils by the U.S. tractor industry. Fluid A, an older generation product, is considered as the high wear oil in full-scale tractor tests although it has never had a problem in field service in over 10 years use. The other three fluids, B, C, and D, are commercial oils that have been in the market for several years. They are considered similar in performance in the full-scale OEM tractor wear test.

Table 5.2.5:   Test Oils – Physical and Chemical Properties

Fluid	Description	Kin Viscosity [cSt] @40 C	@100 C	P	Elements, % wt Zn	S	Ca	Mg
A	UTTO/TF–7	65.9	9.33	0.11	0.08	0.29	0.39	––
B	UTTO/TF–11	57.0	9.05	0.11	0.08	0.35	––	0.31
C	UTTO	78.3	9.78	0.12	0.14	0.43	0.30	––
D	UTTO	52.2	8.82	0.12	0.15	0.27	0.37	––

671

5.2.3 Test Results and Discussion

Prior to each test, the pinion and gear were cleaned with a chlorinated solvent in an ultrasonic bath, dried thoroughly, and weighed to the nearest mg. Each gear was also measured for surface roughness (average of three teeth per gear), Ra [μm], and the involute profile was recorded for three teeth. This information was used to determine the dimensional loss due to wear and serve as confirmation of the weight loss measurement taken in the course of the experiments.

5.2.3.1 Comparison Between Tractor and FZG Test Methods

Results available for the full-scale tractor test show a three-fold increase in wear for Fluid A compared with Fluid B (68 μm vs. 21 μm average wear of the sun pinion gear). These two fluids were first evaluated in the FZG rig with Method A using the two different speeds similar to the full-scale tractor test. The two fluids showed substantial differences in wear gravimetrically, but the trend was the same as that observed in the tractor.

Using typical gravimetric methods Fluid A lost a total of 894 mg over 50 hours while Fluid B lost only 63 mg. The gears were also checked for wear by measuring changes to the involute profile before and after test. The magnitude of the difference between the fluids parallels the wear observed in the tractor (Figure 5.2.5). Of interest here also is the difference in wear rates for each fluid during each part of the test. Fluid A had a very high rate of wear during both parts of the test where as Fluid B exhibited a very low wear rate during Part 2 of the test. This suggests a strong antiwear system for Fluid B and that the wear associated with Fluid A was well beyond what might be considered break-in wear. A breakdown of the results is shown in Table 5.2.6.

Method B represents a simplified version of Method A where the speed is maintained at 0.25 m/s for both parts of the test. The running time is compensated for in each part of the test to insure the same amount of work is done on the gears as in the other methods using the FZG rig. The results with Fluids A and B were essentially the same as with Method A and are summarized in Table 5.2.6. Fluid A lost 936 mg total while Fluid B lost 68 mg. Although severe wear occurred with Fluid A there were no signs of scuffing damage on the gear flanks with either method. Photographs of the pinion gears for Method B with Fluids A and B are shown in Figure 5.2.6. This is consistent with the results from the full-scale tractor test and field service.

These results suggest that under severe operating conditions small differences in film thickness, speed, and load are not going to influence the results significantly. Since there was essentially no difference in Methods A and B the remainder of the study used Method B. These results also suggest that the fluid can play a major role in the overall wear performance.

672

Fluid A

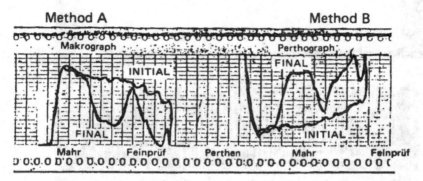

Fluid B

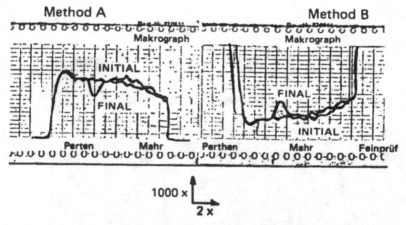

$$1000 \times \quad 2 \times$$

Figure 5.2.5: Tooth Flank Wear — Methods A and B; Fluids A and B

METHOD B

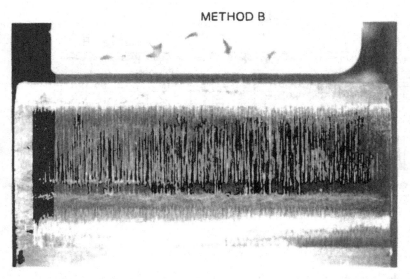

FLUID A

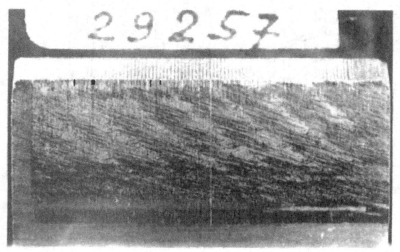

FLUID B

Figure 5.2.6:   Pinion Gears: Method B — Fluids A and B

674

Table 5.2.6:   Comparison of Methods A and B — Low Speed FZG Wear Test

| | Method A | | | Method B | | | Tractor West |
	Phase	Phase	Total	Phase	Phase	Total	Avg Gear Wear
Fluid A							
Wt Loss [mg]	510	384	894	535	401	936	—
Avg Pinion Flank Wear [um]	12	19	19	11	18	18	68
Specific Wear [mg/kWh]	7.04	6.18	6.64	7.38	6.45	6.95	—
Fluid B							
Wt loss [mg]	59	4	63	66	2	68	—
Avg Pinion Flank Wear [um]	5	6	6	8	9	9	21
Specific Wear [mg/kWh]	0.81	0.06	0.47	0.91	0.03	0.51	—

5.2.3.2 Evaluation of Current Generation Fluids (Method B)

Having established that wear differences can be observed in the FZG rig using the simulative methods and that these differences have the same trend as observed in full-scale tractor testing, it was desired to characterize the wear performance of Fluids C and D by the same method. Fluids B, C, and D are typical commercial fluids meeting the requirements of many OEM specifications and known in Europe as Universal Tractor Transmission Oils (UTTO). The information available from OEM testing and field service suggests that these three oils have similar performance characteristics.

The results of the low speed FZG evaluations for the three fluids using Method B are summarized in Table 5.2.7; photographs of the pinion gears are shown in Figure 5.2.7. The results confirm the similarity of the three fluids with respect to their wear performance. Fluid C appears to be slightly different than Fluids B and D in that the wear rate was lower during Part 1 and higher than the others during Part 2 of the test. This suggests that the chemical characteristics controlling the break-in process are different in Fluid C compared with either B or D; a much longer break-in period may be required with this oil before a normal, very low wear rate is established as with most other fluids. A longer duration test would be required to clarify this supposition.

675

Table 5.2.7: Low Speed FZG Wear Test — Method B

	Weight Loss [mg]			Specific Wear [mg/kWh]		
	Phase	Phase	Total	Phase	Phase	Total
Fluid B	66	2	68	0.91	0.03	0.51
Fluid C	37	15	52	0.51	0.24	0.39
Fluid D	46	1	47	0.64	0.02	0.35

METHOD B

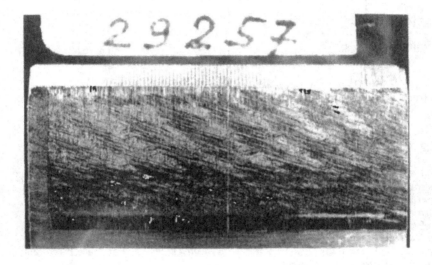

FLUID B

Figure 5.2.7: Pinion Gears: Method B — Fluids B, C, and D, cont.

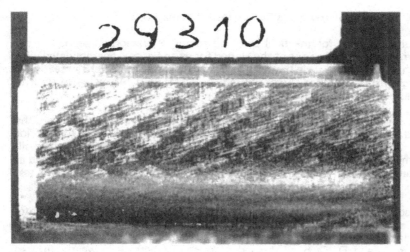

FLUID C

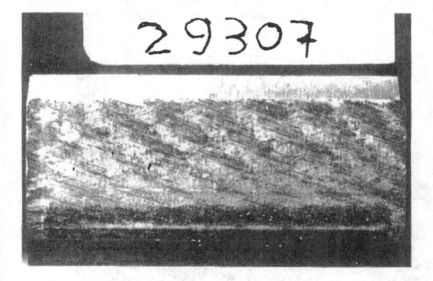

FLUID D

Figure 5.2.7: Pinion Gears: Method B — Fluids B, C, and D

5.2.3.3 Influence of Water as a Contaminant (Method C)

It is quite common for the final drive oil sumps of tractors in service in the southern and western regions of the United States to become contaminated with as much as 1 % water. This has been known to increase the severity greatly, as one might expect. Method C was devised to determine if the increased severity could also be observed in a bench type test.

Except for the reduction in temperature to 80°C, the test conditions were the same as for Method B described in Table 5.2.5. Approximately three hours prior to the start of a test 1 % volume of tap water was added to the test oil. The oil-water mixture was shaken vigorously one time per hour and allowed to stand at room temperature in between. After the oil-water mixture was added to the test gearbox no further water additions were made during the course of the test.

The results with the water contaminant are summarized in Table 5.2.8; photographs of the pinion gears for Fluids B, C, and D are shown in Figure 5.2.8. These data clearly show a significant difference among the wear performance of Fluids B, C, and D. Each oil showed a noticeable increase in wear compared with the non-contaminated test. More interesting is the fact that while the oils are judged to be similar in performance in a non-contaminated situation, the assessment changes significantly when exposed to a common service contaminant.

METHOD C

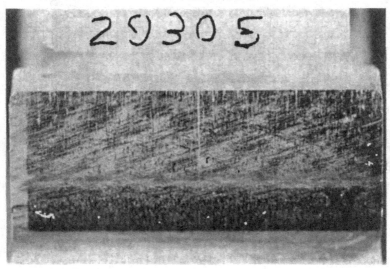

FLUID B

678

FLUID C

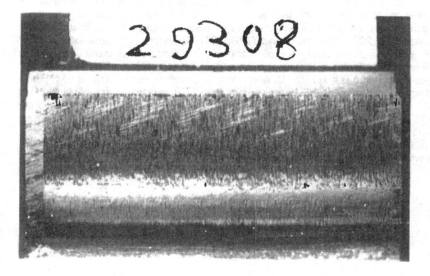

FLUID D

Figure 5.2.8: Pinion Gears: Method C — Fluids B, C, and D, cont.

Table 5.2.8:   Low Speed FZG Wear Test — Method C

	Weight Loss [mg]			Specific Wear [mg/kWh]		
	Phase	Phase	Total	Phase	Phase	Total
Fluid B	95	21	116	1.31	0.34	0.86
Fluid C	438	454	982	6.05	7.30	6.63
Fluid D	273	139	414	3.77	2.24	3.06

One can speculate on two possible causes for the high wear in the presence of water. First, water, being a highly polar molecule, preferentially adsorbs on the gear surface and excludes the antiwear compounds from getting to the surface. The antiwear compounds used in these fluids are Zinc Dialkyl Dithiophosphates (ZDP). Since the water does not possess any load-carrying capacity per se the result is very poor wear control. The other possibility is that the water reacts with the antiwear or other compounds in the bulk oil solution causing a change in the chemical species which may be less effective than the original.

A post-test analysis of the end of test oil suggests that the water has altered the antiwear system directly or indirectly by affecting one or more of the other components in the additive system. End of test samples of Fluids B and C from Methods B (120°C/no water) and C (80°C/1 % water) were analysed by ^{31}P NMR and compared with new oil samples. The spectra are shown in Figures 5.2.9 and 5.2.10. These data should be considered as semi-quantitative only and used only to establish trends. Under the high temperature conditions of Method B, there is still a measurable amount of the ZDP remaining for each fluid. It would be reasonable to expect that at lower temperatures more of the ZDP should be present, all other factors being equal. The spectra for Method C show significant depletion of the ZDP peak for Fluid C while a relatively large peak exists for Fluid B. These data correspond in general to the wear results — very high wear for Fluid C and low wear for Fluid B.

680

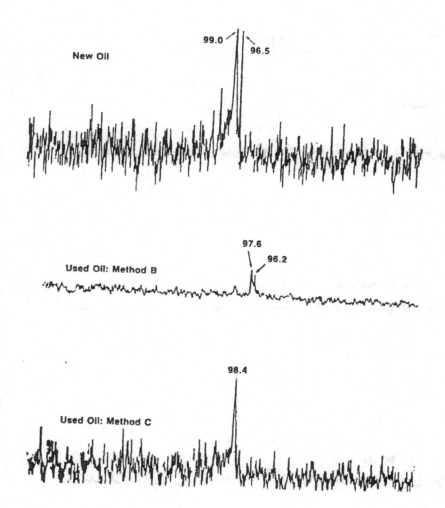

Figure 5.2.9: 31 P NMR Spectra — Fluid B

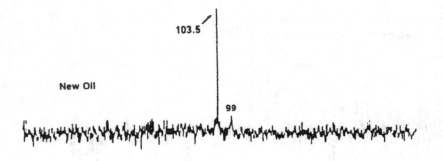

New Oil

103.5

99

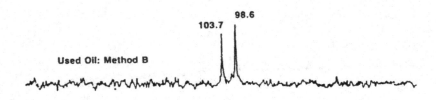

Used Oil: Method B

103.7 98.6

Used Oil: Method C

103 99

Figure 5.2.10: 31 P NMR Spectra — Fluid C

5.2.3.4 Influence of Surface Roughness

It is well known that the surface finish of machine components can have a significant effect on the outcome of wear performance. Compiling the data generated on the aforementioned test program with several additional tests with a different batch of "C" profile gears we find that noticeable differences in wear can be observed as a function of the surface finish of the gears used. Results obtained with Fluids A, B, and C are plotted as a function of the average Ra value ((pinion + gear)/2) of the gearset used in Figure 2.5.11. Separation is maintained between high and low wear fluids but the absolute wear decreases with decreasing Ra values.

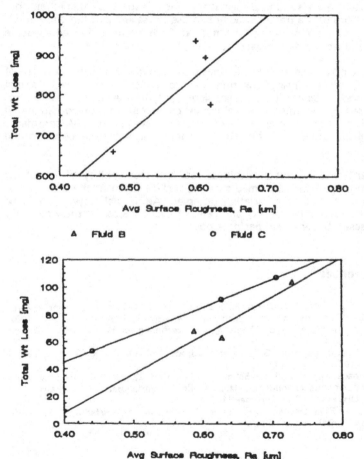

Figure 5.2.11: Low Speed FZG Wear Evaluation-Influence of Surface Roughness

683

5.2.4 Summary

The investigation has shown that a reasonable simulative method can be developed in the FZG test rig if the conditions are chosen on the basis of appropriate calculations. It appears that the use of gear profiles with a lower tip sliding velocity, i.e. "C" type gears, are less likely to produce scuffing during the test. No scuffing occurred during this investigation with "C" type gears whereas previous evaluations with "A" type gears in a similar test resulted in scuffed flanks on many occasions with a variety of oils. Scuffing is not a normal mode of failure with tractor planetary units.

The results with Methods A and B suggest that there is some latitude in reproducing the conditions of the tractor exactly because the severity of the operation is much greater than the sensitivity of the method. Both methods showed adequate separation of the test fluids compared to the full-scale tractor test results.

The use of water as a contaminant is viewed as an extension of the severity of the laboratory test and brings the simulation perhaps a step closer to actual practice. Method C demonstrated that significant differences in wear performance occurred with water contamination compared to non-contaminated conditions. Performance under non-contaminated conditions was considered equivalent for the three oils. This suggests additional work may be warranted in this area.

While sufficient data are not available to establish repeatability of any of the methods discussed it was confirmed that the surface finish of the gears can be a contributing factor to the variation observed. Additional testing should be conducted to establish repeatability and causes of variance before these methods are used to assess lubricant wear performance.

5.2.5 References

(1) Niemann, G. and Winter, H.: Maschinenelemente Band II, Getriebe Allgemein, Zahnradgetriebe — Grundlagen, Stirnradgetriebe. 2. Auflage. Springer Verlag, Berlin 1982.

(2) DIN 3990: Grundlagen für die Tragfähigkeitsberechnung von Gerad- und Schrägstirnrädern. Vorlage 1982.

(3) Dowson, D. and Higginson, G.R.: Elastrohydrodynamic Lubrication. Oxford, Permagon Press, 1966.

(4) Schönnenbeck, G.: Einfluß der Schmierstoffe auf die Zahnflankenermüdung (Graufleckigkeit und Grübchenbildung). Hauptsächlich im Umfangsgeschwindigkeitsbereich 1–9 m/s. Dissertation T.U. München (1983).

(5) DIN 51 354: FZG-Zahnrad-Verspannungs-Prüfmaschine. Prüfverfahren A/8,3/90 für Schmieröle (1984).

Appendix
Symbols and Units

Symbol	Description	Units
a	Center distance	mm
b	Width of contact	m
d	Pitch circle diameter	mm
d_a	Tip circle diameter	mm
d_b	Base circle diameter	mm
d_w	Contact pitch circle diameter	mm
E	Modulus of elasticity	N/mm^2
E'	Reduced modulus of elasticity $= \{0.5^*[(1-\nu_1^2/E_1) + (1-\nu_2^2/E_2)]\}^{-1}$	N/mm^2
F_t	Circumferential force at pitch circle	N
G	Dimensionless material parameter $= \alpha^* \cdot E'$	— —
HRC	Hardness — Rockwell C Scale	
h	Lubricant film thickness	μm
h_{min}	Minimum lubricant film thickness	μm
L	Length of contact	m
M	Torque	N·m
m	Module	mm
m	Viscosity-temperature slope	cSt/°C
n	Rotational speed	rpm
P	Power	kW
P_{Vz}	Power loss at the tooth	kW
Pc	Hertzian pressure at contact circle diameter	N/mm^2
R	Reduced radius of curvature $= r_1 \cdot r_2/r_1 + r_2$	m
Ra	Arithmetic mean roughness	μm
Ss	Temperature dependent scuffing safety factor	— —
T_M	Temperature of gear mass	°C
T_O	Temperature of oil	°C
t	Time	h, min, s

Symbol	Description	Units
U	Dimensionless speed parameter	---
u	Velocity	m/s
v	Circumferential velocity	m/s
v_g	Sliding velocity	m/s
W	Dimensionless load parameter	---
w	Total load	N
X_{Ca}	Tip relief factor	---
X_S	Lubricant factor	---
x	Profile modification	---
Z_B	Single tooth contact factor	---
Z_β	Helical factor	---
Z_E	Elasticity factor	---
Z_E	Overlaß factor	---
Z_H	Zone factor	---
z	Number of teeth	---
α	Normal pressure angle	deg
α_w	Working pressure angle	deg
α^*	Pressure-viscosity coefficient	m^2/N
β	Helix angle	deg
ϵ_α	Profile overlap	---
	Specific film thickness = h/Ra (dimensionless)	---
σ_{HO}	Nominal gear tooth flank pressure	N/mm^2
μ_O	Lubricant viscosity at inlet surface temperature	Pa·s, $N·s/m^2$
λ	Poisson's ratio	---
ν_M	Kinematic viscosity at temperature M	mm^2/s
ν_{40}	Kinematic viscosity at 40°C	mm^2/s
ν_{100}	Kinematic viscosity at 100°C	mm^2/s
ω	Angular velocity	s^{-1}

Indices:

1	Pinion	Z	Sun gear
2	Gear	P	Planet gear
		H	Ring gear

6. Gear Lubrication

6.1 Performance Characteristics of Sulfur-Phosphorus Type Hypoid Gear Oils

S. Watanabe and H. Ohashi
Tonen K. K. Corporate, Saitama, Japan

It is important to predict the performance characteristics of the newly developed automotive gear oils. Using a bench hypoid gear tester, four kinds of the gear tests were carried out to confirm the properties of the new S-P type candidate oils.

Gear protection performance, load carrying properties, peak temperature reduction characteristics and fuel economy improvement were examined by a high temperature and high speed test, modified L-42 test, a peak temperature measurement and a chassis roll test, respectively.

The effectiveness of four additives, olefin polysulfides, phosphorus compounds, ashless dispersants and organo-molybdenum compounds on these performance evaluation is shown. The severity of four tests are compared.

Hypoid gear lubrication in the car rear axle and sulfur-phosphorus additive technology for high performance gear oils are discussed.

6.1.1 Introduction

The conditions under which an automotive gear oil is required to lubricate a passenger car axle are constantly increasing in severity.

Innovations such as the use of smaller and/or lightweight axles, new type transfer gears for a four wheel drive, increasing engine output and newly designed transaxles, all contribute to this increase in performance requirement.

In accordance with automotive manufacturers' requests, the effort on developing new gear oil is under way and it is important to predict the performance characteristics of a candidate oil.

While laboratory tests give some indication of the possible service performance of developed oils, it has been found that, for severe conditions, such as are encountered with automotive hypoid gears, tests on full scale equipment are not only desirable but also necessary.

Hypoid gear lubrication in the passenger car rear axles and sulfur-phosphorus (S-P) additive technology for high performance hypoid gear oils are described in this paper.

The authors will focus the discussion on the following four subjects and report their findings on the effect of S-P additive chemistry upon the performance of hypoid gear oils while these oils are lubricating hypoid gears.

6.1.2 Experimental

6.1.2.1 Axle Gear Tester

An outline of the test rig is shown in Figure 6.1.1. A commercial gasoline engine, for which specifications are described in Table 6.1.1 was used as the power source.

The "T-bar" type rear axle housing was mounted on similar supports to the one specified in the CRC L-42 test.

In order to measure torque with time, strain gauges were fixed at the middle height of the axle support. The detected signals were recorded on an oscillograph through an amplifier.

A thermocouple was installed at the drain plug and a temperature-recording instrument monitored the temperature of the lubricant continuously throughout the test.

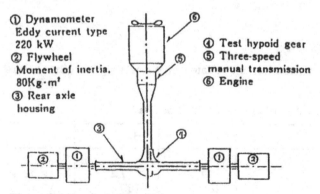

① Dynamometer
  Eddy current type
  220 kW
② Flywheel
  Moment of inertia,
  80Kg·m²
③ Rear axle
  housing

④ Test hypoid gear
⑤ Three-speed
  manual transmission
⑥ Engine

Figure 6.1.1:   Axle Gear Tester

Table 6.1.1:   Power Source

Number of cylinders	Volume of displacement ml	Maximum output kW/rpm	Maximum torque N · m/rpm
6	2563	101/5400	201/3600

6.1.2.2 Test Hypoid Gears

A rear axle assembly of typical Japanese passenger car was used and some details of hypoid gears are shown in Table 6.1.2. The ring gear except one for the modified CRC L-42 test was gas-carburised and treated with manganese phosphate. On the other hand, green gears which were gas-carburised but did not have chemical conversion coating, were used for the load carrying property test.

Table 6.1.2: Specific Details of Hypoid Gear

	Pinion	Ring gear
Number of teeth	10	41
Outside dia., mm	71	167
Face width, mm	52	28
Offset, mm	25.4	
Material	SCM 415	do.
Heat treatment	carburized	do.
Phosphate coating	none	phosphated

Tooth contact pattern was confirmed to be normal for each set of gears and the average values of pinion torque and backlash were measured to within 1.5 − 2.5 Nm and 0.12 − 0.17 mm, respectively.

6.1.2.3 Test Oils

The characteristics of the lubricants evaluated in the test rig are outlined in Table 6.1.3.

Table 6.1.3: Inspection Data of GL-5 Hypoid Gear Oils

	A	B	C
Kinematic Viscosity			
at $100°C$, mm^2/s	17.42	17.48	18.24
at $40°C$, mm^2/s	185.7	189.4	222.1
Viscosity Index	101	99	98
Total Acid No., mgKOH/g	1.6	1.8	1.9
Total Base No.,	1.1	1.6	2.1
Element in Oil			
S, wt %	2.80	2.93	2.72
P, wt %	0.11	0.12	0.11
Cl, wt %	0.02	0.02	0.02
N, wt %	0.08	0.09	0.09

All three oils contained a sulfur-phosphorus (S-P) EP package commercially available for use in hypoid gear oils meeting the API Service GL-5 requirements.

Oil A contained only the GL-5 level of S-P additive package, while oils B and C were blended with the S-P package and an ashless dispersant which is reported to improve the thermal-oxidative stability of gear oils.

Oil B contains a small quantity of dispersant and 10 % more of the EP agent compared with oil A. Oil C contains equal amount of EP agent to oil A and more dispersant than oil B.

6.1.2.4 Procedure

Four kinds of axle gear tests, a high temperature and high speed test, modified L-42, a peak temperature measurement test and a wear test, were carried out using the axle gear tester.

Details of these tests and conditions are described in the respective section.

Fuel economy effect was examined by the chassis roll tests using a Japanese passenger car with the same hypoid gears in the rear axle.

6.1.3 Results and Discussion

6.1.3.1 Gear Protection Performance Under High Temperature and High Speed Conditions

Passenger cars are often driven at very rapid speed on express highways, especially in Europe and some troubles in powertrain lubrications have been reported (1).

When cars with independent suspension rear axles run at speeds of above 120 km/h during the break-in period, the oil temperatures are usually observed to rise up to $160 - 180°C$ (2).

In order to evaluate the service performance of hypoid gear lubricants, two full scale laboratory testing methods, the CRC L-37 and L-42 are well known. The L-37, high torque and low speed test adopt the endurance conditions of 135°C oil temperature and 24 h duration for determining the endurance characteristics of GL-5 gear oil. But these conditions are not severe enough to confirm the abovementioned field performance of hypoid gear oils in high speed and high oil temperature usage.

To clarify the influence of gear oil formulation changes on the hypoid gear failure during high temperature and high speed use, bench axle tests were carried

691

out under the conditions of 4563 rpm input revolution, 14.7 — 137 Nm input torque and 110 — 180°C oil temperature.

Under the conditions of 180°C oil temperature, 4563 rpm pinion speed (equivalent to 120 km/h speed) and 98 Nm torque (road load is about 70 Nm at 120 km/h), the progress of gear distress with time is tested and shown in Table 6.1.4. In the case of oil A, it was observed after 4 h that the center of ring gear teeth turns surface color to grey and white which we named "Whitening".

Table 6.1.4:   Observation of Gear Distress Progress

Durations, h \ Sample Oil	A	B
1	Normal	Wholly Whitening
4	Partly Whitening	Wholly Whitening
8	Partly Whitening	Wholly Whitening
11.5	Worn & Shiny	Seizure
13	Seizure	

180°C, 4563 rpm, 98 Nm

With oil B, the whole ring gear tooth surface changed to grey and white color after 4 h and "Whitening" was noted at 8 h later inspection. After 11.5 h, the oil B test was terminated because of noise and gear tooth seizure.

While with oil A, "Whitening" disappeared before 11.5 h and the gear teeth showed "Worn & Shiny" surfaces, the test was continued till 13 h. At that time gears made noise due to severe gear distress, "Seizure". The surface "Whitening" phenomenon is considered to be the same surface discoloration as "Grey-staining" (3) and "White layer" (4) reported on carburized gears.

The effect of oils on gear distress is compared in Figure 6.1.2 under the conditions of 8 h duration, 168 and 188°C and 14.7 — 137 Nm torque. Vertical axis shows the progress of gear distress observed in these tests. A gear failure, rippling, is located between worn & shiny and seizure.

Use of oil A, B and C resulted in increasing gear distress with increasing test severity as shown in Figure 6.1.2. The authors believe that the sequence of gear distress shown on the vertical axis is logical. Thus, "Whitening" is considered to be a precursor of the well defined gear distress. More details are reported in the paper (5).

The repeatabilities of these experiments were checked in 4 various conditions and are also plotted in Figure 6.1.2 and are thought to be rather good except one test of oil A, 180°C and 137 Nm which results scatter rippling and light ridging.

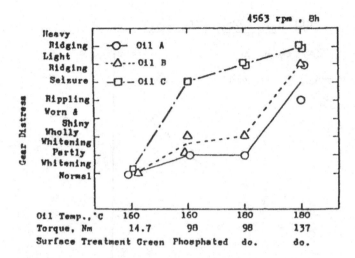

Figure 6.1.2: Gear Distress after 8 hours and under Various Conditions

A clear distinction among the gear protection performances of GL-5 hypoid gear oils at the high temperatures of 160 − 180°C is shown in Figure 6.1.2, and the gear protection performance of oil A in which no ashless dispersant is compounded is the best, while oil C which contains relatively high concentration of dispersant is the worst.

If we suppose that the addition of ashless dispersant produces a harmful effect on the gear protection performance of gear oil at high temperature, then the difference between oil A and B and the lowest protection performance of oil C will be understandable.

The effect of oil temperature on gear distress was examined using oil A, the best gear protectable oil and results are shown in Figure 6.1.3. Two curves of different test conditions are plotted along with oil temperature in Figure 6.1.3. One is 69 Nm torque and 15 h duration, the other is 137 Nm and 8 h. From these curves, it can be seen that oil temperature is the most critical variable.

It has been reported that in laboratory thermal and oxidative stability tests, hypoid gear oils blended with S-P agent and ashless dispersant form less deposits on the catalyst and beaker and display no significant EP elements depletion in the oils compared to the dispersant free analogue (6). Accordingly, dispersant type gear oils have been thought to provide the improved performance in the high temperature. But above-mentioned results indicate that the addition of ashless dispersant to GL-5 gear oil reduces the reaction of EP agent with gear surfaces and lowers the gear protection performance in the high temp. of 160 − 180°C.

693

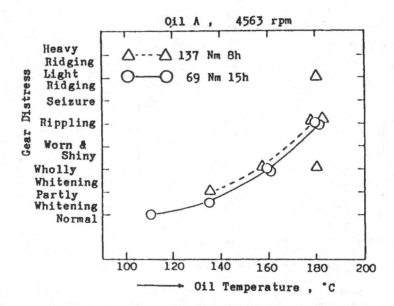

Figure 6.1.3: Effect of Oil Temperature on Gear Distress

Some trials to develop a new oil composition which can retard the progress of gear distress at 180°C were carried out and those results are summarized in Table 6.1.5.

It has been reported that "Grey-staining" is much influenced by oil viscosity (3), but in the tests shown in Table 6.1.5, the high viscosity analogue of oil A does not give a positive effect on gear failure.

Comparing test No. 2 with No. 3 in which each oil contains the S and P component of the S-P additive package respectively, P compounds seem to exhibit better results than olefin polysulfides in those evaluations.

Distinct improvements are demonstrated in test No. 4, 5 and 6 compared with No. 1. The addition of organo-molybdenum compounds exhibits a clear effect on retarding gear failure because Mo compounds are considered to produce MoS_2-like material on the teeth which may reduce friction and frictional heating and may result in a decrease of the gear tooth tempering.

Table 6.1.5: Study on Gear Oil Formulations to Retard Gear Distress

No.	Oil Composition	Gear Observation after 8 h	Gear Observation after 16 h
1	High Viscosity Oil A (SAE 250)	P-Whitening	Rippling
2	Base Oil of A + Sulfur Component	Worn & Shiny	Rippling
3	Base Oil of A + Phosphorus Component	W-Whitening	Worn & Shiny
4	Oil A + MoS$_2$	P-Whitening	P-Whitening
5	Base Oil of A + Soluble Mo	P-Whitening	W-Whitening
6	Base Oil of A + S-P EP Agent + Soluble Mo	P-Whitening	W-Whitening

180°C, 98 Nm, 4563 rpm

6.1.3.2 Load Carrying Properties Under High Speed and Shock Loading

A new high speed and shock load test method (modified CRC L-42) was developed using Japanese commercially produced rear axle hypoid gears. Significance of this test and correlation with CRC L-42 and with data of actual field tests were reported elsewhere (7).

Using modified L-42 test, the influence of base oil viscosity, dosage of an S-P package additive and other additives on load carrying properties of hypoid gear oils was studied.

Increased addition of an S-P EP agent to a low viscosity base oil of 6.5 mm^2/s @ 100°C was examined and the results are shown in Figure 6.1.4. Satisfactory performance was not obtained even with 3.5 times higher than the normal additive dosage defined for API GL-5.

Roles of sulfur and phosphorus components in S-P package additives were also investigated. It was confirmed that olefin polysulfides in S-P package were the key compound determining the high speed shock test performance required for standing the test.

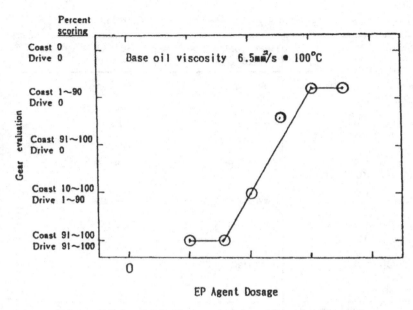

Figure 6.1.4: Effect of S-P EP Agent on Load Carrying Properties

Weight ratio of sulfur to phosphorus in EP agents was found to be important. Pass or failure to modified L-42 can be predicted by the weight ratio. The ratio of more than eleven was required to obtain the successful performance of GL-5 load carrying properties.

In order to exhibit satisfactory load carrying properties, phosphorus components with weak polarity were better than ones with strong polarity.

Addition of ashless dispersant deteriorated the EP performance of S-P EP agents. Passing modified L-42 test results were obtained with the low viscosity base oil blends of 10 mass % viscosity index improvers plus S-P additives.

6.1.3.3 Effect of Hypoid Gear Oils on Axle Break-In Temperature

High lubricant temperatures are generated during the break-in of new differential assemblies, and for a constant speed and load conditions. The break-in of a new axle causes the bulk oil temperature to rise, peak and then decrease to a stabilized temperature (8).

The effect of hypoid oils on the break-in temperatures was studied using a bench axle gear tester which gave good repeatability.

The peak temperature varied with the hypoid gear oil formulation. Typical results are shown in Figure 6.1.5. It was found that certain phosphorus compounds were effective for lowering the peak break-in temperature of axle lubricant.

Figure 6.1.6 shows that the peak temperatures were reduced with increasing dosages of phosphorus compounds.

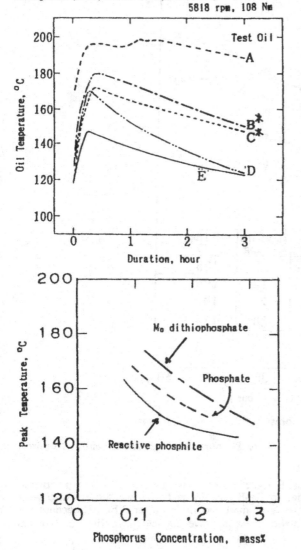

Figure 6.1.5:
Oil Temperature in Axle during Break-In

Figure 6.1.6:
Effect of Phosphorus Component on Peak Temperature

The wear of hypoid gear in the break-in stage was investigated and the effects of a small amount of abrasive particles were compared. The results are shown in Figure 6.1.7. The hypoid gear wear during the initial 8 hours was large, and after then, the wear rate became rapidly low.

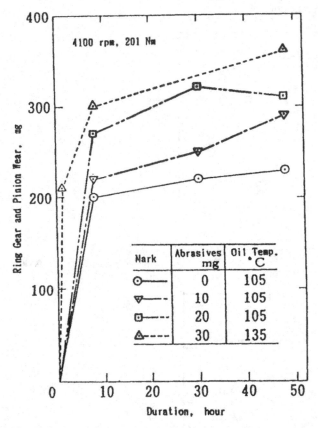

Mark	Abrasives mg	Oil Temp. °C
⊙—	0	105
▽—–	10	105
⊡—––	20	105
△––––	30	135

Figure 6.1.7: Influence of Abrasives on Hypoid Gear Wear

Insolubles produced in tested oil by 8 hours wear test were mainly worn-iron and weighed about 1 gram.

The wear volume of the initial 8 hours was increased by the 10 – 30 mg addition of abrasives, silicone carbide, to GL-5 hypoid gear oil. But the abrasive wear caused by SiC was reduced to very small proportion after 8 hours, because the frictional surfaces ground down those abrasive particles into finer and round ones.

Four commercial organo-molybdenum compounds (OMC) were evaluated with the same molybdenum dosage in GL-5 gear oil. The extent of peak temperature reduction was not associated with M_0 dithiophosphates and M_0 dithiocarbamates but with phosphates and thiophosphates in OMC.

6.1.3.4 Effect of Hypoid Gear Oils on Fuel Economy

The influence of gear oils on fuel saving was investigated using the chassis roll. Chassis dynamometer fuel economy tests with three different viscosity oils were carried out and fuel economy improvements in various driving modes are shown in Figure 6.1.8 with SAE 90 as the reference.

In Figure 6.1.8, it is obvious that the high viscosity oil, 140, consumed more fuel than the reference, 90, but the low viscosity oil, 75 W, consumed less fuel. The fuel economy improvement with the low viscosity oil decreased slightly as the running speed increased.

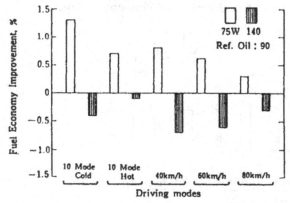

Figure 6.1.8: Influence of Lubricant Viscosity on Fuel Economy Improvement

The temperature rise with different viscosity test oils during chassis tests is shown in Figure 6.1.9. At 60 km/h road load, the temp. of 140 was about 4°C higher than that of 90, and that of 75 W was about 8°C lower.

These oil temp. differences are reasonable because the energy loss of the power transmission resulted in an oil temperature rise and fuel economy improvement with gear oil viscosity reduction is considered to result from decrease of mechanical churning losses in hypoid gears.

Another fuel economy chassis test was carried out with sample oils, 75 W, 75 W + FM and 90. Fuel economy improvements in various driving models were compared in Figure 6.1.10 in which 90 oil was used as the reference. The fuel savings with 75 W + FM were not better than those with 75 W.

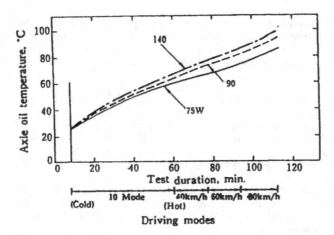

Figure 6.1.9: Oil Temperature in Axle during Fuel Consumption Tests

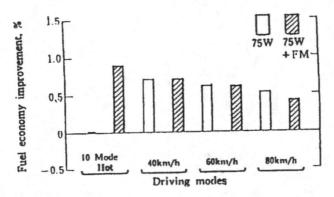

Figure 6.1.10: Influence of Friction Modifier upon Fuel Economy Improvement

In previous friction tests using LFW-1 and 4 Ball Machines, 75 W + FM gave lower friction than 75 W. Under such boundary conditions, friction modifiers must be relied on to reduce friction and to possibly improve fuel economy.

In fuel economy tests in various running modes, the lubrication conditions between hypoid gears were thought to be milder than in laboratory friction tests, so that the difference in the frictional losses between 75 W + FM and 75 W would not result in an improvement of fuel economy.

These facts suggest that the lubrication mode between hypoid gears was dominated mainly by hydrodynamic lubrication under the conditions studied.

700

More detailed information was published in a journal (9).

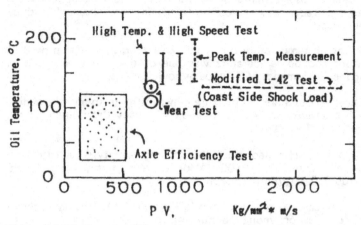

Figure 6.1.11: Severity Comparison between Four Tests

Table 6.1.6: Summary of Additive Effectiveness in Various Tests

Evaluation	S component	P component	Dispersant	M_O dithiophosphate
High Temp. & High Speed	Base	Effective	Harmful	More effective
Modified L-42	More effective		Harmful	No study
Peak Temp.	No study	More Effective	Harmful	No extra effect
Fuel Economy	No study	No study	No study	No extra effect

6.1.4 Conclusions

(1) Hypoid gear lubrications and the additive technology for the higher performance hypoid gear oil than API GL-5 gear oils have been described.

Four kinds of the gear tests using the axle gear tester were carried out to confirm the performance characteristics of the new S-P type candidate product.

The effectiveness of four additives, olefin polysulfides, phosphorus compounds, ashless dispersants and organo-molybdenum compounds are summarized in Table 6.1.6 according to the findings observed in these performance tests.

(2)   The effect of oil temperature on hypoid gear lubrication was proved to be important and oil temperature was one of the critical variables in evaluating the performance of S-P type hypoid gear oils.

Other critical variables are the pressure between the teeth and the sliding velocity. The severity in gear lubrication is usually determined by the pressure-velocity (PV) values.

For the hypoid gears tested, the highest sliding velocity, V and the contact pressure, P are given in 0.00184 (pinion revolution), m/s and 31.5 (pinion torque), 0.5 Kg/mm^2 respectively.

PV values were calculated and plotted in Figure 6.1.11. This figure shows the relative comparison among the four tests.

The fuel economy test is located in the low oil temperature and low PV area, in which the lubrication mode between hypoid gears was dominated by hydrodynamic lubrication.

Various gear distress was observed by the evaluating conditions in the high temperature and high PV area.

6.1.5   References

(1)   Tsuzuki, Y.; Masunaga, K.; Akiyama, K.: Lubricants of Power-Train for Automotive Passenger Cars. J. Japan Soc. Lubri Engers. 28 (1983) 2, 80—86.
(2)   Towle, A.: Some Factors Influencing the Performance and Selection of Drive Line Lubricants. Revue de L'Institut Francais du Petrole 25 (1970) 7—8, 935—956.
(3)   Ariura, Y.; Ueno, T.; Nakanishi, T.: Grey-Staining on the Tooth Surface of Carburized Gears. J. Japan Soc. Lubri. Engers. 24 (1979) 10, 662—669.
(4)   Uetz, H.; Sommer, K.: Investigations of the Effect of Surface Temperatures in Sliding Contact. Wear 43 (1977) 2, 375—388.
(5)   Ohashi, H.; Watanabe, S.: Effect of Hypoid Gear Oils on Gear Distress under High Temperature and High Speed Conditions. JSLE Tokyo 1985. P. 1009—1014. Pro. JSLE Inter. Tribology Conf. JSLE. 8.—10.07.85. Tokyo.
(6)   Zalar, F.V.: A New High Performance Automotive Gear Lubricant. JSLE, Tokyo (1985) P 237—245. Pro. JSLE-ASLE Inter. Lubrication Conf. JSLE 9.—11.06.75. Tokyo.

(7)   Watanabe, S.: Studies on a High Speed and Shock Test for Automotive Gear Lubricants. Finland, Lansi-Savo Oy, 1989. Vol. 4 P 257—262. Pro. 5th Inter. Cong. Tribology, Eurotrib 89, 12.—15.06.89. Helsinki.

(8)   Holubec, Z.M.; Brandow, W.C.: Test Techniques for the Evaluation of Lubricant Effects on Axle Break-in Temperature. SAE Paper 760327, SAE Meeting, Detroit, 1976.

(9)   Watanabe, S.; Ohashi, H.; Hasegawa, Y.: The Influence of Hypoid Gear Oils upon Fuel Economy and Axle Efficiency. J. of JSLE, Inter. Edition 10 (1989) 63—68.

6.2 The Screening of E. P. Oil Formulas by the Use of a New Hypoid Gear Axle Test

G. Venizelos and G. Lassau
Conservatoire National des Art et Métiers, Paris, France
P. Marchand
Institut Francais du Pétrole, Rueil-Malmaison, France

Summary

The evaluation of the performance level of the E.P. additives for hypoid gears remains up to now a very difficult challenge. The optimization of the structure of sulfurized E.P. agents that are the major constituent of packages for hypoid gears is excessively difficult and slow, as it is quite often based on results derived from fleet tests. More versatile and more easily available tests are requested by engineers in order to speed up the optimization process of formulas.

Facing the unsuitability of the classical laboratory mechanical tests in the E.P. lubrication domaine, I.F.P. (Institut Francais du Petrole), together with C.N.A.M. (Conservatoire National des Arts et Metiers), set up a new test with an actual axle. Up to now, tests have been carried out with a Peugeot 505 axle (PC 7 type). The test is based on applying successive torque peaks to a hypoid gear. Torque peaks are created by engaging the clutch of a Diesel engine, thus by motoring it. The procedure is light and short. Following damage, progression is made possible by examining gears through a little window openend in the case wall. Furthermore, several types of axles can be easily fitted to the bench without any substantial modification.

The correlation with CRC-L 42 tests has been established. The test procedure is very promising and its severity can be adjusted in order to suit to various experimental needs.

6.2.1 Introduction

Despite the fact that most European manufacturers of passenger cars turn to the transaxle design, hypoid gears are still widely used in the automotive industries in:

— rear-wheel driven to class cars,
— trucks, bulldozers, payloaders and the like,
— four-wheel driven cars.

The present evolution of these vehicles in our days, leads hypoid gears to heavier duty functions: higher speeds, powers and torques, in a space reduced every day by car designers. The evaluation of the performance level of extreme pressure (E.P.) additives for hypoid gears remains up to now a very difficult challenge. It has been established by various authors that laboratory standard tests (FZG, Timken, . . .) are inapt to evaluate correctly oils of API-GL4 level or higher, while today's commercial products are dominated by oils of GL5 or "GL5 +" level. The gear axle tests induced in these specifications, and particularly the CRC L 37 and L 42 mechanical tests, are old (30 years!) and not really available in Europe, specially the L 42 scoring test.

This unconvenient situation led main users to develop new evaluation procedures:

1. In the United States many efforts were made to bring to date the evaluation method: the L 42 test is the evolution of the CRC L 19 method; its test parts have been changed. New reference oils, representative of the actual formulae are on try (13).

2. In Europe, cooperative programs to develop new procedures have been established, particular in the United Kingdom (4), (12). It seems that these methods are not longer practiced today.

3. Large users of E.P. oils, for instance the French Army, or OEMs, have developed their own methods. None of them has achieved a process which is widely spread or recognized internationally.

4. Research Institutes have tried to develop the standard laboratory tests in order to become discriminating and significant at the GL4 to GL5 + level. The development of a large size FZG machine and its testing process, realized by University of Munich, should be also mentioned (2), (1). This process is under detailed examination by A.S.T.M. and S.A.E., in order to replace the obsolete tests of GL4 classification (13).

Despite these developments, it seems that the evolution of E.P. oils for hypoid gears is rather slow although the need for more multifunctional and more performing packages is greater than ever. The Institut Francais du Pétrole (I.F.P.), in its program of developing new E.P. additives, has decided together with the Conservatoire National des Arts et Métiers (C.N.A.M.), to establish a new screening test to evaluate the scoring resistance of E.P. oils. The axle used for the test is presently equipped by Peugeot parts, but the bench design is such that it can be easily adapted to other axles. It doesn't need any particular substructure because it doesn't use a fired engine.

The principle of the test is to apply successive torque peaks to a hypoid gear axle. The test takes about two hours and it has already allowed to obtain significant results which will be presented in this paper.

6.2.2 Objectives and Test Equipment

6.2.2.1 Recalls on the Principle of the CRC L 42 Test (9) (10)

A simple test equipment permits to apply successive torque peaks to a Dana hypoid gear axle: a fired engine drives the brakes, via the hypoid axle (see Figure 6.2.1) and successively 10 accelerations and 10 decelerations are applied.

The applied torques are as follows:

Test phase	Torque applied to the gear ring (Nm)	Torque applied to the pinion (Nm)
Acceleration	2700	720
Deceleration	1350	360

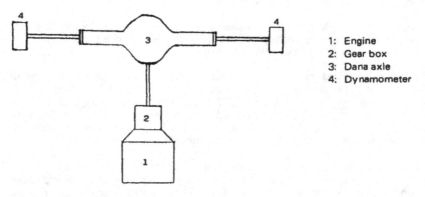

1: Engine
2: Gear box
3: Dana axle
4: Dynamometer

Figure 6.2.1   Principle of CRC-L 42 Test Equipment

In this experiment it has been observed that scoring occurs on the tooth surface in contact during deceleration, corresponding to the reverse gear of a car. The reason is, to our opinion, that the hand of spiral on these hypoid gears is such that the gears tend to take up to backlash under load in this drive.

Although it has been internationally recognized, this procedure presents, to our opinion, major disadvantages, such as: test axle is quite different from the present European hypoid gear axles, test parts are not easily available in Europe and finally it is more complicated to use a fired engine due to cooling, exhaust gases, etc.

6.2.2.2 Objectives

Most hypoid gear axles of European passenger cars are today overhung to the body of the vehicle and are connected to the wheels by Cardan transmissions.
Test equipment should allow to use these kinds of axles. In order to adapt axles from various origins it should have a modulable conception so that the minimum of transformation is necessary when changing axles. To avoid the inconveniences of a fired petrol engine, the method used by Mercedes-Benz for several years, seems a convenient solution: torque peaks are obtained when an inert mass is coupled by clutching to flywheels which are already rotating thanks to an electric motor. Figure 6.2.2 is a schematic view of the test equipment.

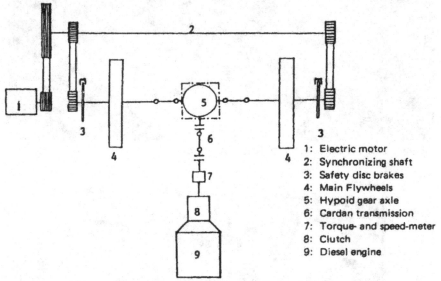

1: Electric motor
2: Synchronizing shaft
3: Safety disc brakes
4: Main Flywheels
5: Hypoid gear axle
6: Cardan transmission
7: Torque- and speed-meter
8: Clutch
9: Diesel engine

Figure 6.2.2: Scope of IFP-CNAM Test Equipment

6.2.2.3 Test Apparatus

Peugeot parts have been chosen as they are widely used by this French manufacturer in the rear axle of its vehicles. Details of the basic test equipment are given in tables 6.2.1 and 6.2.2.

A differential inductive torquemeter is placed in the shaft leading from the gear box to the axle. A heating device (ventilating hot air) is controlled to bring and to keep the testing oil as well as the axle and its carrier, at the desired temperature. Figure 6.2.3 and Figure 6.2.4 show views of the equipment.

707

Figure 6.2.3: General View of Test Installation

Figure 6.2.4: Close-up of Peugeot PC 7 Test Axle

Table 6.2.1:   Main Hardware Specifications

Element	Characteristics
Driving electric motor	Power: 18 kVA
Speed controller	Telemecanique, Altivar
Main flywheels	Diameter: 920 mm
	Total moment of inertia: 52.33 kg $\cdot$ m^2
	(of which flywheels: 47.12 kg $\cdot$ m^2)
Axle type	Peugeot PC7
Engine type	XD2S (Diesel $-$ 2300 cm^3)
Clutch command	Hydraulic
Axle-flywheels connection	Peugeot 505 transmission
Axle-gear box connection	Glaenzer transmission

6.2.2.4 Selection of Test Conditions

a) Sliding velocity:

If the tooth surface of the pinion is in contact with its mate of the ring at the point P, at any instant t, the sliding velocity of pinion (index 1) with respect to the ring (index 2) at P at the instant t, is (1):

$$V_g(1/2; P, t) = \omega_1 \sqrt{(ur_2 \sin \psi_2 - r_1 \sin \psi_1)^2}$$

Table 6.2.2:   Characteristics of the Tested Axle

Type		Peugeot PC7
Axle Gear Reduction ratio	(u)	8/37
Material	ring	20 MC5 gas carburized
	pinion	20 CD4 gas carburized
Surface treatment		phosphation
Teeth:		
Profile		Oerlikon Spirac
Offset		25 mm
Face contact ratio		3.1
Mean pinion radius	r_1	22.3 mm
Mean ring radius	r_2	81.9 mm
Angle of primitive cone	α_1	18.3°
	α_2	71°
Pinion spiral angle	ψ_1	47.1°
Ring spiral angle	ψ_1	30.9°
Pressure angle		
Concave side of pinion	γ'	17.2°
convex side of pinion	γ	20.8°
Principal radii of curvature		
(algebraic value)	R_1'	25 mm
(at mean point)	R_1''	64.8 mm
	R_2'	$-$ 65 mm
	R_2''	∞

Where:

ω_1 : is the rotating speed of pinion,

$u = \dfrac{\omega_2}{\omega_1}$ is the reduction ratio,

ω_2; is the rotating speed of ring,

r_1 and r_2 are the radii at the mean point of pinion and ring respectively,

ψ_1 and ψ_2 are the spiral angles of pinion and ring.

Numerical application using the values of table 6.2.2 gives the sliding velocity in function of pinion's rotating speed:

$$
\begin{aligned}
V_g(1/2; P, t) &= 7.2 \times 10^{-3}\ \omega_1 \ \text{(m/s)(if } \omega_1 \text{ is in rad/s)} \\
&= 0.75 \times 10^{-3}\ \omega_1 \ \text{(m/s)(if } \omega_1 \text{ is in rpm)}
\end{aligned}
$$

So for $\omega_1 = 315\ \text{rad/s}(\cong 3000\ \text{rpm})$ (test conditions)

$$V_g(1/2; P, t) = 2.268\ \text{m/s}$$

b) Hertzian Pressures

In order to calculate Hertzian pressure we follow the method developed by Niemann and used by Naruse (11):

The contact of hypoid gear teeth is geometrically speaking a point contact. Under load it turns to surface contact, which is, according to Hertz, the half of an ellipsoid limited by an ellipse of major axis 2a and minor axis 2b long.

So the maximum pressure P_0 applied to the center of the contact area M, is given by:

$$P_0 = \frac{3}{2}\ \frac{F}{\pi ab}$$

where F denotes the normal load.

We assume that both bodies are homogeneous, elastic, isotropic and that the radii of curvature, defined in the neighborhood of contact, are much greater than the dimensions of contact area, caused by local deformations.

With these conditions and taking values of torque $C_1 + 320$ N.m, Poisson's coefficient $\nu = 0.3$, Young's modulus E = 22000 MPa, principal radii of curvature R'_1, R''_1, R'_2, R''_2, mean radius r_1 of pinion, spiral angle ψ_1 of pinion at M, pressure angle γ on the concave side of the ring (convex side of the pinion) as mentioned on Table 6.2.2, we get:

710

$$F = \frac{C_1}{r_1 \cos \gamma \cos \psi_1} = 70.425 \, C_1$$

in Newtons, with C_1 value in Nm.

The eccentricity (e) of the ellipse can be calculated from the radii of curvature:

$$e^2 = 1 - \frac{B}{3A + 4B}$$

Where:

$$A = \frac{1}{2} \left(\frac{1}{R_1'} + \frac{1}{R_2''} \right) = 2 \times 10^{-2} \; (mm^{-1})$$

$$B = \frac{1}{2} \left(\frac{1}{R_1''} + \frac{1}{R_2'} \right) = 2.37 \times 10^{-5} \; (mm^{-1})$$

so, for the hypoid gear used, $e = 0.99$.

The axes of the ellipse are then:

$$a = \sqrt[3]{\frac{9}{\pi e^2} \frac{1 - \nu^2}{E} \frac{F}{A}} = 5.8 \times 10^{-3} \, F \; (mm)$$

$$b = a \sqrt{1 - e^2}$$

and the maximum Hertzian pressure is:

$$P_0 = \frac{3}{2} \frac{F}{\pi \, ab} = 104.5 \sqrt[3]{F}$$

$$P_0 = 432 \sqrt[3]{C_1}$$

P_0 in MPa; C_1 in Nm

For a torque $C_1 = 320$ Nm applied to the pinion

$$P_0 = 2.954 \; MPa.$$

c) Admissible field of utilisation

Firstly the sliding velocity is limited by the maximum rotating speed of the Diesel engine which is 4500 rpm.

Secondly the torque applied is limited by the mechanical characteristics of the clutch.

The torque applied to the axle is on the other hand limited by safety. Shear pins are used to prevent accidents, for instance in the case of tooth breakage and axle blocking-up.

When combining all these limits, the domain of possible utilization of the test equipment is actually as shown in Figure 6.2.5.

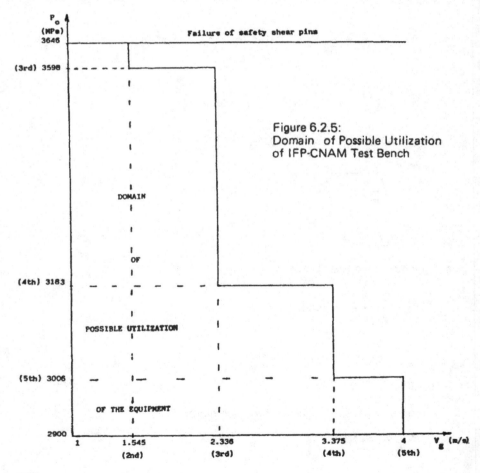

Figure 6.2.5:
Domain of Possible Utilization of IFP-CNAM Test Bench

6.2.3 Test Procedure

The hypoid gear axle and all testing parts of Table 6.2.2 are fitted together following the manufacturer's instructions.

6.2.3.1 Running In

The axle, containing 1.55 l of testing oil is heated up to 90°C. Then the test bench starts to rotate with the Diesel engine clutched out, during 15 min, at a rotating speed of 1000 rpm (ω_1 = 105 rad/s). These is no clutching-in during this phase whose purpose is to clean the gears, to achieve a very light running-in of the mating surfaces and to verify the correct functioning of the test bench.

6.2.3.2 Test

After changing the oil the axle is heated up to 100°C and the electric motor is started up to bring the pinion shaft to a rotating speed of 3000 rpm. The gear box selector being engaged in 4th the Diesel engine is clutched-in for about 2 seconds and then is clutched-out. The driving electric motor re-accelerates the flywheels to the desired speed. A new clutching-in is performed and so on up to a number of 20.

After a series of 20, the bench is stopped and the gear ring teeth are observed through a window opened in the wall of the axle case.

This procedure is repeated by series of 20, up to 80. The total number of times of motoring the Diesel engine may be changed for other objectives. The test procedure is schematized in Figure 6.2.6.

Figure 6.2.7 shows a typical recording of torque as measured between the gear box and the axle during clutching-in.

Once the testing bench is at rest and cooled, the axle is dismantled. The tooth surface condition is rated following the CRC method or by comparison to references.

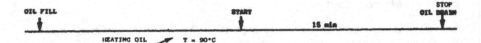

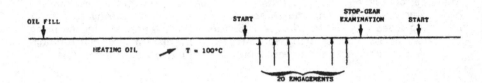

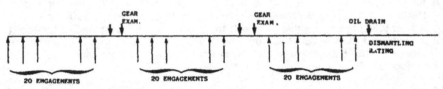

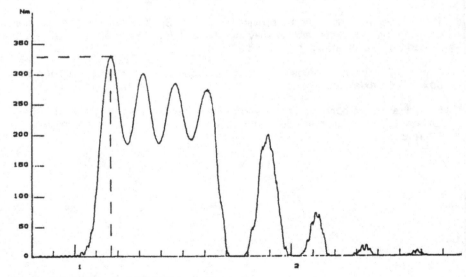

Figure 6.2.6: Chart of Testing Procedure

Figure 6.2.7: Typical Peak Torque Recording

714

6.2.4   Results

6.2.4.1 Severity Level of Test

In this type of procedure there are many ways of adjusting the testing severity. It is obvious that scoring occurs when the oil film is not thick enough to prevent contact between the external envelopes of teeth roughnesses. Thus scoring tendency can be adjusted by choosing the appropriate values of the parameters in Ertel and Grubin's formula:

$$h_0 = 1.18 \; \frac{(V_g \eta_0 \alpha)^{8/11} \; E_r^{1/11} \; R_e^{4/11}}{(W/L)^{1/11}}$$

Where:

h_0:   minimum film thickness in contact area
V_g: sliding velocity
η_0:   dynamic viscosity of the oil at the inlet temperature and atmospheric pressure
α: piezo-viscosity coefficient
E_r: reduced Young's modulus:

$$\frac{1}{E_r} = \frac{1}{2} \left(\frac{1 - \nu_1^2}{E_1} + \frac{1 - \nu_2^2}{E_2} \right)$$

Where:

ν_1 and ν_2 :   Poisson's coefficients of materials
E_1 and E_2 :   Young's modulus of materials
R_e: equivalent radius of curvature;

$$\frac{1}{R_e} = \frac{1}{R_1} + \frac{1}{R_2}$$

W/L: load per unit length.

This formula was obtained from experimental results with two cylinders in contact.

From this formula film thickness seems to be particularly sensitive to two parameters that can be modified in this test: sliding velocity V_g and lubricant viscosity η_0.

The influence of sliding velocity on hypoid gears is quite delicate to establish.

715

Ertel et Grubin's formula does not fully take account of the film thickness evolution at high speeds: thermal effects do reduce the oil film thickness by reducing its viscosity more than the speed does increase the film bearing capacity, as stated by Chang (15).

According to the starting point, the oil film minimum thickness h_0 may increase or decrease when V_g rises.

So the sliding velocity has not been changed, and kept close to service conditions. But oil viscosity has been reduced in this test, in order to achieve a metal-metal contact between the tooth surfaces with torque peaks compatible with the durability of the bench. Figure 6.2.8 shows a typical result.

Figure 6.2.8: Scoring on PC 7 axle gears

The tested additive formulae were all compared in this experience in the same base oil, i. e. 100 NS, having a viscosity of 4.03 mm^2/s at 100°C.

If this choice is unconvenient for instance in case of use of a synthetic base oil several solutions can be achieved, and, if necessary, combined:

— higher torque peaks can be applied, although it may have a detrimental effect on the durability of some bench parts,
— a higher oil temperature can be set,
— a reinforced clutch mechanism can be adapted.

716

6.2.4.2 Correlation with the CRC L 42 Test

As reference oils RGO 105 or RGO 110 are not readily available the test procedure was calibrated using a commercial sulphur-phosphorous E.P. additive A. Results are shown in Table 6.2.3.

Table 6.2.3

Formula	CRC L 42 SAE 90 base oil (pinion scoring)	IFP-CNAM 100 NS base oil (pinion scoring)
(1) 2.5 % A	–	Destruction of tooth surface in contact
(2) 4.25 % A	–	28 % scored
(3) 6.5 % A	2 %	2.2 % scored

Formula (2), although not tested by L 42, is known as having a good API GL4 level and it would score in the L 42 test on about 50 % of the tooth surface in contact. The testing procedure using Peugeot axle gives clearly the difference between GL4 and GL5 levels.

The first favourable comparisons allowed us to test other experimental sulphurized additives, either alone, or in a complete package. Main characteristics of these additives are listed in table 6.2.4.

Table 6.2.4

Mass	Add. B	Add. C
Sulphur content (% mass)	42.3	46
Viscosity at $100°C$ (mm^2/s)	4.35	9.95
Copper corrosion (3 hrs – $121°C$)	1b	2b

The corresponding results are shown on Table 6.2.5.

Table 6.2.5

Composition	Sulphur content (% mass) (1)	CRC L 42 SAE 90 base oil pinion scoring	IFP-CNAM 100 NS base oil pinion scoring
4.03 % add. B	1.71	5 %	0.4 %
3.67 % add. C	1.69	2 %	4.6 %
GL5 experimental package	–	11 %	13 %

(1)   Sulphur from sulphurized additive.

717

6.2.4.3 Conclusions

Despite some differences which we attribute to deviations of repeatability, the correlation between IFP-CNAM and L42 tests is fair and lets us think that this procedure is quite convenient for the screening of E.P. oils of GL5 or lower level. Adjusting the test severity (torque, sliding velocity, temperature) in the admissible field of this testing bench, can allow to test E.P. oils of a higher level.

6.2.5 References

(1)   Winter, H.; Michaelis, K.: Scoring Load Capacity of EP Oils in the FZG L-42 test. SAE paper No. 821183.

(2)   Richter, M.: Grenzen und Möglichkeiten der FZG-Zahnrad-Verspannungsprüfmaschine, insbesondere zur Prüfung von Hypoidgetriebeölen, Teil 6. Schmiertechnik und Tribologie 6 (1972), 177–178.

(3)   Friess, E.W.: Metallurgical Examination of Scored and Ridged Hypoid Gears. SAE paper 7603625.

(4)   Brandow, W.C.; Sands, T.P.; Streets, R.E.: Development of Gear Lubricant Test Technique by the CRC. I.P. Symposium (1964).

(5)   Olszewski, W.F. et al: Evaluation of Gear Lubricant Antiscore Properties. SAE paper 821182.

(6)   Thomas, J.: Design and Manufacture of Spiral Bevel and Hypoid Gears for Heavy-Duty Drive Axles. SAE paper 841085.

(7)   Varin, J.L.: Essais de grippage d'un couple hypoide. Memoire d'ingeniéur CNAM, 7 (1988) 8, Paris.

(8)   Taccoen, D.; Marchand, P.: Calcul des vitesses de glissement et des pressions de contact sur un couple hypoide au cours de l'essai L 37 (1987) Rapport interne.

(9)   Laboratory Performance Tests for Automotive Gear Lubricants Intended for API GL4, GL5 and GL6 Service. ASTM STP 512.

(10) Performance of Gear Lubricants in Axles under High Speed and Shock Loading. Federal Test Method Standard 719b, Method 6507-1.

(11) Haizuka, S.; Naruse, C. et al: Effects of Oil Viscosity and Antiwear Agent on Load Carrying Characteristics of Hypoid Gears. Bulletin of JSME, Vol. 25, No. 205, July 1982.

(12) Hunter, C.E.: Alternative Axles for Hypoid Oil Tests. Automotive Engineer, June–July (1978), 74–77.

(13) Schiemann, L.F. An Overview of ASTM and SAE Gear Lubricant Activities. 3ème Symposium CEC (1989), Paris.

(14) Krupke, E.: Quality Factors for Hypoid Oils measured in Hypoid Gear Axles of Trucks. IP Symposium, 22 November 1964.

(15) Cheng, H.S.: A numerical Solution of the EHD Film Thickness in an elliptical Contact. ASME Trans., J. of Lub. Tech. 82F, P. 155–162, 1970.

(16) Henriot, G.: Bulletin I.E.T. No. 83, Octobre 1983.

6.3 Automatic Transmission Fluids – State of the Art

A. G. Papay
Ethyl Petroleum Additives Division, St. Louis, USA

Summary

Automatic transmission fluids (ATF) are examined in a worldwide context. The function, use and application of ATF, both in the automotive industrial areas, are shown and discussed. The main requirements and test procedures, along with the current original equipment manufacturers (OEM) specifications are reviewed and projections made as to future ones.

The chemistry involved in achieving key ATF properties is examined and a historical overview of its evolution since the early days presented. A comparison of the special capabilities of various chemical approaches is made and pertinent conclusions drawn as to the most promising type of ATF technology.

6.3.1 Introduction

Automatic transmission fluids (ATF) are among the most sophisticated types of lubricants known to the industry. In a sense they constitute the leading edge of lubrication technology. They also amount to a significant volume of product. In 1989 ATF consumption reached 170 million of US gallons in North America (565000 MT), 30 millions (100000 MT) in Europe, and 12 millions (40000 MT) in Japan. The application of ATF includes automatic transmissions and transaxles in passenger cars, trucks, commercial vehicles, and busses. Also included are some manual transmissions, off-highway equipment, tractors, farming implements, construction equipment, industrial machinery, marine equipment, stationary pumps, air compressors, etc. The objective of this paper is to present the state of the art and the main future trends on ATF.

6.3.2 Function and Properties

6.3.2.1 Functions

The ATF functions include those of a fluid coupling medium, energy/torque transmitter, lubricant, and temperature moderator. It transmits hydrodynamic energy in the torque converter, hydrostatic energy in servos and logic circuits, and sliding friction energy in bands and clutches. The ATF is also a heat transfer and temperature control medium.

6.3.2.2 Properties

Viscosity, a very important property, must provide for adequate EHD film at high temperatures with low viscous losses, along with satisfactory low temperature fluidity. Foam control is needed to avoid excessive foaming and air entrainment and for efficient hydraulics, heat removal, and lubrication. Seal compatibility is required to prevent leaks with nitrile, silicone, fluoroelastomers and other seal materials. Corrosion protection, ferrous and nonferrous, is obviously important. Oxidation corrosion affects the life of such parts as cooler fittings brazing, brass bushings and screens, thrust washers etc. Wear control ensures long pump life and protection of moving parts. Thermal/oxidation stability helps to prolong fluid life, and avoid thickening, sludging, additives depletion, and to keep the equipment clean. Friction must be within design parameters. Smooth but sure shifts are desirable, with satisfactory torque transmission under all conditions.

6.3.3 Specifications

6.3.3.1 History

1989 was the 50th anniversary of the automatic transmission. The first major ATF specification, called Type A Suffix A (TASA) was issued in 1957. It was followed by a host of others as shown in Table 6.3.1.

Table 6.3.1

GM		Ford
Type A	1949	—
Type A Suffix A	1957	—
	1959	M2C33-A/B
	1961	M2C33-C/D
DEXRON®	1967	M2C33-F
	1972	M2C33-G (Europe)
DEXRON® II	1973	
	1974	M2C138-CJ
	1981	M2C166-H
	1987	MERCON®

6.3.3.2 Present

In North America the main ATF specifications are DEXRON® II (1) & MERCON® (2). In Europe the main specifications are those of the OEM and include Daimler-Benz 236.6 & 236.7 as well as from ZF, VW, Renk, Voith, Audi, Saab Scania, Volvo and others. In Japan ATF requirements are driven by the OEM who write FF specifications but reveal them only to certain suppliers. The resulting approved fluids are called "genuine" when intended for service fill, but some are also used for FF.

The after market fluids in NA, Europe, and Japan are mostly DEXRON® II, or MERCON®. This simplifies the complex task of satisfying the multitude of vehicles and transmissions in operation around the world.

6.3.3.3 Future

Significant changes and improvements in transmission, especially transaxles have spurred the development of new and more stringent specifications. An example is DEXRON® III, currently under development by GM/Hydramatic with the cooperation of additive suppliers.

6.3.4 Properties Needing Upgrading

Low temperature fluidity: The NA and Japanese OEMs have discussed their desire for better low temperature fluidity. The desired range for service fill ATF is 20,000 to 30,000 cP.

Seal compatibility: It will be extended to VITON and possibly to VAMAC seals. The tests will probably be of an immersion type or an inclusion of seal materials in the new Turbo Hydramatic Oxidation Test (THOT).

Yellow metals corrosion: Improved protection is desired.

Wear protection: The vane pump test (Modified ASTM D2882) is used by Ford for MERCON®. The FZG load stages and perhaps some form of an FZG wear test may be added to the GM requirements. A new test may be required for DEXRON® III fluids, the sprag wear test, which utilizes a sprag clutch from a Hydramatic 440-T4 FWD transaxle; in industrial applications some of the probable wear tests are those in the HF-0 specification and the CRC L-20 gear test.

Oxidation-thermal stability: The THOT test may be upgraded in temperature to 173°C or 183°C (up from 163). The 700-R4 transmission will displace the old THM 350 transmission. We also anticipate that Ford may upgrade the ABOT test in the future.

Friction durability: GM is expected to eliminate the Turbo Hydramatic Cycling test (THCT) and replace it with the Transaxle Cycling Test (TACT) which uses the THM-125C front wheel drive transaxle or the new F-31 4-speed transaxle. The new test is not expected to exceed 9000 cycles, at one cycle per minute, and last about 150 hours. The last 1000 cycles may be run with the axle half-shafts running at a 10 % speed difference. The "differential output mode" will put severe stress on the final drive gears in the transaxle and simulate driving on the "mini-spare" tires. The High Energy Friction Characteristics & Durability (HEFCAD) test will be modified. It will use the actual direct clutch plates from a GM 125-C transaxle. The shift-feel test for GM will be substituted with a procedure using the shape of the torque traces from the TACT test to predict shift feel in a car. This will have the advantage of documenting both friction durability and fluid degradation. A new friction test may be added. This is expected to be a 125-C band test which utilizes the HEFCAD rig but with a band and drum instead of the usual disc clutch (3). The conditions are 100 h, 24000 cycles, 135°C, and 50 cc air per minute injected to induce oxidation.

6.3.5   Chemistry

6.3.5.1   Old Chemistry

The old chemistry of the ATF is shown in Table 6.3.2. It was heavily dependent on metals like Zn, Ca, etc.

Table 6.3.2:   Analysis of Old ATF Packages

	Type A Suffix A	DEXRON®		M2C33F & G	
Element, wt %					
Phosphorus	1.625	0.30	0.045	1.2	0.34
Sulfur	3.75				
Zinc	1.85		0.32	0.42	0.25
Nitrogen		0.75		0.4	0.54
Boron				0.08	
Calcium			0.45		
Barium	13.75				
Specific Gravity	1.12	0.94	0.96	0.93	0.917
Dosage, wt %	6.2	11.0	10.0	8.25	10.9

6.3.5.2   Present Chemistry

The presently utilized chemistry is shown on Tables 6.3.3 and 6.3.4. Table 6.3.3 shows the DEXRON® II ATF additive packages and Table 6.3.4 the MERCON®. All the MERCON® products shown here also meet the DEXRON® II requirements.

722

Table 6.3.3:   Present Chemistry — DEXRON® II Type

Elements, wt %							
Phosphorus	0.17	0.16	0.30	0.45	0.28	0.38	0.30
Sulfur	1.5	1.8	1.3	0.11	0.59	0.66	0.66
Zinc			0.23				
Nitrogen	0.85	0.60	1.03	0.89	0.88	0.95	0.96
Boron	0.18	0.17	0.04	0.15	0.14	0.17	0.14
Calcium		0.73		0.10			
Magnesium			0.05				
Specific Gravity	0.94	0.94	0.94	0.94	0.92	0.93	0.92
Dosage, wt %	9.9	10.9	10.1	11.7	11.7	9.8	10.2

Table 6.3.4:   Present Chemistry — MERCON® Type

Elements, wt %					
Phosphorus	0.16	0.45	0.28	0.38	0.30
Sulfur	1.80	0.11	0.59	0.66	0.66
Nitrogen	0.60	0.89	0.88	0.95	0.96
Boron	0.17	0.15	0.14	0.17	0.14
Calcium	0.73	0.10			
Specific Gravity	0.94	0.94	0.92	0.93	0.92
Dosage, wt %	10.9	11.7	11.7	9.8	10.2

6.3.5.3 Future Chemistry Needs

Changes in transmission design, increase in energy flux through smaller interfaces, extra features added, etc. may being to stress some of the present ATFs to a high degree. This situation has prompted the development of new and more stringent specifications, such as the DEXRON®-III specification. This one is largely an extension of the DEXRON-II specification with several new requirements added. There is also the possibility of a new MERCON® specification with more stringent requirements. To achieve all these new requirements a new chemistry will be required in the ATF packages as most of the present ones are not up to the task. Similar trends are showing also in Europe and Japan.

To understand better the need for a new chemistry and obtain some guidance as to the proper path a comparison of the performance of various formulations, especially in the new areas, will be shown below.

Table 6.3.5: DEXRON® II & MERCON® Performance of Various Chemistries

Chemistry	Zn, P, S, Ca, Ba	Zn, P, S, N, B	P, S, N, B, Ca	P, S, N, B, Ca	P, S, N, B
Antiwear					
PSPT (1) Rating	6 (fail)	3	3	3.5	1
Mod. D2882, wt. loss, mg	320	9	8	9	6
Oxid. Stability					
THOT (Paraf. base)					
Sludge	fail	pass	pass	pass	pass
Cooler corr.	fail	fail	pass	pass	pass
Cu in oil, ppm	400+	252	110	140	23
ABOT (Paraf. base)					
Rating	fail		pass	pass	pass
Friction					
HEFCAD					
Dyn. Nm	128, 133, 125	128, 133, 138	117, 140, 128	124, 140, 138	133, 139, 138
Stat. Nm	131, 130, 129	138, 146, 156	110, 145, 138	123, 140, 141	134, 139, 136
Engag. Time, s.	.63, .63, .70	.65, .60, .61	.66, .59, .74	.66, .58, .60	.61, .60, .60
MERCON® Friction					
Dyn. Nm		136, 133, 137		133, 130, 134	140, 144, 144
Breakaway, Nm		114, 108, 129		113, 94, 107	129, 119, 120
Engag. Time, s		.86, .88, .85		.87, .88, .89	.84, .83

(1) Power Steering Pump test, of DEXRON® II

NOTE: Friction measurements reported at 10, 10000, and 18000 cycles for HEFCAD, and 5, 2000, and 4000 cycles for MERCON

6.3.6 Comparison in Key Properties

6.3.6.1 Comparison in DEXRON® II and MERCON® Requirements

Table 6.3.5 shows how different chemistries can exhibit different levels of performance within the requirements of DEXRON® II and MERCON® specifications.

The DEXRON® II & MERCON® oxidation stability performance of an ATF additive package depends also on the base stock used. It is well known that naphthenic base stocks are generally less stable than their paraffinic counterparts. To assess the ability of different chemistries to cope with the naphthenic factor two ATF fluids were tested in the THOT and the ABOT oxidation tests. Table 6.3.6 shows the results.

Table 6.3.6: Oxidation Performance with Naphthenic Stocks

Chemistry	P, S, N, B, Ca	P, S, N, B
THOT		
Sludge rating	fail	pass
TAN incr.	4.4	2.9
Diff. IR (carbonyl)	0.63	0.54
Cu content, ppm	44	17
Visc. incr. @ 300 h, cSt	9.7	9.1
−40 Brookfield, cP	89,700	55,400
Overall rating	fail	pass
ABOT		
TAN incr. @ 250 h	5.1 (fail)	2.6
IR incr. @ 250 h	.49	.38
Visc. incr. % @ 250 h	90.1	35.4
Visc. incr. % @ 300 h	351.4	122.2
Varnish @ 300 h	fail	pass
Overall rating	fail	pass

It is evident that the fluid with the Calcium (from overbased Ca sulfonate) is not as capable of controlling the naphthenic factor as the ashless fluid is.

6.3.6.2 Comparison in Possible DEXRON® III Properties

Table 6.3.7 shows how different chemistries fare in several tests that are being studied as probable requirements for the coming DEXRON® III specification.

Large S-D values, as seen in the first two fluids, could indicate possible problems with shift-feel quality (5), as well as lower levels of stability. It is obvious from the above data that except for the ashless fluid most of the metal based chemistries will have problems meeting the DEXRON® III requirements. These problems will become more difficult when the TACT procedure becomes available and friction durability under more stringent conditions will take its toll. It is expected that the zinc chemistries due to their low ceiling in thermal/ oxidation stability will have special problems (4).

Table 6.3.7

Chemistry	Zn, P, S, N, B	P, S, N, B, Ca	P, S, N, B
Antiwear			
PSPT, Rating	3 (fail)	3.5 (fail)	1 (pass)
Sprag Wear	pass	fail	pass
Friction			
125-C Band			
Dyn., Nm	160, 144, 159	191, 174, 40	196, 190, 177
Stat., Nm	200, 180, 192	218, 230, 180	218, 212, 204
Engang.Time,s	.48, .57, .53	.48, .50, 1.49	.50, .49, .53
S-D diff. Nm	40, 36, 33	27, 56, 140	22, 22, 27
Comments	pass	fail	pass
		(cracked drum, burned band)	(higher dyn., lower S-D)

NOTE: Torques shown at 20, 12000, and 24000 cycles.

Additional AW			
FZG Load (Stages)	10,11	10,10	11,12
FZG Wear (low speed)			
wt. loss, mg	35	360	5 to 29

6.3.6.3 Comparison in Industrial Requirements

Since ATF is used exensively in non-automatic transmission applications, such as various industrial uses, a comparison in some key industrial tests is desirable. Table 6.3.8 shows the results of the evaluation of different chemistries in several industrial areas.

The results indicate various areas of weakness for two of the chemistries, specifically, the Zn and the Ca-based formulations. Interestingly, the ashless formulation continued to show good performance throughout this study. One of the properties that make ATF so popular with non automotive consumers in North America is the distinct red color that helps to identify the fluid and avoid mix ups. This red color is also important to the automotive users, and to many people the loss of the color is an indication of severe fluid deterioration, a presumption not always correct. Actually Zinc dithiophosphate in the formulation tends to kill the color even after mild thermal decomposition (4). Table 6.3.9 shows the results from a screening test that seems to correlate with the actual observations in passenger car ATF maintenance practice.

Table 6.3.8:   ATF Performance in Industrial Areas

Chemistry	Zn, P, S, N, B	P, S, N, B, Ca	P, S, N, B
Pump Tests			
35VQ25	Pass	pass	pass
P-46 Piston	fail (Shoe wear, cracking of swash plate)	pass	pass
Sundstrand 22, 1 % water			
Flow degradation	fail (heavy)	BL (7,7 %)	pass (0.3 %)
HF-0 Specification			
1000 h sludge	fail	fail	pass
Tractor oil area			
CRC L-20 gear test			
Ridging	trace	heavy	nil
Rippling	trace	heavy	nil
Wear	light	heavy	trace
Rating	pass	fail	pass

Table 6.3.9:   ATF Color Retention Test

Chemistry	Zn,P,S,N,B	P,S,N,B,Ca (0.1 % Ca)	P,S,N,B,Ca (0.73 % Ca)	P,S,N,B
Beaker oven test				
150 C, Steel strip				
1 Day	pale	pale	red	red
7 Days	brown	brown	red	red

The failure of the ZDDP-based fluid to maintain the color is expected. The poor showing of the non-zinc product (0.1 % Ca) is perhaps due to instability of its P-S chemistry which incidentally was found to be of the dithiophosphoric type. The excellent showing of the ashless (non-dithiophosphoric) formulation was also expected given the high level thermal and oxidation stability exhibited in the previous tests and seen in Tables 6.3.5 and 6.3.6. That stability is also confirmed by the consistent durability shown in the friction tests. Another confirmation is seen in a screening test for the IIIE engine sequence test. This test is simple and non-expensive and can be used in screening many types of other lubricants for oxidation stability. It is run at 161°C, with 10 liters of air per hour, in the presence of oxidation catalysts for 64 hours. As an example Table 6.3.10 below presents some comparative data for a number of ATF formulations.

727

Table 6.3.10: Screening Oxidation Test Results for ATFs

Chemistry	Zn,P,S, Ca,Ba	Zn,P,S, N,B	P,S,N, B,Ca	P,S,N, B,Ca	P,S,N,B
Spot Diff.IR (Carb.) State	black tar > 1.00 solid	black tar > 1.00 solid	black tar 0.90 solid	black tar > 1.00 solid	lt. brown 0.15 fluid

6.3.6.4 Japanese ATF Formulations

Japanese automotive manufacturers do not publish ATF specifications but communicate their requirements to their approved lubricant suppliers. They in turn communicate to their chosen additive suppliers their own judgement for a set of requirements appropriate to meet the goal. The OEM service fill approved lubricants are called "genuine" oils. Such genuine fluids are the backbone of the ATF market in Japan. The basic additive technology is usually DEXRON® II or MERCON® with some slight modifications, usually addition of an undisclosed, special friction modifier by the oil company.

As an example, two "genuine" ATFs belonging to two major Japanese automobile manufacturers are shown here and compared to a DEXRON® II/MERCON® fluid of comparable (ashless) chemistry labelled D/M ATF. All three fluids were evaluated in the same tests that include some from the DEXRON® II requirements and some that are typical to Japanese requirements. The latter were run in Japan. Table 6.3.11 shows some of the results of tests run in the USA. The breakaway torque and μ were taken at 0.72 rpm. The coefficient of friction numbers are usually preferred over the torque number by the Japanese OEMs.

Table 6.3.12 shows the results from the tests run in Japan.

There are many important differences between the "genuine" fluids and the closest DEXRON® II/MERCON® counterpart in most of the property areas evaluated. The low ratings in antiwear and oxidation as well as oxidative corrosion, which affects yellow metals performance, could probably be explained as side effects of the addition of the undisclosed special friction modifiers. Formulating is a immensely complex task and any tampering afterwards can upset the delicate balance. The weaknesses seen in the frictional area could perhaps be explained by inadequate oxidation stability.

The Friction Index concept, quantifying the change in coefficient of friction with temperature, is offered by the author for consideration as a possible guide for one aspect of frictional stability.

Table 6.3.11: Performance of "Genuine" Fluids — U.S. Tests

	Genuine A	Genuine B	D/M ATF
Chemistry	P,S,N	P,S,N	P,S,N,B
Antiwear			
PSPT	3.5	6.0	1.0, 1.5
DEXRON® II rating	fair	very poor	very good
Mod. D-2882,wt.loss,mg	559	947	6
MERCON® rating	fail	fail	pass
FZG Load, stages (pass)	8	9	11, 12
FZG Wear (low speed), mg	885	1154	5 to 29
Friction			
HEFCAD			
Dyn. Nm	134, 140, 135	139, 145, 140	133, 139, 138
μ	.141, .145, .140	.142, .150, .145	.140, .145, .148
Stat. Nm	131, 137, 129	135, 143, 135	134, 140, 136
μ	.138, .143, .132	.140, .149, .140	.143, .143, .143
Breakaway, Nm	90, 83, 91	97, 86, 90	90, 100, 115
μ	.110, .085, .097	.109, .090, .090	.129, .111, .120
Engag. Time, s	0.63, 0.60, 0.64	0.62, 0.59, 0.61	0.61, 0.60, 0.60
125C Band test			
Dyn. Nm	182, 176, 110	172, 178, 96	196, 190, 177
μ	.139, .135, .095	.133, .137, .086	.150, .148, .147
Stat. Nm	208, 200, 160	206, 210, 168	218, 212, 204
μ	.152, .149, .128	.157, .154, .131	.158, .158, .158
Engag. Time, s	0.46, 0.50, 0.80	0.47, 0.47, 0.82	0.50, 0.49, 0.50
Rating	fail	fail	pass
Comments	(Drum cracked, Band burned)		Drum & band OK

Table 6.3.12: Performance of "Genuine" Fluids — Japanese Tests

	Genuine A	Genuine B	D/M ATF
Thermal/Oxidation Stab.			
ISO test, 165C, 144h			
Vis.incr. %	17	19	4
Hexane insol. %	0.10	0.10	nil
Cu in oil, ppm	46	36	20
Corrosion			
Acell. Cu Catalyzed Oxidation			
$163°C$, 50h, 10l/h air, Cu			
Cu, wt. loss, mg	26.7	19.7	0.5
Vis. incr. %	128.4	139.9	3.4
Diff. IR (carbonyl)	1.23	0.62	0.10
Cu in oil, ppm	1280	873	88
Friction			
Large LVFA			
427.7 kg, 30 min Br. In			
40 C Run, μ			
1 rpm	0.134	0.120	0.126
100 rpm	0.135	0.132	0.138
80 C Run, μ			
1 rpm	0.105	0.104	0.123
100 rpm	0.132	0.132	0.138
Ratio μ-1/μ-100			
$40°C$	0.99	0.91	0.91
$100°C$	0.80	0.79	0.89
Friction Index			
$100°C$ ratio/$40°C$ ratio	81	87	98

6.3.6.5 Environmental Considerations vs. Formulation

Environmental and health related factors are now the prime shapers of technological approaches in solving problems. In ATF formulations these factors will become more evident as the new generation of products is surfacing. Toxic elements such as barium either as sulfonate or carboxylate will be bad choices. Also zinc as in zinc dialkyldithiophosphate, known for toxicity problems is another questionable choice. Chemistries of poor thermal/oxidation stability, such as dithiophsphoric types, will have problems in the higher temperatures of operation anticipated in the new equipment. Decomposition byproducts with bad odors will accentuate other toxic aspects. Generally, large amounts of metals in formulations are expected to be regarded as a detriment to be avoided. On the other hand practically ashless formulations will be pursued as environmentally more benign and easy to deal with. This shift, however, towards ashless formulations is not expected to be an overly difficult task. As seen above there is already at least one such formulation with excellent results in all tests. Others should be able to join it soon.

6.3.7 Conclusions

Changing technological requirements in the new automatic transmission designs are generating a strong effort worldwide for improved ATFs.

To meet the new requirements new and improved chemistries are called for. Also the new products must offer the potential for more applications in a wide variety of uses, both automotive and industrial.

Comparisons of performance in a variety of areas show metal-based formulations to be generally inferior to those that are not.

An ashless technology can now be developed that addresses both performance and environmental needs for the foreseeable future.

6.3.8 References

(1)   DEXRON® II Automatic Transmission Fluid Specification. General Motors Corporation, Second Edition, 1978.
(2)   MERCON® Automatic Transmission Fluid Specification. Ford Motor Company, 1987.
(3)   SAE Recommended Practives — J1499 February 1987: Band Friction Test Machine (SAE) Test Procedure, Revised February 1987.
(4)   Papay, A.G.: Effect of Chemistry on Performance of Automatic Transmission Fluids. Preprint No. 89-AM-2D-2, STLE Annual Meeting Atlanta, 1989.
(5)   Papay, A.G.; Hartley, R.J.: Frictional Response of Automatic Transmission Fluids in Band and Disc Clutches. SAE Paper 890721, SAE Int'l. Congress, Detroit, 1989.

6.4 Prediction of Low Speed Clutch Shudder in Automatic Transmissions Using the Low Velocity Friction Apparatus

R. F. Watts and R. K. Nibert
Exxon Chemical Company, Linden, USA

Shuddering in automotive drivetrains is most often caused by friction induced stick-slip or dynamic vibration. The prediction of this phenomenon, and its control, are of great interest to equipment designers and lubricant formulators. This paper describes the use of the Low Velocity Friction Apparatus (LVFA) for this purpose. The LVFA is used to model clutch systems and their operating conditions, and from this data predictions of operating conditions under which stick-slip behavior could occur are made. The influence of friction material, mating steel surface, thrust load, temperature and lubricant are all examined.

6.4.1 Introduction

Stick-slip behavior in clutch and brake systems is an undesirable aspect of performance. In the drivetrains of mobile vehicles this phenomenon is observed as squawk in wet clutches, shudder in torque converter clutches and limited slip axles, and chatter in wet brakes. Since the early part of the twentieth century, this "stick-slip" phenomenon has been of interest to researchers, equipment designers and lubricant formulators. In the 1930s Bowden made the first direct measurements of stick-slip behavior by sliding two loaded metal blocks relative to each other (1). Following this initial observation many others have commented on the phenomenon and ways of controlling it (2, 3). Most of the recent literature in this area pertains to the control of this phenomenon in automobile clutches and tractor wet brakes (4, 5, 6).

Stick-slip sliding occurs when the static coefficient of friction exceeds the dynamic coefficient, i. e. when the friction ratio, μ_s/μ_d, exceeds 1.0. Under these conditions the force required to initiate motion is larger than that required to maintain sliding. As increasing force is applied to the movable element it eventually overcomes the static coefficient of friction and begins to move. The acceleration is then opposed by the spring force of the mechanism as well as by friction and viscosity of the lubricant. These decelerating forces cause the velocity to drop, the friction increases and the movable element again stops. This continual acceleration-deceleration is the source of stick-slip vibration (7). The stick-slip phenomenon has been modelled mathematically by Friesen (6) with the same general conclusion, that is: a sufficiently negative slope of the friction-velocity curve leads to self-excited vibration. This relationship is shown in Figure 6.4.1.

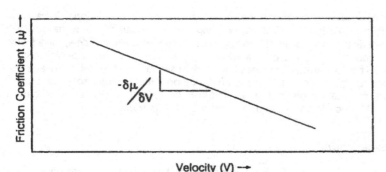

Figure 6.4.1: Friction versus Sliding Speed

A phenomenon closely related to stick-slip is dynamic frictional vibration. The fundamental difference between the two phenomena is that dynamic frictional vibration results from velocity pulsation without actual sticking. If in a decelerating system, friction coefficient is increasing with decreasing speed, $-\delta\mu/\delta V$ (Figure 6.4.1), a self energizing condition exists which amplifies the vibration (7).

Clutch systems used in automatic transmissions are susceptible to both types of phenomena, stick-slip and dynamic frictional vibration. However, truly continuously slipping clutches, such as those employed in some torque converters and differentials, would only be vulnerable to dynamic frictional vibration (8). Whether a system experiences stick-slip conditions or dynamic frictional vibration depends on the relative sliding rates and system elasticity or spring constant.

Many different types of apparatuses have been constructed to investigate these phenomena. Early investigations of stick-slip behavior were conducted using a modified milling machine slideway. Steel surfaces of various finishes could be moved relative to each other in the presence of various lubricants. Coefficient of friction could then be determined under different sliding speeds and applied loads. Stick-slip behavior showed up as vibration in this type of system (3). In the late 1950s the emphasis changed from evaluating steel-on-steel friction behavior to looking at the frictional characteristics of paper or asbestos based clutch linings running against steel plates. Several different apparatuses appeared and were used to evaluate the low speed frictional characteristics of these clutch systems. A modified four ball wear tester was used to measure coefficient of friction at low speeds. The standard specimen cup was replaced by a chamber supported by an air bearing to provide free rotation. A modified spindle, capable of rotating 0.75 inch diameter clutch segments, was fitted to the machine arbor. The system could be loaded by adding dead weights. Temperature was controlled by the cup heater and frictional torque measured by a strain gauge arrangement on the cup (7). About 1960, General Motors reported the construction of an appratus designed specifically for the measurement of

733

friction at low sliding speeds under a wide variety of conditions. This Low Velocity Friction Apparatus (LVFA) was capable of measuring frictional properties of lubricants using a small annulus of clutch material. Speed, load and temperature could be varied over a wide range to simulate field operating conditions (9). This apparatus continues in wide use today. A number of apparatuses similar to the General Motors LVFA but on a much larger scale have been constructed for measuring low speed friction performance on full size clutch plates (10, 11), and even the use of an entire tractor is reported (6).

6.4.2 Background

The most critical aspect of performance in clutch systems operating at very low sliding speeds is elimination of shudder caused by stick-slip and dynamic frictional vibration. Figure 6.4.2 shows three friction coefficients (μ) versus velocity curves. Curve 1 ist that observed for a typical lubricant base oil. As speed is decreased friction coefficient remains constant to a certain point and then increases rapidly. In this area of large $-\delta\mu/\delta V$ shuddering will occur. Curve 2 is representative of a friction modified oil. In this case, as velocity decreases so does friction coefficient (a positive $\delta\mu/\delta V$). This system should be free of shuddering on deceleration. Curve 3 shows a poorly friction modified fluid. Although the performance of this fluid could be generally acceptable, shuddering would occur in the region designated "a" if the clutch system were to operate in or pass through this region. A number of authors have correlated $-\delta\mu/\delta V$ with shudder or squawk in actual clutch operation (4, 5, 6, 9, 11).

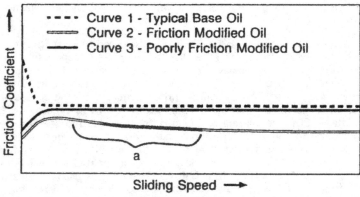

Figure 6.4.2: Friction versus Sliding Speed

For the optimum performance and control of a continuously slipping clutch device, the frictional performance of the system, friction material, mating steel surface, and lubricant must be as constant as possible over a wide range of operating conditions. Many factors determine the frictional performance

734

Table 6.4.1: Fluids studied

Fluid	Viscosities			Elements, ppm							
	KV100. cSt	@0°C, cP	@–40°C, cP	B	Zn	P	S	Ca	Mg	Ba	N
A	8.3	300	37,000	0	920	540	2,600	2,740	30	1,990	100
B	7.1	350	39,500	277	590	580	3,400	0	0	0	900
C	8.1	400	41,500	275	20	180	4,200	820	10	0	700
D	7.9	400	41,000	54	210	290	4,900	0	100	0	120
E	7.6	300	26,000	205	10	310	3,100	0	50	0	1,100

of a clutch system; selection of clutch lining material, the type and finish of the steel surface against which the clutch runs and the lubricant are all critical. Ideally, once a clutch system and lubricant are chosen, friction coefficient would not vary significantly with applied load, temperature or sliding velocity. It is the combined goal of clutch manufacturers and lubricant formulators to achieve this type of performance.

6.4.3 Experimental Results

All experimental results reported were obtained on a Low Velocity Friction Apparatus similar to that described by General Motors (9). The specimen cup was modified so that evaporated liquid nitrogen could be passed through it to extend the range of operating temperatures significantly below ambient.

A numer of combinations of friction materials, fluids, and mating steel surfaces have been studied under a wide range of loads, speeds and temperatures to assess the friction properties of the systems and look for conditions under which clutch induced shuddering could occur. Table 6.4.1 lists the fluids included in this study. Fluids are characterized by their viscometrics and elemental analysis. All of the fluids with the exception of the *Experimental Fluid* are commercially available automatic transmission fluids. Table 6.4.2 lists the friction materials used in the work and gives a brief description of their properties. All of the friction materials are currently used in automatic transmissions. Unless otherwise noted all results were obtained by the standard procedure described in Appendix 1.

Dynamic friction as used in this discussion is measured at a sliding speed of 100 ft/min. Static friction is the value observed just as the system comes to rest, i. e. at zero sliding speed.

Table 6.4.2: Clutch Friction Materials*

Material	Clutch Type	Material** Density	Particles**	Resin** Content
CFM-1	Lock-Up	High	Resin	High
CFM-2	Band	Low	None	Medium
CFM-3	Plate	Low	None	Medium
CFM-4	Plate	Medium	Resin	High
CFM-5	Plate	Low	None	Moderate
CFM-6	Plate	Medium	Graphite	High
CFM-7	Plate	Low	Graphite	Moderate

*   All Materials are Cellulose Based
**   Determined by Optical Microscopy

736

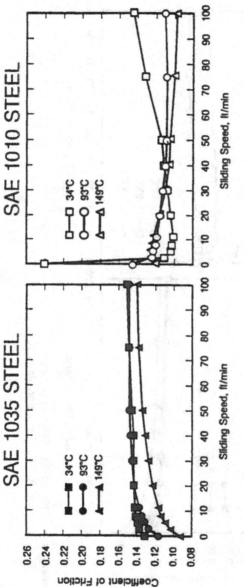

Figure 6.4.3: Effect of Steel Mating Surface
Fluid A, CFM-3 Friction Material

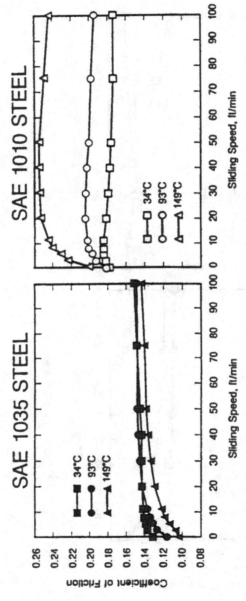

Figure 6.4.4: Effect of Steel Mating Surface
Fluid A, CFM-1 Friction Material

6.4.4 Effect of Mating Steel Surface

The hardness and surface finish of the mating steel surface can greatly affect the overall frictional characteristics of a clutch system. Figures 6.4.3 and 6.4.4 show the different friction response achieved when running Fluid A against two steels of different hardness and surface finish with two different paper based lining materials. The two steels are typical of those used in clutch systems today. The steels referred to as SAE 1035 are generated from a medium carbon steel of Rockwell hardness 24, tumbled to a 10-15 microinch AA (arithmetic average) surface finish. This process produces a hard smooth steel surface. The steels referred to as SAE 1010 are generated from a low carbon steel of Rockwell hardness 8, lapped to a 40 microinch AA surface finish. In contrast to the SAE 1035 steel described above, this process produces a soft rough steel mating surface. Both systems were run-in for 16 hours under a load of 127 psi at room temperature and a sliding speed of 50 ft/min. The friction characteristics of the systems were then evaluated. Both friction materials performed quite well with the SAE 1035 steel, giving somewhat similar results. The CFM-1 friction material gave lower friction coefficients at 34°C than the CFM-3 material and also showed an area of potential shuddering at sliding speeds from 10 to 70 ft/min at 34°C. With the SAE 1010 steel, quite different results were observed with the two friction materials. The CFM-1 material gave reasonable results with the possible exception of 34°C where again, potential for shuddering existed. The CFM-3 material, which is softer and less dense than the CFM-1 material, was totally consumed during the break-in process. The softer lining material was badly abraded by the rougher steel surface and subsequently gave the very poor curves shown. This example highlights the criticalness of matching the friction material to the steel mating surface to achieve good operation over a range of temperatures and to yield a system with good long term durability.

6.4.5 Effect of Friction Material

To illustrate the different frictional characteristics of various clutch lining materials, the seven friction materials shown in Table 6.4.2 were run under standard conditions using Fluid B. The full results of this testing are shown in Appendix 2. None of the lining materials tested produced the extremely high static friction values (0.17 to 0.25) which are sometimes observed with modified resin systems. Neither did they produce extremely low dynamic friction coefficients (0.07 to 0.10) which are observed with very dense materials of low porosity. Systems which exhibit very high static friction coefficients or very low dynamic friction coefficients are predisposed to give stick-slip or dynamic frictional vibration.

Table 6.4.3 shows the extremes of friction performance obtained with Fluid B and the seven friction materials. CFM-1, CFM-3 and CFM-4 gave the most diverse frictional behavior. CFM-1, the most dense material, gave by far the lowest dynamic friction coefficient. This is most likely due to trapping of oil between the friction material and steel surface. Since the dynamic friction coefficient was very low, the friction ratio (μ_s/μ_d) was quite high, 1.034. This high friction ratio would predict poor stick-slip performance. CFM-3, the material of highest porosity, gave a high dynamic friction coefficient, which in turn gave a low friction ratio (μ_s/μ_d). This low friction ratio would predict good stick-slip performance. The third friction material, CFM-4, gave very high static and dynamic friction coefficients which in turn produced a high friction ratio. This system would also be predicted to be susceptible to stick-slip behavior. Figure 6.4.5 shows the friction coefficient versus sliding speed curves for these three friction materials. As discussed, it appears that both CFM-1 and CFM-4 would be capable of giving stick-slip behavior at low sliding speeds when used with Fluid B.

Table 6.4.3: Effect of Friction Material, Fluid B*

Friction Material	μ_s	μ_d**	Friction Ratio
CFM-1	0.122	0.118	1.034
CFM-3	0.122	0.144	0.847
CFM-4	0.152	0.150	1.013

*   100°C, 140 psi
**   at 100 ft/min

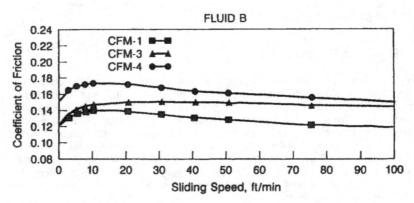

SAE 1035 Steel, 140 psi, 100°C

Figure 6.4.5: Effect of Lining Material, Fluid B

740

6.4.6 Effect of Load

For optimal control, a clutch system would have a constant coefficient of friction under conditions of increasing and decreasing thrust load. This is especially true for continuously slipping clutch devices. To quantify this aspect of performance, model clutch systems can be run on the LVFA under a range of loads and temperatures to determine their response to these changes under actual operating conditions. The thrust load responses of Fluids B, C, D and E were evaluated using CFM-3 run against an SAE 1035 steel disc. Thrust load was varied from 70 psi to 210 psi and friction torques recorded at 0, 40, 100 and 150 degrees C. The full results of this testing are included in Appendix 2.

Static friction coefficient was found to be quite insensitive to thrust load, but quite sensitive to temperature. The temperature sensitivity will be discussed in the next section.

Figures 6.4.6 and 6.4.7 show the dynamic friction coefficients obtained with this system at 0 and 40 degrees C at a sliding speed of 100 ft/min. The effect of changing load was much more pronounced at lower temperatures. In each case the fluids grouped into two types, those that had increasing friction coefficient with increasing load, Fluids C and D, and those with decreasing friction coefficients with increasing load, Fluids B and E. The reason for these two different phenomena is not entirely understood; however, it does not appear to be viscosity related since all of the fluids have quite similar viscosities at 0°C (see Table 6.4.1).

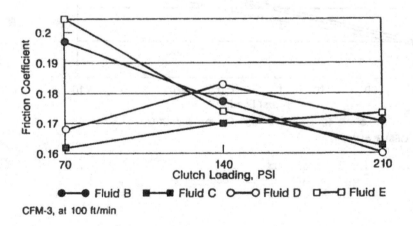

CFM-3, at 100 ft/min

Figure 6.4.6: Dynamic Friction Coefficient
Effect of Load at 0°C

741

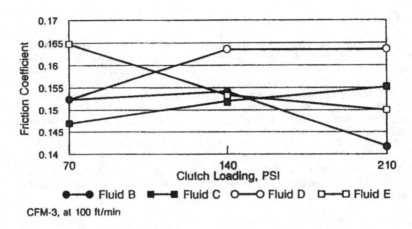

CFM-3, at 100 ft/min

Figure 6.4.7: Dynamic Friction Coefficient
Effect of Load at 40°C

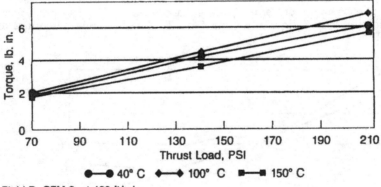

Fluid D, CFM-3, at 100 ft/min

Figure 6.4.8: The Effect of Load
Clutch Torque vs Thrust Load

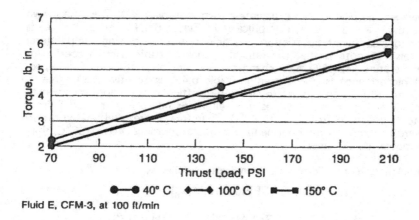

Fluid E, CFM-3, at 100 ft/min

Figure 6.4.9: The Effect of Load
Clutch Torque vs Thrust Load

Figures 6.4.8 and 6.4.9 show the dynamic friction torques obtained in the experiment described above, plotted against thrust load. In this case it is more informative to plot frictional torque rather than friction coefficient, since it is the torque output of the clutch that the driveline actually is subjected to. It is important for torque to vary linearly with thrust load. With Fluid D the slopes of the torque versus thrust load lines vary with temperature, causing them to diverge at high loads. This makes clutch control over a wide range of temperatures more difficult. The clutch is more sensitive to temperature changes at high thrust loads and as a result, the change of torque output with changes in thrust loads is not the same at different temperatures. Fluid E gives a more linear response of transmitted torque to load, and varies less with temperature than Fluid D. The low temperature dependence of Fluid E is evidenced by the tight grouping of the lines in Figure 6.4.9.

6.4.7 Effect of Temperature

Temperature is very critical to friction performance. Since friction modifiers adsorb and desorb from the surfaces of the clutch materials with temperature changes, the frictional performance of a clutch system would be expected to change noticeably with temperature (12). The same experimental data, which was described above in the discussion of loading effects, can be used to examine the effect of temperature on these systems.

Table 6.4.4 shows the variation of static breakaway friction coefficient with temperature. Since static breakaway friction coefficient varies very little with thrust load, the values obtained at all three load levels, 70, 140 and 210 psi, were averaged to obtain the coefficients shown in Table 6.4.4. The data shown in Table 6.4.4 are also shown graphically in Figure 6.4.10. All of the fluids exhibited decreasing static friction coefficient with increasing temperature. What is also of note is that the rate of change of static friction coefficient with temperature ($\delta\mu_s/\delta T$) also changed over the range studied. This aspect of fluid performance is also shown in Table 6.4.4 as Relative $\Delta\mu$/Degree C. In all of the fluids studied, static friction coefficient was more sensitive to temperature at higher temperatures, i. e. all fluids had a higher $\delta\mu_s/\delta T$ from 100°C to 150°C than from 0°C to 40°C. Therefore, more fluid related differences were measures in the intermediate temperature range, 40 – 100°C, than at the temperature extremes.

Table 6.4.4: Effect of Temperature on Static Breakaway**

FRICTION COEFFICIENT (μ_s)

Fluid	0°C	Rel $\Delta\mu$/°C*	40°C	Rel $\Delta\mu$/°C*	100°C	Rel $\Delta\mu$/°C*	150°C
B	0.135	0.20	0.127	0.08	0.122	0.42	0.101
C	0.136	0.15	0.130	0.33	0.110	0.40	0.090
D	0.152	0.27	0.141	0.53	0.109	0.54	0.082
E	0.160	0.17	0.153	0.55	0.120	0.42	0.099

*Relative $\Delta\mu$/Degree C $= \dfrac{|\mu_1 - \mu_2|}{|T_1 - T_2|} \times 1000$

** CFM-3, SAE 1035 Steel

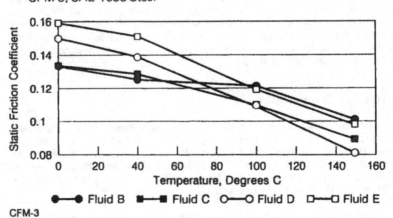

CFM-3

Figure 6.4.10: Static Breakaway Friction Coefficient
Effect of Temperature

Table 6.4.5 shows the variation of dynamic friction coefficient with temperature. This data is also shown graphically in Figure 6.4.11. As with static friction coefficient, dynamic friction coefficients also decrease with increasing temperature. However, contrary to the behavior of static friction coefficient, dynamic friction coefficient rate of change with temperature ($\delta\mu_d/\delta T$) is greater at low temperatures. This behavior may be in part accounted for by viscometric effects. This observation is consistent with the fact that Fluid E has the lowest change in viscosity over the temperature range studied and also the lowest change in dynamic friction coefficient.

Table 6.4.5: Effect of Temperature on Dynamic Friction**

FRICTION COEFFICIENT (μ_D)

Fluid	0°C	Rel Δμ/°C*	40°C	Rel Δμ/°C*	100°C	Rel Δμ/°C*	150°C
B	0.177	0.58	0.154	0.17	0.144	0.24	0.156
C	0.170	0.45	0.152	0.03	0.150	0.20	0.140
D	0.186	0.65	0.160	0.10	0.154	0.02	0.153
E	0.174	0.53	0.153	0.18	0.142	0.08	0.146

*Relative $\Delta\mu$/Degree C $= \dfrac{|\mu_1 - \mu_2|}{|T_1 - T_2|} \times 1000$

** CFM-3, SAE 1035 steel, 100 ft/min, 140 psi thrust load

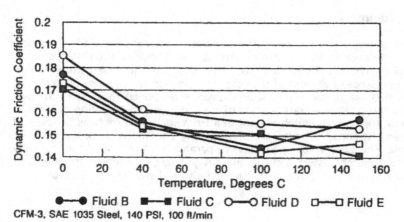

CFM-3, SAE 1035 Steel, 140 PSI, 100 ft/min

Figure 6.4.11: Dynamic Friction Coefficient
                Effect of Temperature

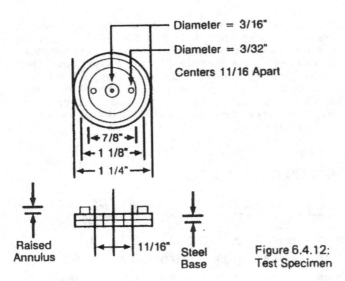

Diameter = 3/16"

Diameter = 3/32"

Centers 11/16 Apart

7/8"

1 1/8"

1 1/4"

Raised
Annulus

11/16"

Steel
Base

Figure 6.4.12:
Test Specimen

The rate of change of friction coefficient with temperature is a critical parameter for prevention or prediction of stick-slip behavior. If the two coefficients are changing at different rates, even though they are moving in the same direction, potential exists for the friction ratio (μ_s/μ_d) to change from a value of less than one to greater than one. The change of friction ratio to a value greater than one would move the system into an area of potential shuddering behavior.

6.4.8 Summary

Stick-slip behavior in clutch systems is undesirable. Occurrence of this phenomenon is dependent on the three materials that comprise the clutch system: the friction material, the mating steel surface and the lubricant present. Further, the presence or absence of stick-slip type phenomena depends on operating variables such as temperature, sliding speed and applied load. The data presented in this paper illustrate that all of these design and operating parameters are critical to having a clutch system free from stick-slip problems. In most of the systems studied, at some set of operating conditions, stick-slip behavior is possible. The Low Velocity Friction Apparatus is a very useful tool for fully evaluating a clutch system early in the design stage, or a lubricant early in its development, to insure that the system will be free of stick-slip type problems.

6.4.9 References

(1) Bowden, F.P.; Leben, L.; Tabor, D.: "The Sliding of Metals, Frictional Fluctuations, and Vibration of Moving Parts", Engineer **168**, 214 (1939).

(2) Blok, H.: "Fundamental Mechanical Aspects of Boundary Lubrication", SAE **46**, 54—68 (1940).

(3) Merchant, M.E.: "Characteristics of Typical Polar and Non-Polar Lubricant Additives Under Stick-Slip Conditions", Lubrication Eng., **2**, 56—61 (1946).

(4) Rogers, J.J.; Haviland, M.L.: "Friction of Transmission Clutch Materials as Affected by Fluids, Additives and Oxidation", SAE 194A (1960).

(5) Haviland, M.L.; Goodwin, M.C.; Rodgers, J.J.: "Friction Characteristics of Controlled-Slip Differential Lubricants", SAE 660778 (1966).

(6) Friesen, T.V.: "Chatter in Wet Brakes", SAE 831318 (1983).

(7) Sprague, S.R.; Cunningham, R.G.: "Frictional Characteristics of Lubricants", Ind. & Eng. Chem., **51**(9), 1047—1050 (1959).

(8) Hiramatsu, T.; Akagi, T.; Yoneda, H.: "Control Technology of Minimal Slip-Type Torque Converter Clutch", SAE 850460 (1985).

(9) Haviland, M.L.; Rodgers, J.J.: "Friction Characteristics of Automatic Transmission Fluids as Related to Transmission Operation", ASLE Preprint 60AM 6A-1 (1960).

(10) Haviland, M.L.; Goodwin, M.C.; Rodgers, J.J.: "New Apparatus Measures Friction Characteristics of CSD Lubricants", SAE Journal **75**(9), 66—71 (1967).

(11) Miyazaki, M.; Hoshino, M.: "Evaluation of Vibration-Preventive Properties of Lubricating Oils in Wet Friction Plates and Retention of Such Properties Using a Friction Tester", SAE 881674 (1988).

(12) Papayf, A.G.: "Oil-Soluble Friction Reducers-Theory and Application", Lubrication Eng., **39**(7), 419—26 (1983).

APPENDIX 1: STANDARD TEST PROCEDURE

Test Specimens:

Clutch friction material with a machined annulus of 1.125 inch O.D., 0.875 inch I.D., and mean diameter 1.00 inch was adhesively bonded to a steel backing disc by the manufacturer using a method representative of production processing (see Figure 6.4.3). SAE 1035 steel discs of 1.50 inch diameter were stamped from separator steel stock and tumble finished to 10-15 microinch A.A. surface roughness finish.

Test Procedure:

After ultrasonically cleaning the steel specimen and test machine fixtures in heptane, the specimens were assembled in the tester and surfaces broken-in for one hour at the given test load (70, 140 or 210 psi) at a sliding speed, from 0 — 100 ft per minute, were then recorded at each temperature. Friction was measured by a dead weight calibrated load cell. Heating of the fluid was accomplished by imbedded cartridge heaters. Coefficient of friction data reported were the average of the plots obtained during acceleration to 100 ft/min and then deceleration back to rest. Static friction was measured after deceleration and was the average of four measurements made by turning the drive shaft by hand.

Error:

Standard deviation for the static friction measurement is 0.006 and 0.004 for the dynamic friction measurement.

APPENDIX 2: TEST RESULTS

LVFA Static and Dynamic Friction Data with Different Fluid-Clutch Friction Material Combinations

Fluid	Clutch Friction Material	Thrust Load (PSI)	0 ft/min (Static)				100 ft/min (Dynamic)			
			0°C	40°C	100°C	150°C	0°C	40°C	100°C	150°C
B	CFM-3	70 140 210	.136 .142 .138	.122 .132 .127	.120 .122 .124	.106 .102 .095	.197 .177 .160	.152 .154 .142	.138 .144 .145	.150 .156 .152
C	CFM-3	70 140 210	.130 .134 .144	.134 .130 .127	.108 .106 .117	.086 .088 .096	.162 .170 .173	.147 .152 .155	.150 .150 .153	.139 .140 .135
D	CFM-3	70 140 210	.150 .154 .152	.140 .135 .148	.101 .108 .119	.084 .080 .083	.168 .182 .170	.152 .164 .164	.141 .155 .156	.134 .134 .136
E	CFM-3	70 140 210	.160 .159 .159	.156 .153 .151	.122 .118 .120	.098 .098 .100	.204 .174 .162	.165 .153 .150	.148 .142 .142	.146 .146 .143
B	CFM-1	140	-	.129	.122	.116	-	.116	.118	.146
B	CFM-2	140	-	.130	.123	.098	-	.118	.116	.144
B	CFM-4	140	-	.144	.152	.127	-	.140	.150	.167
B	CFM-5	140	-	.148	.143	.105	-	.150	.150	.161
B	CFM-6	140	-	.148	.131	.104	-	.143	.144	.158
B	CFM-7	140	-	.156	.132	.102	-	.142	.136	.156

749

7. Lubricant Influence on Ceramic and Seal Materials

7.1 Relation Between Surface Chemistry of Ceramics and Their Tribological Behaviour

7.2 Effect of Hydrogenation Degree of HSN on Various Lubricating Oil Additives Resistance

7.3 Electro-Chemical Investigation of Deposit Formations on Mechanical Seal Surfaces for Diesel Engine Coolant Pumps

7.1 Relation Between Surface Chemistry of Ceramics and Their Tribological Behaviour

T.E. Fischer and W.M. Mullins
Stevens Institute of Technology, Hoboken, USA

Abstract

Despite their chemical stability against chemical attack and corrosion over a wide temperature range, ceramics are chemically quite active when subjected to friction and wear; these tribochemical reactions influence the friction coefficient as well as wear mechanisms and wear rates. Oxide ceramics react differently to water vapor than covalently bonded materials. The latter experience accelerated oxidation that causes smooth surfaces and reduced wear by microfracture. The resultant oxide layers are often lubricious and permit friction coefficients as low as 0.1 under favorable circumstances. Oxide ceramics suffer chemisorption-induced embrittlement and increased wear rates in humid ambients. Depending on the strength of their interaction with water, which in turn depends on their electronic structure, oxides react differently with liquid water. Silicon oxide (formed on silicon nitride or carbide) dissolves tribochemically; alumina reacts to form lubricating stable hydroxides, zirconia remains inert. The interaction with hydrocarbons is less well known. Paraffins and other "inert" hydrocarbons form boundary lubricants on ceramics, but not on metals. Hydrocarbons also cause tribochemical attack of zirconia, where they increase the wear rate by an order of magnitude despite reducing friction coefficients from 0.7 to 0.1. These phenomena are discussed in terms of the surface chemistry of these materials.

7.1.1 Introduction

Early in the exploration of ceramics for tribological service, it was found that the chemical environment has a major influence on the friction of these materials in lubricated and unlubricated contact and on their wear resistance (1—3). An increase in atmospheric humidity can decrease the wear rate of one ceramic by as much as two orders of magnitude (2, 4) or increase the wear of another by similar amounts (5). Obviously, this behavior is related to the chemical processes. This interaction takes several forms, depending on the material, the environment and the mechanical parameters; it can consist in the formation of surface coatings that decrease wear (4), in a purely chemical form of wear (by dissolution in the liquid environment (6, 7)), in chemisorption and boundary lubrication effectiveness (4, 5), and in chemically induced cracking that increases wear rates (5). The rates of these chemical reactions are very much influenced by simultaneous friction; because of this interaction of friction and chemistry, these reactions are called "tribochemical" (8, 9). They have been observed in all materials subjected to friction (10–17), but they are particularly pronounced in the tribological behavior of ceramics; their understanding is a prerequisite for the successful application of these materials in tribological service and can form a rational basis for the development of synthetic lubricants or lubricant additives suited for them.

In what follows, we shall describe the observed chemical effects on the friction and wear of ceramics; in doing so, we shall consider the effects of humidity and water as well as those of some model hydrocarbon lubricants. We shall then review broadly the state of knowledge of the physical chemistry of ceramic surfaces, and finally, we shall, where it is already possible, explain the former in view of the latter.

7.1.2 Chemical Effects of Water on Ceramic Tribology

7.1.2.1 Tribochemical Reaction of Ceramics with Water

It has been discovered quite early that the ambient humidity has a pronounced effect on the wear of silicon nitride and other ceramics (1–3). A more detailed look at this phenomenon (4) revealed that not only the amount of wear, but the wear mechanism itself is modified by humidity. Silicon nitride, for example, wears rapidly when sliding in dry argon; if the environment contains various amounts of water vapor, the wear rate decreases by as much as two orders of magnitude (Fig. 7.1.1). The minimum wear rate is reached at 90 % relative humidity in argon and 50 % relative humidity in air. Under these conditions, the wear scar is much smoother than after sliding in dry gases and it is covered with an amorphous silicon oxide which is probably strongly hydrated. It is known that the oxidation rate of this material is increased up to one thousandfold by the presence of humidity in the air (18). However, oxidation of silicon nitride proceeds measurably only above 1,000 K (19). Obviously, the rate of oxidation of this material is accelerated by the simultaneous action of friction. How this occurs exactly has not been determined yet. In the wear of steel, we have been able to identify the tribochemical mechanism (16, 17): the change in the oxidation kinetics of steel from a logarithmic to parabolic law during friction and the increase in its rate by several orders of magnitude results from a large increase in the density of diffusion paths caused by the extensive plastic deformation due to wear. In the case of silicon nitride (4), the experimental evidence is not as easy to interpret and introduction of diffusion paths by plastic deformation is unlikely, since wear occurs mostly by microfracture (4). One can speculate that the reaction is accelerated because the hydroxide formed on the surface is continuously removed and fresh surface is exposed by friction, but clear experimental evidence for any one mechanism is still lacking.

Humidity Effect on Wear Rate (Si_3N_4/Si_3N_4 , Load = 1kg, U = 1mm/sec 20°

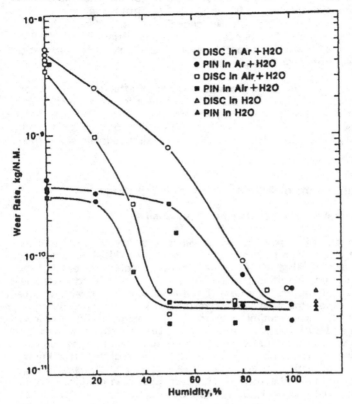

Figure 7.1.1: Wear Rates of Silicon Nitride Sliding at Low Speed in Inert (Dry Argon) and in Chemically Active Environments (Humid Argon and Air)

The friction coefficient of silicon nitride is very nearly the same in dry and in humid environments (4). Thus the total mechanical forces acting on the contacting surfaces are the same. Why then is wear not increasing but decreased when chemical attack is added to the friction force? The answer lies in the different topographies of the surfaces: in humid environment, the tribochemical oxidation occurs at the asperities, consequently, the contacting surfaces are smoother and the load and friction forces are distributed over wider areas than in the absence of tribochemistry. In other words, the local stresses responsible for micro-cracking are reduced by the tribochemical oxidation (20).

When silicon nitride is sliding in water at room temperature, wear produces ultraflat surfaces by molecular dissolution of the material (6). The dissolution occurs by the oxidation of the materials to form water soluble silicic acid (21). The tribochemical nature of this reaction is responsible for the very smooth surfaces obtained: the reaction occurs only at contacting asperities. The smoothness of the surfaces is such that hydrodynamic lubrication (with attendant near-zero friction) is obtained when the silicon nitride surfaces slide in water at a velocity of only 6 cm/s despite an average bearing load pressure of 30 MPa. We estimate that the hydrodynamic water film, and with it the roughness of the surfaces it separates, has a thickness of less than 10 nanometers.

Similar experiments with silicon carbide also produce dissolution of the material, but hydrodynamic lubrication is not obtained because the material is mechanically weaker than silicon nitride and suffers local fracture as well as dissolution.

7.1.2.2 Formation of Lubricious Oxides

At elevated temperature, sliding in humid air reduces friction as well as wear in a limited range of load and sliding velocities (22) (Fig. 7.1.2). A combination of scanning Auger spectroscopy and scanning electron microscopy revealed that the surfaces are covered by a very smooth layer of silicon oxide. As the severity of sliding (i.e. the product of load and sliding velocity) is increased beyond a certain value, friction and wear are high and the wear surfaces are rough. Apparently, the passage from low to high friction is the result of a competition between the kinetics of formation of the lubricious oxide layer and its wear.

Lubricious oxides have recently been formed on silicon nitride by preoxidation (23). In order to achieve low friction (friction coefficients as low as $\mu = 0.05$ have been obtained), it is necessary to prepare a very smooth and flat surface to avoid friction and wear by ploughing. This is achieved by friction in water (7) as described above. Subsequent oxidation in air for a few minutes produces a surface which presents a low friction coefficient.

Gates, Hsu and Klaus (24) have recently shown that sliding in water causes a decrease of the friction coefficient of alumina against itself from $\mu = 0.6$ to $\mu = 0.25$. (The friction coefficient of silicon nitride is not lowered by water when the speeds are so low that hydrodynamic lubrication is not operative). This lowering of friction is attributed by the authors to the formation of stable aluminum hydroxides which are modified to a layered structure (i.e. trihydroxide-bayerite) by the frictional shear stresses.

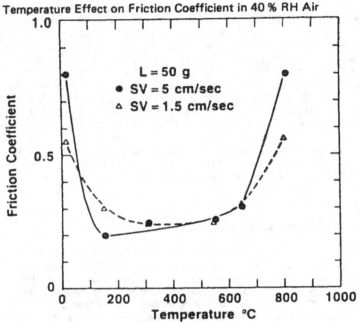

Temperature Effect on Friction Coefficient in 40 % RH Air

L = 50 g
● SV = 5 cm/sec
△ SV = 1.5 cm/sec

Figure 7.1.2: Friction Coefficients of Silicon Nitride Sliding at Moderate Speeds and Loads in Humid Air as a Function of Temperature: Lubrication by Tribochemical Oxide

7.1.2.3 Chemically Induced Fracture in Oxide Ceramics

Oxide ceramics, such as alumina and zirconia, are obviously incapable of undergoing the oxidation reactions that are responsible for the decrease of wear in silicon nitride. This different chemistry manifests itself in friction by a difference in its reaction to the environment and to lubrications: Wallbridge, Dowson and Roberts (25) have reported that wear of alumina sliding in water is higher than in air. With zirconia (5), water increases the wear rate by an order of magnitude over wear in dry nitrogen (Fig. 7.1.3). Even the humidity of room air causes an increase in the wear rate that is almost as large. This increase in wear occurs by intergranular fracture; we are in the presence of chemisorption embrittlement. According to Widerhorn and Michalske and their coworkers (26, 27), this phenomenon occurs by the attach of the bonds between neighboring metal and oxygen ions at a crack tip by water; it is related to the well-known tendency of oxide ceramics to form hydroxylic surfaces (21).

Wear of ZrO$_2$ in Various Lubricants

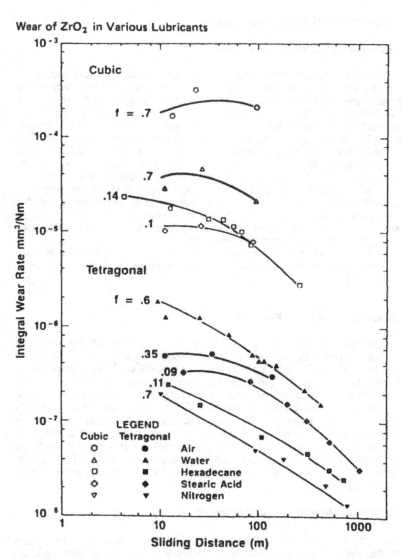

Figure 7.1.3: Friction and Wear of Brittle (Cubic) and Tough (Tetragonal) Zirconium Oxide Doped with Yttria in Various Environments. The Friction Coefficients f are Labelled Next to the Curves

757

7.1.3 Interaction with Hydrocarbons

7.1.3.1 Boundary Lubrication by Paraffins

It is well known that polar molecules such as fatty acids, alcohols and esters adsorb strongly enough onto metallic surfaces to form a coherent monolayer that decreases the adhesion between the rubbing surfaces and produces a friction coefficient $\mu = 0.1$ at low velocities where hydrodynamic lift is not capable of separating the surface. This phenomenon is known as "boundary lubrication" (28). In the case of metals, nonpolar hydrocarbons such as paraffins and the molecules of lubricant basestocks do not act as boundary lubricants: at low enough velocities ($v = 1$ mm/s), the friction coefficient is as high as $\mu = 0.6$ in the presence of these fluids. In the case of all ceramics known to us, paraffins act is boundary lubricants, with friction coefficients in the neighborhood of $\mu = 0.12$ (Fig. 7.1.3). We do not have a proven explanation for this behavior; it may be related to the fact that silicia and alumina powders are used as catalysts for cracking and isomerization of hydrocarbons in the petroleum industry (29). It is to be expected that the acid sites on the surface of ceramics, strong enough to break carbon-carbon bonds at elevated temperatures, are capable of adsorbing the molecules at room temperature. This phenomenon is worthy of further study.

7.1.3.2 Adsorption—Induced Fracture

Adsorption-induced fracture occurs also in the presence of hydrocarbon lubricants. Pure paraffin decreases friction of zirconia to $\mu = 0.11$ as described above. With metals, such a decrease in friction decreases wear by several orders of magnitude. In the case of zirconia (5), paraffin causes an increase in the wear rate of about 50 % over the wear rate in dry nitrogen despite the large reduction in friction from 0.7 to 0.1. When sliding occurs in paraffin with 0.5 % stearic acid, which is a classic boundary lubricant, the friction coefficient decreases further to 0.09 but wear increases by another factor of 3. Electron microscopy of the wear scar reveals chemical attack of the grain boundaries and intergranular fracture as the cause of this increase in wear. In fact, attack of the material by the stearic acid is also observed, but much less pronounced, on the unworn surface.

7.1.4 Chemical Properties of Ceramics

Ceramics are characterized by a large gap between their valence band (which constitues the Highest Occupied Molecular Orbitals, HOMO) and their conduction band (or Lowest Unoccupied Molecular Orbital, LUMO). For this reason, chemical interaction with the environment is dominated by electron transfer, and most ceramics are best considered as solid acids or bases according to Lewis acid-base theory. We recall that a Lewis base donates electronic charge and a Lewis acid accepts electronic charge in the formation of the product or adduct.

758

7.1.4.1 MgO, Al$_2$O$_3$, SiO$_2$

These three base oxides have similar electronic structure (30–34), consisting of a split valence band; the lower band is in the nature of atomic 2s orbitals of oxygen, the upper band consists of 2p oxygen orbitals with very little cationic characteristics. The conduction band is nearly entirely cationic 3s and 3p in nature. The band gaps of these materials are among the largest measured, from 7.5 to 9.5 eV (35). These characteristics classify these oxides as very hard acids and bases.

The elctronic structure of water is very similar to that of these oxides; it has valence and conduction band structure with a 7.5 to 8.0 eV energy gap (36). The reactions of water with these surfaces occurs with sufficient charge transfer (37) to induce the dissociation of adsorbed water molecules into protons and hydroxyls. The surface charge of the oxide attracts the oppositely charged ions in solution, which forms a diffuse space charge region around the surface (38). All of these oxides from stable hydroxides and have pH-dependent solubilities in water which range from 10^{-3} for silica and 10^{-4} for alumina.

Reactions with highly polar functional groups such as —OH, —COOH and —NH$_2$, which tend to dissociate or form polar "hydrogen" bonds (39–42). These reactions are similar to those one would expect to find with water; as a general rule of thumb, if water will dissolve in the compounds, they will tend to wet and spread onto the oxides.

The surfaces of these oxides are known to catalyze many ring opening reactions when the ring is small and highly polar (29). The general tendencies of alumino-silicates to catalyze alkene production, cracking and isomerizations are attributed to the products of these reactions.

7.1.4.2 SiC, Si$_3$N$_4$

Silicon carbide and silicon nitride are narrow band gap materials (E_g = 2.8 eV for SiC and 0.5 eV for Si$_3$N$_4$ (35). They can be considered as semiconductors. In addition, both materials have strong tendencies towards nonstoichiometry, which further reduces the effective band gap.

Both of these compounds react slowly with air or water (18, 19) to form the more chemically stable oxynitride or oxide surface coatings, which have the chemical properties mentioned above. Also, oxides are commonly added as tougheners and sintering aides to bulk silicon nitride and carbide, which makes the grain boundary oxide phases even more prevalent. Consequently these materials have surface chemistries in air and solubilities in water that are nearly identical to that of silica.

7.1.4.3 ZrO$_2$

The band gap of zirconia is 0.5 eV, much smaller than that of alumina and water, but the structure of the bands is expected to be similar. Because of this, ZrO$_2$ is a much softer acid base than alumina or the oxide formations on Si$_3$N$_4$. The aqueous surface chemistry of zirconia is very well known (43, 44); ZrO$_2$ is considered to be very weakly acidic in water. Measurements are complicated by the specific adsorption of softer complex anions onto the surface, such as NO$_3^-$ and ClO$_4^-$, which produce considerable surface charge double layers even though no strong reaction with water has taken place.

Surface reactions with water tend to be simple dissociation of the oxide to form Zr^{4+} and ZrO^{2+} in solution. As the concentrations of these increase to the order of 10^{-8}, complexations to $(Zr_4(OH)_8(H_2O)_{16})^{8+}$ and eventually recrystallization of the oxide occurs. No zirconium hydroxides are known to occur. These reactions are typical of the weak interaction with water and would seem to preclude any possibility for environmentally stimulated fracture, the proposed wear mechanism for ZrO$_2$ in water. But wear testing (5) indicates some form of interaction between water and the ZrO$_2$ that is not presently well understood.

Since pure zirconia undergoes several destructive phase transformations during cooling from high temperature, structural zirconia is often doped with several percent yttria to stabilize the high temperature phase. A fraction of this stabilizer segregates to the grain boundaries of the material. It is possible that water reacts with the Y$_2$O$_3$ stabilizer, which has a slightly wider band gap (5.6 eV) and is much more reactive with water than ZrO$_2$. Many times, also, second-phase small particles are used to further toughen the material. These particles are often alumina or yttria precipitates, these can react with water to the detriment of the mechanical properties of the zirconia.

Unlike the aqueous chemistry, the organometallic chemistry of zirconia is very rich (42, 43), with zirconium compounds acting as Ziegler-Natta catalysts for polymerization and causing olefin hydrozirconation reactions which graft carbonyls onto alkenes to produce carboxylic acids. In tribological conditions, the potential exists for similar types of oxidation and catalysis reactions of organic lubricants with the surfaces or with wear debris to form complex oxidized oligomers. The wear behavior of zirconia in hydrocarbon lubricants reflects this reactivity.

7.1.5 Discussion

We have seen that the surface chemistry of ceramics plays a decisive role in their tribological behavior. In most cases, the chemical influence of the environment of lubricant is tribochemical in that the chemical reaction or attack is strongly influenced by the mechanical phenomena that occur simultaneously. It appears that the chemical interaction takes four major forms, depending on the reactivity of the ceramic toward the environment.

The weakest interaction (apart from total inertness) is simple adsorption. If the adsorbed molecule is large enough, it can provide boundary lubrication. We have seen that it is a particularity of ceramics that they absorb relatively inert hydrocarbons such as paraffins strongly enough for this boundary lubrication to occur. This is not observed in the case of metals.

At the other extreme is strong reaction or dissolution. In that case the environment attacks or corrodes the material generally. This attack can be accelerated by friction. We have encountered this example in the case of lubrication of zirconia by stearic acid: attack on the grain boundaries leads to loss of individual grains everywhere on the material, but this loss is greatly magnified in the wear scar.

Two different forms of tribochemical interaction occur at intermediate reactivities. The interaction of water with silicon nitride, alumina and zirconia will serve to illustrate this phenomenon. Silicon oxide is relatively the most reactive ceramic to water. This reactivity is not sufficient to cause general dissolution at room temperature. However, the tribochemical activation of the oxidation of silicon nitride to silica and of the subsequent dissolution causes removal of the silicon nitride where rubbing occurs. OH^- radicals interact strongly enough with silicon to generate a soluble silicic acid. The interaction of water with alumina is weaker and causes two different tribochemical interactions: the formation of stable hydroxides on the surface of alumina which cause a reduction in friction, and an increase in the wear rate which probably takes the form of stress corrosion cracking. The weakest interaction occurs with zirconia. In this case the reaction occurs only where it is mechanically stimulated by large mechanical stresses. This direct mechanical activation occurs only at crack tips. Even in this case, the reaction does not result in tribochemical dissolution as in the case of silicon nitride, but merely weakens the Zr—O bond and lowers the threshold for propagation. We note, of course, that this adsoprtion stimulates fracture is well known for silica and glass that are subjected only to mechnical stresses in the absence of friction. Silica thus presents the scientifically interesting problem of being susceptible to tribochemical dissolution as well as of adsorption—induced fracture. During friction, dissolution produces smooth surfaces that distribute the contact and friction forces and decrease the stresses, thus decrease wear by fracture. This is exactly the opposite effect from adsorption-induced fracture. It would be interesting to find out which of these two mechanisms predominates, i.e., whether water increases or decreases the wear rate. This has not been investigated to date. For alumina and zirconia, the situation is simple: neither of these materials is dissolved in water and adsorption-induced fracture dominates and causes the observed increases in wear rate. Let us, however recall that zirconia may be an anomaly: its reactivity towards water is so low that even adsorption-induced fracture is surprising on theoretical grounds. It remains to be verified whether zirconia or a grain-boundary phase is responsible for the observed embrittlement.

In the case of hydrocarbon lubricants, our knowledge is much more anecdotal and our understanding is very limited. Yet precisely the tribochemical effects of hydrocarbon lubricants are the phenomena that require the most pressing development since they determine the performance of ceramic elements in machinery that is lubricated by conventional oils. A knowledge of ceramic tribochemistry with hydrocarbons is also necessary as a theoretical basis for the development of synthetic lubricants for these materials, especially for high-temperature service. Our measurements with zirconia have shown the importance of these phenomena: certain hydrocarbons increase its wear rate over that of unlubricated sliding despite the fact that they provide low friction.

7.1.6 References

(1) Tabor, D.: Proc. Roy. Soc. A251 (1959) 378.
(2) Shimura, H.; Tsuya, Y.: Effects of atmosphere on the wear rate of some ceramics and cermets. Wear of Materials 1977, ed. V. Ludema, ASME, New York, p. 452.
(3) Suzuki, K.; Sugita, T.: Characteristics of Magnesium Oxide Single Crystals Polished with Vibrational Sliding in Water. Wear of Material, 1981, S.K. Rhee, A.S. Ruff and K.C. Ludema (eds.) ASME, New York, p. 518.
(4) Fischer, T.E.; Tomizawa, H.: Interaction of Microfracture and Tribochemistry in the Friction and Wear of Silicon Nitride. Wear, 106 (1985) 29.
(5) Fischer, T.E.; Anderson, M.P.; Jahanmir, S.; Salher, R.: Friction and Wear of Tough and Brittle Zirconia in Nitrogen, Air, Water, Hexadecane and Hexadecane containing Stearic Acid. War Materials 1987, K. Ludema, ed. ASME, New York, p. 257—266, also: Wear (1988) 133—148.
(6) Sugita, T.; Ueda, K.; Kanemura, Y.: Material-Removing Mechanism of Silicon Nitride During Rubbing in Water. Wear 97 (1984) 1—8.
(7) Tomizawa, H.; Fischer, T.E.: Friction and Wear of Silicon Nitride and Silicon Carbide in Water: Hydrodynamic Lubrication at Low Sliding Speed Obtained by Tribochemical Wear. ASLE Transactions 30 (1987) 41.
(8) Jahanmir, S.; Fischer, T.E.: Friction and Wear of Silicon Nitride Sliding in Air, Water, Hexadecane and Hexadecane Containing Stearic Acid. STLE Trans. 31 (1988) 32.
(9) Heinicke, G.: 1984. Tribochemistry, Carl Hanser Verlag, Munich.
(10) Thiessen, P.A.; Meyer, K.; Heinicke, G.: "Grundlagen der Tribochemie", Akademie Verlag, Berlin (1967), p. 15.
(11) Sakurai, T.; Kato, K.: Study of Corrosivity and Correlation Between Chemical Reactivity and Load-Carrying Capacity of Oils Containing Extreme Pressure Agents. ASLE Trans. 9 (1966) 77.
(12) Sakurai, T.: Role of Chemistry in the Lubrication of Concentrated Contacts. J. Lubr. Technol. 103 (1981) 473—485.
(13) Habeeb, J.J.; Stover, W.H.: The Role of Hydroperoxides in Engineer Wear and the Effect of Zinc Dialkyldithiophosphates. ASLE Trans. 30 (1987) 419—26.
(14) Willermette, P.A.; Kandah, S.K.; Siegel, W.P.; Chase, R.E.: Influence of Molecular Oxygen on Wear Protection by Surface-Active Compounds. ASLE Trans 26 (1983) 523.
(15) Spedding, H.; Watkins, R.C.: Antiwear Mechanisms of ZDDP. Tribology 15 (1982) 9, ibid. 15 (1983) 15.

(16) Quinn, T.F.: Theory of Oxidative Wear, Tribology Intl. 16 (1983) 257, ibid. 16 (1983) 305.

(17) Sexton, M.D.; Fischer, T.E.: The Mild Wear of 52100 Steel. Wear 96 (1084) 17—30. Fischer, T.E.; Sexton, M.D.: The Tribochemistry of Oxidative Wear. In: Physical Chemistry of the Solid State: Applications to Metals and their Compounds, P. Lacombe, ed., Elsevier (1984) p. 97.

(18) Singhal, S.C.: Effect of Water Vapor on the Oxidation of Hot-Pressed Silicon Nitride and Silicon Carbide. J. Am. Ceram. Soc. 59 (1976) 81.

(19) Kiehle, A.J.; Heung, L.K.; Gielisse, P.J.; Rockett, T.J.: Oxidation Behaviour of Hot Pressed Si_3N_4. J. Am. Ceramic Soc. 58 (1975) 17.

(20) Fischer, T.E.: Tribochemistry Ann. Rev. Mater. Sci. 18 (1988) 303.

(21) Iler, R.K.: The Chemistry of Silica, John Wiley & Sons, New York (1979).

(22) Tomizawa, H.; Fischer, T.E.: Friction and Wear of Silicon Nitride at 150 to 850°C. ASLE Transactions 29 (1986) 481—488.

(23) Fischer, T.E.; Liang, H.; Mullins, W.M.: Tribochemical Lubricious Oxides in Silicon Nitride. Mat. Res. Soc. Symp. Proc. 140 (1989) 339.

(24) Gates, R.S.; Hsu, S.M.; Klaus, E.E.: Tribochemical Mechanism of Alumina with Water. Tribology Transactions 32 (3) (1989) 357—363.

(25) Wallbridge, N.; Dowson, D.; Roberts, E.W.: The Wear Characteristics of Sliding Pairs of High-Density Polycrystalline Aluminium Oxide Under Both Dry and Wet Conditions. Wear of Materials 1983, K.C. Ludema ed. ASME, New York.

(26) Wiederhorn, S.M.; Freiman, S.W.; Fuller, E.R.; Simmons, C.J.: Effects of Water and Other Dielectrics on Crack Growth. J. Materials Sci. 17 (1982) 3450—78.

(27) Michalske, T.A.; Bunker, B.C.: Slow Fracture Model Based on Strained Silicate Structures. J. Appl. Phys 56 (1984) 2686—93.

(28) Bowden, F.P.; Tabor, D.: Friction and Lubrication of Solids. Clarendon Press, Oxford (1964).

(29) Knozinger, H.; Ratnasamy, P.: Catalytic Aluminas: Surface Models and Characterization of Surface Sites. Catal. Rev.-Sci. Eng. 17 (1978) 31.

(30) Tossel, J.A.: Electronic Structures of Mg, Al and Si in Octahedral Coordination with Oxygen From SCF X-Alpha Molecular Orbital Calculations. J. Phys. Chem. Sol. 36 (1975) 1273—80.

(31) Tossel, J.A.: Electronic Structures of Silicon, Aluminium and Magnesium in Tetrahedral Coordination with Oxygen From SCF-X-Alpha Calculations. J. Am. Chem. Soc. 19 (1975) 3840—4.

(32) Ciraci, S.; Batra, I.P.: Electronic Structure of Alpha-Alumina and Its Defect States. Phys. Rev. B 28 (1983) 982—92.

(33) Batra, I.P. Electronic Structure of Alpha Al_2O_3. Solid State Phys. 15 (1982) 5399—5410.

(34) Ching, W.Y.: Theory of Amorphous SiO_2 and SiO_x. Phys. Rev. B. 26 (1982) 6610—21.

(35) Strehlow, W.H.; Cook, E.L.: Compilation of Compound Semiconductors and Insulators. J. Phys. Chem. Ref. Data 2(1) (1973) 163—99.

(36) Pitzner, R.M.; Merfield, D.P.: Minimum Basis Wavefunctions for Water. J. Chem. Phys. 52(9) (1970) 4782—7.

(37) Mullins, W.M.: The Effect of Fermi Energy on Reaction for Water with Oxide Surfaces. Surface Science 217 (1989) 459—467.

(38) Onoda Jr., G.Y.; Casey, J.A.: Surface Chemistry of Oxides in Water. Ultrastructures Process. Ceram. Glasses. eds. L.L. Hench, D.R. Ulrich, Wiley, NY (1984).

(39) Tormey, E.S.: The Adsorption of Glyceryl Esters at the Alumina/Toluene Interface. PhD. Thesis, Massachussetts Institute of Technology (1982).

(40) Bolger, J.C.: Acid Base Interactions Between Oxide Surfaces and Polar Organic Compounds. J. Electrochem. Soc. 128 (1981) 82.

(41) Bolger, J.C.; Michaels, A.S.: Interface Conversion for Polymer Coatings, eds. P. Weiss, P.G. Cheever, American Elsevier (1968).

(42) Fowkes, F.W.: Characterization of Solid Surfaces by Wet Chemical Techniques. Industrial Applications of Surface Analysis, ACS Symposium Series, 199 (1982).

(43) Ray, K.S.; Kahn, S.: Electrical Double Layer at Zirconium Oxide-Solution Interface. Indian J. of Chem. 13, 1975, 577.

(44) Regazzoni, A.E.; Blesa, M.A.; Maroto, A.G.J.: Interfacial Properties of Zirconium Dioxide and Magnetite in Water. J. Colloid Interface Sci. 91, 1983, 560.

7.2 Effect of Hydrogenation Degree of HSN on Various Lubricating Oil Additives Resistance

M. Oyama, H. Shimoda, H. Sakakida and T. Nakagawa, Nippon Zeon Co. Ltd., Tokyo, Japan

It has been reported that HSN exhibits excellent resistance to various lubricating oils such as engine oil, power steering oil and gear oil. However, the effects of hydrogenation degree of HSN on the resistance to various lubricating oil additives are not studied.

In this paper, the effects of hydrogenation degree (unsaturation degree) of HSN on various lubricating oils and lubricating oil additives were presented.

HSN exhibits excellent resistance to lubricating oils. Especially, the fully saturated NBR is little influenced by any of the commercial oils tested.

The lower the iodine value of HSN, the less became the deterioration of HSN by anti-wear agents, detergents, antioxidants and extreme pressure additives. No HSN is affected by dispersants and viscosity index improvers.

The N, Mg, Ca, P, Cl and S elements of the lubricating oils were measured. The S and P content in the lubricating oils affects the resistance of partially saturated NBR to them. However, a fully saturated NBR exhibits good resistance to these oils.

7.2.1 Introduction

Highly saturated Nitrile Elastomer (HSN) is a new oil resistant rubber produced by selective hydrogenation of NBR.

Since our first introduction of HSN in 1984 to the rubber industry, it has found many applications in synchronous belts, various O-rings, static and dynamic seals, gaskets, fuel and oil hoses, diaphragms, etc., because of its excellent properties.

It is reported that HSN exhibits better resistance than NBR to heat, lubricating oil additives, steam and hydrogen sulfide, and has superior physical properties.

The classification system of SAE J200 is intended for use with rubber products for automotive applications. The typical HSN compound can be classified as type D and class H (DH), indicating the heat resistance of 150°C and less than 30 % volume swell in ASTM # 3 oil as shown in Figure 7.2.1. The classification of HSN, however, can cover the heat resistance ranges of class H to class K of 30 to 10 % maximum swell by selecting the appropriate hydrogenation levels and acrylonitrile contents, and by using various compounding techniques.

The temperature in the engine compartment has become higher in recent years due to the use of high powered engines and reduction of space weight. As a result, rubber parts have come to require better heat resistant properties.

In addition, new lubricating oils have been developed for longer maintenance-free life and higher efficiency. The rubber parts for oil delivery system require the higher resistance to lubricating oil additives and oxidized (sour) oil which will aggressively deteriorate the elastomer at high temperatures.

It is reported that HSN shows excellent heat, oil additive and oxidized oil resistances (1—5). In this paper, the effects of unsaturation degree of HSN on various lubricating oils and lubricating oil additives will be presented.

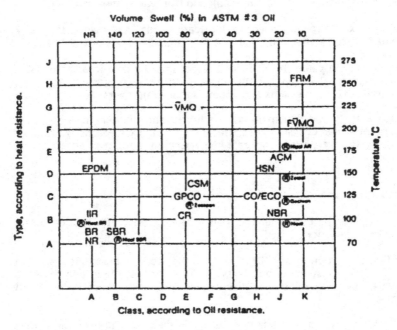

Figure 7.2.1: Classification of HSN by SAE J200

7.2.2 Experimental

7.2.2.1 Test Samples and Physical Tests

Zetpol 2000L, 2010L, 2020L, and 2030L listed in Table 7.2.1 were tested to evaluate their resistances to lubricating oils and additives resistances. Regular NBR N–31 was used for comparison with HSN.

Table 7.2.1: Commercial grade of Zetpol

Grade	Bound ACN %	Iodine Value (g/100g)	Mooney Viscosity (ML$_{1+4}$ 100 C/212 F)
Zetpol 1020	44	25	78
Zetpol 1010	44	10	85
Zetpol 2030L	36	58	57.5
Zetpol 2020	36	28	78
Zetpol 2020L	36	28	57.5
Zetpol 2010	36	11	85
Zetpol 2010L	36	11	57.5
Zetpol 2000	36	4	85
Zetpol 2000L	36	4	65

The formulations and the original properties are shown in Table 7.2.2. The physical tests were made according to JIS K6301, in which dumbbell pattern 3 was used for the original properties and immersion testing.

7.2.2.2 Immersion Test for Commercial Lubricating Oils

Table 7.2.3 shows the 10 types of commercial oils used for the immersion tests made at 150°C for hours.

7.2.2.3 Lubricating Oil Additive Resistance

The effects on various lubricating oil additives were evaluated by adding the additives at the concentrations shown in Table 7.2.4 to ASTM #2 oil at 150°C for 240 hours.

Table 7.2.2:   Test Formula & Original Properties

	HSN Zetpol 2000L	HSN Zetpol 2010L	HSN Zetpol 2020L	HSN Zetpol 2020L	HSN Zetpol 2030L	NBR Nipol N—31
Polymer	100	100	100	100	100	100
Zinc Oxide	–	–	–	5	5	5
Stearic Acid	–	–	–	1	1	1
FEF (N-550)	40	40	40	40	40	40
Peroximon						
F-40	6	6	6	–	–	–
TMPT	2	2	2	–	–	–
Sulfur	–	–	–	0.5	0.5	0.5
Acc. TMPT	–	–	–	2	2	2
Acc. MBT	–	–	–	0.5	0.5	–
Acc. CBS	–	–	–	–	–	1.0
ZnMBI	1	1	1	–	–	–
Agerite						
Resin D	1	1	1	–	–	–
IPPD	–	–	–	1	1	1
Permanax OD	–	–	–	1	1	1
Total	150	150	150	151	150	151.5
Curing Condition						
Temp. (°C)	170	170	170	160	160	160
Time (min)	25	25	25	20	20	20
Original Properties						
Tensile (MPa)	24.9	24.8	24.9	25.1	25.1	21.5
Elongation(%)	470	440	340	500	420	420
Modulus						
100 % (MPa)	1.6	1.7	1.8	1.7	1.9	1.5
Hardness						
(Shore A)	70	70	70	71	71	67

Table 7.2.3: List of Commercial Lubricating Oils

Nisseki	Exxon
1. PAN-XX 10W-30 (SG)	2. SUPER FLO Z 5W-50 (SF/CD)
3. Hidiesel S-3 low (CD)	4. D-3 10W-30 (CD)
5. Super Hiland 46	6. Unipower SQ 46
7. Pan Automatic D-2	8. Automatic Fluid
9. Gearlube EHD-90	10. Gear oil GX 80W-90

Table 7.2.4 List of Oil Additives and its Concentration used

Type of additives	Main chemical composition	Concentration*
Dispersant	Polyalkenyl succinimide	10
	Polyalkenyl succinic ester	10
Detergent	Calcium sulfonate	10
	Calcium phenate	10
Antioxidant	Dialkyl Zinc dithiophosphate	5
Viscosity index improver	Polyalkyl methacrylate	10
Extreme Pressure additives	Dialkyl phosphoric ester	10
	Sulfurized olefines	10
Additive Package	Package for SE oil	10
	Package for SF oil	10
	Package for SG oil	10
	Package for GL—6	10
	Package for Dexron II	10

*: g/100 ml of ASTM #2 oil

7.2.2.4 The Determinations of N, Mg, Ca, P, Cl, and S Elements in the Lubricating Oils Tested

The S and P contents of oils were measured by ion-exchange chromatography. The N contents of oils were measured by the Kjeldahl-Nessler method. On the other hand, the Mg and Ca content was measured by atomic absorption spectrophotometry.

7.2.3 Results and Discussions

7.2.3.1 Resistance of Lubricating Oils

The resistances of NBR N–31, Zetpol 2020L and Zetpol 2000L to commercial lubricating oils such as engine oil, diesel engine oil, hydraulic fluid, auto-transmission fluid, gear oil were evaluated at 150 °C for 240 hours. The changes in the elongation at break (EB) and tensile strength (TB) of N–31, Zetpol 2020L and Zetpol 2000L are shown in Figures 7.2.2 to 7.2.4, respectively. N–31 was deteriorated by these oils and fluids. But Zetpol showed excellent resistance to these oils and fluids. Especially, Zetpol 2000L showed the excellent resistance. It seems that the difference in these resistances between Zetpol and NBR is attributable to the unsaturation degrees of these elastomers.

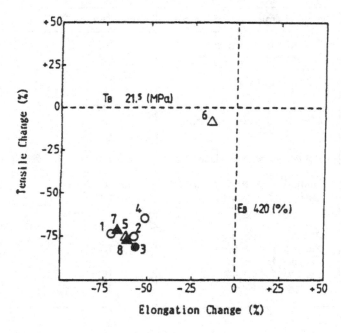

Figure 7.2.2: Lubricating Oil Resistance of N–31
○: Engine oil,
●: Diesel Engine oil,
△: Hydraulic fluid,
▲: ATF,
□: Gear oil (Numbers in figure are mentioned in Table 7.2.3)

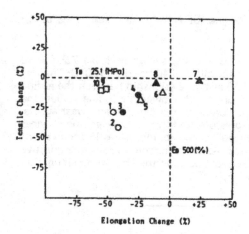

Figure 7.2.3: Lubricating Oil Resistance of Zetpol 2020L
  O: Engine oil,
  ●: Diesel Engine oil,
  △: Hydraulic fluid,
  ▲: ATF,
  □: Gear oil (Numbers in figure are mentioned in Table 7.2.3)

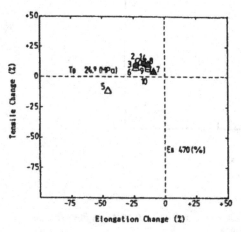

Figure 7.2.4: Lubricating Oil Resistance of Zetpol 2000L
  O: Engine oil,
  ●: Diesel Engine oil,
  △: Hydraulic fluid,
  ▲: ATF,
  □: Gear oil (Numbers in figure are mentioned in Table 7.2.3)

771

7.2.3.2 The Effects of Unsaturation Degree of HSN on the Resistance to Oil Additives

The resistances of HSN and NBR to oil additives as shown in Table 7.2.4 were tested by adding oil additives to ASTM #2 oil for 240 hours at 150°C. The test results are shown in Figure 7.2.5 to 7.2.10 as the relationship between the iodine value and the changes in elongation at break. The lower the iodine values, the less become the changes in elongation at break in the cases of anti-wear agents, detergents, antioxidants and extreme pressure additives. However, the changes in elongation at break are not influenced by the iodine values in the cases of dispersants and viscosity index improvers. As concerns the resistance to additive packages, the lower the iodine value, the less become the changes in elongation at break, except additive package for ATF.

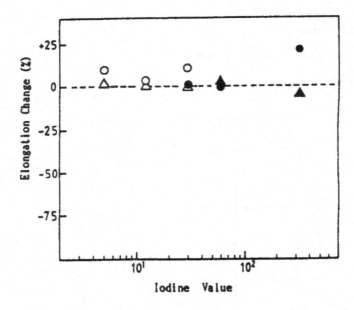

Figure 7.2.5: Effect of Unsaturation Degree on the Oil Additives Resistance
    ○ ●: Polyalkylsuccinimide,
    △ ▲: Polyalkylsuccinic ester (Open symbols are peroxide cured samples and closed are sulfur cured)

772

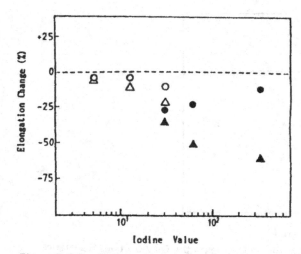

Figure 7.2.6: Effect of Unsaturation Degree on the Oil Additives Resistance
○ ●: Calcium sulfonate,
△ ▲: Calcium phosphate (Open symbols are peroxide cured samples and closed are sulfur cured)

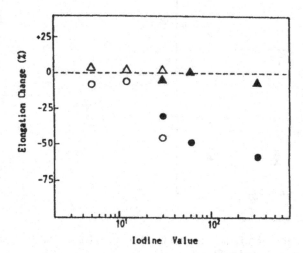

Figure 7.2.7: Effect of Unsaturation Degree on the Oil Additives Resistance
○ ●: Dialkyl Zinc dithiophosphate,
△ ▲: Polyalkylmethacrylate (Open symbols are peroxide cured samples and closed are sulfur cured)

773

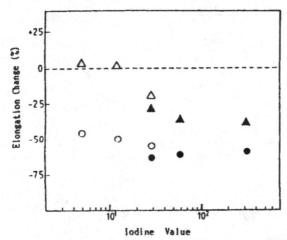

Figure 7.2.8: Effect of Unsaturation Degree on the Oil Additives Resistance
○ ● Dialkylphosphoric ester,
△ ▲: Sulfurized olefines (Open symbols are peroxide cured samples and closed are sulfur cured)

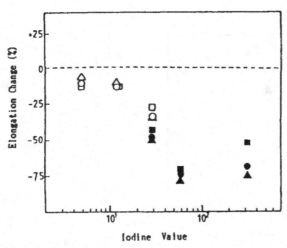

Figure 7.2.9: Effect of Unsaturation Degree on the Oil Additives Resistance
○ ●: Additive package for SE oil,
△ ▲: Additive package for SF oil,
□ ■: Additive package for SG oil
(Open symbols are peroxide cured samples and closed are sulfur cured)

774

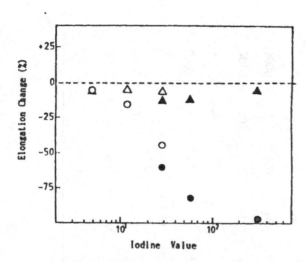

Figure 7.2.10: Effect of Unsaturation Degree on the Oil Additives Resistance
○ ●: Additive package for Gear oil GL—6,
△ ▲: Additive package for ATF Dexron II
(Open symbols are peroxide cured samples and closed are sulfur cured)

7.2.3.3 Analysis of Additives in Lubricating Oil

It is difficult to determine the structures and contents of the additives in the commercial lubricating oils. Therefore, the N, Mg, Ca, P, Cl and S elements in the lubricating oils were measured as shown in Table 7.2.5. The test results are shown in Figure 7.2.11 as the relationship between the contents of these elements and the resistances to the lubricating oils of Zetpol 2020L. The N, Mg, Ca and Cl contents show less effect than the S and P content on the elongation changes of Zetpol 2020L. The P and S derivatives in the lubricating oils deteriorated the Zetpol 2020L, and the higher the S and P content of the oils, the greater become elongation change. The S derivatives in commercial lubricating oil are commonly calcium sulfonate, sulfurized olefins and dialkyl zinc dithiophosphate. On the other hand, the P derivatives are calcium phenate, dialkyl phosphoric ester and dialkyl zinc dithiophosphate. The elongation change of Zetpol 2020L is evidently affected by calcium sulfonate, sulfurized olefines, calcium phenate, dialkyl phosphoric ester and dialkyl zinc dithiophosphate.

Table 7.2.5: Elements Analysis of Commercial Lubricating Oils

Commercial lubricating oils	N	Mg	Ca	Element (ppm) P	Cl	S
Pan XX 10W–30 (SG)	1315	7	1090	270	110	16300
SUPERFLO Z 5W–50 (SF/CD)	–	–	–	–	–	–
Hidiesel S–3 (CD)	459	11	2520	490	110	6000
D–3 10W–30 (CD)	658	1850	1	20	100	6400
Super Hiland 46	–	ND	2	120	ND	–
Unipower SQ 46	130	5	61	130	ND	2800
Pan Automatic D–2	1036	ND	18	28	320	2900
Automatic Fluid	1065	54	11	110	38	3800
Gearlube EHD–90	683	ND	ND	310	260	21700
Gear OIL GX 80W–90	485	ND	ND	400	19	27700

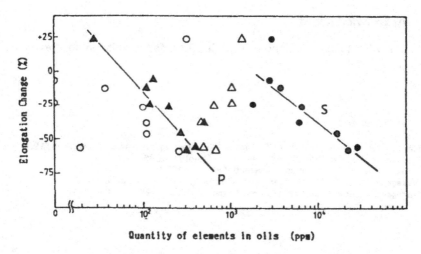

Figure 7.2.11: Effect of Elements on the Commercial Lubricating Oil Resistances
 o: Cl
 ●: S
 △: N
 ▲: P

The resistance of the elastomers to the commercial lubricating oil showed its dependence on unsaturation degree of elastomers. The elongation changes of NBR are greatly affected, but the elongation change of Zetpol 2000L, a fully saturated NBR, is not so affected. It is considered that the S derivatives in the oils will cause HSN and NBR to the crosslinking reactions.

7.2.4 Conclusions

1. HSN exhibits excellent resistance to lubricating oils. Especially, Zetpol 2000L is a little influenced by any of the commercial oils tested.

2. The lower the iodine value of HSN, the less become the deterioration of HSN by anti-wear agents, detergents, antioxidants and extreme pressure additives. No HSN is affected by dispersants and viscosity index improvers.

3. The N, Mg, Ca, P, Cl and S elements of the lubricating oils were measured. The S and P content of the lubricating oils affect the resistance of Zetpol 2020L to them. However, a fully saturated Zetpol 2000 exhibits good resistance to these derivatives.

777

7.2.5 Acknowledgment

The authors wish to thank Nippon Zeon Ca., Ltd. for permission to present this paper.

7.2.6 References

(1) Hashimoto, K. et al.: 127th ACS Rubber Div. Meeting, Los Angeles, April 24, 1985
(2) Hashimoto, K. et al.: 128th ibid., Cleveland, October, 1985
(3) Todani, Y. et al.: 132nd ibid., Cleveland, October, 1987
(4) Nakagawa, T. et al.: Int. Rubber Conf., Goteborg, 1986. Conf. paper, p. 380
(5) Nakagawa, T. et al.: Kautsch. Gummi, Kunstst., vol. 24, 1989, p. 395–399

7.3 Electro-Chemical Investigation of Deposit Formations on Mechanical Seal Surfaces for Diesel Engine Coolant Pumps

H. Hirabayashi, K. Kiryu, K. Okada, A. Yoshino and T. Koga
Eagle Industry Co. Ltd., Okayama-ken, Japan

Summary

Serious leakage problems as a result of deposit formations on mechanical seal surfaces were experienced in actual operation of heavy duty diesel engines. Concerning the cause of these troubles, the action of Cu, Fe, Ca and P ions are concluded to be responsible for deposit formations based upon some chemical analyses. In order to confirm this conclusion, basic laboratory tests were carried out.

From the results of the reproducible tests, it was clarified that the deposit formation on sealing surfaces was closely related to the electro-chemical properties of carbon materials and was promoted by the self-catalytic reactions. In this paper, the countermeasures are also discussed in electrochemical aspects.

7.3.1 Introduction

End-face type mechanical seals are commonly used in automotive coolant water pumps, and are critical to the performance and reliability of engines, requiring effective sealing performance and long operating life (1) (2).

However, several types of leakage problems as a result of deposits adhering to mechanical seal surfaces have been oberserved and some of them had been already investigated (3) (4). Those investigations reported that deposit formations were closely related to additives in engine coolants.

Recently, an increase in the number of serious leakage problems due to deposit formations were experienced in actual operation of heavy duty diesel engines. This increase is considered to be a result of higher engine coolant temperature for improving the thermal efficiency of engines.

As results of various analyses, it was clarified that the deposits were mainly composed of Cu, Fe, Ca, and P elements.

Corresponding to different deposit formation modes, i.e. calcium phosphate deposition, iron compound substrative film formation, metallic copper deposition with the Fe substratum on alumina mating rings and Cu compound formation also with it on carbon seal rings, the effects of concentrations of Fe and Cu anions and PO_3 cation were mainly investigated, where Cu and Fe were generally originated from metallic parts of coolant systems, and where

Ca and P were included in coolant water additives. In order to confirm this conclusion, basic laboratory tests were carried out in electrochemical aspects.

The test results sufficiently reproduced the actual deposition phenomena and were discussed in connection with pH and differences in electrical potential according to the Pourbaix's diagrams under equilibrium conditions (5) (6). Furthermore, dynamic discussions were developed applying the Heterocoagulation theory (7) (8).

7.3.2 Investigations of Failures

At first, the authors investigated the mechanical seals which had face leakages in field service due to deposits adhering to the seal faces. These seals had been applied to heavy duty diesel engines for motor trucks.

The results of the investigations are described as follows:
Table 7.3.1 shows the traveling conditions. Fig. 7.3.1 shows surface as nearly full circular bands. The thickness varied from several micrometers to 30 micrometers.

On the other hand, the deposition on mating ring surfaces were also observed under the certain conditions. However, the deposition was excepted from this investigation because an incidence of this failure was very low in the practical market.

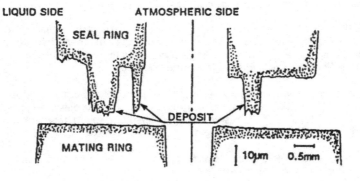

Figure 7.3.1: Surface Profiles of Mechanical Seal with Deposits

Table 7.3.1: An Example of Traveling Condition with Leakage Failure

Classification	Motor Truck
Engine	Heavy Duty Diesel Engine
Running Distance	178,000 km
Running Period	13 months
Coolant	Long Life Coolant

Fig. 7.3.2 shows the secondary electron micrograph of the deposits which adhered to the seal ring surfaces. Furthermore, it is clarified by the X-ray mapping shown in Fig. 7.3.3 that the deposition was mainly composed of Fe, P and Ca elements. Moreover, as the results of the MAXRD (Micro area X-ray diffraction) and XPS (X-ray photoelectron spectroscopy) analyses to identify chemical compounds of deposits, it was confirmed that the deposits were mainly amorphous. For an example, the peak of the MAXRD survey spectra of the deposits shown in Fig. 7.3.4 described a broad shape showing an amorphous structure, so that the chemical compounds could not be identified.

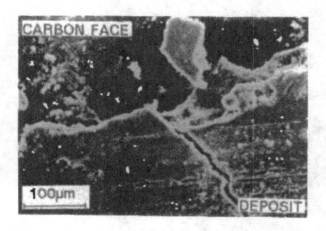

Figure 7.3.2: The Secondary Electron Micrograph of Deposits

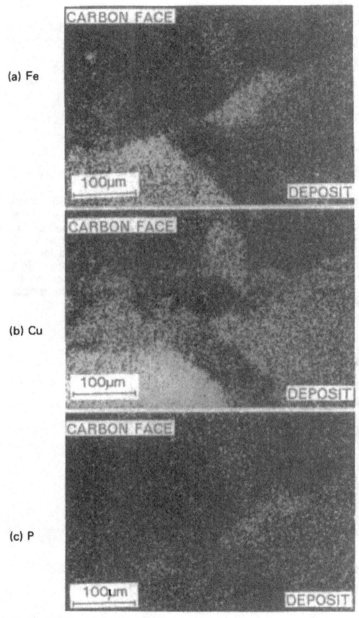

(a) Fe

(b) Cu

(c) P

Figure 7.3.3: The X-ray Mapping of Deposits

782

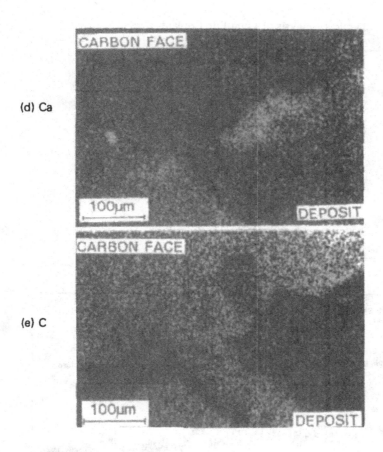

(d) Ca

(e) C

Figure 7.3.3:   The X-ray Mapping of Deposits

Table 7.3.2 shows the chemical analyses of an unused long life coolant applied for the same type of heavy duty diesel engines whick sometimes generate leakage failures. The coolant contained a large volume of phosphate as major inorganic rust preventative, and had no trace of amine and silicate present.

Table 7.3.2:   Chemical Analysis of Long Life Coolant

Color	Green
Concentration	100 %
pH	7.9
PO_4^{-3}	0.81 wt%
NO_3^{-2}	0.09 wt%
SiO_3^{-2}	0.00 wt%
K^+	0.00 wt%
Na^+	0.08 wt%

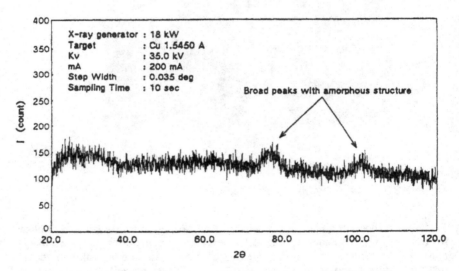

Figure 7.3.4:   The MAXRD Survey Spectra of Deposits

7.3.3 Experiments

7.3.3.1 Mechanical Seals

The test seals are the same as the water pump seals used in the field service operation as shown in Fig. 7.3.5. Fig. 7.3.6 shows a sectional view of a water pump with a mechanical seal for heavy duty diesel engines.

Table 7.3.3 lists the physical properties of sealing materials, where the carbon A generates the deposit formation on seal ring surfaces and the carbon B is no problem in field service operations.

Table 7.3.3: Physical Properties of Sealing Materials

	Seal Ring CARBON A CARBON B		Mating Ring
Material	Phenoric Resin Bonded Carbon		Al_2O_3 93 %
Hardness	100 HsD	89 HsD	93 HR
Gravity	1.46	1.64	3.57
Strength	90 MPa	130 MPa	360 MPa

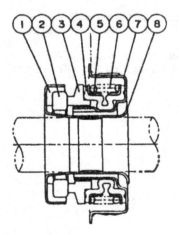

1—CUP GUSKET:NBR
2—MATING RING:ALUMINA CERAMIC
3—SEAL RING:CARBON
4—BELLOWS:NBR
5—SPRING HOLDER:STAINLESS STEEL
6—COIL SPRING:STAINLESS STEEL
7—CARTRIDGE:STAINLESS STEEL
8—SLEEVE:STAINLESS STEEL

Figure 7.3.5: Mechanical Seal to be Tested

785

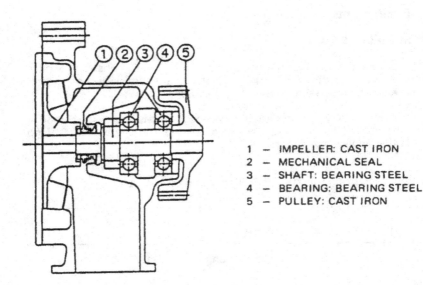

1 — IMPELLER: CAST IRON
2 — MECHANICAL SEAL
3 — SHAFT: BEARING STEEL
4 — BEARING: BEARING STEEL
5 — PULLEY: CAST IRON

Figure 7.3.6: A Sectional View of a Water Pump with a Seal

7.3.4.2 Test Procedure

Fig. 7.3.7 shows an exterior view of the test equipment and Table 7.3.4 describes the test conditions.

The test fluid shown in Table 7.3.5 was compounded in due consideration of the chemical analysis results of the deposits in Fig. 7.3.3 and the long life coolant in Table 7.3.2.

Table 7.3.4: Test Condition

Shaft Speed	4.000 rpm
System Fluid Pressure	100 kPa
Fluid Temperature	90°C
Running Time	400 hrs

Figure 7.3.7:   An Exterior View of the Test Equipment

Table 7.3.5:   Test Liquid

PO_4^{-3}	0.80 wt%
Ca^{+2}	0.05 wt%
Fe^{+3}	0.30 wt%
Cu^{+2}	0.10 wt%
pH	9

7.3.4   Experimental Results

The changes of leakage rates are plotted against the running time in Fig. 7.3.8. All the test pieces of the carbon A materials resulted in abrupt increases of leakage over 200 hrs.

Fig. 7.3.9 shows the representative surface profiles of mechanical seals after the test. The circumferential deposition was observed on the sealing surfaces of the carbon A similarly to the seals which generated excessive leakage due to the deposition on the sealing surfaces experienced in the actual market.

787

Based upon results of analyses, i.e. EPMA (Electron prove X-ray micro analysis). MAXRD and XPS, it was clarified that the deposition was mainly composed of Fe, Cu, P, Ca and its structure was recognized to be amorphous.

On the other hand, all of the carbon B materials generated very little depositions.

From these results, it was considered that the excessive leakage due to the deposition in the actual market could be reproduced in these experiments.

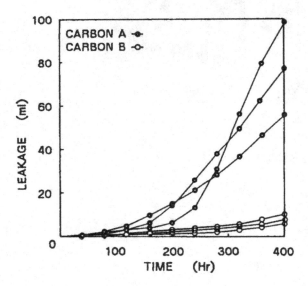

Figure 7.3.8: The Changes of Leakage Rate

788

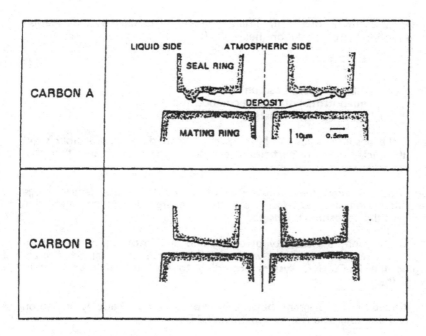

Figure 7.3.9: The Representative Surface Profiles after the Test

7.3.5 Discussions

As it was considered that the electrochemical properties of carbon materials affected to build up depositions on seal ring surfaces, the authors measured a Zeta-potential and surface electric charge.

Table 7.3.6 shows the Zeta-potential measured by the flow-potential method and the amount of the static surface electric charges of the carbon materials A and B.

Comparing both the carbon materials, the following facts were clarified:
The carbon material A which showed the tendency to promote the deposition resulting in leakage was characterized by a larger absolute value of the electric charge as well as the Zeta-potential. The carbon material B shows the quite opposite result.

Consequently, it is considered that the electrochemical properties of carbon materials, i.e. the amount of the potential energy due to the interaction between metal ions (Fe, Cu) and the sealing surfaces of carbon rings, influence the coagulation phenomena of the deposits.

789

The potential energy E, due to the interaction will be obtained by the equation (1) related to the hetrocoagulation theory (7) (8).

$$E = K (\zeta_1{}^2 + \zeta_2{}^2)$$ (1)

Where, ζ_1 : Zeta-potential of carbon material
ζ_2 : Zeta-potential of deposit composition
K : Proportionality constant

However, the absolute value of ζ_2 may be estimated to be negligibly small because the surface areas of particles of deposition are much smaller than that of rubbing surfaces of the carbon rings.

Consequently, the larger the absolute value of ζ_1, and the surface electric charge of the carbon ring are, the larger the potential energy due to the interaction becomes and the deposition builds up.

On the other hand, the stability of oxides and double oxides of Fe and Cu elements can be well explained with the aid of the potential/pH diagram concerning the Fe-Cu-H_2O system according to the Pourbaix's equilibirum theory (5) (6).

It is confirmed by this diagram that Fe_3O_4, Fe_2O_3, Cu_2O, $FeCuO_2$ and so on are stable in the test coolant shown in Table 7.3.5.

Furthermore, as the activation energy for the progress of the electrochemical reaction can be assumed to be very small, the reduction from the deposition to the form of metal ions is considered to be attributed to the self-catalytic reaction. This reaction generally progresses in the solution showing the unstable thermodynamic characteristics, where electrons are generated as a result of the catalytic reaction on the surface of the initially nucleated deposition.

In this experiment, as soon as the hypophosphorous acid ions contact the elements which belong to the VIII group of the periodic table under the definite condition, these metal elements accurate dehydrogenation of H_2O as the following catalytic reaction;

$$[H_2PO_2]^- + H_2O \rightarrow H[HPO_3]^- + 2H$$ (2)

$$Fe^{2+} + 2H \rightarrow Fe + 2H^+$$ (3)

The hypophosphorous acid is generated by the hydrolysis of the sodium hypophosphite, and the equilibrium of the solution due to production of sodium hydroxide shifts to increase alkalinity.

Accompanied by this, the following chain reaction is considered to progress;

$$Fe + 2H_3PO_4 \rightarrow Fe(H_2PO_4)_2 + H_2 \uparrow \tag{4}$$
$$3Fe(H_2PO_4)_2 \rightarrow Fe_3(PO_4)_2 \downarrow + 4H_3PO_4 \tag{5}$$

In this connection, the authors used three kinds of materials for the coolant containers, i.e. stainless steel, aluminium coated with PTFE and acrylic resin, to investigate the possibility of the above reactions. In this experiment, the highest rate deposition was observed in the case of using the stainless steel container. As a result, it was considered that the above reactions accelerated the deposit formation.

From the above discussion, the countermeasures to excessive leakage due to the deposit formation on the seal surfaces are considered to be effective as follows;

a. Decreasing the potential energy due to the interaction.
  ⇒ Decreasing the absolute value of the Zeta-potential and surface potential energy of carbon materials

b. Removing the factors to promote the self-catalytic reaction.
  ⇒ Suppressing the generation of hydrogen
  ⇒ Preventing the elution of metal ions to coolant

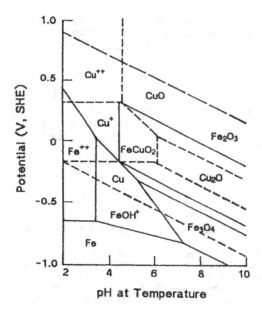

Figure 7.3.10: The Potential-pH Diagram for the Fe-Cu-H$_2$O System at 423 K (150°C)

Table 7.3.6: Electrochemical Properties of Carbon Material A and B

	CARBON A	CARBON B
Height of Deposition	27 μm	0 μm
ZETA—Potential	−61 mV	−33 mV
Surface Electric Charge	− 0.27 nC	− 0.08 nC

7.3.6 Conclusions

1) The deposit formation on sealing surfaces of water pump mechanical seals is closely related to the electrochemical phenomena.

⇒ The electrochemical properties inherent to carbon materials, i.e. the Zeta-potential and the surface electric charge, are responsible for the initial nucleation of the deposits.
⇒ The growth of the depositions is promoted by the self-catalytic reaction which is accelerated by the generation of hydrogen.

2) The countermeasures are considered as follows;

⇒ Decreasing the absolute value of the potential energy due to the inter-action between the deposition particles and the carbon materials.
⇒ Removing the factors to promote the self-catalytic reaction.

7.3.7 Acknowledgement

The authors wish to express their sincere appreciation to Dr. F. Hirano, Professor Emeritus of Kyushu University, for his valuable advice during this investigation, and to thank the directors of Eagle Industry Co., Ltd. for permission to publish this paper.

7.3.8 References

(1) Matsushima, A.: Guide to Automotive Water Pump Seals, SAE (1978) Paper No. 780404.
(2) Kiryu, K.; Fukahori, K.; Matsumoto, S.; Shimomura, T.; Hirabayashi, H.: A Status of Sealing Performance of End-Face Type Seals for Water Pumps of Automotive Engines in Japan, SAE (1988) Paper No. 880303.
(3) Kiryu, K.; Tsuchiya, K.; Yonehara, Y.; Shimomura, T.; Koga, T.: An Investigation of Deposits Formation and Sealing Surfaces of Water Pump End Face Seals, STLE, Lubr. Eng. (1988) Vol. 45, 1, 49—55.

(4)   Kiryu, K.; Tsuchiya, K.; Shimomura, T.; Yanai, T.; Okada, K.; Hirabayashi, H.: The Effect of Coolant Additives and Seal Composition on Performance of Water Pump Seals of Automotive Engines, SAE (1989) Paper No. 890609.

(5)   Pourbaix, M.; Zoubov, N.; Muylder, J.V.: Atlas Dequilibres Electrochimiques a 25°C, Gauthier-Villars & Ceditier Paris (1963).

(6)   Cubicciotti, D.: Pourbais Diagrams for Mixed Metal Oxide Chemistry of Copper in BWR Water, Corrosion (1988) Vol. 44, No. 12, 875–880.

(7)   Devereux, O.F. & Bruyn, P.L.: Interaction of Plane Parallel Double Layers. The Mit Press (1963), Cambridge, Mass.

(8)   Hogg, R.; Healy, T.W.; Fuerstenau, D.W.: Trans., Faraday Soc. (1966) 62, 1638.

Index

794

The Authors

Section 1.1
T.W. Bates
Shell Research Ltd., Chester,
Great Britain
M.A. Vickars
Esso Research Centre, Abingdon,
Great Britain

Section 1.2
J.A. Spearot
General Motors Research Laboratories
Warren, USA

Section 2.1
P. Daucik, T. Jakubik, N. Pronayova
and B. Zuzi
Slovak Technical University
Bratislava, Czechoslovakia

Section 2.2
H.H. Abou el Naga and S.A. Bendary
MISR Petroleum Co.
Cairo, Egypt

Section 2.3
H.H. Abou el Naga, M.M. Mohamed
and M.F. el Meneir
Research Centre, MISR Petroleum Co.
Cairo, Egypt

Section 2.4
K. Rollins, M. Taylor, J.H. Scrivens
and A. Robertson
ICI Wilton Materials Research Centre
Wilton Middlesborough, Great Britain
H. Major
VG Analytical Ltd.
Whythenshane, Great Britain

Section 2.5
R.L. Shubkin and M.E. Kerkemeyer
Ethyl Corporation
Baton Rouge, USA
D.K. Walters and J.V. Bullen
Ethyl Petroleum Additives Ltd.
Bracknell, Great Britain

Section 3.1
C. Kajdas
Technical University at
Radom, Poland

Section 3.2
S. Korcek and M.D. Johnson
Ford Motor Company
Dearborn, USA

Section 3.3
J.M. Georges, J.L. Loubet, N. Alberola
and G. Meille
Ecole Centrale de Lyon, France
H. Bourgognon, P. Hoornaert
and G. Chapelet
Centre de Recherche Elf Solaize
Saint-Symphorien d'Ozon, France

Section 3.4
S.P. O'Connor
BP Chemicals
Hull, Great Britain
J. Crawford
Adibis, Redhill, Great Britain
C. Cane
Adibis, Hull, Great Britain

Section 3.5
G. Deak, L. Bartha and J. Proder
Veszprem University of Chemical
Engineering, Hungary

798

Section 3.6
K. Endo and K. Inoue
Nippon Oil Company Ltd.
Yokohama, Japan

Section 3.7
L. Bartha and J. Hancsok
Veszprem University of Chemical
Engineering, Hungary
E. Bobest
Komarom Petroleum Refinery
Komarom, Hungary

Section 3.8
M.F. Morizur and O. Teysset
Institut Francais du Pétrole
Rueil-Malmaison, France

Section 3.9
D. Wei, H. Song and R. Wang
Research Institute of Petroleum
Processing
Beijing, P.R. China

Section 3.10
J. Dong, G. Chen and F. Luo
Institute of Logistics Engineering
Chongquing, P.R. China

Section 3.11
G.S. Cholakov, K.G. Stanulov
and I.A. Cheriisky
Higher Institute of Chemical
Technology
Sofia, Bulgaria
T. Antonov
Petrochemical Combine
Pleven, Bulgaria

Section 3.12
M. Born, J.C. Hipeaux, P. Marchand
and G. Parc
Institut Francais du Pétrole
Rueil-Malmaison, France

Section 3.13
G. Monteil, A.M. Merillon
and J. Lonchampt
Peugeot S.A.
Voujeaucour, France
C. Roques-Carmes
ENSMM
Besancon, France

Section 3.14
H. Bourgognon and C. Rodes
Centre de Recherche Elf Solaize
Lyon, France
C. Neveu and F. Huby
Rohm & Haas European Operations
Paris, France

Section 3.15
Y. de Vita, I.C. Grigorescu
and G.J. Lizardo
Intevep S.A.
Caracas, Venezuela

Section 4.1
A. Quilley
Adibis BP Chemicals (Additives) Ltd.
Redhill, Great Britain

Section 4.2 Part I + Part II
S.L. Aly, M.O.A. Mokhtar, Z.S. Safar,
A.M. Abdel-Magid, M.A. Radwan
and M.S. Khader
Cairo University
Cairo, Egypt

Section 4.3
J.R. Nanda, G.K. Sharma, R.B. Koganti
and P.K. Mukhopadhyay
Indian Oil Corporation Ltd.
Faridabad, India
R.M. Sundaram
Ministry of Railways
Lucknow, India

Section 4.4
T.W. Selby
Savant Inc.
Midland, USA
T.J. Tolton
Dow Corporation
Freeland, USA

Section 4.5
C.D. Neveu
Rohm & Haas European Operations
Paris, France
W. Böttcher
Röhm GmbH Chemische Fabrik
Darmstadt, Germany

Section 4.6
P.G. Carress
Adibis — BP Chemicals (Additives) Ltd.
Redhill, Great Britain

Section 4.7
D.C. Roberts
Esso Petroleum Co. Ltd.
Abingdon, Great Britain

Section 4.8
P. Tritthart, F. Ruhri and W. Cartellieri
AVL-List Ges.m.b.H.
Graz, Austria

Section 4.9
J.A. McGeehan and E.S. Yamaguchi
Chevron Research Company
Richmond, USA

Section 4.10
A. Zakar
Hungarian Hydrocarbon Institute
Szazhalombatta, Hungary
G. Borsa
Danube Refinery
Szazhalombatta, Hungary

Section 4.11
P. van Donkelaar
Greentech Research sprl
Essen, Belgium

Section 4.12
D. Moura and J.-P. Legeron
Cofran Research Sarl
La Rochelle, France

Section 4.13
D. Kenbeek and G. van der Waal
Unichema International
Gouda, The Netherlands

Section 5.1
D.J. Needle
Smallman Lubricants Ltd.
West Bromwich, Great Britain

Section 5.2
B.M. O'Connor
The Lubrizol Corporation
Wickliffe, USA
H. Winter
Technical University Munich
Munich, Germany

Section 6.1
S. Watanabe and H. Ohashi
Tonen K.K. Corporate
Saitama, Japan

Section 6.2
G. Venizelos and G. Lassau
Conservatoire National des Art et
Métiers
Paris, France
P. Marchand
Institut Francais du Pétrole
Rueil-Malmaison, France

Section 6.3
A.G. Papay
Ethyl Petroleum Additives Division
St. Louis, USA

Section 6.4
R.F. Watts and R.K. Nibert
Exxon Chemical Company
Linden, USA

Section 7.1
T.E. Fischer and W.M. Mullins
Stevens Institute of Technology
Hoboken, USA

Section 7.2
M. Oyama, H. Shimoda, H. Sakakida
and T. Nakagawa
Nippon Zeon Co. Ltd.
Tokyo, Japan

Section 7.3
H. Hirabayashi, K. Kiryu, K. Okada,
A. Yoshino and T. Koga
Eagle Industry Co. Ltd.
Okayama-ken, Japan

Printed in the United States
by Baker & Taylor Publisher Services